Harald Böttger · Valerij V. Bryksin

Hopping Conduction in Solids

Physical Research

Volume 1

Hopping Conduction in Solids

by
Harald Böttger/Valerij V. Bryksin

Hopping Conduction in Solids

by

Harald Böttger/Valerij V. Bryksin

With 107 Figures and 1 Table

Akademie-Verlag Berlin
1985

Authors:

Professor Dr. sc. Harald Böttger
Technische Hochschule „Otto von Guericke“
Magdeburg, Sektion Mathematik und Physik

Dr. sc. Valerij Vjaceslavovič Bryksin
A. F. Ioffe Physical-Technical Institute,
Academy of Sciences of the USSR, Leningrad

Erschienen im Akademie-Verlag Berlin, DDR-1086 Berlin, Leipziger Straße 3—4

Lizenznummer: 202 · 100/429/85
Printed in the German Democratic Republic
Gesamtherstellung: VEB Druckhaus „Maxim Gorki“, 7400 Altenburg
Einband: Karl Salzbrunn
LSV 1184
Bestellnummer: 763 411 5 (6863)

08800

Preface

Phonon-assisted hopping motion of particles between spatially distinct locations is one of the basic transport mechanisms in solids. This mechanism is effective in various materials such as doped semiconductors, amorphous and glassy solids, organic systems, transition metal oxides, superionic materials, and metal hydrides. In particular, it dominates in low-mobility materials.

Historically, the problem of hopping conduction of electrons was first raised towards the end of the fifties, when the low mobility observed in transition-metal oxides became a challenging problem. From the beginning of the seventies, the investigation of the hopping mechanism has been much stimulated by the increasing interest in disordered solids. Hopping conduction is viewed as an important proof of the existence of localized electronic states in disordered solids, which is at present one of the major problems in solid state theory. More recently, the study of hopping transport has been much stimulated by the possibility of using superionic materials and metal hydrides for energy storage.

In recent years, the theory of hopping transport has been characterized by considerable progress, which is based on the application of modern analytical and numerical methods, and which has provided a deeper understanding of transport phenomena in solids.

In this book, we describe the present state of theory of the hopping transport of small polarons in crystals, electrons in disordered systems, and atoms in materials such as superionics and metal hydrides. We develop the theory of hopping transport from a unified point of view by starting from first principles. All steps of the theory are described in detail so that the book may also serve as introductory textbook into the theory of hopping transport. On the other hand, the main results are also derived in an intuitive manner so that the book may also be useful for experimentalists who are not very interested in mathematical details. The theoretical results are illustrated by experimental findings.

The book is organized as follows:

Chapter I contains an overview of ideas and concepts on which modern hopping theory is based. This chapter also contains a compilation of basic elements of the theory of hopping transport such as the electron-phonon Hamiltonian, the small-polaron canonical transformation, and a derivation of the rate equation needed for obtaining the hopping current.

Chapter II is devoted to hopping motion of small polarons in crystals. Furthermore, this chapter also contains a concise description of tunnelling transport of small polarons.

Chapter III is devoted to hopping transport of electrons in disordered systems. In this chapter special emphasis is laid on the problem of localized electron states (Section 4.), the application of ideas of the percolation theory (Section 5.2.), and the application of the effective-medium approximation (Section 6.4.). Furthermore, this chapter also contains a description of the so-called dispersive transport in disordered solids (Section 6.8.).

Chapter IV is devoted to hopping transport of atoms in solids. This chapter also includes a concise description of continous diffusion of atomic particles (Sections 10.3. and 10.4.).

Chapter V, added in proof, gives an overview of results which have been obtained since the book was prepared for publication.

Some topics with more formal character (a diagram technique for strong electron-phonon coupling; rate equation in the presence of a magnetic field; relationship between percolation problems and the Potts model) are given in three appendices.

It should be noted that in order to bring the book up-to-date we also included some problems where no full consensus is evident from the present literature.

The list of references is not complete. It contains, above all, only more recent papers. For older papers, the reader is referred to the review-articles included in the list of references (e.g. Böttger and Bryksin, 1976a, b; 1982).

Finally, we are indepted to Mrs. Trautmann, Akademie-Verlag, for efficient cooperation, and to Mrs. Böttger (the wife of H. B.), who gave us enormous help with the preparation of the manuscript.

H. Böttger, V. V. Bryksin

Contents

Chapter I. Basic Elements of the Theory of Hopping Transport

Chapter II. Small Polarons

Chapter IV. Classical Hopping Transport

10. Diffusive atomic motion

Chapter V. Recent Developments

Chapter I.
Basic Elements of the Theory of Hopping Transport

1. Introduction and overview

Phonon-assisted hopping transitions of particles between spatially distinct locations are a phenomenon encountered in a diverse variety of solid state systems. For instance, in weakly doped and compensated semiconductors the phenomenon of impurity conduction arises from hopping of electrons between impurity sites. Furthermore, in crystals with a narrow conduction band and strong electron-phonon interaction (e.g. orthorhombic sulphur) a new quasi-particle, the small polaron, is formed, which at sufficiently high temperatures moves through the crystal by hopping transitions from site (atom) to site. Another hopping phenomenon is the atom (ion) hopping between lattice sites which are only fractionally occupied with atoms. In this case, depending on the atomic mass, the form of the local potential well, and temperature, the elementary hopping process may be either a thermally activated jump over the potential barrier separating neighbouring sites ("over-barrier hopping") or phonon-assisted tunnelling through the barrier ("under-barrier hopping"). Atom hopping is an important transport mechanism, for example, in ionic crystals, metal hydrides, or on surfaces of solids. The hopping of ions or electrons within spatially localized regions represents a well-known dielectric loss mechanism.

Furthermore, the existence of low-temperature thermal anomalies in certain materials (e.g. amorphous solids, superionic conductors, some ferroelectrics) suggests that in these materials there are atoms or groups of atoms having two or more possible equilibrium positions between which phonon-assisted tunnelling transitions can occur (Anderson et al., 1972, Philipps, 1972, cf. Gilroy and Philipps, 1981, and references therein). Similar transitions (phonon-assisted rearrangements of magnetic clusters) are discussed for spin glasses (Anderson et al., 1972, Villain, 1977, cf. Herlach, 1983, and references therein). Another example of hopping motion in solids is the hopping diffusion of Frenkel excitons in molecular crystals (Gutmann and Lyons, 1967), which is directly analogous to that of small polarons. Furthermore, phonon-assisted energy transport in inhomogeneously broadened spin systems (Hsu and Powell, 1975, Holstein et al., 1976, Lyo et al., 1978) may also be viewed as a hopping phenomenon. Recently, phonon-assisted hopping of electrons between soliton bound states has been discussed as possible dominant conduction mechanism in lightly doped quasi one-dimensional Peierls systems such as polyacetylene (Kivelson, 1981, 1982), and hopping motion of solitons pinned by disorder has been predicted by Larkin and Lee (1978). In the book at hand we restrict ourselves to hopping conduction in solids, i.e. hopping transport of electrons, small polarons and atoms (ions), whereby the emphasis is on the former two problems.

In the case of hopping transport, charge carrier motion is quite differently affected by the interaction with phonons than in the case of band-like transport. While in the latter case the interaction with phonons acts as the mechanism limiting the mobility of the carriers, in the former one it is the mechanism rendering possible charge carrier motion. Note that phonon-assisted hopping is not the only hopping mechanism. Hopping transport can also be due to absorption of photons ("photon-induced hopping") and absorption of energy from excited spins ("ESR-induced hopping").

A necessary prerequisite of hopping transport is the existence of spatially localized charge carriers. Different mechanisms of charge carrier localization are known:
(i) Localization in the small-polaron model (Tyablikov, 1952, Yamashita and Kurosawa, 1958, Holstein, 1959). As mentioned above, in crystals with a narrow conduction band and strong interaction with phonons a new quasi-particle, the small polaron, is formed, the size of which is smaller than the interatomic spacing. Owing to the lattice translational symmetry, in such a system a localized state is quasi-stationary and occurs only at sufficiently high temperatures where the time between consecutive hops is shorter than the decay time of a small-polaron wave packet located at a site. At low temperatures, however, the interaction with phonons leads only to a renormalization (narrowing) of the conduction band and the small-polaron motion is band-like.
(ii) Localization in disordered systems according to Anderson (1958). Because of the lack of long-range order, in disordered systems stationary localized one-particle states may exist, the amplitude of which decreases with increasing distance from a localization center. In the absence of electron-phonon interaction, electrons in localized states do not contribute to conductivity. In the presence of interaction with phonons, however, hopping transitions may occur between localized states (Miller and Abrahams, 1960).
(iii) Localization of elctrons in a strong electrical field (Wannier, 1960). Using a one-band approximation, one finds that in narrow-band semiconductors by a strong electrical field the stationary electron states become localized in the direction of the field. Electron-phonon interaction can produce hopping transitions between these states (Bryksin and Firsov, 1970a, Maekawa, 1970).

Furthermore, in the presence of a magnetic field, localization may occur in the plane perpendicular to the magnetic field due to Landau quantization. For high enough fields, for which the period of the cyclotron motion is smaller than the collision time, the transport perpendicular to the magnetic field is hopping-like and described by the formula of Ţiţeica (1935). Localization may also be due to strong interelectronic correlation. The latter mechanism is supposed to be effective, for instance, in magnetite (cf. Mott, 1979). In the case of atom hopping, classical behaviour (over-barrier hopping) may be expected for heavy particles (e.g. ions in ionic conductors), while for light particles (e.g. hydrogen in metal hydrides or positive muons in metals), quantum effects (under-barrier hopping or even band-like motion) may be important at sufficiently low temperatures. With fair success, methods of the small-polaron theory have been applied to quantum diffusion of atoms (Flynn and Stoneham, 1970, Kagan and Klinger, 1974).

In what follows, we take a quick trip through the history of the investigation of the hopping transport. First, let us consider hopping of small polarons. Towards the end of the fifties, materials with low carrier mobility (less than 10^{-1} cm/Vs) became a challenging problem. It turned out that the classical transport theory, which is based on the assumption of a band mechanism of transport and which makes use of the Boltzmann equation, is unsuitable for studying such materials. The first long step towards a theoretical understanding of low mobility materials was taken by Holstein (1959), who formulated a simple consistent model for the small-polaron problem and used it to study the conductivity.

Holstein (1959) pointed out that strong electron-phonon interaction can cause "self-trapping" of an electron, so that the electron cannot move unless accompanied by its induced lattice distortion (polaron formation). Assuming the electronic overlap term (being proportional to the resonance integral J) in the total Hamiltonian to be small, he calculated the transition rates by perturbation theory (in this case the transitions are said to be "non-adiabatic"). Two kinds of transitions appear: (i) diagonal transitions leaving the number of phonons unaltered and (ii) nondiagonal transitions in which the number of phonons is changed. At sufficiently low temperatures, the diagonal transitions are dominant and give rise to band-like motion. In this temperature range the nondiagonal transitions determine the lifetime of the polaron-band states. At about half the Debye temperature, T_D, the energy uncertainty is equal to the band width so that the band picture loses its meaning. At higher temperatures the transport is by hopping associated with the nondiagonal transitions. In the high-temperature limit, $T \gg T_D$, multiphonon processes (i.e. one-electron transitions with simultaneous absorption and emission of many phonons) determine the transition rate, which is therefore an exponentially rising function of temperature.

The next step towards a theoretical understanding of the small-polaron problem was taken by Lang and Firsov (1962, 1963), who introduced the small-polaron canonical transformation and presented a consequent theory of the conductivity in the small-polaron model by starting from the Kubo formula. In particular, these authors elucidated the appropriate expansion parameter of the perturbation-in-J approach. The first correct investigation of the Hall effect in the small-polaron model was performed by Friedman and Holstein (1963), who used an intuitive concept of classical hops in a magnetic field. The Hall effect was shown to require account of at least three neighboured sites, while the drift mobility is determined by two-site hops. The first investigation of the Hall mobility by means of the Kubo formula was performed by Firsov (1963). It is interesting to note that, in contradiction to the band-transport regime, in the hopping regime the Hall and drift mobilities appreciably differ from each other. This statement does not only hold for the small-polaron model, but it is also valid for hopping in disordered systems. Accordingly, the charge-carrier concentration cannot be obtained from a measurement of the hopping-Hall effect. The Hall mobility in the transition regime between the hopping and the band regimes was studied by Bryksin and Firsov (1974a), and in the polaron-band regime by Friedman (1963).

Kudinov and Firsov (1966) took an important step towards a general transport theory which is valid for ordered systems with electron-phonon interaction of arbitrary strength. In the limit of a small resonance integral and strong interaction with phonons, the theory of these authors describes small-polaron transport. Kudinov and Firsov (1965) pointed out that in small-polaron systems the conductivity can be conveniently investigated by using the site (Wannier) representation for the electronic subsystem (except in the low temperature regime).

Lowest-order perturbation-in-J treatment of the transition rates fails in the following two situations: The first of these is when the resonance integral J is so large that the carrier can be viewed as always adjusting to the instantaneous positions of the atoms. This situation (in which the transitions are said to be "adiabatic") was studied by Holstein (1959), Emin and Holstein (1969) and Emin (1971a). The second situation is when successive hops of a carrier are correlated with each other. The latter problem was considered by Emin (1970, 1971a, b).

Recently, Holstein and coworkers studied in a series of papers (Holstein 1978a, b, 1981, Scher and Holstein, 1981, Arnold and Holstein, 1981) small-polaron hopping from the stand-point of a reaction-rate approach, in conjunction with a two-site model. This so-called quantal occurrence-probability approach provides a physically transparent derivation of the hopping rates for non-adiabatic as well as adiabatic hopping and, unlike the common formalism of multiphonon transitions, it allows non-linear electron-phonon interaction and/or vibrational anharmonicity to be accounted for (cf. Christov, 1982).

In the small-polaron model, intraband absorption of light (light-induced hopping) was studied by Reick (1963) and Klinger (1963b), optical transitions between two polaron bands and between a broad valence band and a polaron band by Eagles (1963, 1966), optical transitions between deep atomic levels and a polaron band by Kudinov and Firsov (1964), and resonant Raman scattering in the small-polaron model by Böttger et al. (1978). The first comprehensive experimental investigation of a low-mobility material, TiO_2, was performed by Bogomolov et al. (1967). Investigations of drift mobility, Hall mobility, infrared absorption, and thermoelectric power suggest that in this material the charge carriers are small polarons. There exists now quite convincing evidence for small polaron transport in a number of materials.

Note that if the lattice is soft enough to allow sufficient local deformation to overcome near-neighbour Coulomb repulsion, pairing of polarons at a single site or near-neighbour sites, i.e. formation of bipolarons, should occur (Chakraverty and Schlenker, 1976). The theory of small bipolarons has been developed by Alexandrov and Ranninger (1981). There is some indication that bipolarons may exist, for example, in polydiacetylen (Movaghar and Cade, 1982).

Let us now turn to hopping conduction in disordered solids. The investigation of this problem began with the discovery of impurity hopping conduction in doped and compensated semiconductors such as germanium and silicon (Fritzsche and Cuevas, 1960, Mott and Twose, 1941) and was very much stimulated by the theoretical analysis of Miller and Abrahams (1960), who pointed out that the dc (direct current)-hopping conductivity can be reduced to the problem of calculating the current in a random-resistor network governed by Kirchhoff's laws. In this network the conductance between any two sites is proportional to the corresponding hopping probability. Miller and Abrahams suggested that the dc current is governed by certain one-dimensional paths connecting the electrodes. This idea proved to be very useful for averaging the resistor network. However, the concrete procedure of determining these paths proposed by the latter authors overestimates the resistance of the network.

According to Miller and Abrahams, the hopping probability $W_{mm'}$ between impurity sites (states) m and m' is described by a product of a tunnelling term and a phonon term

$$W_{mm'} \sim \exp\left(-2\alpha\,|\boldsymbol{R}_{mm'}| - |\varepsilon_{mm'}|/kT\right). \tag{1.1}$$

Here $\boldsymbol{R}_{mm'} = \boldsymbol{R}_m - \boldsymbol{R}_{m'}$ and $\varepsilon_{mm'} = \varepsilon_m - \varepsilon_{m'}$, where $\boldsymbol{R}_m$ and ε_m, respectively, denote the position and the energy of the mth site; α is the reciprocal Bohr radius of an impurity state and T the temperature. At sufficiently high temperatures, the second exponential in (1.1) is small compared with the first one, so that the spread of the hopping probabilities and, accordingly, the current are governed by the random spatial distribution of the impurities. This is the situation of the so-called "R-percolation" ("R-hopping"), where the dc conductivity is determined by nearest-neighbour hops and exhibits an activated form.

At low temperatures the hopping transport will eventually move to within a few kT of the Fermi level. Under these conditions one will observe the so-called variable-range hopping, a mechanism first suggested by Mott (1968, 1969a, b) to cover the situation where the energy between non-nearest neighbour hopping sites is a function of their spatial separation. According to Mott, in this so-called "R-ε-percolation" ("R-ε-hopping") situation, the dc conductivity is governed by hops for which the exponent in (1.1) becomes minimal. By optimization of the hopping probability one obtains the famous Mott "$T^{-1/4}$ law"

$$\sigma \sim \exp\{-(T_0/T)^{1/4}\}, \tag{1.2}$$

where T_0 is some constant proportional to the density of states at the Fermi level.

The problem of averaging the network of Miller and Abrahams has attracted much interest. Application of methods of the percolation theory to this problem was first proposed by Ambegaokar et al. (1971), Shklovskii and Efros (1971), and Pollak (1972). This approach has proved to be very effective and is now widely used. It allows the correct exponential behaviour of the dc hopping conductivity to be predicted. Another effective method for studying hopping phenomena is the effective-medium approximation, which was first applied to random resistor networks by Kirkpatrick (1971, 1973).

Interest in the ac (alternating current) hopping properties began with the work of Pollak and Geballe (1961), who studied ac conductivity in doped and compensated silicon and performed a theoretical analysis of the data obtained by using a two-site (pair) approximation. Such approximation is only valid for sufficiently high frequencies, where during half a period of the external field, electrons can hop solely between nearest-neighbour sites. TANAKA and FAN (1963) and MOTT (1970) investigated the "phononless" ac conductivity (photon-assisted hopping) by using a pair approximation. In like manner, HOLSTEIN (1961) calculated the ac Hall mobility in a disordered system by taking into account only three nearest-neighbour sites. — At lower frequencies, during half a period of the field, the electrons can hop through clusters of sites, whose size increases with decreasing frequency ("multiple hopping") such as to lead to the resistor network of Miller and Abrahams in the limit as the frequency goes to zero. Cluster approaches to ac hopping conductivity were proposed by BÖTTGER et al. (1979a, b, c) and ZVYAGIN (1978, 1980). The basic idea of these approaches is to apply methods of the percolation theory to the ac hopping problem.

A calculation of the current or the Hall mobility in a large frequency range requires knowledge of the equation of motion of the density matrix. Furthermore, after solving this equation, some averaging procedure must be performed. Such an equation of motion, written in the site representation, was proposed by MOTT and TWOSE (1961), BRENIG et al. (1971, 1973a), and BUTCHER (1972), and derived on the basis of first principles by ZVYAGIN and KEIPER (1972) and BÖTTGER and BRYKSIN (1975c). POLLAK (1974) and KIRKPATRICK (1973) pointed out that the problem of solving this equation (rate equation) is equivalent to the problem of calculating the charge distribution in a random-resistor network with capacitors, which in the limit as the frequency goes to zero passes into the network of Miller and Abrahams.

The development of the theory of hopping transport in disordered systems has been much stimulated by the work of SCHER and LAX (1973), who proposed a stochastic description of the hopping motion on the basis of the continuous-time random-walk (CTRW) model. This approach was proved to be correct by KLAFTER and SILBEY (1980b). It allows the conductivity to be described in a broad frequency range. Furthermore, it allows the time dependence of the transit current in time-of-flight experiments to be studied, in particular, in disordered solids where dispersive (non-Gaussian) transport appears due to random hopping or trapping. Note that the frequency dependence of the conductivity and the dispersive transport are intimately related problems.

There is some indication that polaronic effects play an important role in certain disordered solids. Such effects were first theoretically investigated by SCHNAKENBERG (1968) by using the polaron canonical transformation mentioned above. ELLIOTT (1977, 1978) suggested that in certain amorphous materials the dominant conduction process is hopping of bipolarons.

It is interesting to note that the small-polaron Hamiltonian is a special case of the Hamiltonian generally used to describe hopping transport in disordered systems. In the small-polaron model as well as in disordered hopping systems the resonance integral J is in general assumed to be small so that a perturbation-in-J treatment becomes appropriate. Accordingly, the theoretical methods employed to study hopping motion in both systems bear a number of common features. Making use of methods of the small-polaron theory, electron transport in disordered systems can be investigated in the presence of electron-phonon interaction of arbitrary strength. Such a theory was developed by Böttger and Bryksin (1975c). It is applicable to disordered as well as ordered systems and it is valid for electrical fields of arbitrary strength. Unlike other theories, it also accounts for nondiagonal elements of the density matrix and, therefore, renders it possible, to investigate transport properties without an a priori assumption of localized states, too. This theory was generalized such as to describe the hopping-Hall effect in disordered systems (Böttger and Bryksin, 1977a, b, c, d).

In crystalline semiconductors, charge carrier localization by strong electrical fields, and thus the occurrence of the Stark ladder, is due to reflection of the carriers on the boundaries of the Brillouin zone (Wannier, 1960, Hacker and Obermair, 1970). If the spacing between the levels of the Stark ladder is larger than their uncertainty due to interaction with phonons, the conductivity is due to phonon-assisted hops between the levels of the Stark ladder. In this case a resonance situation occurs for optical phonons, which gives rise to an oscillation of the current in dependence of the field ("electro-phonon resonance"; Bryksin and Firsov, 1970a, 1971a, 1972b, Maekawa, 1970).

In the small-polaron model, the hopping conductivity in strong electrical fields was first investigated by Efros (1967). Within the framework of this model it was pointed out by Bryksin and Firsov (1970a) that, by raising the field, the mobility decreases in the tunnelling regime but increases in the hopping regime, which suggests an enhanced tendency to localization due to a strong field. The field dependence of the conductivity in disordered hopping systems was studied by Böttger and Bryksin (1979, 1980) on the basis of the general formalism proposed by Böttger and Bryksin (1975). For not too strong electrical fields, the conductivity was shown to decrease with increasing field. This behaviour markedly differs from earlier theoretical predictions.

Note that there is yet a multitude of interesting aspects of hopping transport which we have not mentioned in this brief overview but which will be described in detail in the following sections. We close this section by giving an overview of review articles and monographs dealing with problems of hopping transport. Theoretical and experimental aspects of small polarons have been reviewed by Appel (1968), Mott and Austin (1969), Firsov (1974), Emin (1973, 1975), and Böttger and Bryksin (1976a, b) (theoretical) and Spear (1974) (experimental). The hopping transport in disordered systems is the subject of the reviews by Shklovskii (1972a), Bonch-Bruevich et al. (1973), Shklovskii and Efros (1975), Böttger and Bryksin (1976a, b, 1982), Banyai and Aldea (1979), Zvyagin (1980), and Butcher (1980) and the monograph by Shklovskii and Efros (1979). The basic ideas of the theory of electronic states and charge transport in disordered solids are described in the monographs by Mott and Davis (1979) and Bonch-Bruevich et al. (1981), where in the former monograph a great deal of experimental work done in this field is reviewed, too. Aspects of atom hopping are considered, for example, in the reviews by Dieterich et al. (1980) and Alexander et al. (1981).

2. General formalism

2.1. The Hamiltonian for a hopping system

A theoretical investigation of the transport in a hopping system requires a model Hamiltonian which renders it possible to account for the essential features of the specific system. The Hamiltonian must include the electron subsystem, the phonon subsystem, the electron-phonon interaction and the interaction with the external fields applied. In the case of a disordered system, at least the Hamiltonian of the electron subsystem must contain parameters which allow effects of disorder to be accounted for.

2.1.1. The electron Hamiltonian

On studying hopping transport the electron subsystem is in general described in terms of the tight-binding approximation. Furthermore, for the sake of simplicity, each atom (site) is assumed to possess only a single energy level, which implies the assumption that interband transitions (in ordered systems) or transitions between different levels of an atom (in disordered systems) do not affect the charge transport.

The one-electron wavefunction of the electron subsystem, $\Psi(\boldsymbol{r})$, is expressed as a linear superposition

$$\Psi(\boldsymbol{r}) = \sum_m a_m \varphi_m(\boldsymbol{r} - \boldsymbol{R}_m) \tag{2.1.1}$$

of the atomic electron wavefunctions, $\varphi_m(\boldsymbol{r} - \boldsymbol{R}_m)$, each localized about a particular site and depending on the position vector $\boldsymbol{R}_m$ and, possibly in discordered systems, the site index m of that site. The equations for the coefficients a_m can be obtained by substitution of (2.1.1) into the Schrödinger equation of the system. Although the atomic wavefunctions located at different sites are not really orthogonal, in practice the corresponding overlap integrals are assumed to be small and in general completely neglected, which is in accordance with the assumptions appropriate to the case of tight-binding. In the case of ordered systems, the problem of nonorthogonality of the functions $\varphi(\boldsymbol{r} - \boldsymbol{R}_m)$ (in such systems the index m is lacking) can be formally by-passed by using Wannier functions for $\varphi_m(\boldsymbol{r} - \boldsymbol{R}_m)$ instead of atomic wavefunctions. On the other hand, in disordered systems in the energy range of interest all states are supposed to be localized due to disorder. If the localization radius of the wavefunctions is less than the interatomic spacing (which is the case, e.g., in weakly doped semiconductors) the states are localized at the lattice (impurity) sites. Therefore, in the case of disordered systems, $\varphi_m(\boldsymbol{r} - \boldsymbol{R}_m)$ may be viewed as localized state having its localization center at $\boldsymbol{R}_m$. Consequently, the neglection of the overlap integrals between functions $\varphi_m(\boldsymbol{r} - \boldsymbol{R}_m)$ located at different sites seems to be a reasonable approximation.

With regard to the absence of a spin index at the functions $\varphi_m(\boldsymbol{r} - \boldsymbol{R}_m)$ in (2.1.1), note that this implies that each site can be occupied at most by a single electron.

In other words, the Hubbard repulsion energy is supposed to be strong enough to prevent doubly occupied states (sites) (see below).

In what follows, we use the formalism of the second quantization. To this end, we introduce the operators a_m^+ (creation operator) and a_m (annihilation operator) which, respectively, create and annihilate electrons in the state $\varphi_m(\boldsymbol{r} - \boldsymbol{R}_m)$ located at the site m. These operators satisfy the common Fermi anticommutation relations

$$[a_{m'}^+, a_m]_+ = \delta_{mm'}, [a_{m'}^+, a_m^+]_+ = [a_{m'}, a_m]_+ = 0, \tag{2.1.2}$$

where

$$[A, B]_+ = AB + BA.$$

Let us assume that the one-electron potential may be written as a sum of atomic potentials, so that in the one-electron approximation the Hamiltonian of the electron subsystem becomes

$$\mathcal{H}_e = -\frac{\hbar^2}{2m_e} \Delta + \sum_m U_m(\boldsymbol{r} - \boldsymbol{R}_m), \tag{2.1.3}$$

where m_e is the electron mass, Δ the Laplace operator, and $U_m(\boldsymbol{r} - \boldsymbol{R}_m)$ the contribution of the mth atom to the potential.

Taking into account that the functions $\varphi_m(\boldsymbol{r} - \boldsymbol{R}_m)$ are solutions of the equations

$$\left[-\frac{\hbar^2}{2m} \Delta + U_m(\boldsymbol{r} - \boldsymbol{R}_m)\right] \varphi_m(\boldsymbol{r} - \boldsymbol{R}_m) = \varepsilon_m \varphi_m(\boldsymbol{r} - \boldsymbol{R}_m),$$

we obtain the following expression for the electron Hamiltonian written in the representation of the second quantization:

$$\mathcal{H}_e = \sum_m \varepsilon_m a_m^+ a_m + \sum_{m' \neq m} J_{m'm} a_{m'}^+ a_m, \tag{2.1.4}$$

where

$$\left.\begin{aligned} \varepsilon_m &= \int d\boldsymbol{r} \varphi_m^*(\boldsymbol{r} - \boldsymbol{R}_m)\, \mathcal{H}_e \varphi_m(\boldsymbol{r} - \boldsymbol{R}_m), \\ J_{m'm} &= \int d\boldsymbol{r} \varphi_{m'}^*(\boldsymbol{r} - \boldsymbol{R}_{m'})\, \mathcal{H}_e \varphi_m(\boldsymbol{r} - \boldsymbol{R}_m). \end{aligned}\right\} \tag{2.1.5}$$

Here ε_m is the electronic energy of the (isolated) mth atom, and $J_{m'm}$ the resonance integral between the sites m' and m. $J_{m'm}$ is determined by the overlap of the wavefunctions located at neighbouring sites. In the theory of hopping transport, the resonance integral is a small quantity. Nevertheless, it must be accounted for, because just it causes the charge transfer from site to site.

For an ordered (crystalline) system, the quantity ε_m is independent of m and may be made equal to zero, which defines the zero point of the energy scale. Furthermore, in this case the resonance integral $J_{m'm}$, which depends on $\boldsymbol{R}_m - \boldsymbol{R}_{m'}$, is not a random quantity, because the position vectors $\boldsymbol{R}_m$ form a regular lattice. Therefore, in (2.1.4) we may pass over from the site to the Fourier (momentum) representation (i.e. from the Wannier to the Bloch representation) as

$$a_m = N^{-1/2} \sum_{k} a_k \exp(\mathrm{i}\boldsymbol{k}\boldsymbol{R}_m),$$

$$a_k = N^{-1/2} \sum_{m} a_m \exp(-\mathrm{i}\boldsymbol{k}\boldsymbol{R}_m). \tag{2.1.6}$$

In this way we obtain

$$\mathcal{H}_e = \sum_{k} \varepsilon(\boldsymbol{k})\, a_k^+ a_k, \tag{2.1.7}$$

where

$$\varepsilon(\boldsymbol{k}) = \sum_{m} J(\boldsymbol{R}_m) \exp(\mathrm{i}\boldsymbol{k}\boldsymbol{R}_m). \tag{2.1.8}$$

Here N is the total number of sites in the system. The Hamiltonian (2.1.7) has the common form of a band Hamiltonian, and $\varepsilon(\boldsymbol{k})$ (2.1.8) is the familiar tight-binding expression for the $\boldsymbol{k}$-dependence of the energy.

On the other hand, in disordered systems ε_m and/or $J_{m'm}$ are random quantities. In the case of the so-called on-diagonal disorder, only ε_m is random, while in that of the so-called off-diagonal disorder, $J_{m'm}$ is random. In actual systems such as doped semiconductors, in general, both ε_m and $J_{m'm}$ are random quantities. A survey of typical disorder effects on electron states and charge transport will be given in Sections 4. and 5.4. Among the major disorder effects are the localization phenomenon and the related hopping mechanism.

In the literature there are different definitions of the term localization (see below). According to one definition there is no dc (direct current) conductivity at zero temperature if the Fermi level is in the region of localized states. Accordingly, a Hamiltonian of the form (2.1.4) may be used to calculate the dc conductivity only if the charge carries are not localized, because in the absence of electron-phonon interaction, by definition, only in this case it provides a finite conductivity. In this case the transport is not due to hopping, but due to nearly free propagation or diffusion, governed by scattering on irregularities of the system. Within the framework of the Hamiltonian (2.1.4), hopping can occur only in the presence of an alternating electrical field. This hopping phenomenon is the so-called photon-induced hopping (cf. Section 6.). Consequently, an investigation of hopping transport at arbitrary frequencies, including the dc limit, requires that we account for electron-phonon interaction.

2.1.2. *Electron-phonon interaction*

Within the framework of the tight-binding formalism presented in Section 2.1.1., an expression for the Hamiltonian of the electron-phonon interaction, $\mathcal{H}_{e-ph}$, can be obtained as follows: We replace in the "atomic" potential $U_m(\boldsymbol{r} - \boldsymbol{R}_m)$ (cf. (2.1.3)) the vector $\boldsymbol{R}_m$ by $\boldsymbol{R}_m + \boldsymbol{u}_m$ in order to account for the small vibrational displacements $\boldsymbol{u}_m$ of the mth atom around its equilibrium position $\boldsymbol{R}_m$. Then we expand $U_m(\boldsymbol{r} - \boldsymbol{R}_m - \boldsymbol{u}_m)$ in powers of $\boldsymbol{u}_m$ up to linear terms and employ the formalism of the second quantization to express $\boldsymbol{u}_m$ by phonon creation and annihilation operators. It is interesting to note that in disordered systems the disorder can also strongly affect the vibrational behaviour of the system (cf., e.g., BÖTTGER, 1983). However, so far only little work has been done to examine the effect of disorder-affected vibrational modes on electron transport phenomena, because this is a rather complex mathematical problem.

In what follows, for the sake of simplicity, we assume that the vibrational subsystem is not affected by disorder. One may argue that this approximation should be valid at least in the important case of impurity conduction in doped semiconductors, because in this case the number of impurity atoms, between which hopping takes place, is much less than the number of the regularly arranged host atoms, which are expected to determine the basic vibrational behaviour of the system. However, even in this case, under certain conditions, the impurities can cause local vibrational modes, which can essentially affect the hopping transport (cf. AHMED and BLACKMAN, 1980a, b).

Using the common quantization procedure for lattice vibrations (see, e.g., KITTEL, 1963, BÖTTGER, 1983), we find that

$$\mathcal{H}_{\text{e-ph}} = \sum_{m\boldsymbol{q}} a_m^+ a_m \hbar\omega_{\boldsymbol{q}} \{u_m(\boldsymbol{q})\, b_{\boldsymbol{q}} + u_m^*(\boldsymbol{q})\, b_{\boldsymbol{q}}^+\}. \tag{2.1.9}$$

Here $\omega_{\boldsymbol{q}}$ is the frequency of the phonon with the wave vector $\boldsymbol{q}$ ($\boldsymbol{q}$ is assumed also to include the index of the vibrational branch), and $b_{\boldsymbol{q}}^+$ and $b_{\boldsymbol{q}}$, respectively, are the phonon creation and annihilation operators satisfying the Bose commutation relations

$$[b_{\boldsymbol{q}'}, b_{\boldsymbol{q}}^+]_- = \delta_{\boldsymbol{q}',\boldsymbol{q}}, \; [b_{\boldsymbol{q}'}^+, b_{\boldsymbol{q}}^+]_- = [b_{\boldsymbol{q}'}, b_{\boldsymbol{q}}]_- = 0. \tag{2.1.10}$$

The quantity $u_m(\boldsymbol{q})$ is related to the dimensionless electron-phonon coupling constant, γ, as

$$u_m(\boldsymbol{q}) = (2N)^{-1/2}\, \gamma_m(\boldsymbol{q}) \exp(-\mathrm{i}\boldsymbol{q}\boldsymbol{R}_m). \tag{2.1.11}$$

In disordered systems the quantity γ may in general depend on the site index m. The Hamiltonian (2.1.9) is widely used in the theory of transport phenomena. For more details on the derivation of this so-called Fröhlich Hamiltonian, the reader is referred to the original papers of HOLSTEIN (1959) and LANG and FIRSOV (1962).

We must yet add to the Hamiltonian (2.1.9) the Hamiltonian of the vibrational subsystem

$$\mathcal{H}_{\text{ph}} = \sum_{\boldsymbol{q}} \hbar\omega_{\boldsymbol{q}}(b_{\boldsymbol{q}}^+ b_{\boldsymbol{q}} + 1/2). \tag{2.1.12}$$

In the special case of an ordered system, the electron operators in the Hamiltonian (2.1.9) may be expressed by the Fourier representation (2.1.6), which yields

$$\mathcal{H}_{\text{e-ph}} = \sum_{\boldsymbol{k},\boldsymbol{q}} \frac{\hbar\omega_{\boldsymbol{q}}}{\sqrt{2N}} \left(a_{\boldsymbol{k}}^+ a_{\boldsymbol{k}-\boldsymbol{q}} \gamma^*(\boldsymbol{q})\, b_{\boldsymbol{q}}^+ + a_{\boldsymbol{k}}^+ a_{\boldsymbol{k}+\boldsymbol{q}} \gamma(\boldsymbol{q})\, b_{\boldsymbol{q}}\right). \tag{2.1.13}$$

This is the familiar form of the Fröhlich Hamiltonian.

It should be noted that in (2.1.9) we have neglected terms proportional to $a_{m'}^+ a_m$, where $m' \neq m$. Strictly speaking, such a neglection may be justified only if the overlap between the atomic wavefunctions located at neighbouring sites is weak, so that the corresponding coupling constant $\gamma_{mm'}$ is small compared with γ_m. In the systems under consideration this requirement is fulfilled, but note that in the literature the model Hamiltonian (2.1.13) is also used in cases where the overlap is not small (for instance, on deriving the Boltzmann equation for scattering of electrons in broad energy bands). On account of nondiagonal terms, in (2.1.13) the coupling constant also depends on the electron momentum $\boldsymbol{k}$.

In the absence of external fields, the total Hamiltonian of the electron-phonon system, $\mathcal{H}$, is composed of the contributions (2.1.4), (2.1.9), and (2.1.12):

$$\mathcal{H} = \mathcal{H}_{\text{e}} + \mathcal{H}_{\text{ph}} + \mathcal{H}_{\text{e-ph}}. \tag{2.1.14}$$

2.1.3. *Interaction with electrical and magnetic fields*

Let us now consider the interaction with an external electrical field $\boldsymbol{E}$ and a magnetic field $\boldsymbol{H}$. Assuming not too high electrical fields, then the material parameters γ, $\boldsymbol{J}$, $\omega_{\boldsymbol{q}}$ etc. do not depend on $\boldsymbol{E}$, and the electrical field gives rise to a Hamiltonian

$$\mathcal{H}_E = -\boldsymbol{DE} \tag{2.1.15}$$

in addition to the Hamiltonian (2.1.14). Here $\boldsymbol{D}$ denotes the operator of the dipole moment

$$\boldsymbol{D} = e \sum_{mm'} \boldsymbol{d}_{m'm} a_{m'}^+ a_m, \tag{2.1.16}$$

where

$$\begin{aligned} \boldsymbol{d}_{m'm} &= \int d\boldsymbol{r}\, \boldsymbol{r} \varphi_{m'}^*(\boldsymbol{r} - \boldsymbol{R}_{m'})\, \varphi_m(\boldsymbol{r} - \boldsymbol{R}_m) \\ &= \boldsymbol{R}_m \delta_{m'm} + \int d\boldsymbol{r} \boldsymbol{r} \varphi_{m'}^* \left(\boldsymbol{r} - \frac{\boldsymbol{R}_m{}' - \boldsymbol{R}_m}{2}\right) \varphi_m \left(\boldsymbol{r} + \frac{\boldsymbol{R}_{m'} - \boldsymbol{R}_m}{2}\right), \end{aligned} \tag{2.1.17}$$

and $e(\equiv -|e|)$ is the electronic charge.

By neglecting the nondiagonal elements with respect to the site indices, (2.1.16) reduces to

$$\mathcal{H}_E = -e\boldsymbol{E} \sum_m \boldsymbol{R}_m a_m^+ a_m. \tag{2.1.18}$$

The expression (2.1.18) is exact if $\varphi_{m'}(-\boldsymbol{r}) = \varphi_m(\boldsymbol{r})$, which holds in ordered systems with inversion symmetry, where the Wannier states have this property. In what follows, we assume that (2.1.18) is a reasonable approximation for disordered solids, too.

Let us now turn to the Hamiltonian in the presence of an external spatially constant magnetic field $\boldsymbol{H}$ (HOLSTEIN, 1961). On account to the magnetic field, the electron Hamiltonian (2.1.3) reads

$$\mathcal{H}_{\text{e}} = \frac{1}{2m_{\text{e}}} \left(\boldsymbol{p} - \frac{e}{c} \boldsymbol{A}(\boldsymbol{r})\right)^2 + \sum_m U_m(\boldsymbol{r} - \boldsymbol{R}_m), \tag{2.1.19}$$

where $\boldsymbol{p} = -\mathrm{i}\hbar\nabla$ is the momentum operator and $\boldsymbol{A}(\boldsymbol{r}) = \frac{1}{2}[\boldsymbol{H}\times\boldsymbol{r}]$ the vector potential.

In the presence of a homogeneous magnetic field, an important property of the wavefunction is the formal loss of the translational symmetry, even in the case of an ordered system. This property is reflected by a phase factor in the wavefunction which depends on the magnetic field. To see this, consider a lattice translation vector $\boldsymbol{R}_0$ with the property $\sum_m U_m(\boldsymbol{r}+\boldsymbol{R}_0-\boldsymbol{R}_m) = \sum_m U_m(\boldsymbol{r}-\boldsymbol{R}_m)$, i.e. the points $\boldsymbol{r}$ and $\boldsymbol{r}+\boldsymbol{R}_0$ are physically equivalent. Replacing in (2.1.19) $\boldsymbol{r}$ by $\boldsymbol{r}+\boldsymbol{R}_0$, we find that the Hamiltonian remains unchanged if one introduces the gauge transformation $\Psi(\boldsymbol{r}) \to \Psi(\boldsymbol{r})\exp\left\{\mathrm{i}\frac{e}{\hbar c}\boldsymbol{A}(\boldsymbol{r})\,\boldsymbol{R}_0\right\}$, where $\Psi(\boldsymbol{r})$ denotes the eigenfunction of the Hamiltonian.

Just as in the absence of a magnetic field (cf. Section 2.1.1.), we now assume that the eigenstates of the Hamiltonian (2.1.19) may be represented as a superposition of local atomic wavefunctions $\varphi_m(\boldsymbol{r},\boldsymbol{R}_m\,|\,\boldsymbol{H})$, which are now solutions of the equations

$$\left\{\frac{1}{2m_\mathrm{e}}\left(\boldsymbol{p}-\frac{e}{c}\boldsymbol{A}(\boldsymbol{r})\right)^2 + U_m(\boldsymbol{r}-\boldsymbol{R}_m)\right\}\varphi_m(\boldsymbol{r},\boldsymbol{R}_m\,|\,\boldsymbol{H}) = \varepsilon_m(\boldsymbol{H})\,\varphi_m(\boldsymbol{r},\boldsymbol{R}_m\,|\,\boldsymbol{H}). \tag{2.1.20}$$

These functions will obviously not have the form $\varphi_m(\boldsymbol{r}-\boldsymbol{R}_m\,|\,\boldsymbol{H})$. However, introducing the gauge transformation

$$\varphi_m(\boldsymbol{r},\boldsymbol{R}_m\,|\,\boldsymbol{H}) = \chi_m(\boldsymbol{r}-\boldsymbol{R}_m\,|\,\boldsymbol{H})\exp\left\{\mathrm{i}\frac{e}{\hbar c}\boldsymbol{A}(\boldsymbol{r})\,\boldsymbol{R}_m\right\}, \tag{2.1.21}$$

we find, instead of (2.1.20), the equation

$$\left\{\frac{1}{2m_\mathrm{e}}\left(\boldsymbol{p}-\frac{e}{2c}\boldsymbol{H}\times(\boldsymbol{r}-\boldsymbol{R}_m)\right)^2 + U_m(\boldsymbol{r}-\boldsymbol{R}_m)\right\}\chi_m(\boldsymbol{r}-\boldsymbol{R}_m\,|\,\boldsymbol{H}) = \varepsilon_m(\boldsymbol{H})\,\chi_m(\boldsymbol{r}-\boldsymbol{R}_m\,|\,\boldsymbol{H}), \tag{2.1.22}$$

the solution of which clearly depends only on the difference $\boldsymbol{r}-\boldsymbol{R}_m$.

Proceeding further as in Section 2.1.1., we obtain the following expression for the electron Hamiltonian in the presence of a magnetic field

$$\mathcal{H}_\mathrm{e} = \sum_m \varepsilon_m(\boldsymbol{H})\,a_m^+a_m + \sum_{m\neq m'} J_{m'm}(\boldsymbol{H})\,a_{m'}^+a_m \tag{2.1.23}$$

where

$$\left.\begin{aligned}
\varepsilon_m(\boldsymbol{H}) &= \int \mathrm{d}\boldsymbol{r}\chi_m{}^*(\boldsymbol{r}-\boldsymbol{R}_m\,|\,\boldsymbol{H})\,\mathcal{H}_\mathrm{e}\chi_m(\boldsymbol{r}-\boldsymbol{R}_m\,|\,\boldsymbol{H}),\\
J_{m'm}(\boldsymbol{H}) &= \left[\exp\left\{\frac{\mathrm{i}e}{\hbar c}\boldsymbol{A}(\boldsymbol{R}_m)\,\boldsymbol{R}_{m'}\right\}\right]\\
&\quad\times\int \mathrm{d}\boldsymbol{r}\chi_{m'}^*(\boldsymbol{r}-\boldsymbol{R}_{m'}\,|\,\boldsymbol{H})\,\mathcal{H}_\mathrm{e}\chi_m(\boldsymbol{r}-\boldsymbol{R}_m\,|\,\boldsymbol{H}).
\end{aligned}\right\} \tag{2.1.24}$$

The dependence of ε_m on $\boldsymbol{H}$ describes the Zeeman effect. It must be accounted for primarily on studying the magnetic susceptibility of an Anderson-localized system. In what follows, however, we may ignore this dependence.

The dependence of the wavefunctions χ_m on $\boldsymbol{H}$ is important for studying the magnetoresistance, because the magnetic field affects the transition probability between neighbouring sites via a deformation of these functions.

Finally, let us turn to the phase factor appearing in the resonance integral (2.1.24). For the sake of simplicity, we assume non-degenerate s-functions as zero-field local atomic functions. In this case, first-order terms with respect to the magnetic field disappear from (2.1.22). For a study of the Hall effect, higher-order corrections to χ_m are irrelevant and may be ignored. This means that the hopping-Hall effect is entirely governed by the phase factor of the resonance integral (2.1.24). Accordingly, for a study of the Hall effect, the resonance integral (2.1.24) may be written as

$$J_{m'm}(\boldsymbol{H}) = J_{m'm} \exp\left\{\frac{ie}{2\hbar c}\, \boldsymbol{H}[\boldsymbol{R}_m \times \boldsymbol{R}_{m'}]\right\}, \tag{2.1.25}$$

where $J_{m'm}$ is the resonance integral in the absence of $\boldsymbol{H}$, which is defined in (2.1.5).

It should be noted that a phase factor appears only in those terms of the Hamiltonian which are nondiagonal in the site indices, since $\boldsymbol{A}(\boldsymbol{R}_m)\, \boldsymbol{R}_m = 0$. Therfore, a phase factor does occur neither in the Hamiltonian (2.1.9) of the electron-phonon interaction, nor in the Hamiltonian (2.1.18), which describes the interaction with the external electrical field.

2.2. The polaron canonical transformation

A perturbation treatment of any physical problem requires an appropriate zeroth order Hamiltonian. For instance, in the case of the band transport in the presence of a weak electron-phonon interaction, one may choose the Hamiltonian of the system without electron-phonon interaction as the zeroth-order Hamiltonian. However, such a choice is reasonable only for crystalline systems, where the electron Hamiltonian can be readily diagonalized[1]). On the other hand, for disordered systems, a diagonalization is in practice infeasible, because it would be nothing else but the solution of the localization problem. However, note that in (disordered) hopping systems the resonance integral $J_{m'm}$ must be sufficiently small. Accordingly, for such systems, it is tempting to use the nondiagonal part of the electron Hamiltonian, which is proportional to $J_{m'm}$, as the perturbation and the remainder (i.e. the diagonal part of the electron Hamiltonian, the electron-phonon interaction, and the phonon Hamiltonian) as the zeroth-order Hamiltonian. The latter Hamiltonian describes the behaviour of an electron that is located at a site and interacting with

---

[1]) For crystals, the Hamiltonian (2.1.23) of the electron subsystem is diagonal in the momentum representation. In the presence of a magnetic field, the translational symmetry can be restored by using the Wigner representation for the density matrix, whereby one works with a nondiagonal zeroth order Hamiltonian (LEVINSON, 1969, BRYKSIN and FIRSOV, 1973).

the phonon subsystem. For sufficiently strong electron-phonon interaction, the atoms in the neighbourhood of the given electron are considerably displaced from their equilibrium positions which they would have in the absence of the interaction with the electron, i.e. the electron is surrounded by a polaron "cloud". About the new equilibrium positions the atoms perform small thermal vibrations. This situation was first theoretically analysed by HOLSTEIN (1959) by examining a simple one-dimensional model.

A very useful tool for a perturbation-in-J treatment of an electron-phonon system described by the Hamiltonian (2.1.14) is the polaron canonical transformation (LANG and FIRSOV, 1962), which will be described in the following. To do this, let us introduce the canonically transformed Hamiltonian

$$\tilde{\mathscr{H}} = \exp(-S)\, \mathscr{H} \exp(S). \tag{2.2.1}$$

Here $\exp(-S)$ is a unitary operator, where

$$S = \sum_{mq} a_m^+ a_m [b_q^+ u_m^*(q) - b_q u_m(q)]. \tag{2.2.2}$$

In what follows, we need knowledge of the transformation behaviour of the operators $\tilde{b}_q = \exp(-S)\, b_q \exp(S)$ and $\tilde{a}_m = \exp(-S)\, a_m \exp(S)$. To study this behaviour, we introduce the operator $b_q(\lambda) = \exp(-\lambda S)\, b_q \exp(\lambda S)$, so that $b_q(1) = \tilde{b}_q$. Differentiating $b_q(\lambda)$ with respect to λ, and carrying out the commutation of the operators S and b_q, then, by means of (2.1.10), we find that

$$\frac{\mathrm{d}}{\mathrm{d}\lambda}\big(b_q(\lambda)\big) = -\sum_m a_m^+ a_m u_m^*(q).$$

Integrating this relation over λ with the "initial condition" $b_q(0) = b_q$, we obtain

$$b_q(\lambda) = b_q - \lambda \sum_m a_m^+ a_m u_m^*(q).$$

Making λ equal to unity in this equation, we get the result we have looked for:

$$\tilde{b}_q = b_q - \sum_m a_m^+ a_m u_m^*(q). \tag{2.2.3}$$

Furthermore, taking the Hermitian conjugate of (2.2.3), we get

$$\tilde{b}_q^+ = b_q^+ - \sum_m a_m^+ a_m u_m(q). \tag{2.2.3a}$$

Analogously, introducing $a_m(\lambda) = \exp(-\lambda S)\, a_m \exp(\lambda S)$, differentiating this expression with respect to λ, and evaluating by means of (2.1.2) the commutator $[S, a_m]$, etc., we obtain

$$\tilde{a}_m = a_m \exp\left\{\sum_q [b_q^+ u_m^*(q) - b_q u_m(q)]\right\}, \tag{2.2.4}$$

$$\tilde{a}_m^+ = a_m^+ \exp\left\{-\sum_q [b_q^+ u_m^*(q) - b_q u_m(q)]\right\}. \tag{2.2.4a}$$

It holds that

$$\tilde{\mathcal{H}}(\{b_{\boldsymbol{q}}, b_{\boldsymbol{q}}^{+}, a_m, a_m^{+}\}) = \mathcal{H}(\{\tilde{b}_{\boldsymbol{q}}, \tilde{b}_{\boldsymbol{q}}^{+}, \tilde{a}_m, \tilde{a}_m^{+}\}).$$

Therefore, by means of (2.2.3) to (2.2.4), the explicit form of the canonically transformed Hamiltonian $\tilde{\mathcal{H}}$ can be readily found. Taking for the Hamiltonian $\mathcal{H}$ the sum of the electron Hamiltonian (2.1.23), the phonon Hamiltonian (2.1.12), the electron-phonon interaction (2.1.9), and the interaction with the electrical field (2.1.18), then we get

$$\tilde{\mathcal{H}} = \mathcal{H}_0^E + \mathcal{H}' + \mathcal{H}_{ee}, \tag{2.2.5}$$

where $\mathcal{H}_0^E$ is the new zeroth-order Hamiltonian, which also includes the electrical field,

$$\mathcal{H}_0^E = \mathcal{H}_{ph} + \sum_m \left(\varepsilon_m - E_p(m)\right) a_m^{+} a_m - e\boldsymbol{E} \sum_m \boldsymbol{R}_m a_m^{+} a_m. \tag{2.2.6}$$

The quantity $E_p(m)$ labels the polaron shift at site m

$$E_p(m) = \sum_{\boldsymbol{q}} \hbar\omega_{\boldsymbol{q}} \left|u_m(\boldsymbol{q})\right|^2 = \frac{1}{2N} \sum_{\boldsymbol{q}} \hbar\omega_{\boldsymbol{q}} \left|\gamma_m(\boldsymbol{q})\right|^2 \tag{2.2.7}$$

which describes the lowering of the energy when an electron is localized at site m owing to a polarization displacement of the atoms surrounding it. In disordered systems, the polaron shift (2.2.7) changes from site to site if the electron-phonon coupling constant γ depends on m. Such a dependence may be important when the strength of the local field strongly fluctuates. In what follows, a dependence of γ and E_p on m will be ignored. Thus, in (2.2.6) a constant energy shift E_p occurs, which may be neglected.

The Hamiltonian $\mathcal{H}'$ has the form

$$\mathcal{H}' = \sum_{m \neq m'} J_{m'm}(\boldsymbol{H})\, a_{m'}^{+} a_m \hat{\Phi}_{m'm}, \tag{2.2.8}$$

where $\hat{\Phi}$ is the multiphonon operator

$$\hat{\Phi}_{m'm} = \exp\left\{\sum_{\boldsymbol{q}} \left[b_{\boldsymbol{q}}^{+}\left(u_m^{*}(\boldsymbol{q}) - u_{m'}^{*}(\boldsymbol{q})\right) - b_{\boldsymbol{q}}\left(u_m(\boldsymbol{q}) - u_{m'}(\boldsymbol{q})\right)\right]\right\}. \tag{2.2.9}$$

In the theory of hopping transport, $\mathcal{H}'$ plays the role of the perturbation.

Finally, in (2.2.5) the term

$$\mathcal{H}_{ee} = -\sum_{\boldsymbol{q}, m' \neq m} \hbar\omega_{\boldsymbol{q}} u_m^{*}(\boldsymbol{q})\, u_{m'}(\boldsymbol{q})\, a_{m'}^{+} a_{m'} a_m^{+} a_m \tag{2.2.10}$$

describes the indirect attractive interaction between electrons located at different sites due to the virtual exchange of phonons (this mechanism is analogous to that giving rise to Cooper pairs in superconductors). We recall that in our treatment the intrasite electron repulsion of the Hubbard type has been assumed to be so strong that any atom may be occupied by no more than one electron only. If we desist from this assumption, we must ascribe to each electron operator a spin index σ, in addition to the site index m. Then, the Hamiltonian $\mathscr{H}_{ee}$ takes the form

$$\mathscr{H}_{ee} = -\sum_{m\sigma\boldsymbol{q}} \hbar\omega_{\boldsymbol{q}} |u_m(\boldsymbol{q})|^2 a^+_{m\sigma}a_{m\sigma}a^+_{m\bar{\sigma}}a_{m\bar{\sigma}} - \sum_{\boldsymbol{q},m\neq m',\sigma\sigma'} \hbar\omega_{\boldsymbol{q}} u^*_m(\boldsymbol{q})\, u_{m'}(\boldsymbol{q})\, a^+_{m\sigma}a_{m\sigma}a^+_{m'\sigma'}a_{m'\sigma'}, \tag{2.2.11}$$

where the second term on the right-hand side corresponds to (2.2.10), while the additional first term describes the phonon induced attraction between two electrons with opposite spins σ and $\bar{\sigma}$ located at the same site.

On considering effects of electron-electron interaction, we must still add the direct electron-electron interaction to (2.2.11). This requires that in (2.2.11) we replace

$$\sum_{\boldsymbol{q}} \hbar\omega_{\boldsymbol{q}} u^*_m(\boldsymbol{q})\, u_{m'}(\boldsymbol{q}) \quad \text{by} \quad \sum_{\boldsymbol{q}} \hbar\omega_{\boldsymbol{q}} u^*_m(\boldsymbol{q})\, u_{m'}(\boldsymbol{q}) - v_{m'm},$$

where $v_{m'm}$ is the matrix element of the Coulomb interaction. Note that, according to (2.2.4), the canonical transformation does not change the form of the density operator at site m, $a^+_m a_m$, so that this transformation does not affect the Hamiltonian of the Coulomb interaction.

Thus, the interaction via virtual phonon exchange and the Coulomb repulsion are competing mechanisms. If the sign of the difference $\left(E_p(m) - v_{mm}\right)$ is positive, then there is a tendency to pairing of electrons with opposite spin directions, otherwise the intrasite electron-electron interaction is repulsive. If kT is less than the energy required to break up such an electron pair, then practically all the charge carriers are paired, which would lead to interesting consequences such as the absence of an electron-spin resonance signal and the lack of a paramagnetic susceptibility of the electron system (ANDERSON, 1975, STREET and MOTT, 1975).

In what follows, in general we assume the Hubbard repulsion to be so strong that a double occupation of sites does not occur. Furthermore, we neglect the intersite electron-electron interaction. Accordingly, we may drop the spin index at the electron operators and ignore the term $\mathscr{H}_{ee}$ in (2.2.5).

The form of the Hamiltonian (2.2.5) is very appropriate for studying hopping transport. By means of this Hamiltonian, the following hopping systems can be investigated: (i) small polarons in crystalline materials, (ii) disordered systems with a weak electron-phonon interaction, and (iii) disordered systems with a strong electron-phonon interaction (small polarons in disordered solids). In principle, this Hamiltonian may be used also to examine systems where no localization occurs (e.g., small polarons in crystals at low temperatures, or disordered systems at energies above the so-called mobility edge (see below)). However, such a situation requires a specific treatment, which markedly differs from that used for systems with strongly localized states. Therefore, problems of delocalized (extended) states are outside the scope of this book.

2.3. The expression for the current and its relation to the Kubo formula

For systems with a strong tendency to localization, the site representation is appropriate, and to calculate the charge transport, it is suitable to start from an expression for the current that is valid for electrical fields of arbitrary strength. The latter statement is true even in the case that one is only interested in the Ohmic behaviour. In this case it is suitable at first to derive the transport equation for an arbitrary field $\boldsymbol{E}$, and only after that to linearize this equation with respect to $\boldsymbol{E}$, though a consequent application of the Kubo formula would, of course, also provide correct results (Böttger and Bryksin, 1976f). Hence, the starting point for the transport theory developed here will be the expression for the current density $\boldsymbol{j}$ in the presence of an electrical field of arbitrary strength.

To derive an expression for the current, we must define the conditions of turning on the electrical field. For times $t < 0$, let the electrical field be absent. Thus, for $t < 0$, the system is in the thermodynamic equilibrium described by the Hamiltonian $\mathcal{H}$ for $\boldsymbol{E} = 0$, which we label $\mathcal{H}^{(0)}$. A magnetic field may be included in $\mathcal{H}^{(0)}$, because such a field does not cause a current, i.e. it does not give rise to a non-equilibrium state. Note that at sufficiently large times after turning on the electrical field the behaviour of the system does not depend on the initial state of the system, i.e. on the choice of the Hamiltonian $\mathcal{H}^{(0)}$, provided that the theory is formulated correctly.

At $t = 0$, the electrical field is suddenly switched on, so that at $t > 0$, the system is described by the Hamiltonian $\mathcal{H}$. At $t > 0$, first relaxational currents occur (transition regime), but after a short time the current becomes stationary (stationary regime). In what follows, we will in general study only the current in the stationary regime.

With aid of the usual relation between the current density and the dipole moment, the current density operator $\hat{\boldsymbol{j}}$ can be expressed as

$$\hat{\boldsymbol{j}}(t) = \frac{1}{\Omega} \frac{\mathrm{d}\boldsymbol{D}(t)}{\mathrm{d}t}, \tag{2.3.1}$$

where

$$\boldsymbol{D}(t) = U^+(t)\, \boldsymbol{D} U(t),$$

$$U(t) = \exp\left\{-\frac{\mathrm{i}}{\hbar} \int_0^t \mathrm{d}t' \mathcal{H}(t')\right\}.$$

Here $U(t)$ is the usual time-evolution operator (cf., e.g., Zubarev, 1971) and Ω the volume of the system. By neglecting the nondiagonal elements[1]) with respect to the

[1]) In the limit as $t \to \infty$ (in the stationary regime of the current) the nondiagonal elements of $\boldsymbol{D}$ with respect to the site indices can be shown (Firsov, 1975) to give no contribution to the current, because the corresponding matrix elements $\boldsymbol{d}_{m'm}$ have a finite range, while $\boldsymbol{R}_m$ may assume any large values in an infinite system.

site indices, the operator of the dipole moment, $\boldsymbol{D}$, becomes

$$\boldsymbol{D} = e \sum_m \boldsymbol{R}_m a_m^+ a_m . \tag{2.3.2}$$

On averaging $\boldsymbol{j}$ (2.3.1) over the thermodynamic equilibrium distribution governed by $\mathcal{H}^{(0)}$, and using for $\boldsymbol{D}$ the expression (2.3.2), we find that the current density $\boldsymbol{j}$ is given by

$$\boldsymbol{j}(t) = \frac{e}{\Omega} \sum_m \boldsymbol{R}_m \mathrm{d}\varrho_m(t)/\mathrm{d}t . \tag{2.3.3}$$

Here ϱ_m is the mth diagonal element of the density matrix. This quantity describes the time evolution of the thermodynamically averaged concentration of the charge carriers at site m after turning on the electrical field. It is defined as

$$\varrho_m(t) = \langle U^+(t)\, a_m^+ a_m U(t) \rangle , \tag{2.3.4}$$

where $\langle \ldots \rangle$ indicates the thermodynamic average over the equilibrium distribution:

$$\langle A \rangle \equiv Tr\{\mathrm{e}^{-\beta(\mathcal{H}^{(0)} - \mu \hat{N})} A\} / Tr\{\mathrm{e}^{-\beta(\mathcal{H}^{(0)} - \mu \hat{N})}\} . \tag{2.3.5}$$

Here μ is the chemical potential, $\beta = 1/kT$, and $\hat{N} = \sum_m a_m^+ a_m$ is the operator of the total particle number. $\hat{N}$ commutates with $\mathcal{H}$ as well as with $\mathcal{H}^{(0)}$ (i.e. the total particle number is conserved in the system).

The expression for the current (2.3.3) may be used to describe various transport phenomena such as transport in weak or strong electrical fields, in the presence or absence of a magnetic field, for weak or strong coupling with phonons, etc. The problem of evaluating the current will be studied in the following section.

Let us now point out that in the linear approximation with respect to the electrical field (i.e. with respect to $\mathcal{H}_E$) the relation (2.3.3) passes into the well-known Kubo formula. To this end, we decompose $\mathcal{H}(t)$ as $\mathcal{H}(t) = \mathcal{H}^{(0)} + \mathcal{H}_E(t)$ and expand the evolution operator $U(t)$ in a power series with respect to $\mathcal{H}_E$ up to the linear term by using the Dyson expansion

$$U(t) \approx \mathrm{e}^{-\frac{\mathrm{i}}{\hbar}\mathcal{H}^{(0)}t} \left\{ 1 - \frac{\mathrm{i}}{\hbar} \int_0^t \mathrm{d}t_1 \mathrm{e}^{\frac{\mathrm{i}}{\hbar}\mathcal{H}^{(0)}t_1} \mathcal{H}_E(t_1)\, \mathrm{e}^{-\frac{\mathrm{i}}{\hbar}\mathcal{H}^{(0)}t_1} \right\} . \tag{2.3.6}$$

Inserting (2.3.6) into (2.3.4), neglecting the zeroth-order term (because for $\boldsymbol{E} = 0$, we have $\boldsymbol{j} = 0$), and accounting for (2.1.15), after some simple manipulations, we find that

$$j_\mu(t) = \frac{\mathrm{i}}{\hbar\Omega} \int_0^t \mathrm{d}t' \langle \dot{D}_\mu(t')\, D_\nu - D_\nu \dot{D}_\mu(t') \rangle\, E_\nu(t - t') ,$$

where μ, ν label the Cartesian components ($\mu, \nu = x, y, z$), and $\dot{\boldsymbol{D}}(t) = \mathrm{d}\boldsymbol{D}(t)/\mathrm{d}t$. Assuming $\boldsymbol{E}(t) = \boldsymbol{E} \exp(\mathrm{i}\omega t)$ and, accordingly, $\boldsymbol{j}(t) = \boldsymbol{j}(\omega) \exp(\mathrm{i}\omega t)$, then the conductivity tensor $\sigma_{\mu\nu}(\omega)$ $\left(j_\mu(\omega) = \sigma_{\mu\nu}(\omega)\, E_\nu\right)$ is given by

$$\sigma_{\mu\nu}(\omega) = \frac{\mathrm{i}}{\hbar\Omega} \lim_{t\to\infty} \int\limits_0^t \mathrm{d}t' \mathrm{e}^{-\mathrm{i}\omega t'} \langle \dot{D}_\mu(t')\, D_\nu - D_\nu \dot{D}_\mu(t')\rangle. \tag{2.3.7}$$

Here the limit $t \to \infty$ was taken in order to exclude the transition regime of the current. The relation (2.3.7) is Kubo's linear response formula in the dipole moment representation (cf., e.g., ZUBAREV, 1971). By integration by parts, the right-hand side of (2.3.7) can be expressed by correlators of the current.

Concluding this section, let us draw attention to the fact that the formula for the current (2.3.3) remains valid also if we replace therein $\mathcal{H}^{(0)}$ by $\tilde{\mathcal{H}}^{(0)}$ and $\mathcal{H}$ by $\tilde{\mathcal{H}}$. To see this, under the sign for the trace in the expression for ϱ_m we insert from the left the operator exp $(-S)$ and from the right the operator exp (S). Then, on account of the fact that the operator $a_m^+ a_m$ is not affected by the canonical transformation, we find that the form of $\varrho_m(t)$ remains unchanged, except that $\mathcal{H}^{(0)}$ is replaced by $\tilde{\mathcal{H}}^{(0)}$ and $\mathcal{H}$ by $\tilde{\mathcal{H}}$.

2.4. The equation of motion for the density matrix in the hopping regime

2.4.1. *Perturbation theory*

To examine $\varrho_m(t)$ we use a perturbation expansion which is obtained by expanding in (2.3.4) the operators $U(t)$, $U(t)^+$, and exp $(-\beta\mathcal{H}^{(0)})$ in powers with respect to $\mathcal{H}'$ (cf. (2.2.5)). Note that thereby the electron-phonon interaction is not required to be weak. Using the Dyson expansion, the evolution operator can be represented by an infinite power series as

$$U(t) = U_0(t) \left\{ 1 + \left(-\frac{\mathrm{i}}{\hbar}\right) \int\limits_0^t \mathrm{d}t_1 \mathcal{H}'(t_1) + \left(-\frac{\mathrm{i}}{\hbar}\right)^2 \int\limits_0^t \mathrm{d}t_2 \int\limits_0^{t_2} \mathrm{d}t_1 \mathcal{H}'(t_2)\, \mathcal{H}'(t_1) + \dots \right\}, \tag{2.4.1}$$

where U_0 is the evolution operator for free particles

$$U_0(t) = \exp\left\{ -\frac{\mathrm{i}}{\hbar} \int\limits_0^t \mathrm{d}t' \mathcal{H}_0^E(t') \right\}$$

and

$$\mathcal{H}'(t) = U_0^+(t)\, \mathcal{H}' U_0(t). \tag{2.4.2}$$

$U^+(t)$ and $\exp(-\beta\mathcal{H}^{(0)})$ can be analogously expressed as power series with respect to $\mathcal{H}'$. Thereby, for $\exp(-\beta\mathcal{H}^{(0)})$, in (2.4.2) we must replace t by $-i\hbar\beta$ and $\mathcal{H}_0{}^E$ by $\mathcal{H}_0(\equiv \mathcal{H}_0{}^E$ for $\boldsymbol{E} = 0)$.

Each term of the perturbation series for $\varrho_m(t)$ can be represented by diagrams on a contour C in the complex t-plane (KONSTANTINOV and PEREL, 1960). This contour consists of an upper and a lower horizontal branch, which are separated from each other by a cut on the real axis from 0 to t, and a vertical branch on the imaginary axis from $-i\hbar\beta$ to 0 (cf. Fig. 2.1). The operator $U(t)$ is expanded on the upper branch, $U^+(t)$ on the lower branch and $\exp(-\beta\mathcal{H}^{(0)})$ on the vertical one. At the turning point of the contour (at t) a "clamp" is placed. This clamp corresponds to the operator $a_m^+a_m$.

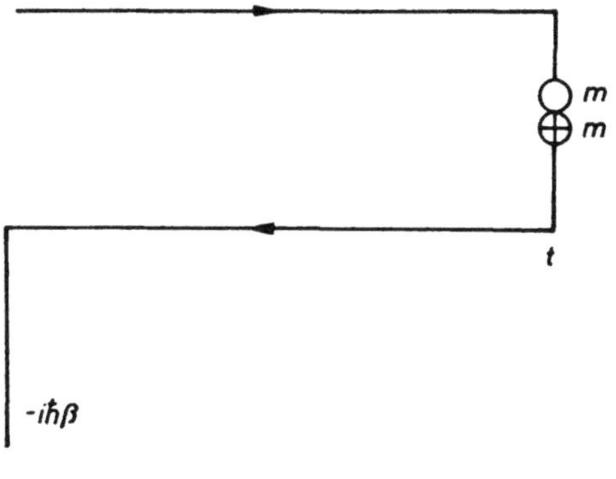

Fig. 2.1. Contour C in the complex time plane. For conveniency, the two horizontal branches of the contour along the cut on the real axis are separated from each other. The arrows indicate the direction of the contour

The perturbation series for $\varrho_m(t)$ corresponds to the sum of all diagrams having an arbitrary number of points placed on any of the three branches of the contour. To each point i at time t_i an operator $\pm\frac{i}{\hbar}\,\mathcal{H}'(t_i)$ is ascribed, where the sign $+$ is associated with a point on the lower horizontal and the vertical branch, and the sign $-$ with a point on the upper horizontal branch. For points on the vertical branch, the evolution operator does not contain the electrical field (see above). The operators must be arranged such that operators belonging to earlier times on the contour (in Fig. 2.1 the arrow on C is directed from earlier to later times) are placed on the right of those belonging to later times. After that it is necessary to integrate over all times t_i on account of the arrangement of the points on the contour. For instance, for two points t_1 and t_2, $t_2 > t_1$, on the upper horizontal branch, one point (t_3) on the lower horizontal branch, and one point (t_4) on the vertical branch, the integration must be performed as

$$\int_0^t dt_2 \int_0^{t_2} dt_1 \int_0^t dt_3 \int_0^{-i\hbar\beta} dt_4 \ldots .$$

The operator expression obtained must still be thermodynamically averaged over the equilibrium distribution (cf. (2.3.5)). Note that in (2.3.5), as usual, contributions from an expansion of the partition function $\langle \exp [-\beta(\mathcal{H}^{(0)} - \mu\hat{N})] \rangle$ with respect to $\mathcal{H}'$ may be ignored if we account only for connected diagrams. As mentioned above, in the stationary regime the current does not depend on the initial condition (cf. the discussion after (2.4.16)), which is given by the concrete form of the Hamiltonian $\mathcal{H}^{(0)}$. Therefore, in particular, we may choose the Hamiltonian of free particles, $\mathcal{H}_0$, (recall that $\mathcal{H}^{(0)} = \mathcal{H}_0 + \mathcal{H}'$) for the Hamiltonian of the initial distribution. For such a choice, the expansion of $\varrho_m(t)$ with respect to $\mathcal{H}'$ does not contain points on the vertical branch of the contour C. This fact much simplifies the diagram technique.

Note that the evolution operator for free particles, $U_0(t)$, commutates with the density operator $a_m^+ a_m$, i.e. it holds that $U_0{}^+ a_m^+ a_m U_0 = a_m^+ a_m$. Owing to this fact, at the clamp of any diagram there appear no factors $U_0{}^+(t)$ and $U_0(t)$, which arise from an expansion of the evolution operator.

On performing the thermodynamic average over the operator product connected with a given diagram, the electron operators are averaged by means of the usual Wick theorem (cf., e.g., THOULESS, 1961). A more complicated problem is to average over the phonon operators, because the multiphonon operators involved in $\mathcal{H}'$ give rise to some technical difficulties. In the theory of small polarons a procedure for averaging such phonon operator products was developed by LANG and FIRSOV (1962) and KUDINOV and FIRSOV (1965, 1966). This procedure may also be applied to disordered systems (BÖTTGER and BRYKSIN, 1975). It will be described in Appendix 1. In the diagram language, the thermodynamic average gives rise to electron and phonon lines connecting all points of a diagram with each other.

The basic feature of the perturbation expansion for the density matrix is the fact that it contains growing powers of the parameter W/ω, where W is the intersite transition probability and ω the frequency of the external field. Accordingly, a perturbation treatment of the density matrix requires that, at least for small ω, an infinite number of diagrams is summed up. It should be stressed that in the limit as $\omega \to 0$ such a summation is needed for any small $\mathcal{H}'$ (W is at least proportional to the second order of $\mathcal{H}'$). In Appendix 1 such a summation will be performed for the density matrix $\varrho_{m'}^{m}(t) = \langle U^+(t)\, a_{m'}^+ a_m U(t) \rangle$, which is a generalization of the density matrix (2.3.4) ($\varrho_m^m \equiv \varrho_m$). It yields the expression (A 1.32), which, differentiated with respect to t, reads

$$\frac{\mathrm{d}\varrho_{m'}^{m}(t)}{\mathrm{d}t} = \frac{\mathrm{i}}{\hbar} [\varepsilon_{m'} - \varepsilon_m - e\boldsymbol{E}(t) (\boldsymbol{R}_{m'} - \boldsymbol{R}_m)] \varrho_{m'}^{m}(t)$$
$$+ \sum_{\{m_i\}} \int_0^t \mathrm{d}t' \{\varrho_{m_2}^{m_1}(t') \overset{(1)}{\mathcal{W}}{}_{m_2 m'}^{m_1 m}(t', t) + \varrho_{m_3 m_4}^{m_1 m_2}(t') \overset{(2)}{\mathcal{W}}{}_{m_3 m_4 m'}^{m_1 m_2 m}(t', t)$$
$$+ \varrho_{m_4 m_5 m_6}^{m_1 m_2 m_3}(t') \overset{(3)}{\mathcal{W}}{}_{m_4 m_5 m_6 m'}^{m_1 m_2 m_3 m}(t', t) + \ldots\}. \tag{2.4.3}$$

Here $\varrho_{m_1' m_2' \ldots m_n'}^{m_1 m_2 \ldots m_n}(t)$ is the many-particle equal-time density matrix

$$\varrho_{m_1' m_2' \ldots m_n'}^{m_1 m_2 \ldots m_n}(t) = \langle U^+(t)\, a_{m_n'}^+ a_{m_n} \ldots a_{m_1'}^+ a_{m_1} U(t) \rangle. \tag{2.4.4}$$

The quantities $\overset{(n)}{\mathcal{W}}$ have the dimension t^{-2}. In the language of the diagram technique, these quantities are irreducible blocks (which do not contain so-called free sections) of the nth order. An important property of the blocks $\mathcal{W}$ is the fact that they remain finite in the limit as $\omega \to 0$.

The relation (2.4.3) does not represent a closed equation for the one-particle density matrix, but it contains many-particle functions, too. Accordingly, generally speaking, an evaluation of $\varrho_m(t)$ requires also equations for the two-, three-, etc. particle density matrices, which can be obtained, at least in principle, by the above diagram technique by replacing at the clamp the one-particle operator $a_{m'}^+ a_m$ by the two-particle operator $a_{m_1'}^+ a_{m_1} a_{m_2'}^+ a_{m_2}$, etc. In this way one gets an infinite hierarchy of connected equations for the one-, two-, etc. particle density matrices. To obtain a solution of this hierarchy, one needs an appropriate decoupling procedure, whose success depends on the intuition of the researcher. In what follows, the decoupling will be carried out already in (2.4.3). Therefore, we do not write down even more complicated equations of the hierarchy.

From a physical point of view, let us now discuss the question, why many-particle functions do affect the behaviour of the one-particle density matrix. To this end, consider a hop of an electron from site m to site m'. This process is determined by the hopping probability between the two sites (which is described by second order blocks) as well as by the occupation probability of the final site (recall that we have assumed that double occupation of sites is excluded). Just the latter fact leads to the occurrence of the two-particle density matrix in the equation for the one-particle density matrix. To see this, note that the two-particle density matrix describes the probability that the initial state is occupied and, at the same time, the final state is unoccupied. Evidently, such correlation is only lacking in the case that the number of electrons (or holes) is much less than the number of sites, i.e. in the case that the Boltzmann statistics holds, where $f_m \ll 1$ (or for holes $1 - f_m \ll 1$). Here f_m labels the equilibrium occupation of site m, which is defined in (A 1.4).

2.4.2. *One-electron approximation*

On averaging products of operators a^+ and a by using the Wick theorem, in the one-electron approximation from all pairing schemes only the following one survives

$$a^+ a a^+ a a^+ \cdots a a^+ a a^+ a .$$

Here each pairing of the type $a_m a_{m'}^+$ is associated with a Kronecker $\delta_{mm'}$ and the single pairing of the type $a_{m'}^+ a_m$ with a factor $f_m \delta_{mm'}$. Therfore, all the many-particle functions (2.4.4) pass into one-particle functions:

$$\varrho_{m_1' \dots m_n'}^{m_1 \dots m_n}(t) = \varrho_{m_n'}^{m_1}(t)\, \delta_{m_1' m_2} \delta_{m_2' m_3} \cdots \delta_{m_{n-1}' m_n} . \tag{2.4.5}$$

Thus, (2.4.3) becomes a closed equation for the one-particle density matrix

$$\frac{\mathrm{d}\varrho_{m'}^m(t)}{\mathrm{d}t} = \frac{\mathrm{i}}{\hbar} \left[\varepsilon_{m'} - \varepsilon_m - e\boldsymbol{E}(t)\,(\boldsymbol{R}_{m'} - \boldsymbol{R}_m)\right] \varrho_{m'}^m(t)$$
$$+ \sum_{m_1 m_2} \int_0^t \mathrm{d}t' \varrho_{m_2}^{m_1}(t')\, \mathcal{W}_{m_2 m'}^{m_1 m}(t', t) , \tag{2.4.6}$$

where

$$\mathscr{W}^{m_1m}_{m_2m'} = \overset{(1)}{\mathscr{W}}{}^{m_1m}_{m_2m'} + \sum_{m_3} \overset{(2)}{\mathscr{W}}{}^{m_1m_3m}_{m_2m_3m'} + \sum_{m_3m_4} \overset{(3)}{\mathscr{W}}{}^{m_1m_3m_4m}_{m_2m_4m_3m'} + \dots \qquad (2.4.7)$$

In the diagram language, the one-electron approximation requires that all points and the clamp must be consecutively connected by electron lines, according to their arrangement on the contour C, such that there is only a single wrong line, the utmost left one. All right lines are associated with a factor unity, instead of $(1 - f_m)$ as usual. The one-electron approximation is in general used in the small-polaron theory.

The second-order diagrams $\overset{(2)}{\mathscr{W}}$ are shown in Fig. 2.2. In what follows, we need only the diagrams where $m' = m$ and $m_1 = m_2$, which determine the two-site hopping probability. The corresponding first order diagrams vanish (see below) and, according to the diagram rules given in Appendix 1., the analytical expression for the second order diagrams reads

$$\overset{(2)}{\mathscr{W}}{}^{m_1m}_{m_1m}(t', t)$$

$$= 2\mathrm{Re}\, \hbar^{-2} \Bigg\{ |\tilde{J}_{mm_1}(\boldsymbol{H})|^2 F_{m_1m}(t' - t) \exp\left\{\frac{\mathrm{i}}{\hbar} \int\limits_t^{t'} \mathrm{d}t''[\varepsilon_m - \varepsilon_{m_1} - e\boldsymbol{E}(t'')\,(\boldsymbol{R}_m - \boldsymbol{R}_{m_1})]\right\}$$

$$- \delta_{mm_1} \sum_{m_2} |\tilde{J}_{mm_2}(\boldsymbol{H})|^2 F_{mm_2}(t' - t) \exp\left\{\frac{\mathrm{i}}{\hbar} \int\limits_t^{t'} \mathrm{d}t''[\varepsilon_{m_2} - \varepsilon_m - e\boldsymbol{E}(t'')\,(\boldsymbol{R}_{m_2} - \boldsymbol{R}_m)]\right\}\Bigg\}.$$

$$(2.4.8)$$

Here we have accounted for $\tilde{J}_{mm_1}(\boldsymbol{H}) = \tilde{J}^*_{m_1m}(\boldsymbol{H})$, where $\tilde{J}$ denotes the resonance integral renormalized by electron-phonon interaction, $\tilde{J}_{m'm}(\boldsymbol{H}) = J_{m'm}(\boldsymbol{H}) \times \exp[-S_{\mathrm{T}}(m', m)]$, where the quantity $S_{\mathrm{T}}(m', m)$ is defined in (A 1.18). The function $F_{m_1m}(t)$ ist related to the function $F^{m_1m_2}_{m_3m_4}(t)$, defined in (A 1.20), by $F_{m_1m} \equiv F^{m_1m}_{mm_1}$. Accordingly, $F_{m_1m}(t)$ is given by

$$F_{m_1m}(t) = \exp\left\{\sum_{\boldsymbol{q}} \frac{|\gamma(\boldsymbol{q})|^2\,[1 - \cos \boldsymbol{q}(\boldsymbol{R}_{m_1} - \boldsymbol{R}_m)]}{N \sinh\left(\frac{1}{2}\,\hbar\omega_{\boldsymbol{q}}\beta\right)} \cos \omega_{\boldsymbol{q}}\left(t + \frac{1}{2}\,\mathrm{i}\hbar\beta\right)\right\} - 1. \qquad (2.4.9)$$

The first term on the right-hand side of (2.4.8) corresponds to the diagrams a) and b) in Fig. 2.2, and the second terms to the diagram c) and d) in Fig. 2.2. These pairs of diagrams are complex conjugate between each other (for $m' = m$ and $m_1 = m_2$). The property of complex conjugation holds if the electrical field is a real quantity. In the case that $\boldsymbol{E}(t)$ is a complex function, at first the sum of two conjugate complex quantities must be inserted into (2.4.8) (to get a real field), which is then rewritten such that only the complex field appears.

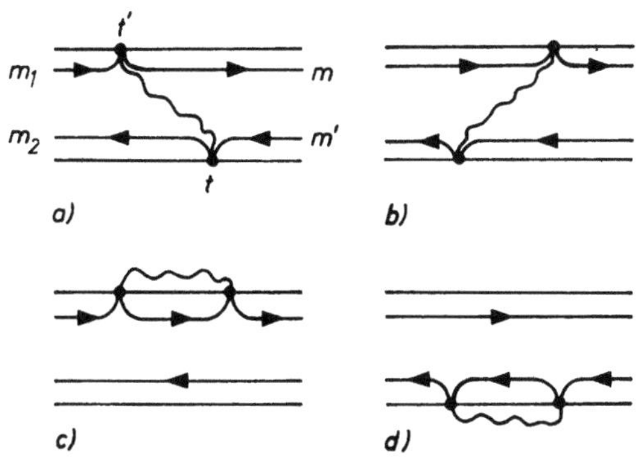

Fig. 2.2. Second-order diagrams in the one-electron approximation describing $\overset{(2)}{\mathcal{W}}{}^{m_1m}_{m_2m'}(t',t)$. The arrows indicate the directions of the electron lines. Wavy lines indicate phonon averages. Diagonal diagrams describe the probability of two-site hops (the diagrams a) and b) describe the incoming part and c) and d) the outgoing one)

The first-order transition probabilities $\overset{(1)}{\mathcal{W}}$ are defined in Appendix 1. in (A1.29).

Note that the expression for the current (2.3.3) contains only diagonal elements, $\varrho_m(t)$, of the density matrix. To determine $\varrho_m(t)$ by means of (2.4.6), we must at first exclude the nondiagonal elements of the density matrix, i.e. $\varrho^m_{m'}$ where $m' \neq m$. In this way we obtain a closed equation for $\varrho_m(t)$. To do this, we consider (2.4.6) for the two special cases $m = m'$ and $m \neq m'$:

$$\frac{\mathrm{d}\varrho_m(t)}{\mathrm{d}t} = \sum_{m_1} \int_0^t \mathrm{d}t' \varrho_{m_1}(t')\, \mathcal{W}_{m_1m}(t', t) + \sum_{m_1 \neq m_2} \int_0^t \mathrm{d}t' \varrho^{m_1}_{m_2}(t')\, \mathcal{W}^{m_1m}_{m_2m}(t', t),$$

$$\frac{\mathrm{d}\varrho^m_{m'}(t)}{\mathrm{d}t} = \frac{\mathrm{i}}{\hbar} [\varepsilon_{m'} - \varepsilon_m - e\boldsymbol{E}(t)\, (\boldsymbol{R}_{m'} - \boldsymbol{R}_m)]\, \varrho^m_{m'}(t)$$

$$+ \sum_{m_1} \int_0^t \mathrm{d}t' \varrho_{m_1}(t')\, \mathcal{W}^{m_1m}_{m_1m'}(t', t) + \sum_{m_1 \neq m_2} \int_0^t \mathrm{d}t' \varrho^{m_1}_{m_2} \mathcal{W}^{m_1m}_{m_2m'}(t', t), \qquad (2.4.10)$$

where $\mathcal{W}_{m_1m} \equiv \mathcal{W}^{m_1m}_{m_1m}$. Using for the nondiagonal probability $\mathcal{W}^{m_1m}_{m_1m'}(t', t)$ the lowest-order expression with respect to J (A1.29) (diagrams with only a single point), then from the second equation of (2.4.10), we find that

$$\varrho^m_{m'} \sim \frac{\tilde{J}_{m_1m}}{\mathrm{i}(\varepsilon_{m'} - \varepsilon_m)} (\varrho_m - \varrho_{m'}). \qquad (2.4.11)$$

According to (2.4.11), for most of the pairs of sites it holds that $\varrho^m_{m'} \sim (\tilde{J}/\Gamma)\, (\varrho_m - \varrho_{m'})$, where Γ is the width of the spread of the site energies. In the present book we are interested in hopping systems, where $\tilde{J}/\Gamma \ll 1$ because Anderson localization requires that $J/\Gamma \ll 1$ (recall that $\tilde{J} < J$, where $\tilde{J}$ is the resonance integral J renormalized due to electron-phonon interaction). Accordingly, in a hopping system it holds that $\varrho^m_{m'} \ll \varrho_{m'}, \varrho_m$, which allows the nondiagonal elements of the density matrix in

(2.4.10) to be neglected. Thus, from (2.4.10) we obtain

$$\frac{\mathrm{d}\varrho_m(t)}{\mathrm{d}t} = \sum_{m'} \int_0^t \mathrm{d}t' \varrho_{m'}(t') \mathscr{W}_{m'm}(t', t). \tag{2.4.12}$$

As mentioned above, first-order diagrams (containing isolated or "empty" sites) do not contribute to the diagonal probabilities $\mathscr{W}_{m'm}$ (cf. (A 1.29)). Therefore, an investigation of hopping transport requires at least account of second-order diagrams. However, we may restrict ourselves to these diagrams only if the symmetric part of the conductivity (i.e. dissipative current) is to be evaluated. However, if we are interested in the antisymmetric part of the conductivity tensor in the presence of a magnetic field (i.e., e.g., in the Hall effect), then such an approach is not sufficient, because (cf. (2.4.8)) the diagonal two-site probabilities $\overset{(2)}{\mathscr{W}}$ are proportional to $|J(\boldsymbol{H})|^2$. According to (2.1.25), in such an expression the magnetic field-dependent phase factor cancels. Therefore, an investigation of the hopping-Hall effect requires account of at least three-site hopping probabilities $\overset{(3)}{\mathscr{W}}$ (HOLSTEIN, 1961). Below, the equation for $\varrho_m(t)$ will be generalized such that it includes also three-site probabilities.

According to (2.4.8), the probability $\mathscr{W}$ satisfies the sum rule

$$\sum_m \mathscr{W}_{m'm} = 0. \tag{2.4.13}$$

It can be shown that this property is preserved for blocks of any order with respect to the resonance integral. This is due to the fact that by transfering the utmost right point in any diagram from the above horizontal branch to the lower one, or vice versa, only the sign of the diagram is changed. The property (2.4.13) allows the quantities $\mathscr{W}_{m'm'}$ to be expressed as

$$\mathscr{W}_{m'm'} = -\sum_{m(\neq m')} \mathscr{W}_{m'm},$$

so that (2.4.12) may be rewritten as

$$\frac{\mathrm{d}\varrho_m(t)}{\mathrm{d}t} = \sum_{m'} \int_0^t \mathrm{d}t' \{\varrho_{m'}(t') \mathscr{W}_{m'm}(t', t) - \varrho_m(t') \mathscr{W}_{mm'}(t', t)\}, \tag{2.4.14}$$

where the sum is over $m'(\neq m)$. The corresponding expression for $\mathscr{W}_{m'm}$ may be readily obtained from (2.4.8) by neglecting the second term on the right-hand side of this formula. We find that

$$\mathscr{W}_{m'm}(t', t) = 2\,\mathrm{Re}\,\frac{|\tilde{J}_{m'm}(\boldsymbol{H})|^2}{\hbar^2} F_{m'm}(t' - t)$$
$$\times \exp\left\{\frac{\mathrm{i}}{\hbar} \int_t^{t'} \mathrm{d}t'' [\varepsilon_m - \varepsilon_{m'} - e\boldsymbol{E}(t'') (\boldsymbol{R}_m - \boldsymbol{R}_{m'})]\right\}. \tag{2.4.15}$$

Recall that in (2.4.15) $\boldsymbol{E}(t)$ must be a real quantity.

Equation (2.4.14) with the two-site probabilities (2.4.15) represents a non-Markovian rate equation for the electron density at site m. This equation in the discrete analog of the continuity equation. The second term on the right-hand side of (2.4.14) describes the electron transfer from site m to neighbouring sites m', and the first term the transfer from neighbouring sites m' to site m. The non-Markovian character of (2.4.14) is in that $\mathrm{d}\varrho_m(t)/\mathrm{d}t$ is not determined only by the balance between incoming and outgoing particles at time t, but by the time integral from 0 to t, i.e. the system has a "memory".

According to (A 1.32), (2.4.14) must be solved on account of the initial condition $\varrho_m(t = 0) = f_m$. In this case, (2.4.14) describes the transition of the system to the stationary state after turning on the electrical field at time $t = 0$. However, we are in general interested in the stationary regime of the current in the limit as $t \to \infty$. As mentioned above, this regime does not depend on the initial state. To study the hopping behaviour in the stationary regime, note that $F(t)$ occurring in (2.4.15) goes to zero at large times t, i.e. there exists a characteristic time scale δt during that this function strongly diminishes. The quantity δt has the meaning of a characteristic time after that the system reaches the stationary regime and the electron and phonon subsystems come into equilibrium with each other (e.g., δt is the time after that a polaron cavity vanishes).

Accordingly, a study of the stationary regime requires that in (2.4.14) we take the limit as $t/\delta t \to \infty$. To do this, in this equation we replace the variable of integration as $t' = t - t_1$. Thus, because $\mathscr{W}(t - t_1, t)$ goes to zero in time $t_1 \approx \delta t$, in the limit as $t/\delta t \to \infty$, in (2.4.14) the upper limit of the integration over t_1 may be set equal to ∞:

$$\frac{\mathrm{d}\varrho_m(t)}{\mathrm{d}t} = \sum_{m_1} \int_0^\infty \mathrm{d}t_1 \{\varrho_{m_1}(t - t_1)\, \mathscr{W}_{m_1 m}(t - t_1, t) - \varrho_m(t - t_1)\, \mathscr{W}_{m m_1}(t - t_1, t)\}\,. \tag{2.4.16}$$

Unlike (2.4.14), equation (2.4.16) describes the system at large times after the disappearance of relaxational processes. However, in the limit of large t, the memory of the initial state is lost. Thus, to solve (2.4.16) we need another additional condition than $\varrho_m(0) = f_m$. Let us assume that $\boldsymbol{E}(t) = \boldsymbol{E} \exp(\mathrm{i}\omega t)$. Then we may choose

$$\varrho_m\left(t + \frac{2\pi}{\omega}\right) = \varrho_m(t) \tag{2.4.17}$$

as new additional condition. Accordingly, the initial condition completely disappeares from the problem in question. This is the fact we have predicted above. In particular, the thermodynamic equilibrium distribution function f_m is lacking now, which appears in all terms of the perturbation expansion of $\varrho_m(t)$ with respect to $\mathscr{H}'$.

Let us now point out that in the limit as $\omega\delta t \ll 1$, (2.4.16) changes into a Markovian rate equation. Note that according to (2.4.17), $\varrho_m(t)$ changes for t of order $1/\omega$, while $\mathscr{W}(t - t_1, t)$ goes to zero in time $t_1 \sim \delta t$. Hence, for $\omega\delta t \ll 1$, in (2.4.16) $\varrho(t - t_1)$ may be taken out of the integral over t_1 after setting t_1 equal to zero in this quantity.

We find that

$$\frac{\mathrm{d}\varrho_m(t)}{\mathrm{d}t} = \sum_{m_1} \{\varrho_{m_1}(t)\, W_{m_1m}(t) - \varrho_m(t)\, W_{mm_1}(t)\}, \tag{2.4.18}$$

where

$$\begin{aligned} W_{m_1m}(t) &\equiv \int_0^\infty \mathrm{d}t_1 \mathscr{W}_{m_1m}(t - t_1, t) \\ &= \frac{|\tilde{J}_{m_1m}(\boldsymbol{H})|^2}{\hbar^2} \exp\left\{\frac{\beta}{2}\,[\varepsilon_{m_1} - \varepsilon_m - e\boldsymbol{E}(t)\,(\boldsymbol{R}_{m_1} - \boldsymbol{R}_m)]\right\} \int_{-\infty}^{\infty} \mathrm{d}t_1 \\ &\quad \times \exp\left\{\frac{\mathrm{i}t_1}{\hbar}\,[\varepsilon_{m_1} - \varepsilon_m - e\boldsymbol{E}(t)\,(\boldsymbol{R}_{m_1} - \boldsymbol{R}_m)]\right\} \\ &\quad \times \left\{\exp\left[\sum_{\boldsymbol{q}} \frac{|\gamma(\boldsymbol{q})|^2\,[1 - \cos \boldsymbol{q}(\boldsymbol{R}_{m_1} - \boldsymbol{R}_m)]}{N \sinh\left(\frac{1}{2}\,\hbar\omega/\beta\right)} \cos \omega_{\boldsymbol{q}} t_1\right] - 1\right\}. \end{aligned} \tag{2.4.19}$$

Because of $t_1\omega \sim \delta t\omega \ll 1$, here we have used the approximation

$$\int_0^{t_1} \mathrm{d}t' E(t - t') \approx t_1 E(t).$$

By means of (2.4.19), it is easy to show that the hopping probability W_{m_1m} has the symmetry property

$$W_{m_1m}/W_{mm_1} = \exp\{\beta[\varepsilon_{m_1} - \varepsilon_m - e\boldsymbol{E}(t)\,(\boldsymbol{R}_{m_1} - \boldsymbol{R}_m)]\}, \tag{2.4.20}$$

which reflects the principle of the detailed balance for hops between sites m_1 and m in an external electrical field. This property plays a very important role in the theory of the hopping transport.

Equation (2.4.18) describes the Markovian itineracy of a charge carrier through a lattice in the presence of an external electrical field, analogous to the Brownian motion in an external (e.g., gravitational) field.

2.4.3. *Hartree-Fock decoupling*

After having studied the one-electron problem, let us now take up again the relation (2.4.3), which accounts for the Hubbard intrasite interelectronic correlation. On using this relation we restrict ourselves to second-order diagrams. Note that for hops close to the Fermi level, the one-electron approximation is not sufficient. But hopping near the Fermi level is a situation often encountered in actual disordered materials. To study this hopping situation, we reduce (2.4.3) by taking into account that for hopping correlators of the electron density $n_m = a_m^+ a_m$ are important and averages of operators a_m^+ and $a_{m'}$ ($m \neq m'$) may be ignored. Accordingly, in the two-particle matrix $\varrho_{m_3m_4}^{m_1m_2}$ the operator $a_{m_4}^+ a_{m_2} a_{m_3}^+ a_{m_1}$ under the sign of the trace may be replaced by

$$\delta_{m_4m_2}\delta_{m_3m_1}n_{m_1}n_{m_2} + \delta_{m_4m_1}\delta_{m_3m_2}n_{m_1}(1 - n_{m_2}).$$

To study hopping, we use this replacement and account for $n_m n_m = n_m$. Thus, by a procedure analogous to that employed in the case of the one-electron approximation, we find for $\varrho_m(t)$ an equation of the form (2.4.16) (or (2.4.18) in the Markovian limit) where on the right-hand side ϱ_m is replaced by $\varrho_m - \varrho_{mm_1}^{mm_1}$ and ϱ_{m_1} by $\varrho_{m_1} - \varrho_{mm_1}^{mm_1}$.

According to (2.4.4), the two-particle function $\varrho_{mm_1}^{mm_1}(t)$ represents the equal-time correlator of the particle densities at sites m and m_1 at time t. Using the simplest approximation, the Hartree-Fock approximation, $\varrho_{mm_1}^{mm_1}(t)$ becomes

$$\varrho_{mm_1}^{mm_1}(t) = \varrho_m(t)\,\varrho_{m_1}(t), \qquad (m_1 \neq m). \tag{2.4.21}$$

It should be noted that the Hartree-Fock approximation fails to describe correlated hops, where the electron "feels" whether the site it will hop to is vacant or occupied. Owing to this approximation, the prohibition of an electron transfer to occupied neighbouring site is replaced by a finite transition probability, which decreases with increasing average occupation of the neighbouring site.

By means of (2.4.21), the non-Markovian rate equation becomes

$$\frac{\mathrm{d}\varrho_m(t)}{\mathrm{d}t} = \sum_{m_1} \int_0^\infty \mathrm{d}t_1 \{\mathscr{W}_{m_1m}(t - t_1, t)\,\varrho_{m_1}(t - t_1)\,\big(1 - \varrho_m(t - t_1)\big)$$
$$- \mathscr{W}_{mm_1}(t - t_1, t)\,\varrho_m(t - t_1)\,\big(1 - \varrho_{m_1}(t - t_1)\big)\}. \tag{2.4.22}$$

In the Markovian limit, (2.4.22) changes into

$$\frac{\mathrm{d}\varrho_m(t)}{\mathrm{d}t} = \sum_{m_1} \{\varrho_{m_1}(t)\,\big(1 - \varrho_m(t)\big)\,W_{m_1m}(t) - \varrho_m(t)\,\big(1 - \varrho_{m_1}(t)\big)\,W_{mm_1}(t)\}. \tag{2.4.23}$$

In the case of the Boltzmann statistics, i.e. for $\varrho_m \ll 1$ (electrons) or $1 - \varrho_m \ll 1$ (holes), (2.4.22) and (2.4.23), respectively, pass into (2.4.16) and (2.4.18), which have been obtained by using the one-electron approximation.

Concluding this section, let us discuss the range of application of the Markovian rate equation. As mentioned above, the non-Markovian behaviour is governed by the parameter δt, the time during which the electron distribution at a site reaches the equilibrium with the phonon subsystem. This time is of order a reciprocal phonon frequency, i.e. $\delta t \approx 10^{-12}$ s. Accordingly, (2.4.18) and (2.4.23) are valid for frequencies of the external electrical field up to the infrared region. An investigation of optical properties requires use of the non-Markovian equations (2.4.16) or (2.4.22).

2.5. The rate equation for weak electrical fields

2.5.1. *The rate equation for arbitrary frequencies*

In this section we study the rate equation (2.4.22) in the linear approximation with respect to the external electrical field $\boldsymbol{E}$. Setting in (2.4.22) $\boldsymbol{E}$ equal to zero, and taking into account that for $\boldsymbol{E} = 0$, ϱ_m does not depend on time, we find that

$$\sum_{m_1} \{\varrho_{m_1}^{(0)}(1 - \varrho_m^{(0)})\, W_{m_1m}^{(0)} - \varrho_m^{(0)}(1 - \varrho_{m_1}^{(0)})\, W_{mm_1}^{(0)}\} = 0. \tag{2.5.1}$$

Here the superscript (0) denotes the zeroth-order term with respect to $\boldsymbol{E}$ of the corresponding quantity. The probability $W^{(0)}$ is defined in (2.4.19). In obtaining $W^{(0)}$ we have accounted for that $\mathscr{W}(t', t)$, involved in (2.4.19), does depend only on $t' - t$ (cf. (2.4.15)).

On account of the symmetry property (2.4.20) of the transition probabilities, it is easy to see that $\varrho_m^{(0)} = f_m$. Thus, in the absence of an electrical field, the density matrix can be obtained by making use only of the principle of the detailed equilibrium, i.e. without the knowledge of the explicit form of the transition probabilities. This is a basic property of dissipative systems, where the transition to the termal equilibrium is due to interaction with the heat bath.

Let us now study the rate equation in the linear approximation with respect to $\boldsymbol{E}$. We assume that the time dependence of the electrical field is given by

$$\boldsymbol{E}(t) = \boldsymbol{E} \exp(\mathrm{i}\omega t), \tag{2.5.2}$$

and write $\varrho_m(t) = \varrho_m^{(0)} + \varrho_m^{(1)}(t)$, where $\varrho_m^{(1)}(t)$ denotes the first-order term in the expansion of $\varrho_m(t)$ with respect to $\boldsymbol{E}$. For convenience, instead of $\varrho_m^{(1)}(t)$, we introduce a new quantity $\delta\mu_m$ having the dimension of an energy:

$$\varrho_m^{(1)}(t) = \beta\delta\mu_m f_m(1 - f_m) \exp(\mathrm{i}\omega t). \tag{2.5.3}$$

Hence, linearizing (2.4.22) with respect to $\boldsymbol{E}$ and using $\varrho_m^{(0)} = f_m$, we obtain the following equation for $\delta\mu_m$:

$$\begin{aligned} \mathrm{i}\omega f_m(1 - f_m)\, \delta\mu_m &= \sum_{m_1} e\boldsymbol{E}(\boldsymbol{R}_m - \boldsymbol{R}_{m_1})\, \Gamma_{m_1m}(\omega) \\ &\quad + \sum_{m_1} \{\delta\mu_{m_1}\tilde{\Gamma}_{m_1m}(\omega) - \delta\mu_m\tilde{\Gamma}_{mm_1}(\omega)\}, \end{aligned} \tag{2.5.4}$$

where

$$\begin{aligned} \tilde{\Gamma}_{m_1m}(\omega) = f_{m_1}(1 - f_{m_1}) \int_0^\infty \mathrm{d}t\, \mathrm{e}^{-\mathrm{i}\omega t} \{\mathscr{W}_{m_1m}^{(0)}(-t)\,(1 - f_m) \\ + \mathscr{W}_{mm_1}^{(0)}(-t)\, f_m\}, \end{aligned} \tag{2.5.5}$$

$$\left.\begin{aligned} \Gamma_{m_1m}(\omega) &= 2f_m f_{m_1} \exp\left[\frac{\beta}{2}(\varepsilon_m + \varepsilon_{m_1} - 2\mu)\int\limits_0^\infty \mathrm{d}t\, \mathrm{e}^{-\mathrm{i}\omega t}\right. \\ &\times \int\limits_{-1/2}^{1/2} \mathrm{d}\lambda\, \frac{|J_{m_1m}(\boldsymbol{H})|^2}{\hbar^2} \cos\left[\frac{(\varepsilon_{m_1} - \varepsilon_m)}{\hbar}(t + \mathrm{i}\hbar\beta\lambda)\right] \\ &\times \left\{\exp\left[\sum_{\boldsymbol{q}} \frac{|\gamma(\boldsymbol{q})|^2\,[1 - \cos \boldsymbol{q}(\boldsymbol{R}_{m_1} - \boldsymbol{R}_m)]}{N \sinh\left(\frac{1}{2}\hbar\omega_{\boldsymbol{q}}\beta\right)} \cos \omega_{\boldsymbol{q}}(t + \mathrm{i}\hbar\beta) - 1\right\}.\right. \end{aligned}\right\} \qquad (2.5.6)$$

Here $\mathscr{W}^{(0)}_{m_1m}(t) = \mathscr{W}_{m_1m}(t + t', t')$ at $\boldsymbol{E} = 0$, which does not depend on t' (cf. (2.4.15)).

Using (2.5.3), (2.5.4), and the expression for the current (2.3.3), we find that in the linear approximation with respect to $\boldsymbol{E}$ the current is given by a sum of two terms

$$\boldsymbol{j}(\omega) = \boldsymbol{j}_1(\omega) + \boldsymbol{j}_2(\omega), \qquad (2.5.7)$$

$$\boldsymbol{j}_1(\omega) = \frac{e^2\beta}{2\Omega} \sum_{mm'} (\boldsymbol{R}_m - \boldsymbol{R}_{m'})\,[\boldsymbol{E}(\boldsymbol{R}_m - \boldsymbol{R}_{m'})]\,\Gamma_{m'm}(\omega), \qquad (2.5.8)$$

$$\boldsymbol{j}_2(\omega) = \frac{e\beta}{2\Omega} \sum_{mm'} (\boldsymbol{R}_m - \boldsymbol{R}_{m'})\,\{\delta\mu_{m'}\tilde{\Gamma}_{m'm}(\omega) - \delta\mu_m\tilde{\Gamma}_{mm'}(\omega)\}. \qquad (2.5.9)$$

The quantity $\delta\mu_m$ describes the charge redistribution at the mth site caused by the external field. It may be identified with the field-induced local change of the chemical potential at this site. In particular, in an ordered system a charge redistribution does not occur, i.e. $\delta\mu_m = 0$, since in the presence of lattice translational symmetry, $\delta\mu_m = 0$ solves (2.5.4). Accordingly, for an ordered system, we have $\boldsymbol{j}_2 = 0$, which allows a closed expression for the conductivity $\sigma_{\mu\nu}$ to be obtained from (2.5.8):

$$\sigma_{\mu\nu}(\omega) = \frac{e^2\beta}{2\Omega} \sum_{mm'} (\boldsymbol{R}_m - \boldsymbol{R}_{m'})_\mu\,(\boldsymbol{R}_m - \boldsymbol{R}_{m'})_\nu\,\Gamma_{m'm}(\omega). \qquad (2.5.10)$$

The evaluation of the conductivity by means of (2.5.10) is not a difficult problem. This problem will be studied in the following chapter within the framework of the small-polaron model in crystals.

The contribution $\boldsymbol{j}_2(\omega)$ to the current is proportional to $\delta\mu_m$ and is encountered only in the presence of a spatial dispersion. The occurrence of this contribution may be understood as follows. In applying a field, the charges are spatially redistributed and the new distribution differs from the equilibrium distribution f_m. Therefore, diffusive currents occur and try to restore the initial distribution. The diffusive contribution to the current is directed against the Ohmic one and diminishes therefore the total current. Note that a similar situation is encountered in systems with macroscopic spatial inhomogeneities, as for instance, in the case of p-n transitions. In the latter systems, the current is phenomenologically given by $\boldsymbol{j} = \sigma\boldsymbol{E} - D\,\nabla n$, where D is the diffusion coefficient and ∇n the gradient of the charge carrier concentration, which may be related to the gradient of the chemical potential.

Now the diffusive contribution is shown to decrease with increasing frequency, so that at sufficiently high frequencies it may be neglected. Let be $\omega\tau \gg 1$, where τ is a characteristic time between consecutive hops from site to site, for which $\tau \sim \tilde{\Gamma}^{-1}$ holds. In this case equation (2.5.4) may be solved by iteration with respect to $\tilde{\Gamma}/\omega$. The zeroth order term of such an iteration reads $\delta\mu_m = 0$. Inserting $\delta\mu_m = 0$ into the right-hand side of (2.5.4), we obtain the first-order expression

$$\delta\mu_m \approx \sum_{m'} \frac{e\boldsymbol{E}(\boldsymbol{R}_m - \boldsymbol{R}_{m'})}{\mathrm{i}\omega f_m(1 - f_m)} \Gamma_{m'm}.$$

Using this expression for $\delta\mu_m$ in $\boldsymbol{j}_2$ (2.5.9), we find that $j_2 \sim j_1/\omega\tau$, i.e. $j_2 \ll j_1$. Consequently, for high frequencies, $\boldsymbol{j}_2(\omega)$ may be neglected. Thus, the formula (2.5.10) describes the conductivity in two cases: (i) in ordered systems (small-polaron model) in a large frequency range and (ii) in disordered systems at sufficiently high frequencies. However, it is not valid in disordered systems in the low frequency limit, where it would lead to an overstimation of the conductivity. Because in strongly disordered systems $\tilde{\Gamma}$ exhibits a large spatially dispersion, so far it is not possible to give the lower limiting frequency down to that (2.5.10) applies. This problem will be discussed in detail in Section 6.

Note that the iterative solution of (2.5.4) with respect to powers of $1/\omega$ may be shown to be equivalent to a power-series expansion of the density matrix with respect to free sections, which represents an expansion with respect to powers of the Hamiltonian $\mathcal{H}'$. Formula (2.5.10) for σ can be ontained by expanding the Kubo formula (2.3.7) with respect to $\mathcal{H}'$ up to the lowest nonvanishing order (second order).

2.5.2. *The low-frequency limit and the random-resistor network with capacitors*

In this section we study the very important low-frequency limiting case, where $\omega\delta t \ll 1$, i.e. where the Markovian rate equation (2.4.23) applies. In this case, on the right-hand side of (2.5.4), we may set ω equal to zero. From (2.5.5) and (2.5.6), we obtain

$$\begin{aligned} \Gamma_{mm_1}(0) &= \Gamma_{m_1m}(0) = \tilde{\Gamma}_{mm_1}(0) = \tilde{\Gamma}_{m_1m}(0) \equiv \Gamma_{mm_1} = \Gamma_{m_1m} \\ &= f_{m_1}(1 - f_m)\, W^{(0)}_{m_1m} = f_m(1 - f_{m_1})\, W^{(0)}_{mm_1}, \end{aligned} \tag{2.5.11}$$

where $W^{(3)}$ is defined by (2.4.19).
On account of (2.5.11), equation (2.5.4) becomes

$$\mathrm{i}\omega f_m(1 - f_m)\,(U_m + \boldsymbol{E}\boldsymbol{R}_m) = \sum_{m'} (U_{m'} - U_m)\,\Gamma_{m'm}, \tag{2.5.12}$$

where, instead of $\delta\mu_m$, we have introduced the quantity

$$U_m = -\boldsymbol{E}\boldsymbol{R}_m + \delta\mu_m/e, \tag{2.5.13}$$

which has the dimension of an electrical potential.

In the same approximation, the expression for the current, (2.5.7) to (2.5.9), takes the form

$$\boldsymbol{j}(\omega) = \frac{e^2\beta}{2\Omega} \sum_{mm'} (\boldsymbol{R}_m - \boldsymbol{R}_{m'}) \Gamma_{m'm}[U_{m'}(\omega) - U_m(\omega)]. \tag{2.5.14}$$

According to (2.5.13), the quantity $eU_m + \mu$ may be viewed as the electrochemical potential at the site m. The expression for the current (2.5.14) has the form well known from the phenomenological theory of charge transport in spatially inhomogeneous media, where $\boldsymbol{j}$ is proportional to the gradient of the electrochemical potential. We note in passing that the results obtained show that a deviation from the Einstein relation (relating the mobility with the diffusion coefficient at $\omega = 0$) is governed by the dimensionless parameter $\omega\delta t$ such that for $\omega\delta t \to 0$, but arbitrary $\omega\tau$, this relation remains valid.

The set of equations (2.5.12) can be related to the electrical equivalent scheme shown in Fig. 2.3. (Pollak, 1973, Kirkpatrick, 1973). To do this, we introduce the quantities

$$\left.\begin{aligned} Z_{mm'} &= (e^2\beta\Gamma_{mm'})^{-1}, \\ C_m &= e^2\beta f_m(1 - f_m), \end{aligned}\right\} \tag{2.5.15}$$

which have the dimension of a resistance and a capacity, respectively. Using (2.5.15), we can rewrite (2.5.12) as

$$\mathrm{i}\omega C_m(U_m + \boldsymbol{E}\boldsymbol{R}_m) = \sum_{m'} (U_{m'} - U_m)/Z_{m'm}. \tag{2.5.16}$$

It is easy to see, that the latter set of equations represents Kirchhoff's laws for a circuit containing random resistors $Z_{m'm}$, capacitors C_m, and sources of the voltages $-\boldsymbol{E}\boldsymbol{R}_m$ (cf. Fig. 2.3). It terms of the quantities (2.5.15), the current (2.5.14) reads

$$\begin{aligned} \boldsymbol{j}(\omega) &= \frac{1}{2\Omega} \sum_{mm'} (\boldsymbol{R}_m - \boldsymbol{R}_{m'}) (U_{m'} - U_m)/Z_{m'm} \\ &= \frac{\mathrm{i}\omega}{\Omega} \sum_m \boldsymbol{R}_m C_m (U_m + \boldsymbol{E}\boldsymbol{R}_m). \end{aligned} \tag{2.5.17}$$

Note that for sites located close of the surface (electrodes) of the sample, the form of Kirchhoff's law differs from that for sites placed deeply in the volume. This is due to the fact that for the former sites we must require that the electrochemical potential coincides with the electrical one, i.e. at the electrodes we have $\delta\mu_m = 0$.

According to (2.5.17), at $\omega \neq 0$, the current $\boldsymbol{j}(\omega)$ can be obtained by multiplying the current through the mth capacitor by the position vector $\boldsymbol{R}_m$, taking the sum over m, and dividing the expression obtained by the volume Ω of the system. From the invariance of (2.5.17) under a rigid body translation $\boldsymbol{R}_0$ of the system as a whole (i.e. replacing $\boldsymbol{R}_m$ by $\boldsymbol{R}_m + \boldsymbol{R}_0$ in (2.5.17)), we get $\sum_m \delta\mu_m C_m = 0$, which represents the condition of charge neutrality, since $e\beta\delta\mu_m C_m$ is the charge at site m.

In the limit as ω goes to zero, the impedance of the capacitors becomes infinitely large, so that the equivalent scheme in question reduces to a scheme consisting of

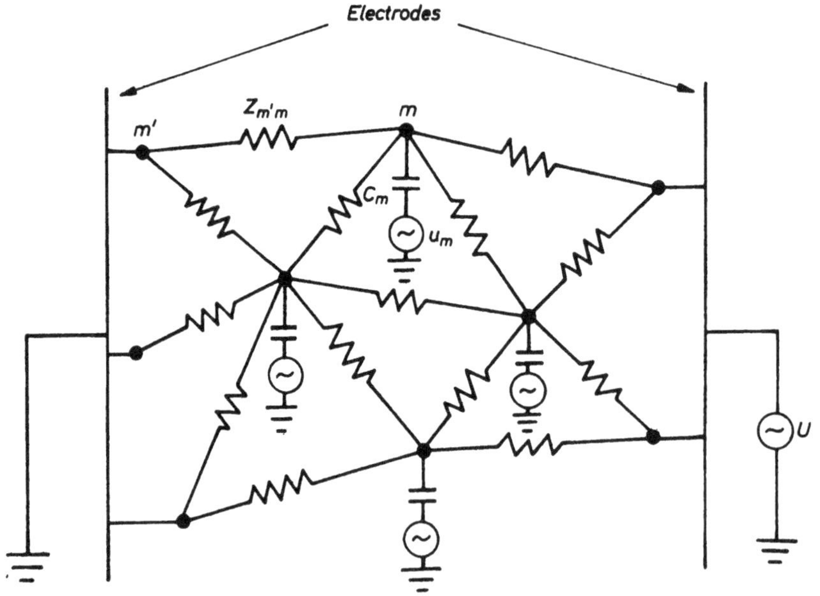

Fig. 2.3. Equivalent scheme for the ac hopping conductivity in disordered systems, which corresponds to Kirchoff's law (2.5.16); $Z_{m'm}$ is the resistor between sites m' and m (cf. (2.5.15)), C_m the capacitor on site m (cf. (2.5.15)), $u_m = -\boldsymbol{E}\boldsymbol{R}_m$, and U is the external potential. The sites placed close to the electrodes are not connected with each other by resistors and are not connected with capacitors

random resistors only, i.e. to a random-resistor network, which was first studied by MILLER and ABRAHAMS (1960). For $\omega \to 0$, the determination of the current by means of (2.5.17) requires first taking the limit $\Omega \to \infty$ and only after that the limit $\omega \to 0$. This procedure can be by-passed by including, for sites close to the electrodes, in the equations of the type (2.5.16), the currents running from the sites into the external circuit. Then, in the limit as $\omega \to 0$, the problem of calculating $\boldsymbol{j}$ reduces to a calculation of the current in the external circuit, I, which must yet be divided by the area S of an electrode. The latter contribution to $\boldsymbol{j}$ must be also accounted for in the case that $\omega \neq 0$. Accordingly, the expression for the current (2.5.17) has to be replaced by

$$\boldsymbol{j}(\omega) = \frac{\mathrm{i}\omega}{\Omega} \sum_m \boldsymbol{R}_m C_m (U_m + \boldsymbol{E}\boldsymbol{R}_m) + \frac{I\boldsymbol{E}}{S\,|\boldsymbol{E}|}, \tag{2.5.18}$$

where the summation is over all sites except those at the surface (in the limit as $\Omega \to \infty$, the exclusion of the sites at the surface is unimportant).

Let us now discuss the Markovian limit of formula (2.5.10), which on account of (2.5.11) reads

$$\sigma_{\mu\nu} = \frac{e^2\beta}{2\Omega} \sum_{mm'} (\boldsymbol{R}_m - \boldsymbol{R}_{m'})_\mu \, (\boldsymbol{R}_m - \boldsymbol{R}_{m'})_\nu \, f_{m'}(1 - f_m) \, W^{(0)}_{m'm} \,. \tag{2.5.19}$$

The relation (2.5.19) is characteristic of hopping transport in ordered systems, because it is proportional to the square of the hopping distance and to the hopping probability $W^{(0)}$. For ordered materials, this relation is valid from the dc limit up to an upper frequency limit, which is determined by the inequality $\omega\delta t \ll 1$. In the case of disordered systems, the relation (2.5.19) may be used at frequencies lying in the interval $\tau^{-1} \ll \omega \ll (\delta t)^{-1}$, which implies the requirement that $\delta t < \tau$, otherwise there is no such interval.

Let us draw attention to the fact that the formula (2.5.19) does not exhibit a frequency dependence. For crystalline materials, this means that (2.5.19) yields the dc conductivity and a frequency dependence of the conductivity can arise only from the parameter $\omega\delta t$. On the other hand, for disordered materials, (2.5.19) describes some high-frequency plateau (which may exist if $\delta t < \tau$), where the conductivity is proportional to the average hopping distance squared and to the hopping probability. To see that (2.5.19) cannot describe the dc limit of disordered systems, consider a simple disordered chain with nearest-neighbour hopping. Suppose that between any two sites the hopping probability is made equal to zero, than the dc conductivity must vanish. However, according to (2.5.19), the conductivity remains finite.

As mentioned above, (2.5.19) overestimates the dc conductivity of disordered systems. This is due to the fact that this formula was obtained on neglecting the diffusive contribution to the current. As will be discussed in detail in Section 6, in disordered systems σ exhibits a specific frequency dependence for $\omega < \tau^{-1}$.

Concluding this section, on the basis of the equivalent scheme from Fig. 2.3, let us explain, why there is no frequency dependent conductivity in ordered materials. Note that in the equivalent scheme of such a material all resistors are equal to each other, and likewise all capacitors. Therefore, the potential U_m is equal to the external potential $-\boldsymbol{E}\boldsymbol{R}_m$. Because of this, at each capacitor the difference of the potentials is equal to zero, so that no current runs through the capacitor. Consequently, the impedance of the system has purely active character and does not depend on the frequency.

2.6. The stochastic interpretation of hopping processes

2.6.1. *The conditional-probability function and its relation to the diffusion coefficient*

In this section the hopping transport in weak electrical fields will not be described by using the one-particle density matrix, but by means of the conditional probability function $P_{m'm}(t)$, which has the meaning of the probability that an electron is at time t on site m if it was placed at time $t = 0$ on site m'. Making use of this function, the hopping motion can be described as an irregular itineracy of the charge carrier from site to site, i.e. as Brownian motion on a random lattice. Note that the formalism of the conditional-probability function is suitable for studying hopping only in the one-electron approximation.

According to its definition, in the absence of an electrical field the quantity $P_{m'm}(t)$ obeys the following equation

$$\frac{\mathrm{d}P_{m'm}(t)}{\mathrm{d}t} = \sum_{m''} P_{m'm''}(t)\, W^{(0)}_{m''m}, \tag{2.6.1}$$

where $W^{(0)}_{m'm}$ labels the hopping probability from site m' to site m at $\boldsymbol{E} = 0$, for which it holds that

$$W^{(0)}_{mm} = -\sum_{m'(\neq m)} W^{(0)}_{mm'}. \tag{2.6.2}$$

Equation (2.6.1) describes Markovian hopping behaviour. To account for non-Markovian processes, this equation may be generalized as follows:

$$\frac{\mathrm{d}P_{m'm}(t)}{\mathrm{d}t} = \sum_{m''} \int\limits_0^t \mathrm{d}t' P_{m'm''}(t')\, \mathscr{W}^{(0)}_{m''m}(t - t'). \tag{2.6.3}$$

Note that $\mathscr{W}^{(0)}$ depends only a single time, since for $\boldsymbol{E} = 0$ or for a time-independent $\boldsymbol{E}$, $\mathscr{W}(t', t)$ (2.4.15) depends on $t' - t$.

According to its definition, the function P satisfies the initial condition

$$P_{m'm}(t)|_{t=0} = \delta_{m'm}. \tag{2.6.4}$$

Equation (2.6.3), which we have obtained here in an intuitive manner, can be derived by means of the diagram technique described in Appendix 1 (Böttger and Bryksin, 1976a), whereby, on account of irreducible blocks of second order, the quantities $\mathscr{W}_{m'm}$ are indeed given by (2.4.15).

It is suitable to study (2.6.3) by using the Laplace transform with respect to time

$$P(s) = \int\limits_0^\infty \mathrm{d}t\, \mathrm{e}^{-st} P(t),$$

by means of which (2.6.3) becomes

$$sP_{m'm}(s) = \delta_{m'm} + \sum_{m''} P_{m'm''}(s)\, W^{(0)}_{m''m}(s), \tag{2.6.5}$$

where

$$W^{(0)}_{m'm} = \int\limits_0^\infty \mathrm{d}t\, \mathrm{e}^{-st} \mathscr{W}^{(0)}_{m'm}(t). \tag{2.6.6}$$

According to (2.6.6), in the Markovian limit, in (2.6.5) $W^{(0)}(s)$ must be replaced by $W^{(0)} = W^{(0)}(s)|_{s\to 0}$. This definition is in agreement with the expression (2.4.19) for $W^{(0)}$.

Introducing the matrix P^{-1}, the inverse of the matrix P (i.e. $PP^{-1} = 1$), then by (2.6.5) we get the expression

$$\left(P^{-1}(s)\right)_{m'm} = s\delta_{m'm} - W^{(0)}_{m'm}(s), \tag{2.6.7}$$

from which we find that

$$\sum_{m''} P_{m'm''}(s)\, W^{(0)}_{m''m}(s) = \sum_{m''} W^{(0)}_{m'm''}(s)\, P_{m''m}(s), \tag{2.6.8}$$

i.e. the matrices P and $W^{(0)}$ are commutable.

The sum rule for the probabilities (2.6.2) (or (2.4.13)) along with equation (2.6.5) lead to the following relation

$$\sum_m P_{m'm}(s) = 1/s, \tag{2.6.9}$$

whose meaning is in that the sum over all probabilities of finding the carrier in the lattice is equal to unity (i.e. equal to $1/s$ in the Laplace representation).
To get still another sum rule for $P_{m'm}(s)$ (now with respect to the first index m') imagine that at the initial moment of time all sites are occupied according to $f_m^B = \exp[-(\varepsilon_m - \mu)/kT]$, the one-electron equilibrium distribution function (Boltzmann distribution function). In this case, the system remains in this state for an infinitely long time, i.e.

$$\sum_{m'} f_{m'}^B P_{m'm}(s) = f_m^B/s. \tag{2.6.10}$$

From (2.6.10) and (2.6.5) we get

$$\sum_{m'} f_{m'}^B W_{m'm}^{(0)}(s) = 0. \tag{2.6.11}$$

This relation is a consequence of the principle of the detailed equilibrium. It reflects the fact that after an arbitrary initial state the system evolves towards the thermodynamic equilibrium. The principle of the detailed equilibrium can be mathematically expressed as

$$P_{m'm}(t)|_{t\to\infty} \to f_m^B. \tag{2.6.12}$$

To find the relation between the current and the function P, first we determine the relation between P and the local change of the chemical potential in the presence of a weak field, $\delta\mu_m$, which was introduced in Section 2.5 (cf. (2.5.3)). This relation is given by

$$\delta\mu_m = -es \sum_{m'} \boldsymbol{E}(\boldsymbol{R}_{m'} - \boldsymbol{R}_m)\, P_{m'm} f_{m'}/f_m, \tag{2.6.13}$$

i.e. $P_{m'm}$ is actually the Green function of equation (2.5.12). Indeed, multiplying both sides of (2.6.5) by $s\boldsymbol{E}(\boldsymbol{R}_{m'} - \boldsymbol{R}_m)\, f_{m'}^B/f_m^B$ and taking the sum over m', we obtain the equation for $\delta\mu_m$ in the form (2.5.12) if we perform the analytical continuation $s \to i\omega$ and set ω equal to zero in the probabilities (i.e. change to the Markovian limit). Thereby, it is necessary to take into account that $f_m \ll 1$, i.e. $f_m \approx f_m^B$.

With the aid of (2.6.13), the conductivity (cf. (2.5.14)) can be expressed by the conditional-probability function as

$$\sigma_{\nu'\nu}(\omega) = s^2 \frac{\beta e^2}{2\Omega} \sum_{mm'} (\boldsymbol{R}_{m'} - \boldsymbol{R}_m)_{\nu'} (\boldsymbol{R}_{m'} - \boldsymbol{R}_m)_\nu\, f_{m'}^B P_{m'm}(s)|_{s=i\omega}. \tag{2.6.14}$$

Consequently, the transport equation for the conditional probability function (2.6.5) along with (2.6.14) determine completely the conduction at low frequencies, where the transport has Markovian character.

Let us now introduce the conception of the diffusion coefficient $D_{\nu'\nu}$ by means of the Einstein relation:

$$D_{\nu'\nu} = \sigma_{\nu'\nu}/(e^2 \, \mathrm{d}n/\mathrm{d}\mu), \tag{2.6.15}$$

where n denotes the concentration of electrons

$$n = \frac{1}{\Omega} \sum_m f_m^B. \tag{2.6.16}$$

Then, according (2.6.14), the diffusion coefficient is related to the conditional-probability function as follows

$$D_{\nu'\nu}(s) = \frac{s^2}{2} \frac{\left\langle \sum_m (\boldsymbol{R}_{m'} - \boldsymbol{R}_m)_{\nu'} (\boldsymbol{R}_{m'} - \boldsymbol{R}_m)_\nu f_{m'}^B P_{m'm}(s) \right\rangle}{\langle f_m^B \rangle}. \tag{2.6.17}$$

Here $\langle \ldots \rangle$ indicates the configuration average, which according to the self-averaging concept for large systems, is defined as

$$\langle A_m \rangle = \frac{1}{N} \sum_m A_m. \tag{2.6.18}$$

Note that the Einstein relation (2.6.15) is usually written only for the static limiting case $s \to 0$. The question, whether or not for $s \neq 0$, the coefficient D is actually a diffusion coefficient occuring in a diffusion equation, is still open.

2.6.2. *Relation between charge transport and diffusion*

Let us now point out that for ordered systems, the diffusion coefficient D (2.6.17) does not depend on the parameter s, so that in this case the relation (2.6.15) corresponds actually to the static limit. In crystalline materials, where f_m does not depend on m and $P_{m'm}$ depends only on $\boldsymbol{R}_{m'} - \boldsymbol{R}_m$, formula (2.6.17) takes the form

$$D_{\nu'\nu} = \frac{s^2}{2} \sum_m P_{m'm}(s) \, (\boldsymbol{R}_{m'} - \boldsymbol{R}_m)_{\nu'} \, (\boldsymbol{R}_{m'} - \boldsymbol{R}_m)_\nu. \tag{2.6.19}$$

Let us pass over to the momentum representation

$$P(\boldsymbol{q}, s) = \sum_m \mathrm{e}^{\mathrm{i}\boldsymbol{q}(\boldsymbol{R}_m - \boldsymbol{R}_{m'})} P_{m'm}(s). \tag{2.6.20}$$

In this representation the expression (2.6.19) reads

$$D_{\nu'\nu} = -\frac{s^2}{2} \frac{\partial^2}{\partial q_{\nu'} \partial q_\nu} P(\boldsymbol{q}, s)|_{\boldsymbol{q}=0}. \tag{2.6.21}$$

The equation for P (2.6.5) can be readily solved. We find that

$$\left.\begin{aligned} P(\boldsymbol{q}, s) &= 1/\big(s - W^{(0)}(\boldsymbol{q})\big), \\ W^{(0)}(\boldsymbol{q}) &= \sum_m W_{m'm}^{(0)} \, \mathrm{e}^{\mathrm{i}\boldsymbol{q}(\boldsymbol{R}_m - \boldsymbol{R}_{m'})}. \end{aligned}\right\} \tag{2.6.22}$$

According to the sum rule (2.5.2), we have $W(\boldsymbol{q})|_{\boldsymbol{q}=0} = 0$. Furthermore, owing to the symmetry of the problem in the absence of a magnetic field, it holds that $W(\boldsymbol{q}) = W(-\boldsymbol{q})$. Thus, from (2.6.21) and (2.6.22), we find that

$$D_{\nu'\nu} = -\frac{1}{2}\,\delta_{\nu'\nu}\,\frac{\partial^2}{\partial q_\nu^{\,2}}\,W^{(0)}(\boldsymbol{q})|_{\boldsymbol{q}=0} = \delta_{\nu'\nu}\,\frac{1}{2}\sum_m (\boldsymbol{R}_m)_\nu^{\,2}\,W_{0m}^{(0)}, \tag{2.6.23}$$

where we have set $\boldsymbol{R}_0 = 0$. Accordingly, in the Markovian limit, the diffusion coefficient for hopping transport in ordered systems is independent of s and has the standard form of a product of the square of the hopping distance and the corresponding hopping probability. On comparing the expressions for D (2.6.23) and $\sigma_{\nu'\nu}$ (2.5.19), we see that the Einstein relation (2.6.15) is indeed fulfilled.

As can be seen from (2.6.22), the quantity D plays the role of the diffusion coefficient in a diffusion equation, since, considering sufficiently large times, in (2.6.22) we may expand $W^{(0)}(\boldsymbol{q})$ in a power series with respect to $\boldsymbol{q}$ up to the second order[1]), which in the $\boldsymbol{r}$ and t representation yields (for an isotropic medium, where $D_{xx} = D_{yy} = D_{zz} = D$)

$$\left.\begin{aligned} &\frac{\partial P(\boldsymbol{r}, t)}{\partial t} = D\Delta P(\boldsymbol{r}, t),\\ &\int \mathrm{d}\boldsymbol{r} P(\boldsymbol{r}, t) = 1. \end{aligned}\right\} \tag{2.6.24}$$

The solution of (2.6.24)

$$P(\boldsymbol{r}, t) = (4\pi Dt)^{-3/2} \exp\{-r^2/(4Dt)\} \tag{2.6.25}$$

describes the customary Gaussian diffusion behaviour.

Consequently, at least in ordered systems, the hopping transport may be viewed as Brownian motion of charge carriers interacting with the heat bath (KUDINOV and FIRSOV, 1966). Thereby, a carrier may be considered as a classical object, irrespective of the quantum-mechanical origin of the hopping probability. Owing to this fact, the formulation of the quantum mechanical analog of the Langevin equation may be performed without difficulties. In the Markovian limit, the two-time correlator of the random forces $\langle f(t')\, f(t)\rangle$ occurring in the Langevin equation is proportional to $\delta(t' - t)$, while in the non-Markovian case this δ-function is smeared out over a time interval δt, which may be thought of as a collision time. The hopping probability plays the role of a reciprocal time characterizing slow friction processes. Note that a division into rapidly changing random forces describing the interaction

---

[1]) Note that such an expansion with respect to $\boldsymbol{q}$ may be performed if $tWX \gg |\boldsymbol{r}|$, where X is the hopping distance, W the hopping probability, t the time of observation, and $\boldsymbol{r}$ the position vector of the particle at time t (we assume that the particle was at $t = 0$ at $\boldsymbol{r} = 0$). This criterion means that the time t is so large that the particle gets from $\boldsymbol{r} = 0$ to $\boldsymbol{r}$ by means of a large number of hopping events and via a large number of possible trajectories. Thereby, the discrete structure of the underlying lattice becomes meaningless so that $\boldsymbol{R}_m$ may be replaced by the continuous position vector $\boldsymbol{r}$.

with the heat bath and slowly changing friction forces is reasonable only if $\delta t W \ll 1$. For more details on the theory of the Brownian motion, the reader is referred to the books by CHANDRASEKAR (1947) and SCHWINGER (1962). The stochastic interpretation of the hopping transport in the small-polaron model is discussed in the book by FIRSOV (1975).

A more difficult problem is the hopping transport in disordered systems. Let us assume that for such systems there exists some diffusion function $P(\boldsymbol{q}, s)$ having the form (2.6.22), where $W^{(0)}(\boldsymbol{q})$ denotes some effective probability. From $P(\boldsymbol{q}, s)|_{q=0} = 1/s$, which results from the normalization $\int \mathrm{d}\boldsymbol{r} P(\boldsymbol{r}, s) = 1/s$, we obtain $W^{(0)}(\boldsymbol{q})|_{q=0} = 0$. Furthermore, from the spatial isotropy of the configuration averaged system we get $W^{(0)}(\boldsymbol{q}) = W^{(0)}(-\boldsymbol{q})$. Thus, for small $\boldsymbol{q}$, we have $W^{(0)}(\boldsymbol{q}) \approx -q^2 D$, and D may termed the diffusion coefficient. Now we suppose that this diffusion coefficient is related to the conductivity by the Einstein relation (2.6.15), i.e. D is given by (2.6.17). Consequently, even in the Markovian limit, D depends on s and, correspondingly, σ depends on ω (cf. Section 2.5.). Accordingly, for disordered systems, after performing the configuration average, the diffusion equation has a non-Markovian form

$$\frac{\mathrm{d}P(\boldsymbol{r}, t)}{\mathrm{d}t} = \int\limits_0^t \mathrm{d}t' D(t-t')\, \Delta P(\boldsymbol{r}, t'), \quad D(t) = \frac{1}{2\pi \mathrm{i}} \int\limits_{-\mathrm{i}\infty+a}^{\mathrm{i}\infty+a} \mathrm{d}s\, D(s)\, \mathrm{e}^{st}. \tag{2.6.26}$$

The solution of (2.6.26) may be written in the $\boldsymbol{r}$-s-representation as

$$P(\boldsymbol{r}, s) = \int\limits_0^\infty \mathrm{d}\tau [4\pi D(s)\, \tau]^{-3/2} \exp\left\{-s\tau - \frac{r^2}{4D(s)\, \tau}\right\}. \tag{2.6.27}$$

Thus, for sufficiently large times, where the dependence of D on s may be neglected, the expression for $P(\boldsymbol{r}, t)$ passes into the customary Gaussian form (2.6.25). However, for sufficiently small times, this dependence may not be neglected, and in a disordered system the carrier packet markedly differs from the Gaussian form (2.6.25). Thereby, the deviation from the Gaussian diffusion behaviour is governed by the frequency dependence of the conductivity (BUTCHER, 1978; cf. Section 6.8.). For more details on the diffusion problem in disordered hopping systems the reader is referred to Section 6., where for some class of disordered systems the function $P(\boldsymbol{q}, s)$ will be introduced rigorously, and where the assumption of the Einstein relation for $\omega \neq 0$ will be discussed.

Chapter II. Small Polarons

3. The small-polaron model; crystalline materials with a narrow conduction band and strong electron-phonon interaction

3.1. The low-mobility problem and the small polaron

The investigation of small polarons has been much stimulated by experimental findings which indicate in a number of materials low mobilities, $u < 1\ \text{cm}^2/\text{Vs}$, which cannot be understood within the framework of the usual band picture. Indeed, by means of the Boltzmann equation one gets $u \sim e\tau/m^*$, where τ is the relaxation time and m^* the effective mass of the charge carrier. Taking into account that $\tau \simeq l/v$ (v — thermal velocity of the carrier, l — mean-free path), we obtain $l \sim m^* uv/e$, which, for $u < 1\ \text{cm}^2/\text{Vs}$, yields $l < a$ (a — lattice constant), which is not a reasonable result (IOFFE, 1956).

On the other hand, as pointed out by HEIKES and JONSTON (1957) such low mobilities may be due to thermally activated jumps of the carriers from site to site. For such processes the diffusion coefficient is given by $D \sim a^2 W$, where a is the hopping distance (and also the lattice constant), and W the hopping probability (cf. (2.6.23)). By means of the Einstein relation $u = eD/kT$, we find that $u \sim u_0 \hbar W/kT$, where $u_0 = ea^2/\hbar$ has the dimension of a mobility and is of order $1\ \text{cm}^2/\text{Vs}$. If the hopping carrier must overcome an energy barrier of the height E_a, then $W \sim \frac{\omega_0}{2\pi} e^{(-E_a/kT)}$, where ω_0 is a characteristic phonon frequency. Thus, we find that

$$u \sim u_0 \frac{1}{2\pi} \frac{\hbar\omega_0}{kT} \exp\left(-E_a/kT\right). \tag{3.1.1}$$

Consequently, for hopping transport, the mobility may be much less than $1\ \text{cm}^2/\text{Vs}$.

However, the concept of hopping transport is suitable only for materials where the charge carriers are spatially localized. Such localization may be due to strong electron-phonon interaction, which causes a displacement of the equilibrium positions of the atoms in the vicinity of the carrier such that the carrier is surrounded by a polaron "cloud". In this case the system lowers its energy by an amount E_p, the polaron shift defined in (2.2.7). On the other hand, the kinetic energy of the carrier in the rigid (non-deformed) lattice attempts to remove the carrier from its site by causing "melting-away" of the wave packet. Since the kinetic energy is approximately equal to the resonance integral J (cf. (2.1.5)), the radius of the polaron is of order aJ/E_p. Accordingly, the formation of a polaron whose radius is small compared to the lattice constant (small polaron) requires that

$$J/E_p \ll 1. \tag{3.1.2}$$

Note that in materials with broad bands and strong electron-phonon interaction, where instead of (3.1.2) it holds that $J/E_p \gg 1$, large (continuous) polarons are formed. For $J/E_p \approx 1$, the polarons are said to have intermediate size.

Owing to the translational symmetry of the lattice the localized states in question are quasi-stationary. Therefore, the hopping transport is realized only if the time between consecutive hops, W^{-1}, is less than the life time of the quasi-stationary state, which is given by $\hbar/\Delta\varepsilon_p$, where $\Delta\varepsilon_p$ is the width of the polaron band (YAMASHITA and KUROSAWA, 1958). Consequently, the criterion for the occurrence of hopping motion in crystals reads

$$\hbar W/\Delta\varepsilon_p \gg 1. \tag{3.1.3}$$

Now we must find $\Delta\varepsilon_p$. If $\Delta\varepsilon_p$ would be of the order of the band width of the rigid lattice, i.e. proportional to J, then, in general, the criterion (3.1.3) could not be satisfied, since for a characteristic band width of the rigid lattice of order 1 eV, the decay time of a wave packet located at a site is of order 10^{-14} s, while W^{-1} is of order 10^{-12} s or greater. The most important feature of the polaron state is the strong renormalization of the band width due to electron-phonon interaction, so that $\Delta\varepsilon_p \ll J$ holds (TJABLIKOV, 1952). This is due to the fact that in the polaron state the band width is not governed only by the overlap between the wavefunctions located at neighbouring sites (i.e. J), as in the case of weak electron-phonon interaction, but also by the overlap between the oscillator wavefunctions of neighbouring sites. Owing to the polarization of the lattice, the oscillator wavefunctions at neighbouring sites have equilibrium positions which are displaced relatively to each other, so that the corresponding overlap is proportional to a small parameter.

To study the polaron band quantitatively, note that the band-like motion is due to quantum mechanical tunnelling between the sites without changing the phonon numbers (diagonal phonon transitions). Accordingly, this motion is governed by the matrix elements of the operator $\mathcal{H}'$ (2.2.8) which are diagonal in the number $n_{\boldsymbol{q}}$ of phonons in the state $\boldsymbol{q}$. Using $\langle n_{\boldsymbol{q}}|\, \hat{\Phi}_{m'm}\, |n_{\boldsymbol{q}}\rangle$ from (A 1.14) and the Weyl identity (A 1.10) (cf. Appendix 1.), we obtain

$$\langle n_{\boldsymbol{q}}|\, \mathcal{H}'\, |n_{\boldsymbol{q}}\rangle = \sum_{mm'} J_{m'm}\, \mathrm{e}^{-S_N(m',m)} a^+_{m'} a_m, \tag{3.1.4}$$

where

$$S_N = \sum_{\boldsymbol{q}} \frac{|\gamma(\boldsymbol{q})|^2}{N} [1 - \cos \boldsymbol{q}(\boldsymbol{R}_{m'} - \boldsymbol{R}_m)]\, (n_{\boldsymbol{q}} + 1/2). \tag{3.1.5}$$

Taking the thermodynamic average, in these relations $n_{\boldsymbol{q}}$ is replaced by the Planck distribution $N_{\boldsymbol{q}}$ (A 1.15), i.e. S_N is replaced by S_T defined in (A 1.18). Diagonalizing now the Hamiltonian (3.1.4) by using the momentum representation, we obtain the usual band Hamiltonian (3.1.4), but where $\varepsilon(\boldsymbol{k})$, defined in (2.1.8), is determined by the renormalized resonance integral $\tilde{J}_{m'm} = J_{m'm} \exp\left(-S_T(m'm)\right)$. Consequently, the width of the polaron band is proportional to $\tilde{J}$, i.e. it is much less than the width of the band in the absence of interaction with phonons. The width of the polaron band decreases with increasing temperature. Owing to the renormalization of the band width due to the diagonal phonon transitions[1]), the criterion (3.1.3) for hopping transport can be fulfilled in materials having, in the absence of electron-phonon interaction, a band width of order 1 eV. Note that the renormalizing factor $\exp(-S_T)$ is intimately related to the well-known Debye-Waller factor describing the decrease of the height of the neutron-scattering peaks with increasing temperature and determining the temperature dependence of the Mössbauer effect etc. (see, e.g., KITTEL, 1963). This relation is due to the fact that the exponents of both these quantities are given by the mean-square displacement of an atom from its equilibrium position.

With decreasing temperature, the hopping probability W rapidly decreases, while the width of the polaron band increases. Accordingly, in the small-polaron model at sufficiently low temperatures the transport becomes band-like, whereby the relaxation time τ satisfies the relation

$$\Delta\varepsilon_p \tau/\hbar \gg 1, \tag{3.1.6}$$

which means that the uncertainty of energy due to scattering processes is less than the width of the energy band, i.e. the Boltzmann equation may be used.

In this book we describe in detail only the hopping transport of small polarons (i.e. the transport at high temperatures). The reader who is interested in the transport of small polarons in a large temperature interval is referred to the review by APPEL (1968) and the book by FIRSOV (1975).

3.2. DC conductivity

3.2.1. *Hopping conduction of non-adiabatic polarons*

The dc (direct current) hopping conductivity of small polarons in crystals may be calculated by starting from formula (2.5.19), which on account of resonance integrals only between nearest neighbours takes the form

$$\sigma_h = eu_0 n(1 - f) \frac{\hbar W}{kT}, \tag{3.2.1}$$

---

[1]) Nondiagonal transitions with respect to the phonons give rise to polaron scattering and hopping motion of polarons. To discriminate between diagonal and nondiagonal transitions, in the Hamiltonian of the system, $\hat{\Phi}$ must be replaced by $(\hat{\Phi} - \exp(-S_T)) + \exp(-S_T)$ (LANG and FIRSOV, 1962), which in the language of the diagram technique means that $P(t)$ is replaced by $(P(t) - 1) + 1$ (cf. (A1.20)), which leads to modified diagrams. BRYKSIN and FIRSOV (1969) formulated the theory of small polarons in terms of diagonal and nondiagonal (with respect to n_q) components of the density matrix by making use of the method of KOHN and LUTTINGER (1957).

where $f = \{e^{-\beta\mu} + 1\}^{-1}$ labels the occupation of a site, $n = Nf/\Omega$ the concentration of the charge carriers, and $u_0 = ea^2/\hbar$ is a constant having the dimension of a mobility (a — lattice constant). The index h at σ_h indicates that the conduction has hopping character.

According to (2.4.19), the hopping probability between nearest-neighbour sites, W, may be written as

$$W = \frac{J^2}{\hbar^2} e^{-2S_T} \int_{-\infty}^{\infty} dt \left\{ \exp\left[\sum_{\boldsymbol{q}} \frac{|\gamma(\boldsymbol{q})|^2 (1 - \cos \boldsymbol{q}\boldsymbol{g})}{N \sinh (\hbar\omega_{\boldsymbol{q}}\beta/2)} \cos \omega_{\boldsymbol{q}} t \right] - 1 \right\}, \tag{3.2.2}$$

where $\boldsymbol{g}$ is the vector directed to a nearest-neighbour site, $|\boldsymbol{g}| = a$, and J is the resonance integral between nearest neighbours.

For $|\gamma|^2/\sinh (\hbar\omega_{op}\beta/2) \gg 1$ (ω_{op} — optical phonon limiting frequency), in (3.2.2) the integration can be approximately performed, since the integrand, $F(t)$, is an oscillating function, which decreases with increasing time. The time dependence of $F(t)$ will be studied in Appendix 1. (cf. (A1.19a)), where it turns out that $F(t)$ goes

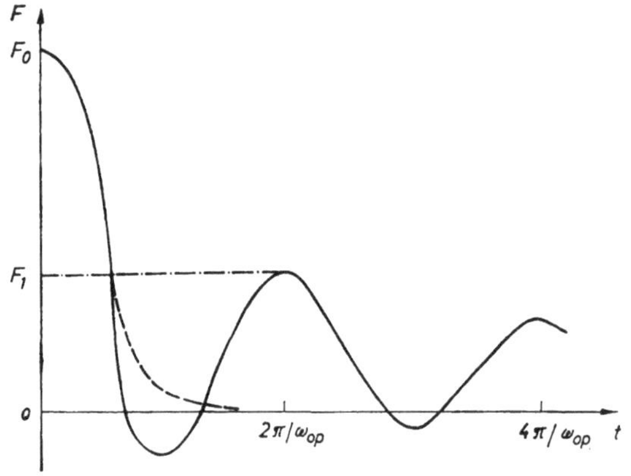

Fig. 3.1. Schematic representation of the time dependence of the function $F(t)$ occurring in (3.2.2) (solid line). The dashed line indicates for $F(t)$ the approximation which is based on an account of the first saddle point only (ω_{op} — optical phonon frequency)

to zero at large times owing to the dispersion of the phonon frequencies. The qualitative time dependence of $F(t)$ is illustrated in Fig. 3.1. We see that $F(t)$ has its maximum value at $t = 0$. At $t \simeq \pi/\omega_{op}$ the function $F(t)$ displays a minimum, and after that, in the interval from π/ω_{op} to $2\pi/\omega_{op}$, it increases, and so on. A very important feature of $F(t)$ is the decrease fo the height of its maximum values with increasing time, i.e. in Fig. 3.1 we see that $F_0 \gg F_1$. This decrease is a consequence of the phonon dispersion $\Delta\omega$. According to (A1.19a), it holds that ($\Delta\omega/\omega_{op} \ll 1$):

$$F_1/F_0 \approx \exp\left\{ -\frac{\Delta\omega}{\omega_{op}} \frac{\gamma^2}{\sinh (\hbar\omega_{op}\beta/2)} \right\},$$

which is a small quantity provided that

$$\frac{\Delta\omega}{\omega_{op}} \frac{\gamma^2}{\sinh (\hbar\omega_{op}\beta/2)} > 1. \tag{3.2.3}$$

If the inequality (3.2.3) is satisfied, then in (3.2.2) the integration with respect to t may be performed by expanding in the exponent $\cos \omega_{\boldsymbol{q}} t$ in a power series with respect to t up to the second-order term. Thereby, the unity, which compensates the divergency for $t \to \infty$, may be neglected. Thus, we find that

$$W \approx \frac{J^2}{\hbar^2} \exp\left(-\frac{E_a(T)}{kT}\right) \int\limits_{-\infty}^{\infty} dt \exp\left(-t^2/(\delta t)^2\right), \tag{3.2.4}$$

where

$$E_a(T) = kT \sum_{\boldsymbol{q}} \frac{|\gamma(\boldsymbol{q})|^2}{N} (1 - \cos \boldsymbol{qg}) \tanh\left(\frac{\hbar\omega_{\boldsymbol{q}}}{4kT}\right), \tag{3.2.5}$$

$$\delta t = \left\{\sum_{\boldsymbol{q}} \frac{|\gamma(\boldsymbol{q})|^2}{2N} (1 - \cos \boldsymbol{qg}) \frac{\omega_{\boldsymbol{q}}^2}{\sinh(\hbar\omega_{\boldsymbol{q}}/2kT)}\right\}^{-1/2}. \tag{3.2.6}$$

Performing the integration in (3.2.4), we obtain the following expression for the two-site hopping probability

$$W = \sqrt{\pi} \frac{J^2 \delta t}{\hbar^2} \exp\left(-E_a(T)/kT\right). \tag{3.2.7}$$

In particular, for $kT > \hbar\omega_{\text{op}}/2$, the activation energy E_a becomes independent of temperature and is given by

$$E_a \equiv E_a(T)|_{T\to\infty} = \sum_{\boldsymbol{q}} \frac{|\gamma(\boldsymbol{q})|^2}{4N} \hbar\omega_{\boldsymbol{q}}(1 - \cos \boldsymbol{qg}). \tag{3.2.8}$$

In this case the characteristic time δt becomes

$$\delta t = \hbar/\sqrt{4E_a kT}, \tag{3.2.9}$$

and the hopping conductivity (3.2.1) takes the form

$$\sigma_h = eu_0 n(1 - f) \frac{\sqrt{\pi}}{2} \frac{J^2}{kT\sqrt{E_a kT}} \exp(-E_a/kT). \tag{3.2.10}$$

Note that the activated temperature dependence of σ_h (3.2.10) is in agreement with the intuitive result (3.1.1). However, we see that the characteristic frequency ω_0 occurring in (3.1.1) is of order $J^2/\hbar\sqrt{E_a kT}$, i.e. ω_0 has no purely vibrational origin. This fact is a consequence of the assumption that the resonance integral J is a small quantity, which allows the probability W to be calculated in lowest (second) order with respect to J. In this case the hopping probability has no purely classical character. As will be shown below, for not too small J, the characteristic frequency ω_0 turns out to be indeed of order ω_{op}, i.e. in this case the hopping probability has classical character.

The characteristic time δt (3.2.9) has the meaning of the duration of a hopping event. According to Section 2., for $\delta t \ll W^{-1}$, the hopping behaviour is Markovian.

With the aid of (3.2.7), the criterion $\delta t \ll W^{-1}$ may be rewritten as $J^2 \ll E_a kT \times \exp(E_a/kT)$, which in the small-polaron model is always satisfied, since $J \ll E_a$ and $kT < E_a$. For small polarons, non-Markovian hopping transport may occur if the inequality (3.2.3) does not hold, so that on calculating the time integral in (3.2.2), one may not restrict oneself to the first saddle point. In this situation, δt turns out to be much larger than (3.2.9). This case was studied by EMIN (1971), who pointed out that for non-Markovian hopping transport the activation energy is smaller than the value given in (3.2.5). In the non-Markovian case, after a hop from site 1 to site 2 the electron-phonon system does not achieve thermal equilibrium before the carrier hops from site 2 to another site. In this case, the second hop is not determined only by its probability, but also by the residence time of the carrier at site 2, i.e. the hop away from site 2 "remembers" the preceding hop. In the presence of memory, the carrier has no time to form the polaron cavity perfectly, which leads to a reduction of the activation energy and, therefore, to an enhancement of the mobility.

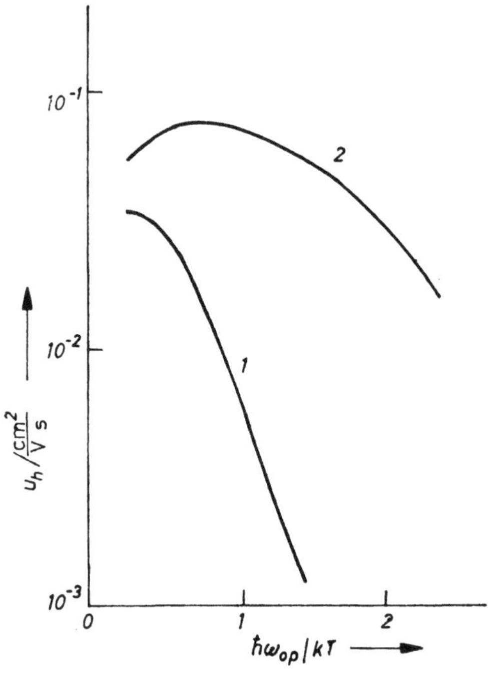

Fig. 3.2. Temperature dependence of the hopping drift mobility u_h in the small-polaron model evaluated by means of formula (3.2.10) (curve 1) for $E_a = 5\hbar\omega_{op}$, $\frac{\sqrt{\pi}}{2} u_0 J^2 / \left(\hbar\omega_{op} \sqrt{E_a \hbar\omega_{op}}\right) = 1\ \mathrm{cm^2/Vs}$; curve 2 indicates this dependence for hopping with memory (EMIN, 1971) for $\Delta\omega/\omega_{op} = 0.1$ (see text)

The temperature dependence of the drift mobility $u_h = \sigma_h / en(1 - f)$, calculated by means of (3.2.10), is shown in Fig. 3.2. This figure also shows the corresponding dependence for the case of non-Markovian transport, which was calculated by EMIN (1971).

As mentioned above, the expression for σ_h (3.2.10) is valid if the resonance integral J is small. For hopping transport of small polarons, the applicability of lowest-order-in-J expressions is governed by the following three dimensionless parameters

$$\eta_1 = J/E_a, \quad \eta_2 = J^2/\left(\hbar\omega_{op}\sqrt{E_a kT}\right), \quad \eta_3 = J^2/(E_a kT). \tag{3.2.11}$$

The parameter η_1 characterizes the size of the polaron. For $\eta_1 \ll 1$, the polaron size is small, which follows from (3.1.2), since, according to (3.2.8) and (2.2.7), $E_p \approx 2E_a$. For $\eta_3 \ll 1$, only hops between nearest-neighbour sites must be taken into account in a formula of the type (2.5.19), which allows the problem in question to be reduced to a two-site one (Lang and Firsov, 1967). Note that this is true for arbitrary values of η_2, although, in particular, the expression for σ_h (3.2.10) holds only if η_2 is small. Because of this, and owing to the fact that at $kT > \hbar\omega_{op}$ the lattice vibrations may be treated classically, the hopping problem can be rigorously solved for arbitrary η_2. Since the two-site model is useful not only for studying the dc conductivity of small polarons, we consider this model in more detail in the following section.

3.2.2. *The two-site model and the adiabatic polaron*

For classical vibrational behaviour, the Hamiltonian of the two-site model reads

$$\mathcal{H} = \frac{M}{2}(v_1^2 + v_2^2) + \frac{M}{2}\omega_{op}^2(x_1^2 + x_2^2) + A(x_1 - x_2)(a_1^+a_1 - a_2^+a_2) + J(a_1^+a_2 + a_2^+a_1). \tag{3.2.12}$$

Here M is the mass of a vibrating ion, x_i the displacement of the ith ion from its equilibrium position, $v_i = \mathrm{d}x_i/\mathrm{d}t$, and A is a constant describing the coupling between the electron and the lattice. We assume that this coupling has short-range character. In (3.2.12) the electron-lattice coupling is proportional to the difference of the displacements of the ions at neighbouring sites, which is in accordance with the fact that in the small-polaron model only the interaction with optical phonons is accounted for. Note that the interaction of electrons with acoustic phonons is considered in the theory of condensons (Toyozawa, 1961).

In the Hamiltonian (3.2.12), phonon dispersion is lacking, i.e. all ions are assumed to vibrate with a single frequency, ω_{op} (Einstein approximation). Accordingly, in this approximation there is no mechanism of energy dissipation, so that it is necessary to introduce such a mechanism ad hoc. A two-site model including phonon dispersion was proposed by Lang and Firsov (1968).

The Hamiltonian (3.1.12) is a simplified two-site version of the general Hamiltonian of an electron-phonon system of the Fröhlich type (2.1.9). In particular, on comparing these two expressions, we see that the coupling constant A is related to the dimensionless coupling constant γ occurring in (2.1.9) (for short-range interaction γ is independent of $\boldsymbol{q}$) by

$$\gamma = 2A(\hbar\omega_{op}^3 M)^{-1/2}. \tag{3.2.13}$$

This relation was obtained by taking into account that, according to the rules of the second quantization,

$$x_m = \sum_{\boldsymbol{q}} (2\hbar/NM\omega_{\rm op})^{1/2} \left(b_{\boldsymbol{q}} \exp (i\boldsymbol{q}\boldsymbol{R}_m) + b_{\boldsymbol{q}}^+ \exp (-\,\mathrm{i}\boldsymbol{q}\boldsymbol{R}_m)\right).$$

Let us now introduce the coordinate $x = x_1 - x_2$, which describes the relative motion of the ions, and the coordinate $X = (x_1 + x_2)/2$ describing the motion of the system as a whole, and the corresponding velocities $v = \mathrm{d}x/\mathrm{d}t$ and $V = \mathrm{d}X/\mathrm{d}t$, too. In the Hamiltonian the terms containing X and V may be ignored, since they do not involve electron operators. Thus, $\mathcal{H}$ takes the form

$$\mathcal{H} = \frac{Mv^2}{4} + \frac{\omega_{\rm op}^2 M}{4} x^2 + Ax(a_1^+a_1 - a_2^+a_2) + J(a_1^+a_2 + a_2^+a_1). \quad (3.2.12\mathrm{a})$$

The drift mobility is calculated on the basis of this Hamiltonian by using the familiar relation between the diffusion coefficient and the hopping probability, $D = a^2W$, and by making use of the Einstein relation.

To elucidate the physical situation, let us adopt the adiabatic approximation. Thus, keeping fixed the coordinate x, by means of the equation of motion $\dot{a}_i = \frac{\mathrm{i}}{\hbar} [\mathcal{H}, a_i]$ and the ansatz $a_i(t) = a_i \exp\left(\frac{\mathrm{i}}{\hbar} Et\right)$, we find that the energy E of the two-site system is given by

$$E_\pm = \frac{\omega_{\rm op}^2 M}{4} x^2 \pm \sqrt{A^2x^2 + J^2}\,. \quad (3.2.14)$$

The adiabatic levels $E_\pm$ (3.2.14) are illustrated in Fig. 3.3. The lower level E_- has two minima, at $x = \pm x_0$, where

$$x_0 = (2A/\omega_{\rm op}M) \left(1 - (1/2)\,(J\omega_{\rm op}^2 M/A^2)^2\right)^{1/2}.$$

Note that according to (3.2.8) and (3.2.13), the activation energy $E_{\rm a}$ is related to A as

$$E_{\rm a} = A^2/\omega_{\rm op}^2 M\,, \quad (3.2.13\mathrm{a})$$

i.e. $E_{\rm a}$ is a purely classical quantity.

By means of (3.2.13a), we get $x_0 = (2A/\omega_{\rm op}M)\left(1 - \frac{1}{4}\eta_1^2\right)^{1/2}$, and since for small polarons, η_1 (3.2.11) is small, we have $x_0 \approx 2A/\omega_{\rm op}M$. For the depth of the wells in in $E_-(x)$, we get $E_{\rm a}\left(1 - \frac{1}{4}\eta_1^2 - \eta_1\right) \approx E_{\rm a}$. At $x = 0$, i.e. at the point of the maximum approach between the levels E_+ and E_-, the difference between E_+ and E_- is equal to $2J$. Thus, a non-zero resonance integral J leads to a splitting of the levels (cf. Fig. 3.3).

Let us now discuss what happens when at the initial moment of time the system is located in the (left) well 1 (cf. Fig. 3.3). Since the depth of the well, $E_{\rm a}$, much exceeds kT, to a good approximation, the system may be viewed as performing small thermal oscillations about the equilibrium position $x = -x_0$, which differs from that in the absence of an electron, $x = 0$, owing to the polarization of the lattice. Due to statistical fluctuations, there is a small, but finite probability that the system reaches $x = 0$ (the point of the quasi-intersection of the levels). In this situation the system can either non-adiabatically pass over from E_- to E_+ and proceed to move along E_+ to the right, or overcome the maximum at $x = 0$ and fall into the well 2. In the former case, sooner or later, the system stops and starts to move back along E_+ from the right to the left. At $x = 0$, the system either non-adiabatically passes over to E_- and falls into the well 1, or proceeds to move to the left along E_+, stops, moves

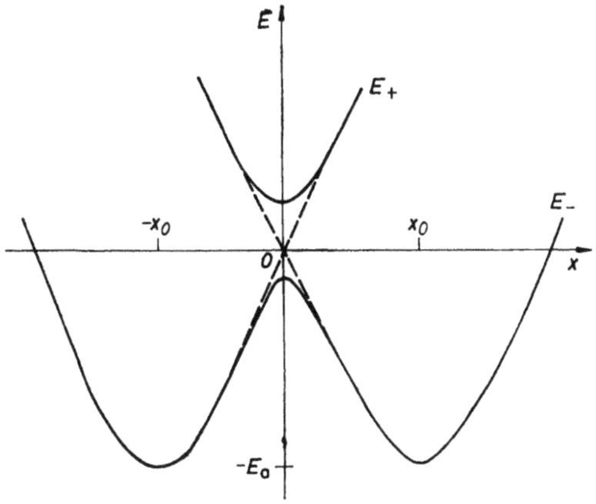

Fig. 3.3. The adiabatic levels E_+ and E_- in the two-site model in dependence of the relative displacement of the nuclei x, evaluated by means of formula (3.2.14). The minimal distance between E_+ and E_- (at $x = 0$) is equal to $2J$ (J — resonance integral). In the absence of interaction ($J = 0$), the levels are indicated by dashed lines

back to the right, and so on. Let P be the probability for a non-adiabatic transition at $x = 0$ from E_- to E_+ or vice versa, and P_t the probability of finding the system in the well 2 as a result of multiple passages through the point of the quasi-intersection of the levels. Taking the sum over the passage events, we find that

$$P_t = (1 - P) + P^2(1 - P) + P^2(1 - P)^3 + \dots + P^2(1 - P)^{2k+1} + \dots$$
$$= 2\,\frac{1 - P}{2 - P}. \tag{3.2.15}$$

On deriving (3.2.15), we assumed that the system, after moving down along E_- (in the well 1 or 2), does not return to the point of the quasi-intersection (i.e. the system thermalizes in a well). This is a reasonable assumption, provided that there is a sufficiently strong friction for the vibrational motion associated with E_-, i.e. an effective mechanism of energy dissipation. As mentioned above, in the two-site model under consideration such a mechanism is lacking, so that energy dissipation must be introduced ad hoc. Recall that for strong friction, i.e. strong phonon dispersion, on calculating the integral (3.2.2), one may restrict oneself to the first saddle point. In this case the hopping transport has Markovian character. On the other hand, for weak friction, the system, after starting from $x = 0$, performs a weakly damped oscillatory motion with a period $2\pi/\omega_{op}$. In this case the hopping transport has non-Markovian character.

Let us now evaluate the transition probability P introduced above. The probability of a quasi-classical transition between adiabatic levels E_+ and E_- is given by (cf., e.g., LANDAU and LIFSHITZ, 1963, § 53)

$$P = \exp\left\{-\frac{2}{\hbar}\,\mathrm{Im}\int\limits_{t'}^{t_0} \mathrm{d}\tau[E_+(\tau) - E_-(\tau)]\right\}, \tag{3.2.16}$$

where E_+ and E_- are defined in (3.2.14), t' is a real time, and t_0 is some complex time, which is defined by the equation $E_+(t_0) = E_-(t_0)$. The time t_0 is complex, because for real values t $\big($i.e. $x(t)\big)$, there is no intersection between the adiabatic levels. We define by $x(t_c) = 0$ the time t_c, which we choose as lower limit in the integral (3.2.16) i.e. $t' = t_c$. Then, we have

$$P = \exp\left\{-\frac{4}{\hbar}\,\mathrm{Im}\int\limits_0^{t_0-t_c} \mathrm{d}\tau\,\sqrt{A^2x^2(\tau + t_c) + J^2}\right\}. \tag{3.2.17}$$

Because the minimum spacing between the adiabatic levels, $2J$, is small, in the time interval between 0 and $t_0 - t_c$ the function $x(\tau + t_c)$ may be expanded in a power series with respect to τ. Preserving only the leading term, then $x(\tau + t_c) \approx v_c\tau$, where v_c is the velocity of the system at $x = 0$ and $t = t_c$.
We find that $t_0 \simeq t_c + \mathrm{i}J/A\,|v_c|$, and after integrating over τ in (3.2.17), we obtain

$$P = \exp\left\{-\frac{\pi J^2}{\hbar A v_c}\right\}. \tag{3.2.18}$$

To get the transition probability W of the system from the well 1 to the well 2, we must multiply $P_t(v_c)$ (3.2.15) by $\delta(x)\,|v_c|$ and the Boltzmann factor $C\exp\left\{-\frac{1}{kT}\right.$ $\left.\times\left[\frac{Mv_c^2}{4} + \frac{\omega_{\mathrm{op}}^2 M}{4}x^2 + Ax\right]\right\}$ (C — normalization factor), and integrate the expression over x and v_c. We find that

$$W = \frac{2}{Z}\int\limits_0^\infty \mathrm{d}v v\, \mathrm{e}^{-Mv^2/4kT}\,\frac{1 - P(v)}{2 - P(v)}, \tag{3.2.19}$$

where Z is a normalization factor

$$Z = \int\limits_{-\infty}^\infty \mathrm{d}x\int\limits_0^\infty \mathrm{d}v \exp\left\{-\frac{1}{kT}\left[\frac{Mv^2}{4} + \frac{\omega_{\mathrm{op}}^2 M}{4}x^2 + Ax\right]\right\}.$$

With the aid of (3.2.19), we obtain the following expression for the hopping drift mobility in the two-site model

$$u_{\mathrm{h}} = u_0 \frac{1}{2\pi} \frac{\hbar\omega_{\mathrm{op}}}{kT} \exp\left(-E_{\mathrm{a}}/kT\right)$$
$$\times 2 \int_0^\infty \mathrm{d}\varepsilon\, \mathrm{e}^{-\varepsilon} \left[1 - \exp\left(-\frac{\pi}{2}\eta_2 \Big/ \sqrt{\varepsilon}\right)\right] \Big/ \left[2 - \exp\left(-\frac{\pi}{2}\eta_2 \Big/ \sqrt{\varepsilon}\right)\right]. \qquad (3.2.20)$$

In the two limiting cases $\eta_2 > 1$ and $\eta_2 \ll 1$, from (3.2.20) we get

$$u_{\mathrm{h}} = u_0 \frac{1}{2\pi} \frac{\hbar\omega_{\mathrm{op}}}{kT} \mathrm{e}^{-E_{\mathrm{a}}/kT} \times \begin{cases} 1 & \text{for} \quad \eta_2 > 1 \\ \pi^{3/2}\eta_2 & \text{for} \quad \eta_2 \ll 1. \end{cases} \qquad (3.2.21)$$

For $\eta_2 > 1$, the probability P for a non-adiabatic transition is small, so that $P_{\mathrm{t}} \approx 1$ (cf. (3.2.15)). Consequently, in this limiting case the system adiabatically overcomes the point $x = 0$ (i.e. the system moves along E_-) on its passage from one well to the other one. The mobility obtained in this case agrees with the expression (3.1.1), which was derived in an intuitive manner. On comparing the two expressions, we find that ω_0 is equal to ω_{op}.

In the opposite limiting case, where $\eta_2 \ll 1$, near $x = 0$, the distance between the adiabatic levels is such small that the electron cannot follow rapid changes of the vibrational subsystem. In this case non-adiabatic transitions are very important, which implies that P is approximately equal to unity. Accordingly, the system on its first passage from the well 1 through the point $x = 0$ in all likelihood skips from E_- to E_+ and on its return from E_+ to E_-, so that it falls back into the well 1. Therefore, the probability of finding the system in the well 2 is very small. It is given by the small probability that on the passage through $x = 0$ a non-adiabatic transition does not occur. It should be noted that the expression for u_{h} obtained in the non-adiabatic limit $\eta_2 \ll 1$ (3.2.21) agrees with the expression (3.2.10), which was derived by the diagram technique described above in lowest order with respect to the resonance integral J. Lang and Firsov (1968) pointed out that the expression for u_{h} (3.2.20) may be obtained by this diagram technique by calculating the irreducible blocks on the basis of an infinite series of diagrams with an arbitrary number of interaction points, but by making use of the two-site model and the high-temperature limit, where $kT > \hbar\omega_{\mathrm{op}}/2$.

The above considerations suggest to call η_2 the parameter of adiabaticity, and the regime, where $\eta_2 > 1$, the adiabatic regime and the opposite regime, where $\eta_2 \ll 1$, the non-adiabatic one.

3.2.3. *Non-hopping transport of small polarons in the low-temperature regime*

As mentioned above, at sufficiently low temperatures the small-polaron transport becomes band-like. In this case, in (2.4.10) the nondiagonal elements of the density matrix with respect to the site indices, $\varrho_{m'}^{m}$, may not be neglected. Accordingly, in the linear case with respect to the electrical field $\boldsymbol{E}$, in addition to the diagonal elements $\delta\mu_m$ (2.5.12), we must introduce nondiagonal elements $\delta\mu_{m'}^{m}$ and obtain a set of coupled equations for $\delta\mu_m$ and $\delta\mu_{m'}^{m}$. Because of this, the conditional-probability function P, which is the Green function of the equation for $\delta\mu$, has now four site indices: $P_{m_2m_4}^{m_1m_3}$ (cf. (2.6.13)). Without describing in detail the corresponding procedure, let us write down the set of equations for $P_{m_2m_4}^{m_1m_3}(s)$, which is a straigthforward generalization of (2.6.5) (KUDINOW and FIRSOV, 1966, BÖTTGER and BRYKSIN, 1976):

$$\left[s + \frac{\mathrm{i}}{\hbar}(\varepsilon_{m_3} - \varepsilon_{m_4})\right] P_{m_2m_4}^{m_1m_3} = \delta_{m_1m_3}\delta_{m_2m_4} + \sum_{m_5m_6} \overset{(0)}{W}{}_{m_2m_6}^{m_1m_5} P_{m_6m_4}^{m_5m_3}. \tag{3.2.22}$$

Recall that the index (0) indicates the absence of an electrical field. Here and in the following we restrict ourselves to the one-electron approximation. $\overset{(0)}{W}{}_{m_2m_4}^{m_1m_3}$ denotes the Laplace transform with respect to the time of $\overset{(0)}{\mathscr{W}}{}_{m_2m_4}^{m_1m_3}(t)$ (2.4.7), where $\overset{(0)}{\mathscr{W}}(t) \equiv \overset{(0)}{\mathscr{W}}(t', t' + t)$ does not depend on t'. In (3.2.22) the term proportional to $(\varepsilon_{m_3} - \varepsilon_{m_4})$ vanishes for crystalline materials because ε_m is independent of m.

KUDINOV and FIRSOV (1966) pointed out that in the absence of a magnetic field the conductivity is related to the four-site conditional-probability function (the kinetics in the presence of a magnetic field and an arbitrary electrical field was studied by BRYKSIN and FIRSOV, 1973) as

$$\sigma_{\nu'\nu} = -\frac{e^2\beta}{2\Omega} \lim_{s\to 0} s^2 \sum_{\boldsymbol{k}} f(\boldsymbol{k}) \left.\frac{\partial^2 P(\boldsymbol{k}, \varkappa, s)}{\partial\varkappa_{\nu'}\partial\varkappa_\nu}\right|_{\varkappa=0}. \tag{3.2.23}$$

Here $f(\boldsymbol{k})$ is the exact (on account of the interaction) thermodynamic equilibrium density matrix in the momentum representation

$$f(\boldsymbol{k}) = Tr\{\mathrm{e}^{-\beta\mathscr{H}^{(0)}} a_{\boldsymbol{k}}^+ a_{\boldsymbol{k}}\}/Tr\{\mathrm{e}^{-\beta\mathscr{H}^{(0)}}\},$$

which is the Fourier transform of $F_m^{m'}$ (A 1.26), $f(\boldsymbol{k}) = \sum_m \mathrm{e}^{\mathrm{i}\boldsymbol{k}\boldsymbol{R}_m} F_{m'}^{m'+m}$ Note that owing to the translational symmetry of the problem, $F_{m'}^{m'+m}$ does not depend on m', so that $f(\boldsymbol{k})$ is diagonal in $\boldsymbol{k}$. In the one-electron case, it is more convenient to use the normalization condition $\sum_{\boldsymbol{k}} f(\boldsymbol{k}) = N_\mathrm{e}$($N_\mathrm{e}$ — total number of electrons in the system) rather than to introduce the chemical potential μ.

The function $P(\boldsymbol{k}, \varkappa, s)$ is the Fourier transform of $P_{m_2m_4}^{m_1m_3}(s)$:

$$P(\boldsymbol{k}, \varkappa, s) = \sum_{m'm_1} P_{m'-m_1m}^{m'+m_1m} \mathrm{e}^{2\mathrm{i}\boldsymbol{k}\boldsymbol{R}_{m_1}} \mathrm{e}^{\mathrm{i}\varkappa(\boldsymbol{R}_{m'} - \boldsymbol{R}_m)}. \tag{3.2.24}$$

$P(\boldsymbol{k}, \varkappa, s)$ is independent of m, owing to the property $P_{m_2+m,m_4+m}^{m_1+m,m_3+m} = P_{m_2m_4}^{m_1m_3}$, which is a consequence of the translational symmetry.

For materials with small energy bands, where kT exceeds the band width, the dependence of f on $\boldsymbol{k}$ may be neglected, since $f(\boldsymbol{k}) \approx (N_\mathrm{e}/N) + 0(\Delta\varepsilon_\mathrm{p}/kT)$. In this limiting case, the relation (3.2.23) changes into a formula of the type (2.6.14) (but which is valid only in the one-electron limit and for an ordered system):

$$\sigma_{\nu\nu'} = \frac{e^2\beta}{2} n \lim_{s\to 0} s^2 \sum_m (\boldsymbol{R}_m)_\nu (\boldsymbol{R}_m)_{\nu'} P_{m0}(s), \tag{3.2.25}$$

where $P_{m'm} \equiv P^{m'm}_{m'm}$, and n is the concentration of electrons.

Hence, equation (3.2.25) is valid down to very low temperatures

$$\frac{\Delta\varepsilon_\text{p}}{kT} \sim \frac{Je^{-S_T}}{kT} \ll 1. \tag{3.2.26}$$

At still lower temperatures, the small-polaron mobility has the standard form $u = e\tau/m^*$, where τ is the inverse scattering rate of the polarons, and $m^* = \hbar^2 \times \exp(S_\text{T})/2Ja^2$ the effective mass in the polaron band. In what follows we restrict ourselves to a discussion of the case (3.2.26).[1]

To calculate the conductivity by means of (3.2.25), we need a closed equation for the diagonal elements $P_{m'm}$. To get such an equation, we rewrite (3.2.22) as follows

$$sP_{m'm} = \delta_{m'm} + \sum_{m_1} \overset{(0)}{W}_{m'm_1} P_{m_1m} + \sum_{m_1 \neq m_2} \overset{(0)}{W}{}^{m'm_1}_{m'm_2} P^{m_1m}_{m_2m},$$

$$sP^{m_1m}_{m_2m} = \sum_{m_3} \overset{(0)}{W}{}^{m_1m_3}_{m_2m_3} P_{m_3m} + \sum_{m_3 \neq m_4} \overset{(0)}{W}{}^{m_1m_3}_{m_2m_4} P^{m_3m}_{m_4m}. \tag{3.2.27}$$

According to Lang and Firsov (1962), all the quantities $W^{m_1m_3}_{m_2m_4}$, where $m_1 \neq m_2$ and $m_3 \neq m_4$, include the small factor $\exp(-S_T)$, except those where $m_1 = m_3$ and $m_2 = m_4$. Retaining in the second equation of (3.2.27) only the latter probabilities, by means of this equation we may express the nondiagonal elements of P by the diagonal ones:

$$P^{m_1m}_{m_2m} \approx (s - \overset{(0)}{W}{}^{m_1m_1}_{m_2m_2})^{-1} \sum_{m_3} \overset{(0)}{W}{}^{m_1m_3}_{m_2m_3} P_{m_3m}.$$

Inserting this expression into the first equation of (3.2.27), we obtain a closed equation for $P_{m'm}$:

$$sP_{m'm} = \delta_{m'm} + \sum_{m_1} \overset{(0)}{\underset{\sim}{W}}_{m'm_1} P_{m_1m}, \tag{3.2.28}$$

where

$$\overset{(0)}{\underset{\sim}{W}}_{m'm} = \overset{(0)}{W}_{m'm} + \sum_{m_1 \neq m_2} \tau(m_1 - m_2)\, \overset{(0)}{W}{}^{m'm_1}_{m'm_2} \overset{(0)}{W}{}^{m_1m}_{m_2m}. \tag{3.2.29}$$

---

[1]) In what follows, for brevity, we use the term tunnelling mechanism for the non-hopping transport in the temperature range (3.2.26). Strictly speaking, the band transport at very low temperatures as well as the (under-barrier) hopping transport are, of course, also due to tunnelling processes.

Here the parameter s was made equal to zero, and

$$\tau(m_1 - m_2) \equiv -1/\overset{(0)}{W}{}^{m_1 m_1}_{m_2 m_2}.$$

Equation (3.2.29) formally agrees with (2.6.5) if we replace there $\overset{(0)}{W}$ by $\overset{(0)}{\underset{\sim}{W}}$. Hence, the conductivity is given by (2.5.19) with the replacement $\overset{(0)}{W} \to \overset{(0)}{\underset{\sim}{W}}$:

$$\sigma_{xx} = \frac{e^2\beta}{2} n \sum_m (\boldsymbol{R}_m)_x^2 \overset{(0)}{\underset{\sim}{W}}_{m0}. \tag{3.2.30}$$

According to (3.2.30) and (3.2.29), in the small-polaron model the mobility u is additively composed of two terms

$$u = u_\text{h} + u_\text{t}, \tag{3.2.31}$$

where the hopping term u_h is described above, and the tunnelling term u_t is given by

$$u_\text{t} = \frac{e\beta}{2} \sum_{m, m_1 \neq m_2} \overset{(0)}{W}{}^{mm_1}_{mm_2} \tau(m_1 - m_2) \overset{(0)}{W}{}^{m_1 0}_{m_2 0} (\boldsymbol{R}_m)_x^2. \tag{3.2.32}$$

In the lowest order with respect to J, for $\overset{(0)}{W}{}^{mm_1}_{mm_2}$ and $\overset{(0)}{W}{}^{m_1 0}_{m_2 0}$, we may use diagrams with a single phononline (cf. (A 1.29), accounting for $\overset{(0)}{W} = \int_0^\infty \mathrm{d}t\, \overset{(0)}{\mathscr{W}}(t_1, t + t_1))$. The quantity $\sum_{mm_1m_2} \boldsymbol{R}_m{}^2 \overset{(0)}{W}{}^{mm_1}_{mm_2} \overset{(0)}{W}{}^{m_1 0}_{m_2 0}$ has the meaning of the square of the polaron velocity $v_x = J\,\mathrm{e}^{-S_T} a/\hbar$. The blocks $\overset{(0)}{W}{}^{m_1 m_1}_{m_2 m_2}$ are negative quantities and $\tau(m_1 - m_2)$ has the meaning of a relaxation time due to friction at the tunnelling transition from site to site.

Consequently, the tunnelling term u_t may be written as

$$u_\text{t} = \frac{e}{kT} v_x{}^2 \tau = u_0 \frac{J\,\mathrm{e}^{-S_T}}{kT} \frac{J\,\mathrm{e}^{-S_T}\tau}{\hbar}, \tag{3.2.33}$$

where $\tau = \tau(m_1 - m_2)$ for $\boldsymbol{R}_{m_1} - \boldsymbol{R}_{m_2} = \boldsymbol{g}$. Owing to the symmetry of the problem, τ does not depend on $\boldsymbol{g}$.

In the lowest order with respect to J, the blocks $W^{m_1 m_1}_{m_2 m_2}$ are described by the two diagrams having two points either on the upper branch of the contour or on the lower branch, wherby the points are connected with each other by a phonon line. Up to the sign, the sum of both these diagrams agrees with the lowest-order-in-J expression for the hopping probability (given by two diagrams having one point on the upper branch and one point on the lower one) obtained above. In this approximation the relaxation time reads (Holstein, 1959)

$$\tau = (\hbar/J^2)\,(4E_\text{a}kT/\pi)^{1/2} \exp\,(E_\text{a}/kT). \tag{3.2.34}$$

In the case that J is not very small, on evaluating τ, blocks with four points must be accounted for. These blocks do not contain the exponentially small factor exp $(-E_a/kT)$, which occurs in the blocks having two points. For four-point blocks, one finds that (LANG and FIRSOV, 1962)

$$\tau \approx (E_a/J)^4 \, \Delta\omega/\omega_{op}^2 \sinh^2 (\hbar\omega_{op}/2kT). \tag{3.2.35}$$

The expression for τ (3.2.35) must be used when the corresponding value turns out to be less than that from (3.2.34).

Thus, in the small-polaron model the drift mobility increases with decreasing temperature due to the decrease of S_T (cf. (A 1.18)). Owing to the additivity of the contributions u_t and u_h, the tunnelling mechanism dominates at low temperatures, but the hopping mechanism at high temperatures. The temperature dependence of the mobility in the small-polaron model is illustrated in Fig. 3.4.

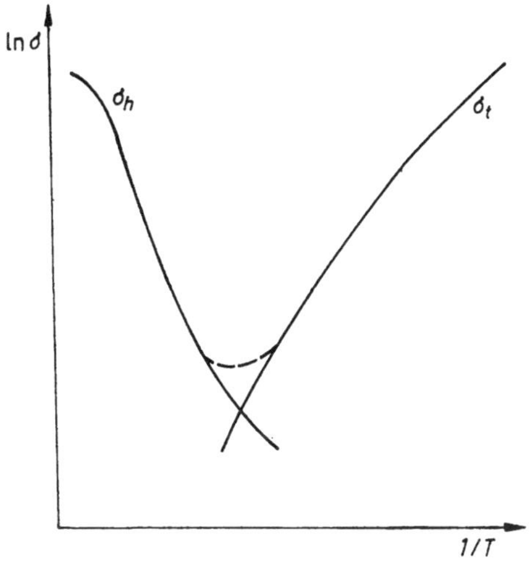

Fig. 3.4. Schematic representation of the temperature dependence of the hopping conductivity σ_h(3.2.1), (3.2.7) and the tunnelling conductivity σ_t(3.2.33). The dashed line indicates the total conductivity in the transition regime. The deviation of the plot ln σ_h versus $1/T$ from a straight line at low temperatures is due to the temperature dependence of the activation energy $E_a(T)$ (3.2.5)

The transition temperature between the tunnelling and the hopping regimes can be obtained by equating u_t and u_h. It approximately agrees with the minimum of the curve $u(T)$ (cf. Fig. 3.4). The concrete form of the curve $u(T)$ strongly depends on the parameters characterizing the system (η_2, η_3, $\Delta\omega/\omega_{op}$, etc.). It will not be discussed here in detail.

Note that the expression for the tunnelling mobility (3.2.33) may be also obtained by means of the standard Boltzmann equation. However, the applicability of (3.2.33) ist not restricted to $\Delta\varepsilon_p\tau/\hbar \gg 1$, the condition for the applicability of a kinetic equation, but this expression also holds for $\Delta\varepsilon_p\tau/\hbar \ll 1$. Accordingly, for this mechanism, the term tunnelling mechanism is more appropriate than the term band mechanism.

Note further that a small polaron can also affect the vibrational behaviour of the system, e.g., it can give rise to local vibrational modes (EAGLES, 1966, BRYKSIN and FIRSOV, 1968).

3.3. Optical properties in the small-polaron model

To study the frequency dependence of the conductivity $\sigma(\omega)$ in the small-polaron model, we use formula (2.5.10), which holds for small resonance integrals J. Recall that $\sigma(\omega)$ is related to the dielectric function $\varepsilon(\omega)$ as

$$\varepsilon(\omega) = \bar{\varepsilon}(\omega) + \frac{4\pi \mathrm{i}\sigma(\omega)}{\omega} \tag{3.3.1}$$

where $\bar{\varepsilon}(\omega)$ denotes the contribution to $\varepsilon(\omega)$ resulting from all other mechanisms, in particular, from the lattice mechanism.

On account of nearest-neighbour hops, according to (2.5.10) and (2.5.6), in the small-polaron model $\sigma(\omega)$ reads

$$\sigma(\omega) = 2en(1-f)\frac{\beta J^2}{\hbar}\,\mathrm{e}^{-2S_T}\int\limits_0^\infty \mathrm{d}t\,\mathrm{e}^{-\mathrm{i}\omega t}\int\limits_{-1/2}^{1/2}\mathrm{d}\lambda$$
$$\times\left\{\exp\left[\sum_{\boldsymbol{q}}\frac{|\gamma(\boldsymbol{q})|^2\,(1-\cos\boldsymbol{gq})}{N\sinh(\hbar\omega_{\boldsymbol{q}}\beta/2)}\cos\omega_{\boldsymbol{q}}(t+\mathrm{i}\hbar\beta\lambda)\right]-1\right\}. \tag{3.3.2}$$

Decomposing $\sigma(\omega)$ into a real and an imaginary part $\sigma = \sigma' + \mathrm{i}\sigma''$, from (3.3.2), we get

$$\sigma' = 2eu_0n(1-f)\frac{J^2}{\hbar^2\omega}\,\mathrm{e}^{-2S_T}\sinh(\hbar\omega\beta/2)\int\limits_{-\infty}^{\infty}\mathrm{d}t\,\mathrm{e}^{-\mathrm{i}\omega t}$$
$$\times\left\{\exp\left[\sum_{\boldsymbol{q}}\frac{|\gamma(\boldsymbol{q})|^2\,(1-\cos\boldsymbol{gq})}{N\sinh(\hbar\omega_{\boldsymbol{q}}\beta/2)}\cos\omega_{\boldsymbol{q}}t\right]-1\right\},$$
$$\sigma'' = 2eu_0n(1-f)\frac{J^2}{\hbar^2\omega}\,\mathrm{e}^{-2S_T}\left\{\int\limits_0^{\hbar\beta/2}\mathrm{d}t\left[\cosh\left(\omega\left(\frac{\hbar\beta}{2}-t\right)\right)-1\right]\right.$$
$$\times\left\{\exp\left[\sum_{\boldsymbol{q}}\frac{|\gamma(\boldsymbol{q})|^2\,(1-\cos\boldsymbol{gq})}{N\sinh(\hbar\omega_{\boldsymbol{q}}\beta/2)}\cosh\omega_{\boldsymbol{q}}t\right]-1\right\}-\int\limits_0^\infty\mathrm{d}t\sinh(\hbar\omega\beta/2)$$
$$\left.\times\sin\omega t\left\{\exp\left[\sum_{\boldsymbol{q}}\frac{|\gamma(\boldsymbol{q})|^2\,(1-\cos\boldsymbol{gq})}{N\sinh(\hbar\omega_{\boldsymbol{q}}\beta/2)}\cos\omega_{\boldsymbol{q}}t\right]-1\right\}\right\}. \tag{3.3.3}$$

In (3.3.3), the integral over t can be calculated by using the saddle-point method. The saddle point is lying on the imaginary axis, say, at $-\mathrm{i}\eta$. Setting at the point $t = -\mathrm{i}\eta$ the derivative of the exponent with respect to t equal to zero, we get the following equation for η:

$$\omega = \sum_{\boldsymbol{q}}\omega_{\boldsymbol{q}}\frac{|\gamma(\boldsymbol{q})|^2\,(1-\cos\boldsymbol{gq})}{N\sinh(\hbar\omega_{\boldsymbol{q}}\beta/2)}\sinh(\omega_{\boldsymbol{q}}\eta). \tag{3.3.5}$$

Here we account only for the first saddle point, because for not too strong dispersion of the optical modes, the contribution from the following saddle points, $-i\eta + \frac{2\pi}{\omega_{op}}$, $-i\eta + \frac{4\pi}{\omega_{op}}$, etc., is small, for the same reason as for the integral in (3.2.2).

For simplicity, let us consider the regime of not too high frequencies ω, where $\eta(\omega)\,\omega_{op} \ll 1$ (for $\omega \to 0$, $\eta \to 0$). Then, expanding in (3.3.5) sinh $(\omega_q \eta)$ in a power series with respect to η, we obtain the following explicit expression for the saddle point:

$$\eta = \frac{1}{2}\,\omega(\delta t)^2. \tag{3.3.6}$$

The parameter δt has the dimension of a time and is defined by (3.2.6).

With the aid of (3.3.6), we find that the condition $\eta\omega_{op} \ll 1$ implies that

$$\omega \ll 2/\omega_{op}(\delta t)^2. \tag{3.3.7}$$

In (3.3.3), we now deform the integration contour in the complex time plane such that it passes through the point $t = -i\eta$ along the path of the steepest descend. After that, in (3.3.3) we expand the exponent in a power series with respect to t about the point $t = -i\eta$ and restrict ourselves to the second order terms in $(t + i\eta)$ (note that the linear term vanishes due to (3.3.5)). The resulting integral can be readily evaluated. We find that

$$\sigma'(\omega) = \sigma_h \frac{\sinh\,(\hbar\omega/2kT)}{\hbar\omega/2kT} \exp\left[-\frac{1}{4}\,\omega^2(\delta t)^2\right], \tag{3.3.8}$$

where σ_h is the dc hopping conductivity given by (3.2.1) and (3.2.7).

For high temperatures, $kT \gg \hbar\omega_{op}/2$, where δt has the form (3.2.9), formula (3.3.8) becomes

$$\sigma'(\omega) = eu_0 n(1-f)\,\frac{\sqrt{\pi}}{2}\,\frac{J^2}{\sqrt{E_a kT}}\,\frac{1-\exp\,(-\hbar\omega/kT)}{\hbar\omega}\,\exp\left\{-\frac{(\hbar\omega - 4E_a)^2}{16E_a kT}\right\}. \tag{3.3.9}$$

Thus, the dependence $\sigma'(\omega)$ has the form of a weakly asymmetric Gaussian peak centred about $\hbar\omega = 4E_a$ and having the half-width $4\sqrt{E_a kT}$ (Reik, 1963, Klinger, 1963). This result can be readily understood in terms of optical transitions between the adiabatic levels in the two-site model shown in Fig. 3.5. Suppose that the electron was initially (before the absorption process) in the well 1 (having its minimum at $x = -x_0$). According to the Franck-Condon principle, the position of the nuclei is not changed during an optical transition, i.e. in Fig. 3.5 such a transition occurs vertically (without changing x) from the level 1 to the level 2. In Fig. 3.5 this transition is indicated by an arrow. Using the expression for the adiabatic levels (3.2.14) (after setting J equal to zero in this formula, i.e. neglecting the level repulsion at $x = 0$, which is irrelevant to the problem in question) we find that the photon energy required to excite an electron from the bottom of the well is equal to $4E_a$, which corresponds to the energy of the maximum of the curve $\sigma'(\omega)$. Since $kT \ll E_a$, the major contribution to the absorption spectrum comes from transitions in the region between the dashed lines in Fig. 3.5. A simple estimation shows that the corresponding photon energies are lying in the interval between $4E_a - 4\sqrt{E_a kT}$ and $4E_a + 4\sqrt{E_a kT}$, i.e. the half-width of the absorption peak is given by $4\sqrt{E_a kT}$, which is in agreement with the result obtained above.

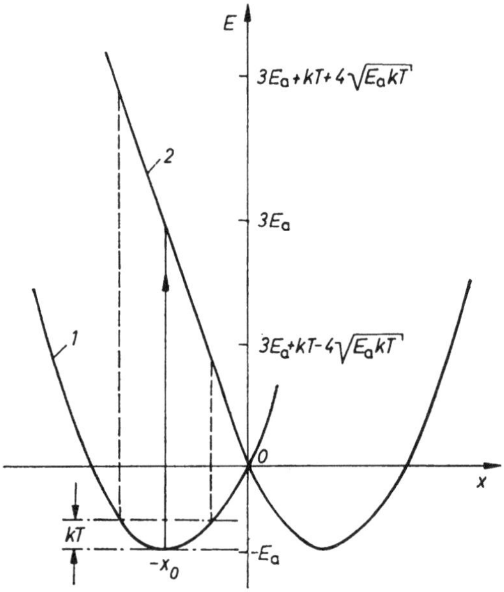

Fig. 3.5. Schematic representation of optical transitions between the adiabatic levels in the two-site model. The level repulsion near the point $x = 0$ is not shown (cf. Fig. 3.3)

It is interesting to note that the most probable optical transition occurs at about $2E_p$ (recall that $2E_a \approx E_p$). This is due to the fact that the energy E_p is required to take the electron out of its polaron cavity at its initial site (site 1) and another energy of this amount to put the electron on site 2, where the polaron cavity is still lacking and the equilibrium positions of the surrounding atoms are still determined by the cavity at site 1.

Let us now turn to the investigation of the imaginary part of $\sigma(\omega)$. For simplicity, we restrict ourselves to the high-temperature limit, where $kT > \hbar\omega_{op}/2$, and to not too high frequencies, where the inequality (3.3.7) holds. In this case, the exponent in (3.3.4) may be expanded in a power series with respect to t up to the second order. We find that

$$\sigma''(\omega) = \sigma_h \frac{2kT}{\hbar\omega} \exp\left(-\frac{\hbar^2\omega^2}{16E_a kT}\right) \left\{\frac{1}{2} \exp\left(\frac{\hbar\omega}{2kT}\right) f\left(\sqrt{\frac{E_a}{kT}} - \frac{\hbar\omega}{4\sqrt{E_a kT}}\right) + \frac{1}{2} \exp\left(-\frac{\hbar\omega}{2kT}\right) f\left(\sqrt{\frac{E_a}{kT}} + \frac{\hbar\omega}{4\sqrt{E_a kT}}\right) - f\left(\sqrt{\frac{E_a}{kT}}\right)\right\}, \quad (3.3.10)$$

where $f(x) = \frac{2}{\sqrt{\pi}} \int_0^x dt\, e^{t^2}$.

For $(4E_a - \hbar\omega)^2 > 16E_akT$ and $E_a > kT$, we can get a simple analytical expression for $\sigma''(\omega)$. Taking into account that for $x \gg 1$, $f(x) \approx e^{x^2}/2x$, from (3.3.10), we obtain

$$\sigma''(\omega) \approx \sigma_h \frac{2(kT)^{3/2} \hbar\omega \exp(E_a/kT)}{(\pi E_a)^{1/2} [(4E_a)^2 - (\hbar\omega)^2]}. \tag{3.3.11}$$

This formula is not valid in the interval between $4E_a - 4\sqrt{E_akT}$ and $4E_a + 4\sqrt{E_akT}$, i.e. in the region where $\sigma'(\omega)$ is peaked. In this region, σ'' rapidly changes from positive to negative values. Note that, unlike $\sigma'(\omega)$, $\sigma''(\omega)$ does not in an activated manner exponentially depend on temperature. The qualitative behaviour of $\sigma'(\omega)$ and $\sigma''(\omega)$, as suggested by (3.3.9) and (3.3.10), is illustrated in Fig. 3.6.

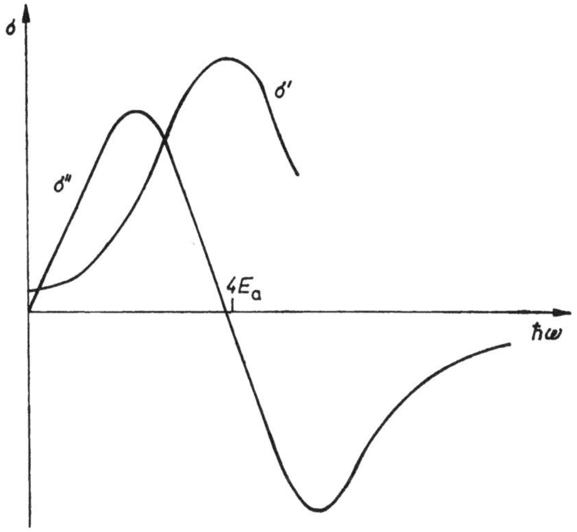

Fig. 3.6. Schematic representation of the real part of the conductivity σ' and the imaginary part σ'' in the small-polaron model (see text)

For arbitrary values of the parameter η_3, the frequency dependence of σ was studied by Firsov (1968), who pointed out that, for $\eta_3 > 1$, the shape of the absorption curve strongly depends on the concrete form of the unrenormalized electron band. For $\eta_3 > 1$, the light induces electron transitions between more remote sites than nearest-neighbour ones. In this case, an electron expelled by the light from its polaron cavity stays in a free state for a rather long time before it will be again localized at some site located far from the original one. This transport mechanism is termed the relay-race mechanism.

In the low-temperature region, where the tunnelling mechanism or the band-transport mechanism is operative, the dependence $\sigma(\omega)$ has a Drude-Lorentz form

$$\sigma(\omega) = \sigma_t \frac{1}{1 + i\omega\tau(\omega)}. \tag{3.3.12}$$

Here the relaxation time τ strongly depends on the frequency, even at rather low frequencies, because for a narrow polaron band, one-phonon scattering processes are forbidden for $\omega = 0$, while for $\omega \neq 0$, such processes may occur.

The optical absorption mechanism considered above may be called the intra-band mechanism, because optical transitions occur within a single electron band. On the other hand, optical transitions between a narrow polaron band and a broad band with weak interaction with phonons were studied by EAGLES (1966), who pointed out that in this case a series of absorption lines should appear at $E_g + l\hbar\omega_{op}$ (E_g — width of the gap between the two bands, l — integer). Furthermore, KUDINOV and FIRSOV (1964) investigated optical transitions between deep atomic levels and a polaron band. ARNOLD and HOLSTEIN (1974) studied the absorption of light in the small polaron model in the presence of an external magnetic field, and AUSTIN (1972) investigated the dependence $\sigma'(\omega)$ in the presence of a strong time-independent electrical field (with regard to the latter paper, see also Section 3.5.).

Let us also draw attention to the papers of BARI (1974), COLL and BENI (1974), and HINKELMANN and REIK (1975), who studied the absorption of light in the small-polaron model in the presence of strong electron-electron correlation of the Hubbard type. In this case, instead of a single peak, there appear three peaks: at $|4E_a - U|$ (transition from a site occupied by two electrons to an unoccupied site), $4E_a$ (transition from a site occupied by a single electron to an unoccupied site, or from a site occupied by two electrons to an unoccupied site), and $4E_a + U$ (transition from a site occupied by a single electron to another site occupied by a single electron), where U denotes the intrasite Hubbard repulsion energy.

3.4. The small polaron in a strong electrical field

3.4.1. *Current-voltage characteristic in the hopping regime*

The small-polaron transport in a strong time-independent electrical field and in the limit of low charge carrier concentrations can be studied by means of the rate equation (2.4.18). Restricting ourselves to small resonance integrals, we may use the expression for the transition probability (2.4.19). In the case of an ordered system, the set of equations (2.4.18) can be readily solved by using the momentum representation, because in this case, $W_{m'm}$ depends only on $m' - m$. In (2.4.18), we take the Laplace transform with respect to time and account for the fact that for a constant field, $W_{m'm}$ does not depend on t. Then we obtain

$$s\varrho_m(s) = f_m + \sum_{m'} \{\varrho_{m'}(s)\, W_{m'm} - \varrho_m(s)\, W_{mm'}\}, \tag{3.4.1}$$

where $f_m = \varrho_m(t)|_{t=0}$. It turns out below, that the stationary current does not depend on the explicit form of the initial distribution f_m. Making use of the momentum representation

$$\left.\begin{aligned} \varrho(\boldsymbol{k}, s) &= \sum_m \varrho_m(s)\, e^{i\boldsymbol{k}\boldsymbol{R}_m}, \\ W(\boldsymbol{k}) &= \sum_m W_{m',m+m'}\, e^{i\boldsymbol{k}\boldsymbol{R}_m}, \end{aligned}\right\} \tag{3.4.2}$$

from (3.4.1), we find that

$$\varrho(\boldsymbol{k}, s) = f(\boldsymbol{k})/\{s + W(0) - W(\boldsymbol{k})\}. \tag{3.4.3}$$

Rewriting now the expression for the current (2.3.3) in the Laplace representation, and passing over from $\varrho_m(s)$ to $\varrho(\boldsymbol{k}, s)$, we obtain

$$\boldsymbol{j}(s) = -\frac{\mathrm{i}e}{\Omega} \nabla_{\boldsymbol{k}} \{s\varrho(\boldsymbol{k}, s) - f(\boldsymbol{k})\}|_{\boldsymbol{k}=0}. \tag{3.4.4}$$

Inserting the expression for $\varrho(\boldsymbol{k}, s)$ (3.4.3), into (3.4.4), we find that

$$\left.\begin{aligned} \boldsymbol{j}(s) &= -\frac{\mathrm{i}e}{\Omega s} f(\boldsymbol{k}) \nabla_{\boldsymbol{k}} W(\boldsymbol{k})|_{\boldsymbol{k}=0}, \\ \boldsymbol{j}(t) &= en \sum_m (\boldsymbol{R}_{m'} - \boldsymbol{R}_m) W_{m'm}. \end{aligned}\right\} \tag{3.4.5}$$

Here we have accounted for that $f(0) = \sum_m f_m = N_e$ (N_e — total number of electrons in the system) and $n = N_e/\Omega$ is the concentration of charge carriers. Thus, as noted above, the current does not depend on the explicit form of the initial distribution.

To illustrate (3.4.5), let us consider a one-dimensional model with nearest-neighbour hopping, where (3.4.5) becomes

$$j = env,$$

with

$$v = a(W_+ - W_-). \tag{3.4.6}$$

Here a is the nearest-neighbour distance, $W_+(W_-)$ the hopping probability along (opposite to) the electrical field E, where for $E \neq 0$, $W_+ \neq W_-$. The quantity v has the meaning of the carrier velocity in the direction of the field.

According to the principle of the detailed equilibrium in the presence of a field (2.4.20), it holds that $W_+/W_- = \exp(eEa/kT)$. Introducing the symmetrized hopping probability $W = \sqrt{W_+W_-}$, then, with the aid of the latter relation, we get $W = W_+ \exp(-eEa/2kT) = W_- \exp(eEa/2kT)$. Thus, (3.4.6) can be written as

$$j = 2ena \sinh(eEa/2kT)\, W. \tag{3.4.7}$$

Since W is a symmetrical function of E, for $E \to 0$, W passes into the nearest-neighbour hopping probability in the absence of a field, and in the linear approximation with respect to the field, (3.4.7) takes the form (3.2.1).

The fact, that in (3.4.5) $\boldsymbol{j}(t)$ does not depend on t, is a consequence of the use of Markovian limit in the rate equation, i.e. (3.4.5) describes $\boldsymbol{j}(t)$ for $t \to \infty$, where the transition processes have already run out.

According to (2.4.19), in (3.4.7), the quantity W is given by

$$W = \frac{J^2}{\hbar^2} \mathrm{e}^{-2S_T} \int_{-\infty}^{\infty} \mathrm{d}t\, \mathrm{e}^{-\frac{\mathrm{i}eEa}{\hbar}t} \left\{\exp\left[\sum_{\boldsymbol{q}} \frac{|\gamma(\boldsymbol{q})|^2 (1 - \cos \boldsymbol{g}\boldsymbol{q})}{N \sinh(\hbar\omega_{\boldsymbol{q}}\beta/2)} \cos \omega_{\boldsymbol{q}} t\right] - 1\right\}. \tag{3.4.8}$$

In this expression, after replacing eEa by $\hbar\omega$, the integral over t agrees with that in $\sigma'(\omega)$ (3.3.3). Therefore, we may use the result (3.3.9), which leads to the following expression for the current in a strong electrical field (EFROS, 1967):

$$j = \sigma_{\mathrm{h}} E \, \frac{\sinh\,(eEa/2kT)}{eEa/2kT} \exp\left[-\frac{(eEa)^2}{16E_{\mathrm{a}}kT}\right]. \tag{3.4.9}$$

According to (3.4.9), beyond the Ohmic region, at fields of order $2kT/ea$, the dependence $j(E)$ exhibits an exponential form $j(E) \sim \exp\,(eEa/2kT)$. When the parameter $eEa/\sqrt{16E_{\mathrm{a}}kT}$ becomes of order unity, the increase of the current slows down. At $eEa = 4E_{\mathrm{a}}$ the dependence $j(E)$ exhibits a maximum, and after that the current decreases with increasing field. Near the maximum, $j(E)$ has a Gaussian shape $\exp\,\{-(eEa - 4E_{\mathrm{a}})^2/(16E_{\mathrm{a}}kT)\}$.

Let us now discuss this result in the two-site model, which also allows its range of applicability to be understood. An electrical field leads to the following additional term in the two-site Hamiltonian (3.2.12a):

$$-e\boldsymbol{E}\boldsymbol{R}_1 a_1^+ a_1 - e\boldsymbol{E}\boldsymbol{R}_2 a_2^+ a_2 = -\frac{1}{2}\, e\boldsymbol{E}\{(\boldsymbol{R}_1 + \boldsymbol{R}_2)\,(a_1^+ a_1 + a_2^+ a_2) + (\boldsymbol{R}_1 - \boldsymbol{R}_2)\,(a_1^+ a_1 - a_2^+ a_2)\}$$

(cf. (2.1.18)). Hence, the adiabatic levels E are given by

$$E_{\pm} = \frac{\omega_{\mathrm{op}}^2 M}{4}\, x^2 \pm \sqrt{\left(Ax + \frac{eEa}{2}\right)^2 + J^2}. \tag{3.4.10}$$

On deriving (3.4.10), we have used $a_1^+ a_1 + a_2^+ a_2 = 1$, which guarantees that one electron is placed on the selected pair of sites.

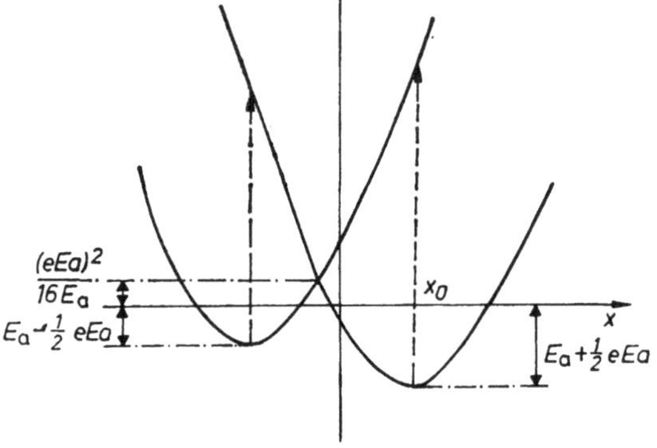

Fig. 3.7. Adiabatic levels in the two-site model in the presence of an electrical field $\boldsymbol{E}$. The level splitting owing to $J \neq 0$ is not shown. The dashed lines with arrows indicate optical transitions (see text)

On account of an electrical field, the adiabatic levels are illustrated in Fig. 3.7, from which we learn that the height of the barrier for a hop from the left polaron well to the right one is given by $E_{\mathrm{a}} - \frac{1}{2}\, eEa + (eEa)^2/16E_{\mathrm{a}}$ (on neglection of the level splitting, i.e. for $J = 0$). Hence, it follows that the hopping probability is proportional to

$$\exp\left\{-\frac{1}{kT}\left(E_a - \frac{1}{2}\,eEa + (eEa)^2/16E_a\right)\right\}$$

in agreement with the result obtained above. At $eEa = 4E_a$, the barrier vanishes, so that at these fields one may not speak about localization at a site. For $eEa > 4E_a$, such a barrier formally again appears, which leads to an exponential decrease of the current. However, it is clear that at such strong electrical fields, the two-site approximation fails. Indeed for sufficiently low barriers, after a transition over the barrier, the system rapidly falls down on the bottom of the well, whereby the friction forces have no time to damp the vibrational process, i.e. the carrier does not achieve thermal equilibrium with the lattice and one may not speak about localization at a site. For low barriers, the carrier has no time to be localized. After a hop, the carrier immediately proceeds ("by inertia") to a third site, and hence to a fourth site and so on. Hence, for $eEa \lessapprox 4E_a$ heating of the carriers occurs due to the external field, and the polaronic localized state dissociates. The threshold value of the field at which the polaron dissociates depends on details of the relaxation processes (friction forces). So far, this value is not known exactly.

Consequently, one may expect that at $eEa/\sqrt{16E_akT} \sim 1$ the exponential increase of the current slows down. At higher fields, $eEa \lessapprox 4E_a$, formula (3.4.9) becomes incorrect. Hence, at fields $eEa > 4E_a$, a decrease of the current should not occur.

For $eEa/4\sqrt{E_akT} \ll 1$ and arbitrary values of $eEa/2kT$, the two-site model may also be used to calculate the current for the adiabatic polaron ($\eta_2 \gg 1$). For such fields, in (3.4.7) we may take the zero-field expression for W, which is given in Section 3.2. Hence, it follows that the expression for the current (3.4.9) corresponds to the non-adiabatic polaron ($\eta_2 \ll 1$). In the adiabatic case ($\eta_2 \gg 1$), the form of formula (3.4.9) is preserved, but σ_h stands for the dc hopping conductivity of the adiabatic polaron. In the field regime in question, the quantity $\exp\{-(eEa)^2/(16E_akT)\}$ may be replaced by unity. Thus, we obtain the following expression for the current in the adiabatic limit:

$$j = \frac{en\omega_{op}}{\pi}\sinh\frac{eEa}{2kT}\exp\left(-E_a/kT\right). \tag{3.4.11}$$

The problem of the adiabatic polaron at arbitrary values of $eEa/\sqrt{16E_akT}$ (but below the dissociation threshold $eEa < 4E_a$) may be studied by the methods described in Section 3.2, but by making use of the adiabatic levels (3.4.10). So far, such a study has not been performed.

Note that typical values of $4E_a$ are between 0.5 and 1 eV, i.e. polaron dissociation may be expected to occur only at very high electrical fields, of order 10^7 V/cm.

In strong electrical fields, the reduction of the potential barrier depends on the hopping distance R_c and is given by $eER_c/2$. Therefore, in the small-polaron model, there is a tendency to an increase of the optimal hopping distance when the field is enhanced, which may essentially affect the current-voltage characteristic (recall that the results obtained above hold for nearest-neighbour hopping). This effect was studied by Bryksin and Firsov (1972), who pointed out that nearest-neighbour hops dominate for $\eta_3 \ll 1$ and $\eta_1^2 \exp(eEa/kT) \ll 1$.

3.4.2. *The small polaron in strong fields at low temperatures and the Stark ladder*

Let us now discuss the current-voltage characteristic for small polarons at low temperatures, where in the case of weak fields the tunnelling (or band) mechanism is operative, i.e. where $\sigma \sim \sigma_t$ (Bryksin and Firsov, 1970a, 1972b). By a procedure similar to that used for evaluating σ_t, one finds that

$$j(E) = \sigma_t E \frac{\tau(E)}{\tau(0)} \left\{1 + \left(\frac{eEa\tau(E)}{\hbar}\right)^2\right\}^{-1}. \tag{3.4.12}$$

Here $\tau(0)$ is the relaxation time in the absence of a field, which is given in (3.2.34) and (3.2.35), and $\tau(E)$ is this quantity for $E \neq 0$. As mentioned in Section 3.2, in lowest order with respect to J, the relaxation time is given by the two diagrams having both interaction points either on the upper or the lower branch of the contour. One gets

$$[\tau(E)]^{-1} = \frac{J^2}{\hbar^2} e^{-2S_T} \cosh \frac{eEa}{2kT} \int_{-\infty}^{\infty} dt \cos\left(\frac{eEa}{\hbar}t\right) \times \left\{\exp\left[\sum_{\boldsymbol{q}} \frac{|\gamma(\boldsymbol{q})|^2 (1 - \cos \boldsymbol{gq})}{N \sinh(\hbar\omega_{\boldsymbol{q}}/2kT)} \cos \omega_{\boldsymbol{q}} t\right] - 1\right\}. \tag{3.4.13}$$

The peculiarity of this expression is in that at low temperatures, when $\sum_{\boldsymbol{q}} |\gamma(\boldsymbol{q})|^2$ $N \sinh(\hbar\omega_{\boldsymbol{q}}/2kT) \ll 1$, $1/\tau$ oscillates with the field, which readily can be seen by expanding the exponential in the integrand in a series with respect to γ. The terms of such a series are proportional to $\delta(\omega_{\boldsymbol{q}_1} + \dots + \omega_{\boldsymbol{q}_n} - eEa/\hbar)$, i.e. they have maxima at $eEa = n\hbar\omega_{op}$, where n is an integer. Hence, it follows that for $eE\tau/\hbar \gg 1$, where $j(E) \sim 1/\tau(E)$ (cf. (3.4.12)), the current-voltage characteristic exhibits oscillations (with maxima given by the resonance condition $n\hbar\omega_{op} = eEa$) superimposed on a function monotonously decreasing with the field (Bryksin and Firsov, 1970a). On account of tunnelling transitions between more remote sites than nearest-neighbour ones, the resonance condition becomes

$$n\hbar\omega_{op} = meEa, \tag{3.4.14}$$

where n and m are integers.

Such oscillating field dependence of the current is not only characteristic of the small-polaron model, but it also occurs in the case of weak electron-phonon interaction and electron bands of finite width. In such systems, the resonance condition (3.4.14) corresponds to transitions between the levels of the Stark ladder (Wannier, 1960) with emission of n phonons (Bryksin and Firsov, 1971a). This phenomenon is termed the electro-phonon resonance. Note that in strong electrical fields, if the distance between the levels of the Stark ladder, eEa, exceeds their width due to collision, $\hbar/\tau$, the current is always proportional to $1/\tau$ (i.e. to the transition probability) and not to τ, as in the case of the customary band transport. Accordingly, in strong fields the charge transport is caused by hops between the levels of the Stark ladder, although in weak fields, it is determined by thė band mechanism. This is due to the fact, that in strong (quantizing) electrical fields the electron states become spatially localized in the direction along the field within a region of the dimension Δ/eE, where Δ is the width of the electron band. Experimentally, the electro-phonon resonance was first observed by MAEKAWA (1970) on ZnS crystals.

A detailed discussion of the hopping transport in quantizing electrical fields is beyond the scope of this book. The reader who is interested in more details of this problem is, e.g., referred to BÖTTGER and BRYKSIN (1976a), SAWAKI and NISHINAGA (1980), BRYKSIN et al. (1981) and references therein.

Some information on the character of the charge transport can be obtained from a study of the absorption of light by small polarons in the presence of an electrical field (AUSTIN, 1972). This problem can be solved analogous to the evaluation of σ' in weak fields (cf. Section 3.3.), because in the presence of a field of arbitrary strength the expression for σ' differs from (3.3.3) only by an additional factor $g_\nu^2 \exp\{e\boldsymbol{E}\boldsymbol{g} \times [(it/\hbar) + (1/2kT)]\}$ in the integrand, by a summation over nearest-neighbours $\boldsymbol{g}$, and by a division by za^2, where z is the coordination number. It turns out that in this case, the conductivity tensor $\sigma_{\nu\nu'}$ becomes anisotropic, such that $\sigma_{xx} \neq \sigma_{yy} = \sigma_{zz}$ (the external field is directed along the x-axis), whereby σ_{yy} and σ_{zz} agree with the corresponding expression obtained for $\boldsymbol{E} = 0$. σ_{xx} is given by a superposition of two Gaussians peaked at $\hbar\omega = 4E_a \pm eEa$. This result can be readily understood by considering the optical transitions (on account of the Franck-Condon principle) between the adiabatic levels illustrated in Fig. 3.7, where the optical transitions are indicated by arrows (cf. Fig. 3.5).

3.5. The Hall effect

3.5.1. *Hall mobility versus drift mobility in the case of hopping transport*

For materials with brand transport and weak coupling with phonons, by means of the Bolzmann equation, one gets the following expression for the Hall coefficient R:

$$R = A/enc, \tag{3.5.1}$$

where n is the concentration of charge carriers, c the velocity of light, and A a numerical factor, which only weakly depends on the type of scattering processes and which is of order unity (see, e.g., the monograph by BLATT, 1957). For scattering by lattice vibrations, one finds that $A = 3\pi/8$, by ionized impurities $A = 315\pi/512$, etc. The expression for R (3.5.1) allows the concentration n to be determined. Furthermore, from the sign of the Hall effect, one can obtain the sign of the charge carriers, i.e. one can distinguish between electron and hole transport.

From a measurement of the conductivity σ_{xx} and the Hall coefficient R one can obtain the carrier mobility u_{H}, which is related to σ_{xx} and R as

$$u_{\text{H}} = cR\sigma_{xx}. \tag{3.5.2}$$

Introducing the antisymmetric component of the conductivity tensor, σ_{xy}, by $j_x = \sigma_{xy}E$ (E is directed along the y-axis), then it holds that

$$R = \frac{1}{H}\frac{\sigma_{xy}}{\sigma_{xx}^2}; \quad u_{\text{H}} = \frac{c}{H}\frac{\sigma_{xy}}{\sigma_{xx}}. \tag{3.5.3}$$

Inserting now the expression for R (3.5.1) into (3.5.2), we find that $u_{\text{H}} = A\sigma_{xx}/en$, i.e. u_{H}, which will be called below the Hall mobility, agrees with the drift mobility up to numerical factor A.

A very important property of the hopping conductivity is in that the Hall mobility, defined in (3.5.2), has nothing in common with the drift mobility, neither with respect to the order of magnitude, nor with respect to the temperature dependence, and often even not with respect to the sign, i.e. A turns out to be strongly temperature dependent and to differ from unity by orders of magnitude sometimes. In the case of hopping transport, the sign of the Hall effect does often not depend on the type of transport, but has electron character. In the literature, this phenomenon is termed the p-n-anomaly (Holstein, 1973).

3.5.2. *The general expression for the Hall current in the hopping regime*

The hopping-Hall effect of small polarons is calculated on the basis of the equation of motion for the density matrix (2.4.3). Nondiagonal elements of the density matrix (with respect to the site indices) are neglected, because we are interested in hopping motion. Since the two-site hopping probabilities do not depend on the magnetic field $\boldsymbol{H}$, a treatment of the Hall effect requires at least account of three-site probabilities in the rate equation. The latter probabilities are proportional to $\overset{(3)}{\mathcal{W}}$ (cf. (2.4.3)). Holstein (1961) was the first who drew attention to the decisive role of three-site hops for the Hall effect in the hopping regime. On account of three-site hops, the rate equation will be derived in Appendix 2. (see also the original paper by Böttger and Bryksin, 1977a). In the Markovian limit, this equation reads

$$\frac{\mathrm{d}\varrho_m}{\mathrm{d}t} = \sum_{m_1} \{\varrho_{m_1}(1-\varrho_m)\, W_{m_1m} - \varrho_m(1-\varrho_{m_1})\, W_{mm_1}\}$$
$$+ \sum_{m_1m_2} \{\varrho_{m_1}(1-\varrho_{m_2})\,(1-\varrho_m)\, W^{(\text{e})}_{m_1m_2m} - (1-\varrho_{m_1})\,\varrho_{m_2}\varrho_m W^{(\text{h})}_{m_1m_2m}. \tag{3.5.4}$$

The equation (3.5.4) is a generalization of the rate equation (2.4.23). As pointed out in Appendix 2., the three-site probabilities $W^{(\text{e,h})}_{m_1m_2m}$ can be calculated with the aid of the diagram technique described above (cf. (A 2.6)).

From the explicit form for $W_{m_1m_2m}$ it follows that these quantities satisfy the following symmetry relations:

$$W^{(\text{e,h})}_{m_1m_2m}(\boldsymbol{H}) = W^{(\text{e,h})}_{mm_2m_1}(-\boldsymbol{H}) \exp\left\{\mp\beta\left(\varepsilon_m - \varepsilon_{m_1} - e\boldsymbol{E}(\boldsymbol{R}_m - \boldsymbol{R}_{m_1})\right)\right\}. \tag{3.5.5}$$

Here the upper sign refers to electrons (indicated by the superscript (e)) and the lower one to holes (h). The relation (3.5.5) may be viewed as a generalization of the condition of the detailed equilibrium (2.4.20) if we consider $W_{m_1m_2m}$ as the probability for a hop from site m_1 to site m via the intermediate site m_2.

Let us now turn to the expression for the current (2.3.3). Inserting $d\varrho_m/dt$, from (3.5.4), into (2.3.3), and keeping only the contribution depending on the magnetic field, we find that

$$\boldsymbol{j}_{\mathrm{H}}(t) = \frac{e}{\Omega} \sum_{mm_1m_2} (\boldsymbol{R}_m - \boldsymbol{R}_{m_1}) \{\varrho_{m_1} (1 - \varrho_{m_2}) (1 - \varrho_m) W^{(e)}_{m_1m_2m}(\boldsymbol{H}) - (1 - \varrho_{m_1}) \varrho_{m_2}\varrho_m W^{(h)}_{m_1m_2m}(\boldsymbol{H})\}. \quad (3.5.6)$$

For reasons of symmetry, under the summation sign in (3.5.6), $\boldsymbol{R}_m$ was replaced by $\boldsymbol{R}_m - \boldsymbol{R}_{m_1}$, which may be done owing to the following symmetry relation for $W_{m_1m_2m}$:

$$W^{(e,h)}_{m_1m_2m}(\boldsymbol{H}) = W^{(e,h)}_{m_1mm_2}(-\boldsymbol{H}). \quad (3.5.7)$$

Since in $W_{m_1m_2m}$ only the linear term with respect to $\boldsymbol{H}$ is accounted for, in (3.5.6) the term containing $\boldsymbol{R}_m$ vanishes.

In ordered materials, at sufficiently large times after turning on the electrical field, the distribution function $\varrho_m(t)$ passes into its thermodynamic equilibrium value $f = \{\exp(-\mu/kT) + 1\}^{-1}$, provided that the condition of charge neutrality of the crystal is fulfilled. Hence, for $t \to \infty$, i.e. in the stationary current regime, we obtain the following closed expression for the current:

$$\boldsymbol{j}_{\mathrm{H}} = \frac{e}{\Omega} f(1 - f) \sum_{mm_1m_2} (\boldsymbol{R}_m - \boldsymbol{R}_{m_1}) W_{m_1m_2m}(\boldsymbol{H}). \quad (3.5.8)$$

Note that this expression does not require solution of the rate equation. On deriving (3.5.8), it was accounted for the fact that in ordered systems for the antisymmetric (with respect to $\boldsymbol{H}$) parts of the hopping probabilities it holds that $W^{(e)}_{m_1m_2m} = W^{(h)}_{m_1m_2m} \equiv W_{m_1m_2m}$, which is a consequence of the rule for passing from $W^{(e)}$ to $W^{(h)}$, given in Appendix 2.

The expression for the Hall current (3.5.8) was first obtained in an intuitive manner by Friedman and Holstein (1963) for the case of the Boltzmann statistics ($f \ll 1$). To realize the physical meaning of (3.5.8), note that the product of the hopping distance ($\boldsymbol{R}_m - \boldsymbol{R}_{m_1}$) and the hopping probability $W_{m_1m_2m}$ represents the carrier velocity.

Consider now three-site hops in a hexagonal crystal in the plane perpendicular to the axis C_3 (cf. Fig. 3.8.). Let the electron be located at site 1. Owing to the fact that in the presence of a magnetic field the hopping probability along the trajectory $1 \to 2 \to 3$ is not equal to that along $1 \to 3 \to 2$, one gets a non-zero average electron velocity perpendicular to the magnetic and the electrical field, i.e. one gets a Hall current. The occurrence of a hopping-Hall current can be also understood in terms of the effect of the external electrical field on the diamagnetic currents induced by the magnetic field (this picture is discussed in detail in Section 8.).

Let us now linearize the expression for the current (3.5.8) with respect to the electrical field. To do this, in (3.5.8), we interchange the summation indices as $m_1 \rightleftarrows m_2$, add the expression obtained to the original one, and multiply the resulting sum by a factor 1/2. Thus, on account of (3.5.5) and (3.5.7), the current becomes (recall that in $W_{m_1m_2m}$ only the contribution being linear in $\boldsymbol{H}$ is accounted for):

$$\boldsymbol{j}_{\mathrm{H}} = \frac{e}{2\Omega} f(1-f) \sum_{mm_1m_2} \boldsymbol{R}_m W_{m_1m_2m}(\boldsymbol{H}) \{1 - \exp[e\beta\boldsymbol{E}(\boldsymbol{R}_{m_1} - \boldsymbol{R}_{m_2})]\}. \quad (3.5.9)$$

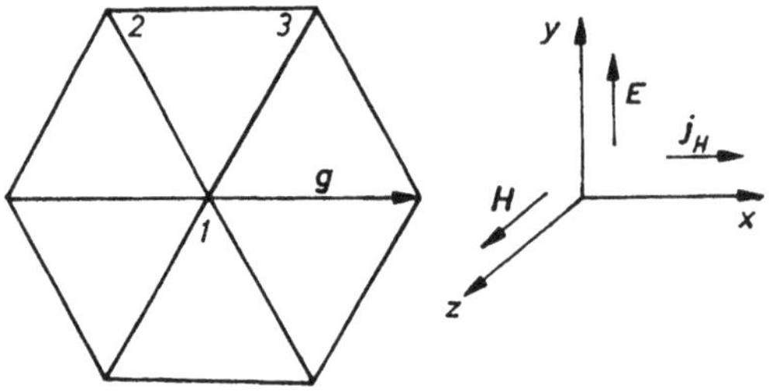

Fig. 3.8. Structure of a hexagonal lattice in the plane perpendicular to the C_3-axis. The directions of the magnetic field $\boldsymbol{H}$, the electrical field $\boldsymbol{E}$, and the Hall current $\boldsymbol{j}_{\mathrm{H}}$ are indicated by arrows ($\boldsymbol{H}$ is perpendicular to the picture plane)

From (3.5.9), it is easy to get the linear approximation with respect to $\boldsymbol{E}$:

$$\boldsymbol{j}_{\mathrm{H}} = \frac{e^2}{2\Omega kT} f(1-f) \sum_{mm_1m_2} \boldsymbol{R}_m[\boldsymbol{E}(\boldsymbol{R}_{m_1} - \boldsymbol{R}_{m_2})] \, W^{(0)}_{m_1m_2m}(\boldsymbol{H}). \quad (3.5.10)$$

Here the superscript (0) indicates the absence of the electrical field. It is interesting to note that (3.5.10) allows the antisymmetric part of the conductivity tensor to be expressed by the hopping probabilities in the absence of a field, i.e. such a formula may be also obtained from the Kubo formula of the linear response theory (for more details on the latter approach, see the book by Firsov, 1975).

Let us now rewrite (3.5.10) such that it only contains differences of the position vectors of the sites. To this end, in (3.5.10), we interchange the site indices as follows: (i) $m \rightleftarrows m_2$, $m_1 \leftrightarrows m_1$, (ii) $m \to m_2$, $m_1 \to m$, $m_2 \to m_1$. Then we add the resulting two expressions to the original one and multiply the sum obtained by a factor 1/3. On account of the fact that, according to (3.5.5) and (3.5.7), at $\boldsymbol{E} = 0$, in $W^{(0)}_{m_1m_2m}$ the interchange of any pair of site indices changes the sign of this quantity, we obtain

$$\boldsymbol{j}_{\mathrm{H}} = \frac{e^2 f(1-f)}{3\Omega kT} \sum_{m_1m_2m} (\boldsymbol{R}_m - \boldsymbol{R}_{m_2}) \, [\boldsymbol{E}(\boldsymbol{R}_m - \boldsymbol{R}_{m_1})] \, W^{(0)}_{m_1m_2m}(\boldsymbol{H}). \quad (3.5.11)$$

This formula describes the Hall current in an ordered hexagonal crystal if the magnetic field $\boldsymbol{H}$ is directed along the C_3 axis (cf. Fig. 3.8).

3.5.3. *The Hall mobility for hopping transport*

Now we use the explicit expression for the hopping probability $W_{m_1m_2m}$, given in Appendix 2. (formula (A 2.6)), and set in this expression $\varepsilon_m = 0$, $\boldsymbol{E} = 0$, $\boldsymbol{R}_m - \boldsymbol{R}_{m'} = \boldsymbol{g}$ ($\boldsymbol{g}$ — vector connecting nearest neighbours). We find that

$$W^{(0)}_{m_1m_2m}(\boldsymbol{H}) = \frac{e\boldsymbol{H}}{2\hbar^4 c} [\boldsymbol{g}_1 \times \boldsymbol{g}_2] \, J^3 I, \quad (3.5.12)$$

where $\boldsymbol{g}_1 = \boldsymbol{R}_m - \boldsymbol{R}_{m_1}$, $\boldsymbol{g}_2 = \boldsymbol{R}_{m_2} - \boldsymbol{R}_{m_1}$, J is the resonance integral between nearest neighbours in the absence of a magnetic field in the plane shown in Fig. 3.8, and I is given by

$$I = e^{-3S_T} \int_{-\infty}^{\infty} dt_1\, dt_2 \left\{ \exp \left[\sum_{\boldsymbol{q}} \frac{|\gamma(\boldsymbol{q})|^2 \, (1 - \cos \boldsymbol{g}\boldsymbol{q})}{2N \sinh (\hbar\omega_{\boldsymbol{q}}\beta/2)} \right. \right.$$

$$\times \left(\cos \omega_{\boldsymbol{q}} \left(t_1 + \frac{i\hbar\beta}{2} \right) + \cos \omega_{\boldsymbol{q}} \left(t_2 + \frac{i\hbar\beta}{2} \right) \right.$$

$$\left. \left. \left. + \cos \omega_{\boldsymbol{q}} \left(t_1 - t_2 + \frac{i\hbar\beta}{2} \right) \right) \right] - 1 \right\}. \qquad (3.5.13)$$

Note that, owing to the symmetry of the problem, the quantity I does not depend on the direction of $\boldsymbol{g}$, but depends only on $|\boldsymbol{g}| = a$.

Inserting (3.5.12) into (3.5.11), after some straightforward transformations, one obtains the following expression for the antisymmetric part of the conductivity tensor:

$$\sigma_{xy} = \frac{3}{4} \frac{e^3 H f(1 - f)\, J^3 \mathcal{N} a^4}{\hbar^4 c k T}. \qquad (3.5.14)$$

Here $\mathcal{N} = N/\Omega$ is the concentration of lattice sites. In (3.5.13), the twofold integral can be evaluated by the saddle-point method. To do this, analogously as in Section 3.2., we assume that the dispersion of the optical phonons is not too small so that only the contribution from the first saddle point may be accounted for. It turns out that this saddle point is lying at $t_1 = -\frac{2}{3} i\hbar\beta$, $t_2 = -\frac{1}{3} i\hbar\beta$. On expanding the exponent in powers of t_1 and t_2 about this point up to the second-order terms, and performing the integration over t_1 and t_2 (a more rigorous, but also more extensive, procedure of evaluating these integrals was given by FRIEDMAN and HOLSTEIN, 1963) we find that

$$I = \frac{4\pi}{\sqrt{3}} \left\{ \sum_{\boldsymbol{q}} \omega_{\boldsymbol{q}}^2 \frac{|\gamma(\boldsymbol{q})|^2 \, (1 - \cos \boldsymbol{g}\boldsymbol{q})}{N \sinh (\hbar\omega_{\boldsymbol{q}}\beta/2)} \cosh (\hbar\omega_{\boldsymbol{q}}\beta/6) \right\}^{-1}$$

$$\times \exp \left\{ -3 \sum_{\boldsymbol{q}} \frac{|\gamma(\boldsymbol{q})|^2 \, (1 - \cos \boldsymbol{g}\boldsymbol{q})}{N \sinh (\hbar\omega_{\boldsymbol{q}}\beta/2)} \sinh (\hbar\omega_{\boldsymbol{q}}\beta/3) \sinh (\hbar\omega_{\boldsymbol{q}}\beta/6) \right\}. \qquad (3.5.15)$$

In the high-temperature limit, where $\hbar\omega_{\text{op}} < 2kT$, we have $I \approx \pi\hbar^2 \exp \left(-\frac{4}{3} \times E_a/kT \right) \Big/ \left(2\sqrt{3}\, E_a kT \right)$ and, according to (3.5.14), σ_{xy} becomes

$$\sigma_{xy} = e\mathcal{N} f(1 - f)\, u_0 \frac{u_0 H}{c} \frac{\pi\sqrt{3}}{8} \frac{J^3}{E_a (kT)^2} \exp(-4E_a/3kT). \qquad (3.5.16)$$

Accordingly, the activation energy for three-site hops is equal to $\frac{4}{3} E_a$, which proves to be less than the activation energy for two uncorrelated two-site hops, $2E_a$. The activation energy for the Hall transport $\frac{4}{3} E_a = \frac{2}{3} E_p$ (recall that $2E_a = E_p$) can be viewed as arising from a configuration (caused by a thermal fluctuation) in which the energy of the occupied site is enhanced by $\frac{2}{3} E_p$, while the energy of each of the two other sites is reduced by $\frac{1}{3} E_p$, so that the levels are equalized and a resonance transition may occur. In this configuration, the three sites are surrounded by polarization clouds of the same kind.

Let us draw attention to an important peculiarity of to the expression (3.5.16): σ_{xy} is a symmetric function of the chemical potential, because $f(1 - f) = \{2 \cosh \mu / 2kT\}^{-2}$. Hence, it follows that the sign of the Hall effect does not change when we displace the chemical potential from a position below the polaron level (electron transport) to a position above this level (hole transport), i.e. for small polarons in hexagonal crystals we find the *p-n*-anomaly mentioned above. Note that for $\mu = 0$, the hopping-Hall effect does not vanish. In the hopping regime, the Hall effect sensitively depends on the symmetry of the crystal lattice. In particular, in cubic crystals, the hopping-Hall effect is governed by four-site hops, because in this case there is no contribution from three-site hops. For four-site hops, however, no *p-n*-anomaly occurs (for more details on the *p-n* anomaly, see Section 8.).

Let us now determine the Hall mobility by means of (3.5.3). To do this, for the conductivity we use the expression (3.2.10), after multiplying it by a factor 3/2, which corresponds to the transition from a cubic to a hexagonal crystal. The latter assertion can be readily verified by means of (2.5.10), by using the concrete geometry of the hexagonal lattice illustrated in Fig. 3.8. Thus, for hexagonal lattices, the hopping-Hall mobility becomes (FRIEDMAN and HOLSTEIN, 1963)

$$u_H = u_0 \sqrt{\frac{\pi}{12}} \frac{J}{\sqrt{E_a kT}} \exp(-E_a/3kT). \tag{3.5.17}$$

According to this result, in hexagonal crystals, the Hall mobility exhibits an activated temperature dependence. The corresponding activation energy is three times smaller than that of the drift mobility. For small polarons, the factor $A = u_H/u_h$ (u_h — drift mobility in the hopping regime) is very large and strongly depends on temperature:

$$A = \frac{2}{3\sqrt{3}} \frac{kT}{J} \exp(2E_a/3kT), \tag{3.5.18}$$

whereas for band transport this factor is close to unity.

Analogous to the conductivity, the Hall mobility of small polarons is governed by the dimensionless parameters η_1, η_2, and η_3 (3.2.11). The results presented above apply to the case where all these parameters are small. EMIN and HOLSTEIN (1969) studied the Hall mobility of the adiabatic polaron ($\eta_2 \gg 1$, but $\eta_1, \eta_3 \ll 1$) by methods analogous to those described in Section 3.2.2., but by using a three-site model instead of a two-site one. For $2kT > \hbar\omega_{op}$, they obtained the following result

$$u_H = u_0 \frac{\hbar\omega_{op}}{kT} \frac{1}{4\pi} \varphi \exp\left(-\frac{\frac{1}{3}E_a - J}{kT}\right), \tag{3.5.19}$$

where φ is a dimensionless function of the parameter J/kT. For $J/kT \gg 1$, φ is equal to unity, and for $J/kT \ll 1$, $\varphi = kT/J$. In the case of the adiabatic polaron, the factor A turns out to be much smaller than for $\eta_2 \ll 1$. For sufficiently small values of the parameter J/kT, A may be less than unity (Austin and Mott, 1969).

As mentioned above, the hopping-Hall effect sensitively depends on the symmetry of the crystal. For cubic crystals, the Hall mobility was calculated in the one-electron approximation by Bryksin and Firsov (1970b, 1971b) and Emin (1971c). At high temperatures, $2kT > \hbar\omega_{op}$, the hopping-Hall mobility in such crystals is given by

$$u_H = u_0 \frac{\sqrt{\pi}}{8} \frac{J^2}{E_a\sqrt{E_a kT}} \frac{\delta^2}{(2-\delta)(\delta-1)^{3/2}} \exp\left\{-\frac{(\delta-1)E_a}{kT}\right\}, \tag{3.5.20}$$

where $\delta = (1 - E_a/4E_a')$, with

$$E_a' = \sum_q \frac{|\gamma(q)|^2}{4N} \hbar\omega_q[1 - \cos q(g + g')]. \tag{3.5.21}$$

Here g and g' are nearest-neighbour vectors lying perpendicular to each other. It turns out that $4/3 < \delta < 2$. In particular, if $|\gamma(q)|^2 \omega_q$ does not depend on q (model with short-range electron phonon coupling), then $\delta = 4/3$.

According to (3.5.20), the assertion that the activation energy of the Hall mobility equals (1/3) E_a does not generally hold. It applies to hexagonal crystals, but to cubic ones only if the electron-phonon interaction has short-range character. Furthermore, (under the same conditions) in cubic crystals the ratio u_H/u_h is smaller than in hexagonal ones, though it remains larger than unity if $\eta_1, \eta_2, \eta_3 < 1$.

3.5.4. *The Hall mobility in a large temperature range including also non-hopping transport*

On studying the drift mobility, we found that below some temperature, the hopping mechanism ceases to dominate, the tunnelling mechanism becomes operative, and at very low temperatures the band mechanism. The situation is similar in the case of the Hall effect. In the band regime, the Hall effect of small polarons was studied by Friedman (1963) by means of the Boltzmann equation. Bryksin and Firsov (1974b) investigated u_H in a large temperature range. They pointed out that in the small-polaron model σ_{xy} consists of a sum of contributions arising from different mechanisms of the Hall transport, analogously as σ_{xx} consists of a hopping and a tunnelling term. For hexagonal crystals, in a large temperature range the expression for u_H reads (BRYKSIN and FIRSOV, 1974b)

$$u_H = u_0 3 \sqrt{\pi} \frac{J}{\hbar\omega_{op}} \frac{\nu}{2\Theta^2} \exp(-\Theta \cosh \alpha)$$
$$\times \frac{1 + \sqrt{2}(\Theta^{3/2}/\nu) \exp(\Theta) + \left(1/3\sqrt{3}\right)(\Theta^3/\nu^2) \exp\left(3\Theta \cosh \frac{\alpha}{3}\right) \Big/ \cosh \frac{\alpha}{3}}{1 + (\Theta^{3/2}/2\nu) \exp(2\Theta)}. \tag{3.5.22}$$

Here $\alpha = \hbar\omega_{op}/2kT$, $\Theta = \gamma/\sinh \alpha$, $\gamma = \sum_{\boldsymbol{q}} |\gamma(\boldsymbol{q})|^2 \frac{1}{2N} (1 - \cos \boldsymbol{gq})$, $\nu = \lambda\gamma^2 \frac{\Delta\omega}{\omega_{op}} \times (E_a/J)^4$, and λ is a number of order unity.

The three terms in the numerator of the fraction in (3.5.22) describe the three different mechanisms of the Hall transport, and the two terms in the denominator the tunnelling (the unity) and the hopping mechanism for σ_{xx}. Let us consider the characteristic temperature regimes of the Hall mobility.

In the regime 1, the Hall mobility is governed by the first terms in the numerator and the denominator of the fraction in (3.5.22). In this regime, the Hall transport has band character, and the effect of the magnetic field on the electron is described by the Lorentz force. The small-polaron kinetics is described by a transport equation of the Boltzmann type, wherby the width of the polaron band is less than kT and $\hbar/\tau$ (τ — relaxation time). Accordingly, in the regime 1, the transport mechanism may be termed the quasi-Lorentzian mechanism. On the temperature scale, this regime follows after that of the band transport in broad bands, where the kinetics may be described by the standard Boltzmann equation. The latter regime is not covered by formula (3.5.22).

The lower limiting temperature of the regime 1, T_{01}, is defined by $kT_{01} = \Delta\varepsilon_p(T_{01})$, where $\Delta\varepsilon_p$ is the width of the polaron band (cf. Section 3.1.). The corresponding upper limiting temperature, T_{12}, can be found by equating the first and the second term in the numerator of (3.5.22), which yields $\Theta_{12} = -\frac{3}{2} \ln \Theta_{12} + \ln\left(\nu/\sqrt{2}\right)$, $\Theta_{12} = \gamma/\sinh(\hbar\omega_{op}/2kT_{12})$.

According to (3.5.22), in this temperature regime, the Hall mobility quickly decreases with increasing temperature, due to the factor $\exp(-\Theta \cosh \alpha)$ (i.e. $\exp(-S_T)$), which is a consequence of the reduction of the width of the polaron band with increasing temperature.

For the drift mobility, the transition from the tunnelling to the hopping mechanism occurs inside the regime of the quasi-Lorentzian Hall transport mechanism, i.e. inside this regime the first and the second terms in the numerator of the fraction in (3.5.22) become equal to each other. The corresponding temperature T_0 is defined by the equation $2\Theta_0 = -\frac{3}{2} \ln \Theta_0 + \ln(2\nu)$. Since, for small polarons, in general we have $\nu \gg 1$, it holds that $T_{12} > T_0$. This means, that for σ_{xx}, the transition to hopping transport occurs at lower temperatures than for σ_{xy}. Consequently, there may exist a broad temperature range within that the drift mobility exhibits an activated temperature dependence while the Hall mobility decreases with increasing temperature. Note that in the temperature range $T_0 < T < T_{12}$ the Hall mobility decreases even more quickly than in the range $T_{01} < T < T_0$.

The regime 2 is governed by the second terms of the numerator and the denominator in (3.5.22). In this regime the Hall transport is due to the tunnelling mechanism. The effect of the magnetic field is not given by the Lorentz force, but by the linear terms with respect to $\boldsymbol{H}$ of the transition probability. The drift transport is due to hopping. The upper limiting temperature of this regime, T_{23}, results form equating the second and the third terms of the numerator in (3.5.22): $2\Theta_{23} = -\frac{3}{2} \times \ln \Theta_{23} + \ln\left(\nu 3 \sqrt{6}\right)$. Here, we have assumed that $\hbar\omega_{op}/6kT_{23} \lessgtr 1$. On passing from the regime 1 to the regime 2, the decrease of u_H with increasing T strongly slows down.

The regime 3 ($T > T_{23}$) is governed by the last terms of the numerator and the denominator. In this regime, the Hall as well as the drift transport exhibit hopping character. The corresponding temperature dependence of u_H was already discussed in detail above.

The temperature dependence of u_H, calculated by means of (3.5.22), is illustrated in Fig. 3.9. For comparison, in this figure, we also illustrate the temperature depen-

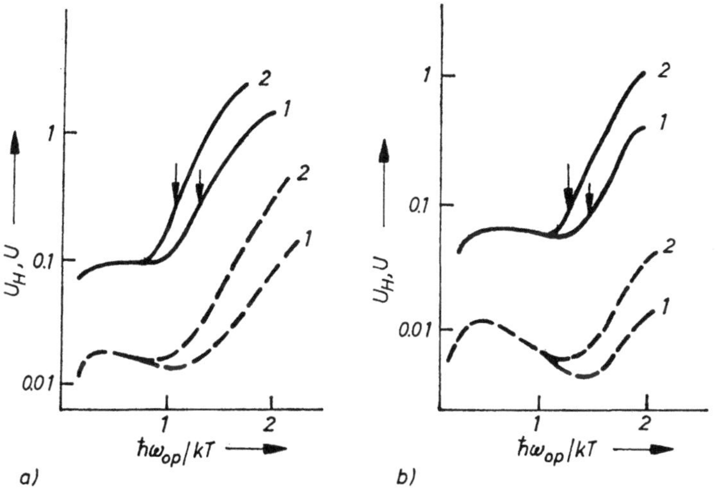

Fig. 3.9. Temperature dependence of the Hall mobility U_H (solid curves) and the drift mobility U (dashed curves) in hexagonal crystals for different values of the parameters γ and ν, evaluated by means of formulae (3.5.22) and (3.5.23). U_H and U are dimensionless quantities: $U_H = \frac{2}{3\sqrt{\pi}} \frac{\hbar\omega_{op}}{J} \frac{u_H}{u_0}$, $U = \frac{1}{3\sqrt{\pi}} \left(\frac{\hbar\omega_{op}}{J}\right)^2 \frac{u}{u_0}$. In a), the curves are evaluated for $\gamma = 3$, and in b), for $\gamma = 4$. The curves 1 correspond to $\nu = 30$, and the curves 2 to $\nu = 90$. The arrows on the curves U_H indicate the position of the minimum of the corresponding curve U. The decrease of U and U_H with increasing T in the high temperature regime (for hopping transport) is due to the pre-exponential factors

dence of the drift mobility, which is the sum of a hopping (3.2.1), (3.2.7) and a tunnelling (3.2.33), (3.2.35) term:

$$u = u_0 6 \sqrt{\pi} \frac{J^2}{(\hbar\omega_{op})^2} \frac{\nu\alpha}{\Theta^2} \exp(-2\Theta \cosh \alpha) \left(1 + \frac{\Theta^{3/2}}{2\nu} e^{2\Theta}\right). \tag{3.5.23}$$

We learn from Fig. 3.9, that there is indeed a temperature range where, with increasing temperature, u_H rapidly decreases, while u increases.

3.6. Thermoelectric power

As pointed out in the preceding section, by examining the Hall effect in the hopping regime, neither the charge carrier concentration nor the sign of the carriers can be determined. The single promising method for obtaining these data is to measure the Seebeck coefficient S. As discussed in Section 9.1., the formulation of the theory of the thermopower in the hopping regime is in general an intricate and so far not completely solved problem.

To study the thermopower of small polarons in the hopping regime, we use the simple (but not rigorous) formula for the Seebeck coefficient (9.1.6)

$$S = -\frac{1}{e} \frac{\langle \varepsilon \rangle - \mu}{T}, \tag{3.6.1}$$

where $\langle \varepsilon \rangle$ denotes the average energy of the charge carriers ($e = |e|$). Since the width of the polaron band is very small, we may assume that the energy transferred by a carrier is equal to the polaron shift $-E_p$. Then, (3.6.1) becomes

$$S = \frac{1}{e} \frac{E_p + \mu}{T}. \tag{3.6.2}$$

Here μ is counted from the unrenormalized polaron band (for $J \to 0$). Taking into account that in the small-polaron model the site occupation f is related to the chemical potential by

$$f = \left\{\exp\left(-\frac{\mu + E_p}{kT}\right) + 1\right\}^{-1},$$

(3.6.2) may be written as (cf. (9.1.8))

$$S = \frac{k}{e} \ln \frac{n}{\mathcal{N} - n}, \tag{3.6.3}$$

where $n = \mathcal{N}f$ is the concentration of electrons ($\mathcal{N}$ — concentration of sites in the lattice).

It is interesting to compare (3.6.3) with the formula for S obtained from the Boltzmann equation for the band transport

$$S = \frac{k}{e} \left(\frac{\mu}{kT} - A\right), \tag{3.6.4}$$

where the term A is due to the kinetic energy of the carriers. For instance, in the case of the Boltzmann statistics, and for a relaxation time not depending on energy, one finds that $A = 5/2$. On comparing (3.6.5) with (3.6.3), we note that for hopping transport, a term arising from the kinetic energy is absent. This is a reasonable result, which formerly, however, was called in question (cf. the review by APPEL, 1968).

Note that according to (3.6.4), in the hopping regime the thermopower changes its sign on passing over from electron transport ($2n < \mathcal{N}$) to hole transport ($2n > \mathcal{N}$), i.e. in this regard, the thermopower behaves as in the case of band transport.

As mentioned in Section 9.1., in formulating the theory of the thermopower in the hopping regime, the major problem is in that the heat flux cannot be defined uniquely. Therefore, the above expression for the Seebeck coefficient may not be viewed as a rigorous result. Furthermore, since during a hopping event the polaron energy increases by an amount E_a, it is a priori not clear, whether a hopping electron transfers the energy $-E_p$ or $-E_p + E_a$. These problems were studied by EFROS (1967), who put forward convincing arguments in favour of (3.6.2). For more details on the hopping thermopower, see Section 9.1.

3.7. Experimental findings

Small polarons may be expected in materials with a narrow electron band (i.e. with a small resonance integral J) and with strong coupling between electrons and optical phonons. Among such materials are transition-metal compounds (characterized by a small overlap between d-orbitals), certain organic semiconductors, molecular solids, rare-gas solids, and possibly some alkali halides. The theory of small polarons is best worked out for the case that the electron correlation may be neglected. Therefore, materials with very low charge-carrier concentration (empty conduction band of the pure material) are most suitable for a comparison of experimental and theoretical results. However, most of the transiton-metal compounds (NiO, CaO, MnO, Fe_3O_4, and others) have a partly filled d-conduction band and exhibit magnetic order and a metal-insulator transition. All these phenomena "screen" small-polaron properties obtained in the one-electron approximation. The problem of hopping transport of small polarons in the presence of magnetic order was studied by HAUBENREISSER (1961) and APPEL (1966). It turns out that on these conditions the hopping probability is lowered due to the interaction of the electron spin with the spin configuration of the ions in the initial and the final states. Therefore, for materials with magnetic order, below the Curie (Néel) point the expression for the hopping mobility (3.2.21) must be multiplied by an additional factor $\frac{1}{2} \times \left(1 - M^2(T)/M^2(0)\right)$, where $M(T)$ is the magnetization of a sublattice.

The polaron transport mechanism was first proposed for NiO (cf. the reviews by APPEL, 1968, and AUSTIN and MOTT, 1969). However, so far the conduction mechanism in NiO is not yet completely understood. In the literature arguments have been put forward in favour of the existence of small polarons in the following materials: MnO, VO_2, $BaTiO_3$, Ti_4O_7, KCl, LiF, Fe_3O_4, $LaCoO_3$, β-$FeSi_2$, solid N_2, O_2, and CO, raregas solids with hole-type conductivity, and certain organic polymers (the corresponding references may be found in the review by BÖTTGER and BRYKSIN, 1976b).

The most comprehensive investigation for establishing the polaron character of charge carriers has been performed on TiO_2 (BOGOMOLOV et al., 1967, KUDINOV et al., 1969). From experimental investigations of the drift mobility, the Hall mobility, the absorption of light, the thermoelectric power, and from a detailed comparison of experimental findings and theoretical predictions (TiO_2 has an empty d-conduction band) one gets quite convincing evidence for small-polaron transport in TiO_2.

Results of investigations of the drift mobility and the Hall effect obtained by BOGOMOLOV et al. (1967) are represented in Fig. 3. 10. The curve of the drift mobility

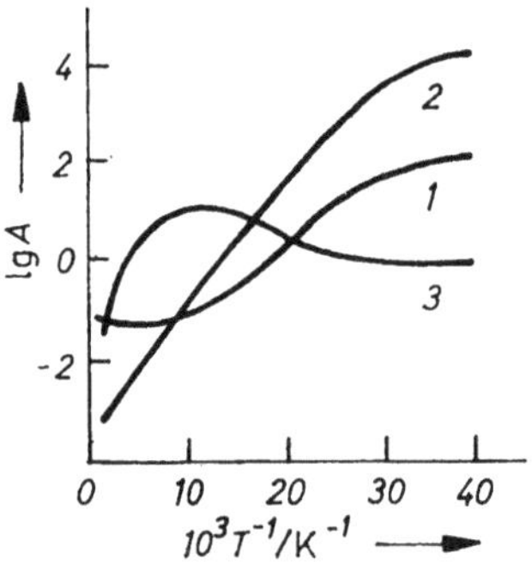

Fig. 3.10. Temperature dependence of dimensionless quantities (A) measured on TiO_2: curve 1, $A = u_{xx}/u_0$; curve 2, $A = u_{xy}/u_0$; curve 3, $A = encR \equiv u_H/u_{xx}$. $u_0 = ea^2/\hbar$ is a quantity having dimension of a mobility

has a minimum at $T = 300$ K, which is characteristic of small polarons (cf. Section 3.2. and Fig. 3.9) and which suggests that the hopping mechanism dominates at $T > 300$ K. According to the estimations performed in the latter paper, in TiO_2, the case of adiabatic small polarons is realized, i.e. $\eta_2 > 1$.

The curves for u_{xy} ($\sigma_{xy} = enu_{xy}$) and the Hall coefficient R show that the Hall mobility strongly differs from the drift mobility and that the ratio u_H/u_{xx} depends on temperature, which is characteristic of small polarons (cf. Section 3.5.). The fact that in the high-temperature regime this ratio becomes less than unity does not contradict the theory, because such a behaviour is in agreement with that expected for adiabatic small polarons ($\eta_2 > 1$). The curve for u_{xy} decreases with increasing temperature within the whole temperature range investigated and does not exhibit a minimum, which at the first sight contradicts the theory of small polarons, all the more since there is an activated increase of u_{xx}. However, as pointed out in Section 3.5., for u_{xy}, the activated bahaviour sets in at higher temperatures than for u_{xx} (cf. Fig. 3.9). Thus, the absence of an activated temperature dependence of u_{xy} only indicates that the transition temperature to an activated behaviour of u_{xy} was not reached in the experiment.

The frequency dependence of the absorption coefficient of TiO_2 doped with Nb, measured by KUDINOV et al. (1969), is represented in Fig. 3.11. This curve well agrees with the theoretical dependence (3.3.9). The temperature dependence of this peak is in good agreement with the theory.

The thermopower measured on TiO_2 by BOGOMOLOV et al. (1967) can be well described by formula (3.6.4). The weak anisotropy observed in the thermopower cannot be understood by means of (3.6.4). This anisotropy possibly indicates a finite width of the polaron band, i.e. a contribution from the kinetic energy of the carriers.

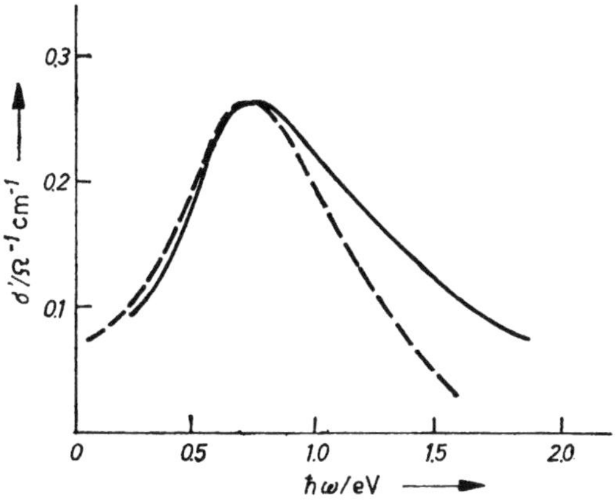

Fig. 3.11. Frequency dependence of the real part of the conductivity, $\sigma'(\omega)$, measured on TiO_2 (solid curve) doped with Nb, for $T = 600$ K. The dashed curve shows results of a calculation performed on the basis of formula (3.3.9) by making use of the following parameters $\hbar\omega_{op} = 0.1$ eV, $kT = 0.05$ eV, $E_a = 0.2$ eV

BARBANEL and BOGOMOLOV (1975) suggested that in TiO_2 the charge carriers are localized in two directions, while in the third direction the carriers move in a narrow band.

An additional argument in favour of the applicability of the small-polaron picture to TiO_2 is the good agreement of the parameters obtained from comparisons of the theory with the results from different experimental methods (cf. the table in the paper by BOGOMOLOV et al., 1967). From such comparisons, the following parameters were obtained for TiO_2: electron-phonon coupling constant $\gamma = 4 \dots 7$ $\left(\gamma = \frac{1}{N} \times \sum_q |\gamma(q)|^2\right)$, activation energy $E_a \approx 0.2$ eV, resonance integral $J \approx 0.1$ eV.

In the past few years the research in the field of small polarons has been mainly concentrated on polarons in organic quasi-one dimensional systems and the phenomenon of polaron pairing (formation of bipolarons) (cf. Section 11.1.).

Chapter III. Hopping Conduction in Disordered Systems

4. Electron states and charge transport in disordered systems

4.1. Effect of disorder on electron states and charge transport

Roughly speaking, there are two broad classes of disordered systems: (i) systems with substitutional, cellular or compositional disorder, and (ii) systems with positional, topological or structural disorder. Compositionally disordered alloys and mixed crystals, but also crystals with substitutional impurities are among the systems of the first class. Glassy and amorphous materials, as well as superionic conductors are among the systems of the second class. It is now a well accepted concept that any kind of disorder can give rise to spatially localized one-particle states.

Localization of electron states due to disorder is one of the most attractive problems in the theory of disordered systems. It is of vital importance for an understanding of transport properties, but it is also an interesting problem as such. Although electron localization has been studied already for more than two decades, many questions, particularly concerning quantitative aspects, are still open.

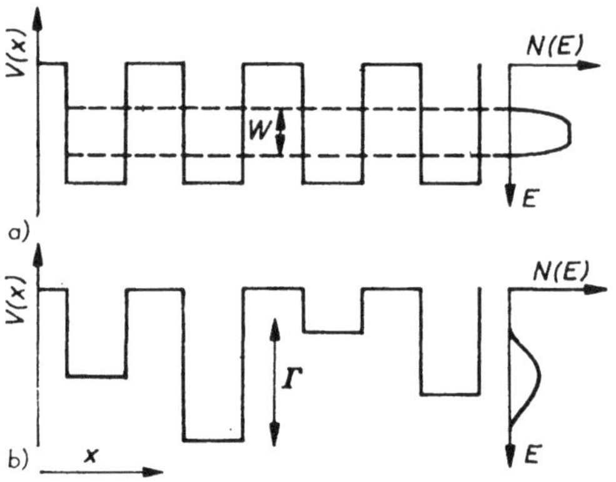

Fig. 4.1. Schematic representation of the potential energy $V(x)$ in a system with diagonal disorder (Anderson model). a) with $\Gamma = 0$; b) with finite value Γ (Γ – width of the spread of the energy levels, W – band width). The density of states $N(E)$ is also shown

Until some few years back, the localization problem was primarily attacked in the manner of ANDERSON (1958) by studying the absence of diffusion in a model with strong disorder, the so-called Anderson model, which is illustrated in Fig. 4.1 (cf. Section 4.2). Towards the end of the seventies, the localization problem has gained new impetus by applying to it ideas of the theory of phase transitions (scaling arguments) (cf. Sections 4.3 and 11.2.1.).

The Anderson model is a tight-binding model with only one state per lattice site. The corresponding Hamiltonian reads (cf. (2.1.4))

$$\mathcal{H} = \sum_{m} \varepsilon_m a_m^+ a_m + \sum_{m \neq m'} J_{m'm} a_{m'}^+ a_m , \tag{4.1.1}$$

where ε_m is the energy of the mth site and $J_{m'm}$ the resonance integral between sites m and m' (note here we drop the index e at the electron Hamiltonian). The resonance integral is assumed to have a constant value between nearest-neighbour sites and to be otherwise zero. The disorder is introduced by making the site energies

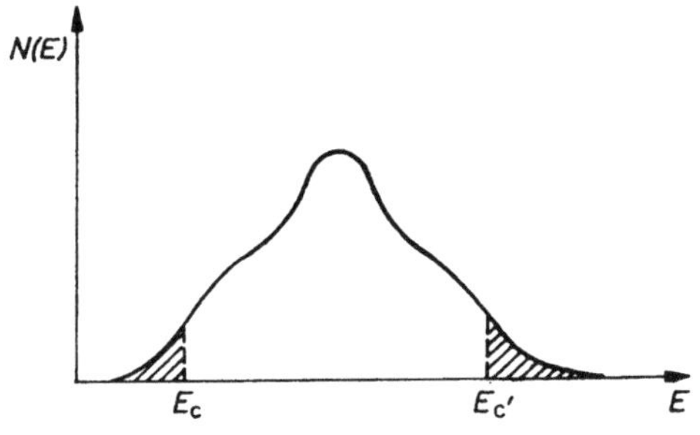

Fig. 4.2. Schematic representation of the density of states $N(E)$ as function of the energy E for a band of a disordered solid. The energies E_c and $E_{c'}$ separate the regions of localized states (shaded) from that of nonlocalized ones

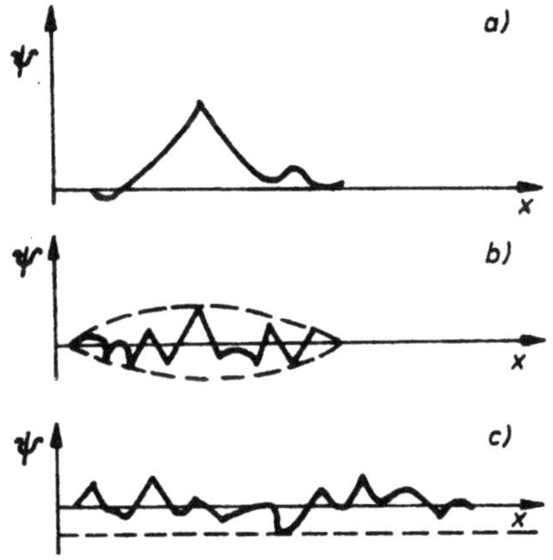

Fig. 4.3. Schematic representation of a strongly localized a), weakly localized b), and nonlocalized (extended) c) wavefunktion

independent random variables. For a uniform distribution of the site energies over a range Γ, the model may be viewed as simulating to some extent properties of systems with (continuous) positional disorder. On the other hand, if ε_m may take only one of two possible values, the Hamiltonian (4.1.1) represents a model for binary alloys.

The effect of disorder on the density of states is schematically shown in Fig. 4.1. We see that disorder removes sharp band edges and other features (van Hove singularities) resulting from lattice periodicity, such that a band with tails occurs. ANDERSON (1958) predicted that for an electron moving in the potential of Fig. 4.1 (b), all states would be localized if Γ/W (W — bandwidth in the absence of disorder) exceeds some critical value, which was later shown to be of order 2 (see, e.g., EDWARDS and THOULESS, 1972).

If Γ/W is less than the critical value, then according to MOTT (1967) the states in the tails are nevertheless localized, and sharp energies E_c and $E_{c'}$ (called the mobility edges) separate the localized and extended states (Fig. 4.2). Localized and extended states are schematically shown in Fig. 4.3. The radial extension of localized wavefunctions tends to infinity as the energy approaches a mobility edge.

Now let us ask what happens when the Fermi energy μ can be made to cross the mobility edge E_c (cf. Fig. 4.2), by varying either the degree of disorder or the density of electrons (in doped crystalline semiconductors the latter can be done by changing the compensation, or in silicon inversion layers by changing the gate voltage). If μ lies well above E_c (but below $E_{c'}$ in Fig. 4.2), one may expect a "weak scattering theory" to be appropriate for describing the electronic conduction (analogously as in liquid metals (ZIMAN, 1961)), i.e. in this case conduction electrons propagate through the system with a mean-free path large compared with the distance between neighbouring atoms.

What happens when μ approaches E_c (from the right in Fig. 4.2) is still a matter of controversy. According to MOTT (1967) (see also MOTT, 1981a) there exists a finite "minimum-metallic conductivity"[1]) (cf. section 11.2.1.).

$$\sigma_{\min} = \text{const}\,\frac{e^2}{\hbar a} \tag{4.1.2}$$

(a — distance between neighbouring atoms), which is typically of some hundred $\Omega^{-1}\,\text{cm}^{-1}$. On the other hand, other authors (e.g. WEGNER, 1979, ABRAHAMS et al. 1979, GÖTZE 1981, PAALANEN and THOMAS, 1983, and references therein) have cast doubt on the existence of such $\sigma_{\min}$. They predicted that the conductivity tends to zero at E_c, as it would in the classical percolation model (see below). For instance, according to ABRAHAMS et al. (1979), in a three-dimensional system the conductivity vanishes near E_c as

$$\sigma \sim \big((E - E_c)/W\big)^s \tag{4.1.3}$$

where s is of order unity (W — bandwidth).

When the Fermi energy lies below the mobility edge (below E_c in Fig. 4.2), so that the states at the Fermi energy are localized, there are two conduction mecha-

---

1) More precisely, MOTT (1976) has argued that a "minimum-metallic conductivity" exists if $\nu d \geqq 2$, but it does not if $\nu d < 2$, where d is the dimensionality of the system and ν the critical exponent of the localization length defined in (4.3.1). The values of ν reported in the literature (cf. SARKER and DOMANY, 1980) vary over a wide range and cannot be used to test Mott's prediction.

nisms: at high temperatures electrons are excited to states above E_c[1]), and at low temperatures conduction is realized by thermally activated hopping from one localized state to another. Both these mechanisms lead to a strong temperature dependence of the conductivity, while for $\mu > E_c$, the conduction by propagation should be nearly independent of temperature. At zero temperature, localized states do not contribute to the dc conductivity.

A disordered system where the Fermi energy lies in the region of localized states, is often called a "Fermi glass". — There are indications that intersite electron-electron interaction may be important in disordered systems. In this case the potential felt by an electron is the combined field (in the sense of the Hartree-Fock approximation) of the random potential of the atoms and of all other electrons. If the latter part dominates (which is believed to be the case, e.g., in $Fe_3O_{4-x}Fe_x$, $x > 0.05$; cf. GRAENER et al., 1979) the system is a "Wigner glass".

Intersite electron-electron interaction can also affect the hopping conduction: (i) for single electron hops via change of the density of one-electron excitations and (ii) by many-electron hops (see Section 5.4.). An account of the effect of electron-electron interaction on the hopping transport is a very intricate problem, which is at present not completely solved.

Of course, the electronic behaviour can be also affected by intrasite electron correlation. An account of this correlation within the framework of the Hubbard model leads to the Hamiltonian

$$\mathcal{H} = \sum_{m\sigma} \varepsilon_m a_{m\sigma}^{+} a_{m\sigma} + \sum_{\substack{mm' \\ \sigma}} J_{m'm} a_{m'\sigma}^{+} a_{m\sigma} + \frac{U}{2} \sum_{m\sigma} n_{m\sigma} n_{m-\sigma}, \tag{4.1.4}$$

where $n_{m\sigma} = a_{m\sigma}^{+} a_{m\sigma}$ is the particle-number operator at site m with spin σ. Note that (4.1.4) does not account for intersite electron-electron interaction.

In the absence of disorder ($\varepsilon_m = E_0$), the Hamiltonian (4.1.4) has been widely used to study metal-insulator transitions (Mott-transition). As qualitatively shown in Fig. 4.4, in this case for $W/U \to 0$ (W — bandwidth for $U = 0$) the single-particle density of states exhibits two sharp peaks, one at E_0 (only one electron at a given site) and another at $E_0 + U$ (two electrons with opposite spin directions at a site). With increasing parameter W/U the two peaks (bands) broaden and at some critical value $(W/U)_c$ they begin to overlap. For the case of one electron per site, at $(W/U)_c$ a metal-insulator transition occurs. For a study of the metal-insulator transition

[1]) The charge transport in extended states sufficiently apart from the mobility edge is generally supposed to be band-like, i. e. due to nearly free propagation interrupted at random by scattering events. For transport in extended states close to the mobility edge where the mean free path is of the order of the lattice constant, MOTT (1976) and COHEN (1970) proposed a model, now called the random-phase model, in which the wavefunction is represented as a linear superposition of atomic wavefunctions with coefficients which have no phase correlation from one site to the next. This model was used by HINDLEY (1970) and HOSHINO and WATABE (1978) to calculate the conductivity, and by FRIEDMAN (1971) and KANEYOSHI (1976) to calculate the Hall mobility. It has been criticized by FRIEDMAN (1980). According to MOTT (1981b), it is correct when the mobility edge lies in midband.

on the basis of the Hamiltonian (4.1.4) see, e.g., MA (1982) and for an account of additional interactions between nearest and next-nearest neighbours in a Hamiltonian of the type (4.1.4) see, e.g., OLES et al. (1983). Note that even in the absence of disorder the Hubbard model (4.1.4) can be solved exactly only for a one-dimensional system.

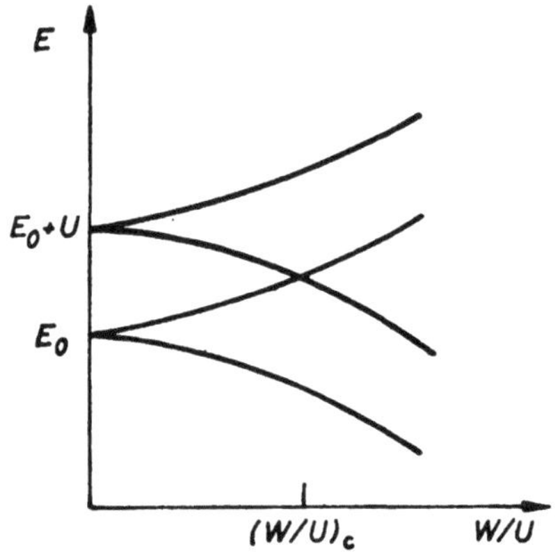

Fig. 4.4. Schematic representation of the metal-insulator transition in the Hubbard model in the absence of disorder (E_0 — site energy, U — Hubbard intra-site repulsive energy, W — band width). For one electron per site, on the left of $(W/U)_c$ the system is an insulator and on the right of it a metal

The density of states of a doped semiconductor (e.g. Si:P), as suggested by experimental studies of charge transport, is schematically shown in Fig. 4.5 (see, e.g., MOTT and DAVIS, 1980, and references therein). It can be qualitatively understood by a tight-binding Hamiltonian of the type (4.1.4): the D^0 (neural donors)-band represents the lower Hubbard band and the D^- (negatively charged donors)-band

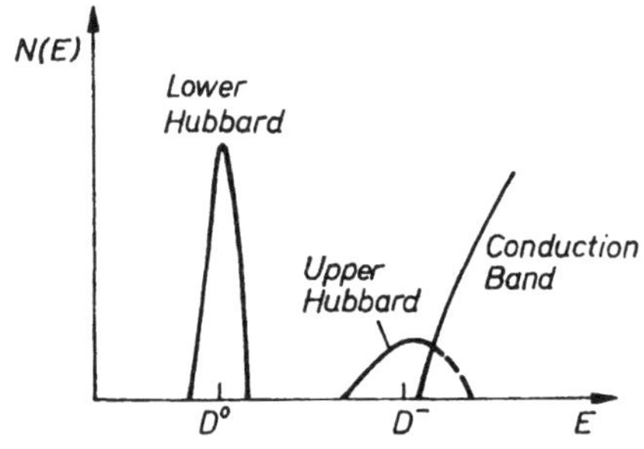

Fig. 4.5. Schematic representation of the density of states $N(E)$ in Si:P for low concentration of donors (see text)

represents the upper Hubbard band. A quantitative study of the density of states and charge transport on the basis of the Hamiltonian (4.1.4) (which would include the problem of localization, the metal-insulator transition etc.), however, has so far not been performed and is, obviously, not practicable at present.

Concluding this concise overview of some aspects of electron states in disordered systems, let us turn to non-crystalline materials, in particular, to non-crystalline semiconductors. The density of states of amorphous silicon, as suggested by experimental studies, is schematically shown in Fig. 4.6. A continuous random network (structure in which all bonds are satisfied, but in which there is no long-range order) is generally used as model for amorphous silicon. The localized band-tail states (in Fig. 4.6 the states between E_c and E_a, and E_v and E_b) are supposed to be due to the lack of long-range order. Accordingly, the position of the mobility edges (E_c and E_v in Fig. 4.6) is an "intrinsic" property of the continous random-network model. On the other hand, the deeper lying gap states (denoted by E_x and E_y in Fig. 4.6) are ascribed to structural defects such as dangling bonds or vacancies. The position of the Fermi energy is determined by the charge distribution in the gap states. It has been suggested that the Fermi energy could be pinned by defect states, such as by two charge states of the same defect, which are separated by a Hubbard energy U and broadened by disorder (for the problem of gap states see, e.g., MOTT, 1980, and references therein).

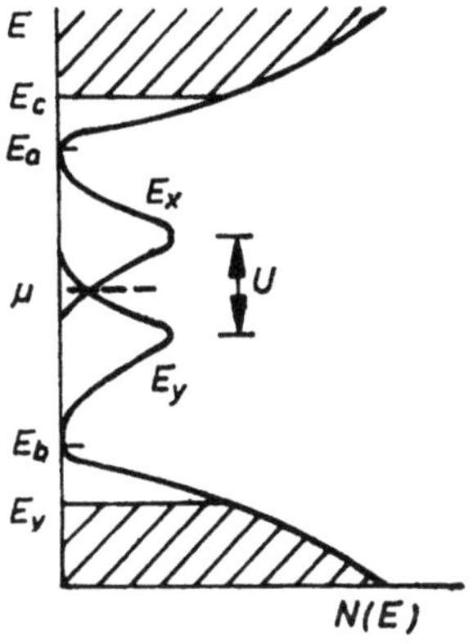

Fig. 4.6. Schematic representation of the density of states $N(E)$ for amorphous silicon (see text); the regions of extended states are shaded

Many glassy and amorphous semiconductors (e.g. chalkogenide glasses) exhibit diamagnetic behaviour, which indicates that spin-pairing must be taking place. ANDERSON (1975) postulated that this may be due to attractive phonon-induced electron-electron interaction, provided that this interaction is strong enough to overcome the repulsive Hubbard U. In this case the electrons experience a negative (attractive) Hubbard energy U_{eff} (cf. Section 2.2.). Of course, such U_{eff} would also effect the charge transport, because in this case the charge carriers are pairs of electrons (bipolarons).

The concept of a negative U_{eff} is now widely used to explain properties of materials with non-bonding (lone-pair) orbitals (e.g. chalkogenide glasses) (STREET and MOTT, 1975, see also MOTT, 1980, SHIMAKAWA, 1982) but also of other materials such as amorphous silicon (ELLIOTT, 1978) and amorphous SiO_2 (SHIMAKAWA and KONDO, 1983).

A first-principles theory of the electron states in non-crystalline solids should be able to explain all the features mentioned above. At present the theory is still far from this goal. So far, only some broad features of the density of states of non-crystalline materials can be theoretically understood. For more details on the behaviour of electrons in disordered solids the reader is, e.g., referred to MOTT and DAVIS (1979), ZIMAN (1979), YONEZAWA (1981), BONCH-BRUEVICH et al. (1981), LIFSHITZ et al. (1982).

4.2. Localization in the Anderson model

Anderson's method (ANDERSON, 1958, see also ANDERSON, 1978) for studying localized states is based on the Hamiltonian (4.1.1). Anderson defines localization as follows: At time $t = 0$ an electron is put on some site, say site 0. If localized states overlap with this site, the electron has a finite probability of being initially in each of these eigenstates, and there is a finite probability p_{00} of rediscovering the electron on this site at $t \to \infty$. On the other hand, if no localized states overlap with site 0, the electron will diffuse away and we have $p_{00} = 0$.

The time averaged probability of finding the electron on site 0 (i.e. in state $|0\rangle \equiv a_0^+ |\Psi_{\text{vac}}\rangle$; $|\Psi_{\text{vac}}\rangle$ — vacuum state) at $t \to \infty$, if it was at $t = 0$ on the same site, may be expressed as

$$p_{00} = \lim_{\varepsilon \to +0} \frac{\varepsilon}{\pi} \int_{-\infty}^{\infty} \mathrm{d}E\, G_0(E + \mathrm{i}\varepsilon)\, G_0(E - \mathrm{i}\varepsilon), \tag{4.2.1}$$

where $G_0(E)$ is the zero diagonal element of the Green function

$$G(E) = [E - \mathcal{H}]^{-1}, \tag{4.2.2}$$

i.e. $G_0(E) = \langle 0| [E - \mathcal{H}]^{-1} |0\rangle$. The expression (4.2.1) can be derived as follows: If the electron is at $t = 0$ in the state $|\Psi(0)\rangle$ and at time time t in the state $|\Psi(t)\rangle$ then the probability of finding the electron at t in the state $|\Psi(0)\rangle$ is

$$|\langle\Psi(t)|\, \Psi(0)\rangle|^2 = \sum_{\substack{nn' \\ (E_n \neq E_{n'})}} |\langle\Psi(0) \mid \Phi_n\rangle|^2\, |\langle\Psi(0) \mid \Phi_{n'}\rangle|^2$$

$$\times \cos \frac{1}{\hbar} (E_n - E_{n'})\, t + \sum_{\substack{nn' \\ (E_n = E_{n'})}} |\langle\Psi(0) \mid \Phi_n\rangle|^2\, |\langle\Psi(0) \mid \Phi_{n'}\rangle|^2, \tag{4.2.3}$$

where $|\Phi_n\rangle$ is the eigenstate of $\mathcal{H}$ and E_n is the corresponding eigenvalue. Taking the time average $\frac{1}{t} \int_0^\infty \mathrm{d}t \ldots$ (to avoid unnecessary oscillations at $t = \infty$) the first term on the right-hand side of (4.2.3) vanishes for $t \to \infty$. The second term may be rewritten as

$$\lim_{\varepsilon \to +0} \frac{\varepsilon}{\pi} \int_{-\infty}^{\infty} \mathrm{d}E \langle\Psi(0)|\, G(E + \mathrm{i}\varepsilon)\, |\Psi(0)\rangle\, \langle\Psi(0)|\, G(E - \mathrm{i}\varepsilon)\, |\Psi(0)\rangle. \tag{4.2.4}$$

This can be readily verified by expanding $|\Psi(0)\rangle$ with respect to $|\Phi_n\rangle$. For $|\Psi(0)\rangle = |0\rangle$, (4.2.4) agrees with (4.2.1).

The expression p_{00} (4.2.1) may be rewritten as

$$p_{00} = \int_{-\infty}^{\infty} dE\, f_0(E), \tag{4.2.5}$$

where

$$f_0(E) = N_0(E)\, Z_0(E) \tag{4.2.6}$$

with

$$Z_0(E) = \lim_{\varepsilon \to +0} \left[1 - \frac{\Delta_0(E + i\varepsilon) - \Delta_0(E - i\varepsilon)}{2i\varepsilon}\right]^{-1} \tag{4.2.7}$$

and

$$N_0(E) = \lim_{\varepsilon \to +0} \frac{i}{2\pi} [G_{00}(E + i\varepsilon) - G_{00}(E - i\varepsilon)]$$
$$= \sum_n |\langle 0| \Phi_n\rangle|^2\, \delta(E - E_n). \tag{4.2.8}$$

Here $\Delta_0(E)$ is the self-energy defined by

$$G_0(E) = [E - \varepsilon_0 - \Delta_0(E)]^{-1}, \tag{4.2.9}$$

and $N_0(E)$ is the (weighted) density of the states which overlap with the site 0. The second line in (4.2.8) was obtained by making use of

$$G(E) = \sum_n \frac{|\Phi_n\rangle \langle\Phi_n|}{E - E_n} \tag{4.2.10}$$

and of the identity

$$\frac{1}{E - E_n \pm i\varepsilon} = \frac{P}{E - E_n} \pm i\pi\delta(E - E_n) \tag{4.2.11}$$

(P — principal part, $\varepsilon \to +0$).

$f_0(E)$ is a positive quantity, since $G_0(E + i\varepsilon) = [G_0(E - i\varepsilon)]^*$. Hence, it may be viewed as the contribution of states with energy E to p_{00}. From (4.2.7) it follows that a branch cut in $\Delta_0(E)$ along the real axis corresponds to extended states, since in this case we have $Z_0(E) = 0$ and thus $f_0(E) = 0$, in spite of $N_0(E) \neq 0$. Localized states correspond to $Z_0(E) \neq 0$ and, of course, $N_0(E) \neq 0$. In the energy range of localized states, the imaginary part of $\Delta_0(E)$ tends to zero as the imaginary part of E tends to zero, except at a finite number of points where δ-function singularities appear. A vanishing of $Z_0(E)$ due to such a singularity implies nothing about localization or nonlocalization of states with energy E, because $N_0(E)$ vanishes, too. Consequently, an eigenstate with energy E is localized only if $\Delta_0(E)$ is analytic at E. Note that the basic difference in the behaviour of the imaginary part of $\Delta_0(E)$ for localized and extended states vanishes, if $\Delta_0(E)$ is configuration averaged, since then the δ-function singularities, mentioned before, will be broadened to continuous nonzero functions. Therefore, one has to evaluate the probability density function $\Delta_0(E)$ instead of taking the configuration average of it.

To study the analytic properties of $\Delta_0(E)$, we expand $\Delta_0(E)$ in a perturbation series and study its convergence. This procedure is based on the assumption that the states of energy E are localized, so that $\Delta_0(E)$ may be expanded in a series, since it is an analytic function. If the convergence of this series can be proved, the original assumption of localized states was correct, otherwise it was wrong and the states of energy E are extended.

To obtain a series expansion of $\Delta_0(E)$ in terms of known quantities, we use the equation for the Green function

$$(E - \varepsilon_m)\, G_{mm'} = \delta_{mm'} + \sum_{m''} J_{mm''} G_{m''m'}, \quad (J_{mm} = 0), \tag{4.2.12}$$

where

$$G_{mm'} = \langle m| \, [E - \mathcal{H}]^{-1} \, |m'\rangle \tag{4.2.13}$$

($|m\rangle \equiv a_m^+ \, |\Psi_{\text{vac}}\rangle$), and express (by iteration) on the right-hand side of this equation the nondiagonal elements of G by diagonal elements. In this manner, by (4.2.9), we obtain

$$\begin{aligned} \Delta_0(E) = {} & \sum_{m'} J_{mm'} g_{m'}(E)\, J_{m'm} \\ & + \sum_{m'm''} J_{mm'} g_{m'}(E)\, J_{m'm''} g_{m''}(E)\, J_{m''m} + \dots, \end{aligned} \tag{4.2.14}$$

where

$$g_m = [E - \varepsilon_m]^{-1}. \tag{4.2.15}$$

The series (4.2.14) may be represented by all paths starting from site 0 and returning to it only at the end. This series has two essential shortcomings: (i) spurious divergencies may arise from a repeated occurrence of sites connected with a small energy denominator; (ii) the series has not the correct analytic properties.

With regard to the latter shortcoming note that

$$G^r_{mm'} = G_{mm'} - G_{mr} G_{rm'} / G_{mm}, \tag{4.2.16}$$

where G^r denotes the Green function of a system with $\varepsilon_r \to \infty$. This relation can be obtained by replacing in (4.2.12) ε_m by $\varepsilon_m' = \varepsilon_m + (\varepsilon_r' - \varepsilon_r)\, \delta_{mr}$, by expressing $G(\dots \varepsilon_r')$ by $G(\dots \varepsilon_r \dots)$, finally by putting $\varepsilon_r' \to \infty$. Equation (4.2.16) shows that the zeros of G_0 (poles of Δ_0) are given by the poles of G^0_m (m is such that $G_{0m} \neq 0$). That is, the poles of Δ_0 do not coincide with the poles of the Green functions (4.2.15) (i.e. with the energies ε_m), as suggested by (4.2.14).

Both the above shortcomings can be removed by resuming the series (4.2.14) in such a way that no repetition of the same site on a path occurs. To realize this, let us rewrite (4.2.12) as

$$\sum_{m''} S_{mm''} G_{m''m'} = \delta_{mm'}, \tag{4.2.17}$$

where

$$S_{mm'} = (E - \varepsilon_m)\, \delta_{mm'} - J_{mm'}. \tag{4.2.18}$$

G_0 may be expressed as

$$G_0 = S^0/S, \tag{4.2.19}$$

where S is the determinant of the matrix **S**, and S^0 is the principal minor, which is obtained from **S** by deleting the 0th row and column.

From (4.2.19) and (4.2.9) we get

$$\Delta_0 = E - \varepsilon_0 - \frac{S}{S^0}. \tag{4.2.20}$$

Analogously, we may define

$$\Delta_{m_i}^{0m_1\ldots m_{i-1}} = E - \varepsilon_{m_i} - \frac{S^{0m_1\ldots m_{i-1}}}{S^{0m_1\ldots m_{i-1}m_i}} \tag{4.2.21}$$

and

$$G_{m_i}^{0m_1\ldots m_{i-1}} = [E - \varepsilon_{m_i} - \Delta_{m_i}^{0m_1\ldots m_{i-1}}]^{-1}, \tag{4.2.22}$$

where $S^{0m_1\ldots m_i}$ is the principal minor obtained from **S** by deleting the 0th, m_1th ..., and m_ith rows and columns. The determinant S may be written in terms of its principal minors (FUJITA and HORI, 1972)

$$S = S_{m_1m_1}S^{m_1} + \sum_{n=2}^{\infty} \sum_{m_2m_3\ldots m_n}' (-1)^{n-1} S_{m_1m_2}S_{m_2m_3} \cdots S_{m_nm_1}S^{m_1m_2m_3\ldots m_n}, \tag{4.2.23}$$

where the prime means that the sum must be taken over all sets of distinct indices $m_2, m_3, \ldots, m_n$, all of which are distinct from m_1. Inserting now (4.2.23) into (4.2.20) and using (4.2.21) and (4.2.22) we obtain the so-called renormalized perturbation expansion (RPE)

$$\Delta_0(E) = \sum_{n=1}^{\infty} \sum_{m_1\ldots m_n}' J_{0m_1}G_{m_1}^0(E)\, J_{m_1m_2}G_{m_2}^{0m_1}(E) \cdots J_{m_{n-1}m_n}G_{m_n}^{0m_1\ldots m_{n-1}}(E)\, J_{m_n0}. \tag{4.2.24}$$

Here the summation over n is over all sets of sites $\{m_1, \ldots, m_n\}$ which form self-avoiding paths starting and ending at site 0.

Analogous to (4.2.16), the poles of $G_{m_i}^{0m_1\ldots m_{i-1}}$ can be shown to agree with the zeros of $G_{m_{i+1}}^{0m_1\ldots m_i}$. Consequently, all poles in (4.2.24) cancel, except those of G_m^0. That is, $\Delta_0(E)$ from (4.2.24) exhibits the correct analytic behaviour. Repeated scattering from the same site, which leads in (4.2.14) to spurious divergencies, is eliminated. The price one pays is the introduction of an additive term $\Delta_{m_i}^{0m_1\ldots m_{i-1}}$ in each of the energy denominators. These terms may be expressed through relations similar to (4.2.24). Repeating this process of eliminating all the unknowns on the right-hand side of (4.2.24), we obtain an infinite continued fraction-like expression in each term of (4.2.24). The convergence of the whole RPE requires (i) the convergence of the latter iteration procedure for expressing the self-energies in the energy denominators and (ii) the convergence of the series (4.2.24) alone (that is, of the so-called renormalized perturbation series (RPS)). So far, studies of the convergence of the RPE are restricted to a study of either the convergence of the iteration procedure (ABOU CHACRA et al., 1973, HEINRICHS, 1977, cf. BREZINI and OLIVIER, 1983) or to the RPS (ANDERSON, 1958, ECONOMOU and COHEN, 1972, LICCIARDELLO and ECONOMOU, 1975, SOUKOULIS and ECONOMOU, 1980). Quite generally, most localization studies on the basis of (4.2.24) involve some assumptions, which cannot be proved rigorously.

By localization studies on the basis of (4.2.24) the following results have been obtained: for on-diagonal disorder (where in the Hamiltonian (4.1.1) the ε_m's are the random quantities) in one- and two-dimensional systems all states are localized; while in three-dimensional systems all states are localized, only if the spread of the energy levels ε_m exceeds some critical value ("Anderson transition"). Below this critical value, in the latter systems extended as well as localized states occur, as qualitatively shown in Fig. 4.2. Similar results have been obtained for purely off-diagonal disorder (in the Hamiltonian (4.1.1) the $J_{m'm}$'s are random). In particular, in a semiconductor all states in an impurity band are localized, if $\alpha\mathcal{N}^{-1/3}$ (α — reciprocal Bohr radius, $\mathcal{N}$ — concentration of impurities) exceeds some value, which equals 6.2 according to FERTIS et al. (1981) and 2.9 according to PURI and ODAGAKI (1981).

Note that Anderson's perturbation-in-J treatment of the localization problem is an approach from the limit of strong disorder, which gives rise to strongly localized states. In this limit localization is rather well understood. More difficult is the approach to the localization problem from the opposite limit of weak disorder by treating the disorder as a perturbation of free electrons (EDWARDS, 1970, ABRAM and EDWARDS, 1972, FREED, 1972, cf. Section 11.2.1.).

4.3. Scaling theory of localization

At large distance r the enveloppe of a localized wavefunction behaves as $\exp(-r/L)$ with the localization length L diverging as energy E approaches the mobility edge E_c from the localized regime (cf. Fig. 4.2). L is expected to diverge at E_c as

$$L \sim |E - E_c|^{-\nu} \tag{4.3.1}$$

(ν is a "critical" exponent), which reminds of the behaviour of the correlation length in a magnetic phase transition, where the correlation length diverges as the temperature T approaches the critical temperature T_c. This suggests us to use ideas of the theory of phase transitions for studying the localization problem.

The two basic ideas of the modern theory of phase transitions are universality and scaling (see also Section 5.2.2.). The principle of universality is based on the conception that the behaviour of a thermodynamic system near a phase transition (i.e. in the critical region) is determined by only the gross features of fluctuations whose size is large compared to the interparticle distance. Because of this, the critical behaviour should be independent of short-range properties such as lattice structure, but only dependent on the spatial dimension. A change of the length scale (by partitioning the lattice into cells and viewing it from a perspective in which these cells are renormalized sites) should not affect the gross features of the large size fluctuations. By establishing a relation (scaling relation) between the corresponding problems with different scales of length, the critical behaviour (critical exponents) can be studied.

The partitioning procedure just described is the basis of the so-called real-space (or position-space) renormalization-group (RG) method. On applying this method to the Anderson localization problem, here we follow DOMANY and SARKER (1979) (see also SARKER and DOMANY, 1981, and references therein). The system is described by the Anderson Hamiltonian (4.1.1)

$$\mathcal{H} = \sum_{m} \varepsilon_m a_m^+ a_m + \sum_{\langle m,m'\rangle} a_{m'}^+ a_m, \tag{4.3.2}$$

where $\langle m, m'\rangle$ are nearest-neighbour sites in a d-dimensional hypercubic lattice, the resonance integral $J_{m'm}$ between sites $\langle m', m\rangle$ is chosen to be a constant $J_{m'm} = 1$ (i.e. the energies are in units of J), and the random energies ε_m have a Gaussian distribution with a width Γ and zero average.

We replace the infinite lattice by a finite one and write the Hamiltonian (4.3.2) as a sum of two parts

$$\mathcal{H} = \mathcal{H}_0 + \mathcal{H}_1. \tag{4.3.3}$$

The lattice is broken into cells, and in $\mathcal{H}_0$ are included all site-diagonal terms and off-diagonal terms of (4.3.2) which connect sites m, m' within the same cell. The remaining off-diagonal terms are included in $\mathcal{H}_1$. The eigenvalues $\mathcal{E}_\alpha$ and normalized eigenstates $|\Phi_\alpha\rangle$ of $\mathcal{H}_0$ are found by exact diagonalization, with ε_m taken randomly. They are assumed to be known.

If a cell contains b^d sites, then there are b^d states per cell. We are interested in localization properties of a state with a pre-chosen energy E (near the band center $E = 0$). From each cell we pick out one state, the state whose energy is closest to E. These states are included in the subspace D of the Hilbert space in which $\mathcal{H}$ operates. Using a method based on the Brillouin-Wigner perturbation theory (see, e.g., THOULESS, 1972) the Hamiltonian may be mapped onto a new Hamiltonian $\mathcal{H}'$ characterized by a new width Γ' of the distribution of the random site energies.

From $(\mathcal{H} - E)\,|\Psi\rangle = 0$, $(\mathcal{H}_0 - \mathcal{E}_\alpha)\,|\Psi_\alpha\rangle = 0$ and $|\Psi\rangle = \sum_\alpha a_\alpha\,|\Phi_\alpha\rangle$ we get

$$(E - \mathcal{E}_\alpha)\, a_\alpha = \langle\Phi_\alpha|\, \mathcal{H}_1\, |\Psi\rangle. \tag{4.3.4}$$

Introducing the projection of $|\Psi\rangle$ onto D, $|\Psi_D\rangle = P_D\,|\Psi\rangle$, where $P_D = \sum_{\alpha\in D} |\Phi_\alpha\rangle\langle\Phi_\alpha|$, and the Green function

$$G = \sum_{\alpha\notin D} (E - \mathcal{E}_\alpha)^{-1}\, |\Phi_\alpha\rangle\langle\Phi_\alpha| = (E - \mathcal{H}_0)^{-1}\,(1 - P_D),$$

then by means of (4.3.4) we get

$$|\Psi\rangle = |\Psi_D\rangle + G\mathcal{H}_1\,|\Psi\rangle. \tag{4.3.5}$$

Defining the operator $\hat{\Omega}$ by $|\Psi\rangle = \hat{\Omega}\,|\Psi_D\rangle$, and V_D by $V_D = \mathcal{H}_1\hat{\Omega}$, then with the aid of (4.3.5) we find that $\hat{\Omega} = 1 + G\mathcal{H}_1\hat{\Omega}$ and therefore

$$V_D = \mathcal{H}_1 + \mathcal{H}_1 G V_D. \tag{4.3.6}$$

Thus, because of $V_D |\Psi_D\rangle = \mathcal{H}_1 |\Psi\rangle$, we have

$$(E - \mathcal{E}_\alpha)\, a_\alpha = \langle \Phi_\alpha | \, V_D \, |\Psi_D\rangle,$$

which implies that (cf. (4.3.4))

$$\mathcal{H}_D |\Psi_D\rangle = E |\Psi_D\rangle, \tag{4.3.7}$$

where

$$\mathcal{H}_D = \mathcal{H}_0 + V_D. \tag{4.3.8}$$

So far, the procedure has been rigorous. To make it practicable, we expand V_D in powers of $\mathcal{H}_1$. If we account for the first-order term in $\mathcal{H}_1$ only, i.e. if $V_D \approx \mathcal{H}_1$, we obtain

$$\mathcal{H}_D = \sum_\alpha \mathcal{E}_\alpha a_\alpha^+ a_\alpha + \sum_{\langle \alpha, \alpha' \rangle} V_{\alpha\alpha'} a_\alpha^+ a_{\alpha'}, \tag{4.3.9}$$

where $\langle \alpha, \alpha' \rangle$ are nearest neighbour cell indices and $V_{\alpha\alpha'} = \langle \alpha | \, \mathcal{H}_1 \, |\alpha'\rangle$. Note that in the new Hamiltonian $\mathcal{H}_D$ besides the diagonal elements, $\mathcal{E}_\alpha$, the off-diagonal elements, $V_{\alpha\alpha'}$, are random, too. The latter randomness is belived to be unimportant for states near the band centre (cf. Fertis et al., 1981, and Puri and Odagaki, 1981) and is ignored here, i.e. in (4.3.9) $V_{\alpha\alpha'}$ is replaced by a uniform V_{eff} (taken as average of the modulus of $V_{\alpha\alpha'}$). To put $\mathcal{H}_D$ (4.3.9) in the form of the Hamiltonian (4.3.2), we divide (4.3.9) by V_{eff}. Furthermore, we have to shift the origin of energy by $\langle \mathcal{E}_\alpha \rangle$, the average value of $\mathcal{E}_\alpha$. For general E, we have $\langle \mathcal{E}_\alpha \rangle \neq 0$. But in particular for $E = 0$, we get $\langle \mathcal{E}_\alpha \rangle = 0$, because for a symmetric distribution of site energies, the energy E maps onto E', and $-E$ onto $-E'$, and thus $E = 0$ onto $E' = 0$. Hence, our RG transformation maps the eigenvalue problem $\mathcal{H} |\Psi\rangle = E |\Psi\rangle$ onto a new one, $\mathcal{H}' |\Psi_D\rangle = E' |\Psi_D\rangle$, such that all lengths are rescaled by a factor $b > 1$. The new width Γ' o the random site energy distribution and the new eigenvalue E' respectively are given by the recursion relations

$$\Gamma' = (\langle \mathcal{E}_\alpha^2 \rangle - \langle \mathcal{E}_\alpha \rangle^2)^{1/2} / V_{\text{eff}}, \tag{4.3.10}$$

$$E' = (E - \langle \mathcal{E}_\alpha \rangle) / V_{\text{eff}}, \tag{4.3.11}$$

and the scaling equation for the localization length reads

$$L' = L(\Gamma', E') = b^{-1} L(\Gamma, E). \tag{4.3.12}$$

For $\Gamma = \infty$, we have $L = 0$, and thus $L' = 0$, which implies that $\Gamma' = \infty$. Furthermore, for $\Gamma = 0$ we have $\Gamma' = 0$, since the RG transformation cannot produce any randomness. Hence, in the (Γ, E)-plane the points $(\infty, 0)$ and $(0, 0)$ are fixed points, i.e. each of these points is mapped onto itself by the RG transformation. According to (4.3.12), a fixed point implies either $L = L' = 0$ (point $(\infty, 0)$, see below) or $L = L' = \infty$ (point $(0, 0)$, see below).

The line $E = 0$ (in the (Γ, E)-plane) is invariant under RG transformation, since (4.3.11) yields $E' = E = 0$. On this line it is sufficient to study solely the recursion relation (4.3.10), which yields $\Gamma' = f(\Gamma)$. The latter relation describes how Γ scales under the RG transformation. As shown in Fig. 4.7 this relation has an unstable fixed point (for $d = 3$), which corresponds to a critical randomness Γ_c (point $(\Gamma_c, 0)$) above which the system flows by iteration to the localized ($\Gamma \to \infty$) regime (large randomness), while for $\Gamma < \Gamma_c$ it flows to the extended ($\Gamma \to 0$) regime (small randomness). Since repeated iteration decreases the localization length (according to (4.3.12), $L' = b^{-1}L < L$), for $\Gamma < \Gamma_c$ (scaling toward the point of absent disorder) L cannot be finite, so that the state is extended. On the contrary, for $\Gamma > \Gamma_c$ (scaling toward infinite disorder) L is finite, i.e. the state is localized. Consequently, Γ_c represents the critical randomness of the Anderson transition (all states localized). For more details on localization studies by the method at hand, the reader is referred to SAKER and DOMANY (1981). Note that according to HILLEBRAND and NIEMINEN (1981) the results of this method depend crucially on the truncation of the perturbation series for V_D (cf. (4.3.6)).

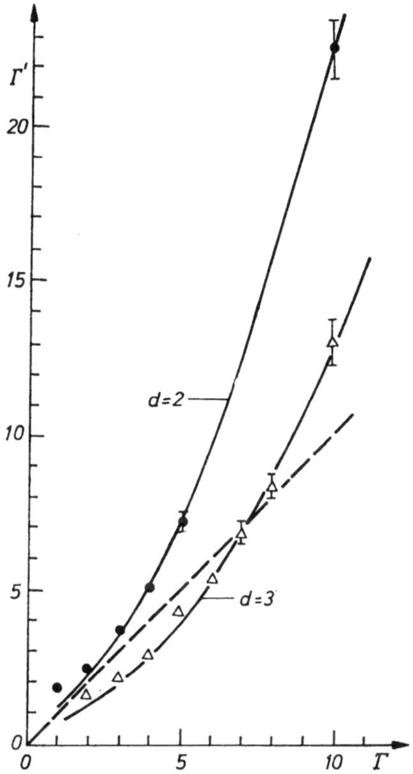

Fig. 4.7. Γ' versus Γ in two and three dimensions for $E = 0$. The dashed line denotes $\Gamma' = \Gamma$. In two dimensions, no fixed point is found, while in three dimensions, there is a fixed point at $\Gamma_c \approx 7.0$ (after SARKER and DOMANY, 1981)

Now let us turn to the scaling theory of localization of ABRAHAMS et al. (1979), which uses the previous idea of THOULESS (1974) that the ratio of ΔE, the sensitivity of energy levels to boundary conditions, to $\mathcal{W}$, the energy level separation, may be taken as localization criterion.

For an exponentially localized state the energy change ΔE should be exponentially small, when the boundary conditions are changed (from periodic to antiperiodic, say). On the other hand for extended states this change should be much larger, roughly of the order of the level spacing. If the state is localized, $\Delta E/\mathscr{W}$ decreases as the size of the system is increased, but it does not decrease, if the state is extended.

For extended states, $\Delta E/\mathscr{W}$ may be shown to be related to the conductance (cf. THOULESS, 1980, 1982) as follows: Consider an electron wavepacket, initially located at the center of a system of linear dimension l, with width many-free paths across but much less than l. It takes time τ of order l^2/D (D — diffusion constant) to reach the boundary. After this time the motion of the wavepacket depends strongly on the boundary conditions. The quantity τ is related to ΔE by

$$\Delta E \tau \approx \hbar . \tag{4.3.13}$$

Using the Einstein relation for the conductivity, $\sigma = e^2 D \mathscr{N}(\mu) = e^2 D/\mathscr{W} l^d$ ($\mathscr{N}(\mu)$ — density of states at the Fermi energy, l^d-volume of a d-dimensional hypercube of side l), we find the conductance

$$G = \sigma l^{d-2} = (e^2/\hbar)\,(\Delta E/\mathscr{W}) . \tag{4.3.14}$$

In the theory of ABRAHAMS et al. (1979) the scaling argument is as follows. For a given energy, we define a generalized dimensionless conductance

$$g(l) = \Delta E(l)/\mathscr{W}(l) . \tag{4.3.15}$$

Now we imagine fitting b^d hypercubes together to form a larger block of size bl. ABRAHAMS et al. (1979) assume that the relevant quantity determining the new g is the old one. From this assumption it follows that there is a RG equation for $g(bl)$ given by

$$g(bl) = f\big(g(l), b\big) , \tag{4.3.16}$$

i.e. $g(bl)$ depends on l only through the dependence of $g(l)$. A differential recursion relation for g can be obtained, by putting in (4.3.16) $b = 1 + \varepsilon$, where ε is infinitesimal, and expanding both sides of (4.3.16) about $b = 1$. Then

$$\frac{\mathrm{d} \ln g(l)}{\mathrm{d} \ln l} = \frac{1}{g(l)} \left.\frac{\partial f(g, b)}{\partial b}\right|_{b=1} \equiv \beta[g(l)] \tag{4.3.17}$$

is the scaling equation for $g(l)$.

Let us consider the scaling function $\beta(g)$. For small g (localized limit), where $g \sim \mathrm{e}^{-l/L}$, we have

$$\lim_{g \to 0} \beta(g) = \ln g , \tag{4.3.18}$$

and for large g (extended limit), where (cf. (4.3.18)) $g \sim l^{d-2}$, we have

$$\lim_{g \to \infty} \beta(g) = d - 2 . \tag{4.3.19}$$

ABRAHAMS et al. (1979) assume that β for all g is smoothly and monotonically connecting (4.3.18), (4.3.19) (see Fig. 4.8). At $d = 3$, there is a fixed point g_c (mobility edge) of (4.3.17) determined by $\beta(g_c) = 0$. This fixed point is unstable, since by starting at the microscopic scale l_0 at an energy E such that $g_E(l_0) \equiv g_0 > g_c$, the system scales to larger values g at large l (metallic behaviour), while for $g_0 < g_c$ it scales to small values g (exponential localization). For $d = 2$, has no zero and is always negative, i.e. for $d = 2$ (and also $d = 1$), all states are localized.

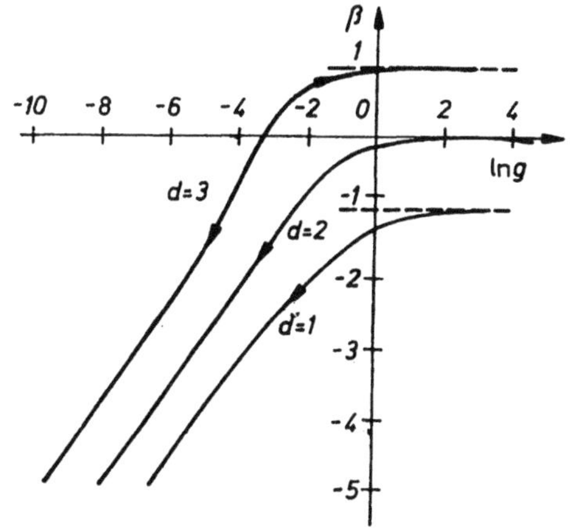

Fig. 4.8. The scaling function $\beta(g(l)) = d \ln (g/d) \times \ln l$ versus $\ln g$ is shown for dimensions $d = 1$, 2, 3 (after VOLLHARDT and WÖLFLE, 1982)

For $d = 3$ the behaviour about the mobility edge can be studied by linearizing $\beta(g)$ near g_c and integrating from a microscopic l_0, g_0. Then

$$\beta = \mathrm{d} \ln g/\mathrm{d} \ln l \approx (1/\nu g_c)\,(g - g_c), \tag{4.3.20}$$

where $(\nu g_c)^{-1}$ denotes slope of β at g_c. Thus, we have

$$\mathrm{d}g/(g - g_c) = (1/\nu)\, \mathrm{d} \ln l, \tag{4.3.21}$$

and by integrating (4.3.21) from g_0, l_0 we find that

$$l/l_0 = \big((g - g_c)/(g_0 - g_c)\big)^{\nu}. \tag{4.3.22}$$

First, consider the regime of extended states ($g_0 > g_c$). For size l large compared to the mean-free path a conductivity independent of l may be defined, which (according to (4.3.14), (4.3.15)) is given by

$$\sigma = \frac{e^2}{\hbar}\, g(l)\, l^{2-d}. \tag{4.3.23}$$

From (4.3.22) and (4.3.23), we get (with $g_c < g$)

$$\sigma \approx A(e^2/\hbar)\, l_0^{2-d} \varepsilon^{(d-2)\nu}, \tag{4.3.24}$$

where A is of order unity and $\varepsilon = (g_0 - g_c)/g_c \sim (E - E_c)/E_c$ measures the energy to the mobility edge (note that $g_0 = g_E(l_0)$, $g_c = g_{E_c}(l_0)$). Hence, according to (4.3.24), for $d = 3$, the conductivity goes continuously to zero at the mobility edge E_c, i.e. σ does not exhibit a "minimum-metallic conductivity" (cf. Sections 4.1. and 11.2.1.).

Now consider the localized regime ($g_0 < g_c$). From (4.3.22) we obtain (with $g \ll g_c$)

$$g \approx \frac{g_c}{\nu} \left(1 - \frac{l}{l_0} |\varepsilon|^\nu\right) \approx \frac{g_c}{\nu} \, e^{-\frac{l}{l_0}|\varepsilon|^\nu}, \tag{4.3.25}$$

i.e. the localization length L becomes (cf. (4.3.1))

$$L = l_0 \, |\varepsilon|^{-\nu}, \tag{4.3.26}$$

so that ν is also the exponent of the localization length.

The scaling function $\beta(g)$ has been calculated numerically and analytically (see, e.g., LEE and FISHER, 1981, and VOLLHARDT and WÖLFLE, 1982; cf. Fig. 4.8). Note that there are also doubts about the validity of the scaling approach of ABRAHAMS et al. (1979), which is based on the assumption that a one-parameter scaling function (cf. (4.3.16)) exists (cf. SARKER, 1983, MAC KINNON, 1983). There are conflicting theoretical results, and an experimental test (DAVIES et al., 1983, cf. SOONPAA and SCHWALM, 1984) of this approach for two-dimensional transport in silicon inversion layers has been found to be inconsistent with a one-parameter scaling assumption. For further aspects of localization, in particular, of the scaling theory of localization, the reader is referred to ANDERSON et al. (1980), ANDERSON (1981), and to the monograph on Anderson localization edited by NAGAÓKA and FUKUYAMA (1982), see also Sections 5.4. and 11.2.1.

5. DC hopping conductivity

5.1. Basic formulae and the problem of averaging the random-resistor network

5.1.1. *Formulation of the problem*

To study the dc (direct current) hopping conductivity in disordered systems in the presence of a weak electrical field, we use the $\omega = 0$ limit of the rate equation (2.5.12) and the expression for the current (2.5.14). Note that the rate equation was obtained by neglecting the off-diagonal elements in the equation of motion of the density matrix, which implies the assumption of strong localization, i.e. localization on atomic sites (impurity sites in the case of weakly doped semiconductors).

The two basic equations for the dc hopping transport may be written as

$$\sum_{m'} (U_{m'} - U_m) \, Z_{m'm}^{-1} = 0, \tag{5.1.1}$$

$$\boldsymbol{j} = \frac{1}{2e\Omega} \sum_{mm'} (\boldsymbol{R}_m - \boldsymbol{R}_{m'}) \, (U_{m'} - U_m) \, Z_{m'm}^{-1}. \tag{5.1.2}$$

Here

$$Z_{mm'} = (e^2 \beta \Gamma_{mm'})^{-1} \tag{5.1.3}$$

is a quantity with dimension of a resistance, $\Gamma_{mm'}$ is the symmetrized hopping probability between sites m and m', $\beta = 1/kT$, $\boldsymbol{j}$ is the current density, Ω is the volume of the system and $\boldsymbol{R}_m$ denotes the position of the mth site. U_m is the internal (electrical) potential at the site m. According to (2.5.13), eU_m is additively composed of the potential of the external field, the change of the chemical potential (due to the charge redistribution by the external field) and, strictly speaking, the self-consistent potential caused by all other electrons. As mentioned in Section 2.5.2., the expression $eU_m + \mu$ (μ — chemical potential) may be viewed as the electrochemical potential at site m.

Note that for an ordered system (small-polaron model), equation (5.1.2) suffices for calculating the current, since (owing to the lattice periodicity) in this case, U_m is simply given by $-\boldsymbol{E}\boldsymbol{R}_m$, the potential of the external field (cf. (2.5.10)).

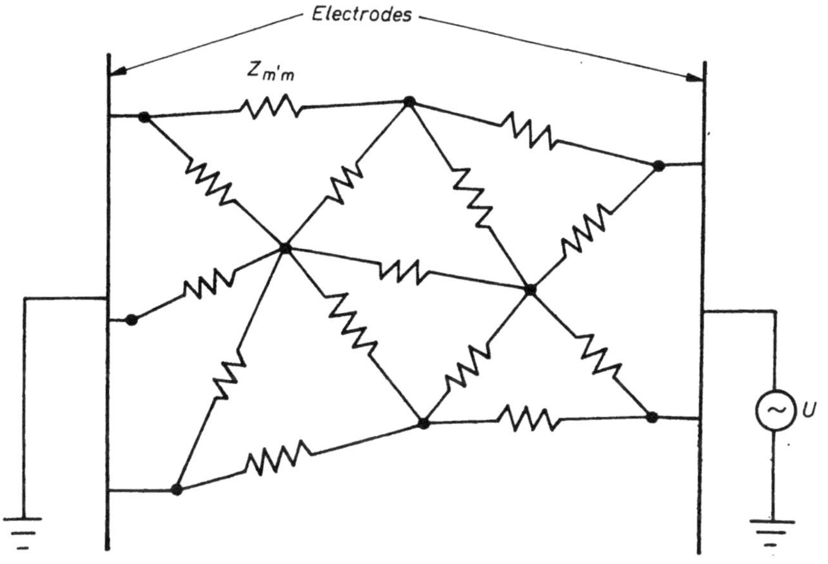

Fig. 5.1. Schematic representation of the random-resistor network of Miller and Abrahams (1960) ($Z_{m'm}$ — resistance between sites m' and m)

On considering $Z_{m'm}$ as resistance between the sites m' and m and U_m as voltage on site m, the problem of calculating the current by equations (5.1.1) and (5.1.2) becomes equivalent to that of calculating the current in a random-resistor network (Miller and Abrahams, 1960) (see Fig. 5.1). Equation (5.1.1) represents Kirchhoff's law for site m, and (5.1.2) describes the current through the boundaries (electrodes) of the sample (i.e. (5.1.2) may be replaced by Kirchhoff's law for the external circuit). The latter fact follows from an inspection of (5.1.2) by taking into account that (5.1.1) holds only for internal sites of the specimen, but not for sites at the electrodes (cf. Section 2.5.2.).

The network problem in question may be studied either numerically (by computer simulation of systems with some hundreds or thousands sites) or analytically. Owing to the large number of sites in actual systems (of order 10^{20}), analytical methods include a configuration-averaging procedure. Hence, in disordered systems a calculation of the dc hopping conductivity requires determination of the hopping probability and performing a configuration average, while in ordered systems there is no need for averaging.

5.1.2. *Expressions for the dc hopping probability*

According to (2.5.15), (2.5.11), and (2.4.19), the hopping probability, or the resistance $Z_{mm'}$, between sites m and m' is given by

$$Z_{mm'}^{-1} = e^2\beta f_{m'}(1 - f_m) \frac{1}{\hbar^2} |J_{m'm}|^2 \exp\left[\frac{\beta}{2} \varepsilon_{m'm} - 2S_T(m \quad m)\right]$$

$$\times \int_{-\infty}^{\infty} dt\, e^{-\frac{i}{\hbar}\varepsilon_{m'm}t} \left\{\exp\left[\sum_{q} \frac{|\gamma(q)|^2 [1 - \cos q(R_{m'} - R_m)]}{N \sinh(\hbar\omega_q\beta/2)} \cos \omega_q t\right] - 1\right\}. \tag{5.1.4}$$

Here $f_m = [e^{\beta(\varepsilon_m - \mu)} + 1]^{-1}$ is the Fermi function, $J_{m'm}$ the resonance integral between sites m' and m, ε_m the energy of site m, $\varepsilon_{m'm} = \varepsilon_{m'} - \varepsilon_{m'}$, ω_q the frequency of the phonon with wave vector q (the branch index is droped), $\gamma(q)$ the electron-phonon coupling constant and S_T is defined in (A 1.18).

For weak interaction with phonons ($|\gamma|^2 \ll 1$) and only one-phonon processes, by expanding the exponential in (5.1.4) and retaining the first two terms, we find that

$$Z_{mm'}^{-1} = Z_0^{-1} e^{-2\alpha|R_{m'm}|}/8 \cosh\left[\frac{\beta}{2}(\varepsilon_{m'} - \mu)\right] \cosh\left[\frac{\beta}{2}(\varepsilon_m - \mu)\right] \sinh\left[\frac{\beta}{2}|\varepsilon_{m'm}|\right], \tag{5.1.5}$$

where

$$Z_0^{-1} = \frac{2\pi e^2\beta J_0^2}{N\hbar^2} \sum_q |\gamma(q)|^2 [1 - \cos q(R_{m'} - R_m)]\, \delta(\omega_q - |\varepsilon_{m'm}|/\hbar). \tag{5.1.6}$$

Here we have used $J_{mm'} = J_0 \exp(-\alpha |R_{m'm}|)$, $(R_{mm'} = R_m - R_{m'})$, which implies the assumption of s-like "impurity" wavefunctions with Bohr radius α^{-1}.

For low temperatures ($kT \ll |\varepsilon_{m'} - \mu|$, $|\varepsilon_m - \varepsilon_\mu|$, $|\varepsilon_{m'} - \varepsilon_m|$), the expression (5.1.5) becomes (Ambegaokar et al., 1971)

$$Z_{mm'}^{-1} = Z_0^{-1}\, e^{-2\alpha|R_{m'm}| - \frac{\beta}{2}(|\varepsilon_{m'} - \mu| + |\varepsilon_m - \mu| + |\varepsilon_{m'} - \varepsilon_m|)}. \tag{5.1.7}$$

In principle, Z_0 depends on the configuration, too. However, the dependence of Z_0 on $R_{mm'}$ and $\varepsilon_{mm'}$ is weak (as powers of these quantities) and may in general be ignored.

Note that also for strong electron-phonon interaction (i.e. in presence of polaron effects) at low temperatures the expression (5.1.4) may be brought into the form

(5.1.7). To this end, we must expand the exponential in (5.1.4) in a power series. At low temperatures, it suffices to account for phonon emission processes only cf. EMIN, 1974, which renders it possible to sum up this series again. The result of such a procedure is (5.1.7), with Z_0 given by

$$Z_0^{-1} = e^{2\beta} \frac{J_0^2}{\hbar^2} e^{-S_0(m',m)} \int_{-\infty}^{\infty} dt \, e^{-\frac{1}{\hbar}|\varepsilon_{m'm}|t} \times \left[\exp\left\{\sum_q \frac{1}{N} |\gamma(q)|^2 [1 - \cos q(R_{m'} - R_m)] e^{i\omega_q t}\right\} - 1\right]. \quad (5.1.8)$$

Here $S_0(m, m') = S_T(m, m')|_{T=0K}$. It is not hard to see that Z_0 in (5.1.8) depends only weakly on $R_{mm'}$, if $|R_{mm'}| \gg a$ (a — lattice constant of the underlying host lattice) and just so on $\varepsilon_{mm'}$, if $|\varepsilon_{mm'}|/\hbar\omega_D \ll 1$ (ω_D — Debye frequency), provided that the parameter $|\gamma|^2$ is not too large. Consequently, at low temperatures, also in presence of strong interaction with phonons, the resistance $Z_{mm'}$ in the network of Miller and Abrahams has the form (5.1.7).

5.1.3. The network of Miller and Abrahams

The problem of averaging the random-resistor network was first attacked by MILLER and ABRAHAMS (1960), who suggested that the overall resistance of the network is governed by certain one-dimensional critical resistor chains connecting both ends (electrodes) of the network. Furthermore, they assumed that hops occur only between nearest-neighbour sites and that the optimal current path is determined solely by the mutual position of sites (i.e. by the quantities $R_{mm'}$), but not by the energies of the sites. That is, they assumed that the spread of the resistances arises from the factor $\exp(-2\alpha |R_{mm'}|)$ in (5.1.7). The average conductivity σ may then be written as

$$\sigma = \varrho_3^{-1} e^{-\varepsilon_3/kT}, \quad \varrho_3 = \varrho_0 e^{f(\mathcal{N})}. \quad (5.1.9)$$

Here $\mathcal{N}$ is the concentration of sites (impurities) involved in hopping transport, ϱ_0 a quantity with dimension of a specific resistance, which only weakly depends on $\mathcal{N}$ and kT (as powers of these quantities), and ε_3 a characteristic activation energy.

The quantity $f(\mathcal{N})$ depends on the choice of the critical current paths. Miller and Abrahams assumed these paths to be formed by going from site to site via the smallest resistance in the forward direction. However, as observed by many authors (AMBEGAOKAR et al., 1971, SHKLOVSKII and EFROS, 1971, POLLAK, 1972, etc.) such an assumption does not yield the most effective current paths. The paths of Miller and Abrahams pass across regions where the nearest-neighbour distance appreciably exceeds the average one (cf. Fig. 5.2), which leads to $f(\mathcal{N}) = c\alpha^{3/2}\mathcal{N}^{-1/2}$, where c is some numerical factor. However, such "hard" hops are relatively seldomly encountered in the network and may be by-passed by the current (cf. Fig. 5.2).

It can readily be seen that the expression $f(\mathcal{N})$ of Miller and Abrahams corresponds to that obtained by averaging the resistance. To this end, let us consider a hard hop connected with a spherical cavity of lacking sites. The probability for such a cavity with radius R is given by the Poisson distribution and is equal to $\exp\left(-\frac{4\pi}{3}\mathcal{N}R^3\right)$. The hard hop determines the overall resistance of a resistor chain, which is therefore proportional to

$$\int \mathrm{d}R\, \mathrm{e}^{2\alpha R}\, \mathrm{e}^{-\frac{4\pi}{3}\mathcal{N}R^3} \tag{5.1.10}$$

The product of the exponentials in (5.1.10) has a sharp maximum at $R_m = (\alpha/2\pi\mathcal{N})^{1/2}$, which determines the resistance of a typical chain, so that the overall resistance becomes proportional to $\exp(2\alpha R_m)$, i.e. $f(\mathcal{N}) = \sqrt{\frac{2}{\pi}}\,\alpha^{3/2}\mathcal{N}^{-1/2}$.

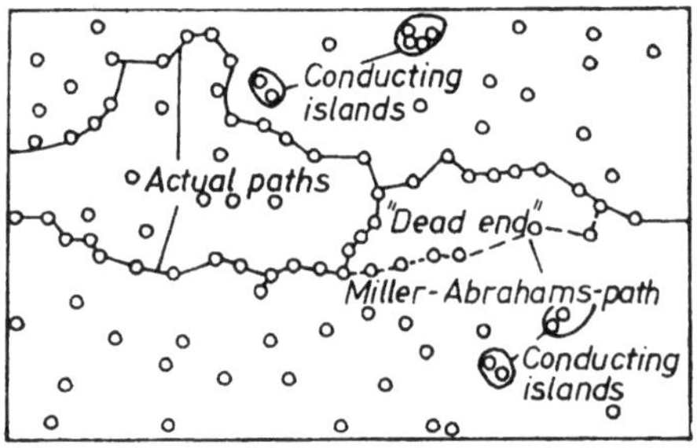

Fig. 5.2. Schematic representation of the actual current paths in a hopping system and of the current path according to Miller and Abrahams, which does not by-pass "dead ends". The conducting islands do not contribute to the dc conductivity, but they play an important role for the ac conductivity (after Brenig et al., 1973b)

Besides pairs of sites with large spacing, pairs with small spacing are likewise unimportant for the dc hopping conductivity, since the latter pairs cannot form macroscopic current paths. Consequently, the major contribution to the overall resistance of the network arises from pairs of sites with a spacing of the order of the mean one, i.e.

$$f(\mathcal{N}) = \eta\alpha\mathcal{N}^{-1/3}, \tag{5.1.11}$$

where η is a number of order unity. An expression like (5.1.11) was first derived by Twose (see Mott and Twose, 1961) by means of an effective-medium method, who found that $\eta = 0.9$.

The percolation theory represents a very effective method for studying the current flow through the network of Miller and Abrahams. By this method the current through the network can be shown to be governed by chains with resistors between neighbouring sites with spacing of the order of the average one. In particular, the number η in (5.1.11) can be found by this method. The percolation theory and its application to the dc hopping problem will be the subject of Section 6.

Let us now turn to the temperature dependence occurring in (5.1.9). For a density of (localized) states with a sharp maximum, such a dependence can readily be obtained. In this case hopping occurs either near the Fermi level or near the maximum of the density of states. Hopping near the Fermi level dominates at very low temperatures and will be discussed in the next section. Hopping near the maximum of the density of states, ε_0, dominates at higher temperatures. In the latter case we may put $\varepsilon_m \approx \varepsilon_{m'} \approx \varepsilon_0$ in (5.1.7), so that the temperature exponent in (5.1.9) becomes $\exp(-\beta |\varepsilon_0 - \mu|)$, or

$$\varepsilon_3 = |\mu - \varepsilon_0| . \tag{5.1.12}$$

The situation in question is illustrated in Fig. 5.3, where Fig. 5.3a) corresponds, for instance, to weakly compensated semiconductors and Fig. 5.3b) to strongly compensated ones.

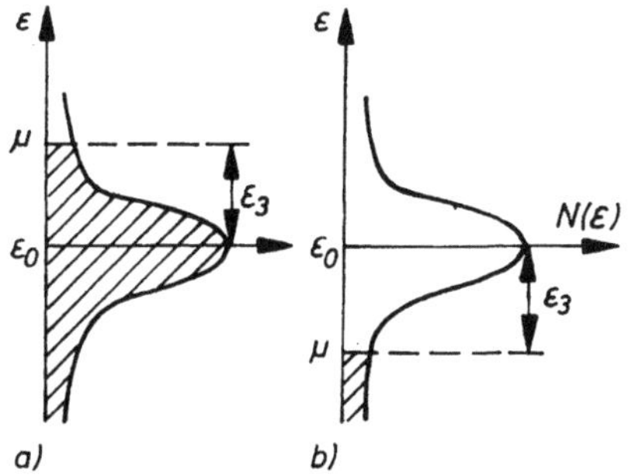

Fig. 5.3. Schematic representation of the density of states, $N(\varepsilon)$, and the position of the Fermi level, μ, in a) weakly compensated and b) strongly compensated semiconductors. The regions occupied by electrons are shaded

For weakly doped semiconductors, ε_3 has been investigated in detail. Such an investigation was first performed by MOTT (1956) for the case weak compensation by considering ionized donors (energy levels placed above the Fermi level) in the vicinity of negatively charged acceptors. The Coulomb interaction between a donor and an acceptor raises the energy level of the donor relative to the level of an isolated donor by an amount of order $e^2\mathcal{N}_D^{1/3}/\varkappa$ ($\varkappa$ — dielectric constant, $\mathcal{N}_D$ — donor concentration). Assuming coincidence of the maximum of the density of states with the energy of an isolated donor, one obtains

$$\varepsilon_3 = Ce^2\mathcal{N}_D^{1/3}/\varkappa , \tag{5.1.13}$$

where C is a number of order unity. A formula of the type (5.1.13) holds for any compensation $K = \mathcal{N}_A/\mathcal{N}_D$ ($\mathcal{N}_A$ — acceptor concentration), but C depends on K. This is due to the fact that ε_3 is governed by the Coulomb interaction either between donors and acceptors or between donors. The analytical expression for C at weak compensation ($K \ll 1$) reads (EFROS et al., 1972)

$$C = 0.99(1 - 0.3K^{1/4}) \tag{5.1.14}$$

and at strong compensation $(1 - K) \ll 1$ (SHKLOVSKII and EFROS, 1971)

$$C = \nu/(1 - K)^{1/3} , \tag{5.1.15}$$

where ν is a number of order unity. For more details on calculations of ε_3 see, e.g., SHKLOVSKII and EFROS, 1979 (Chapter 9).

Concluding this section it should be noted that an activated temperature dependence of $\sigma(T)$ may be absent. The conditions for its presence are $\varepsilon_3 > kT$ (in the opposite case the temperature dependence of the pre-exponential factor may play the decisive role) and $\Gamma < kT$, i.e. it is necessary that $\varepsilon_3 > \Gamma$.

5.1.4. *Mott's law for the variable-range hopping regime*

The expression (5.1.1) for the conductivity is based on the assumption of nearest-neighbour hops. This is a reasonable assumption, if the temperature exponent in (5.1.7) is unimportant. Hence, (5.1.9) holds if $\Gamma/kT \ll \alpha\mathcal{N}^{-1/3}$ (MOTT and DAVIS, 1979). At low temperatures, however, this exponent plays an important role. As suggested by MOTT (1968, 1969), in this case an electron prefers to hop to a more remote site than to the nearest-neighbour one, in order to reduce the energy required for the hop ("variable-range hopping"). For hops near the Fermi level, where $\varepsilon_{m'} - \mu < 0$, $\varepsilon_m - \mu > 0$ (or vice versa), (5.1.7) becomes

$$Z_{mm'}^{-1} = Z_0^{-1}\, \mathrm{e}^{-2\alpha|\boldsymbol{R}_{mm'}|}\, \mathrm{e}^{-\beta|\varepsilon_{m'm}|}. \tag{5.1.16}$$

According to Mott, in this situation the conductivity is determined by hops giving rise to a minimal exponent in (5.1.16). The minimization is performed with the aid of the relation

$$\frac{4\pi}{3} R^3 \int\limits_{\mu}^{E+\mu} \mathrm{d}E' N(E') = 1, \tag{5.1.17}$$

which represents the condition that on the average one pair of sites with an energy difference in the interval 0 to E is present in the volume $(4\pi/3)\, R^3$ ($N(E)$ — density of states). Assuming that $N(E)$ depends only weakly on E near the Fermi level, one may express E by R by means of (5.1.17). Inserting the expression resulting for E into the exponent (5.1.16) (instead of $|\varepsilon_{m'm}|$), we find the following exponent: $2\alpha R + 3\beta/4\pi R^3 N(\mu)$. Minimizing this expression we obtain the hopping distance

$$R_{\min} = [9\beta/8\pi\alpha N(\mu)]^{1/4}. \tag{5.1.18}$$

Using (5.1.18), from (5.1.16) we find Mott's law for variable-range hopping[1])

$$\sigma = \sigma_0 \exp\left[-\left(\frac{T_0}{T}\right)^{1/4}\right], \quad T_0 = c_3{}^4\alpha^3/kN(\mu), \tag{5.1.19}$$

which describes the conductivity at low temperatures. Here c_3 is a number given in Table 5.1. The pre-exponential factor σ_0 cannot be obtained by the procedure under consideration.

This procedure is valid only for sufficiently low temperatures. Indeed, $R_{\min}$ should not be less than the average spacing between the atoms. Replacing on the left-hand side of (5.1.18) $R_{\min}$ by $\left[\frac{3}{4\pi}\mathcal{N}^{-1}\right]^{1/3}$, we obtain the temperature T_c up to which Mott's law holds:

$$T_c = \frac{1}{k} \frac{9}{8\pi\alpha\mathcal{N}(\mu)} \left(\frac{4\pi}{3}\mathcal{N}\right)^{4/3}. \tag{5.1.20}$$

---

[1]) An account of the correlation between the distances and energy differences of localized states, due to the splitting in energies arising from the interaction between the states via the resonance integral, does not change the $T^{1/4}$-law at very low temperatures. At higher temperatures, a slight deviation from this law may be expected (ORTUNO, 1981).

Table 5.1. The values of the numbers c_3 for the three-dimensional case (cf. (5.1.19)) and c_2 for the two-dimensional case (Mott's law reads here $\sigma = \sigma_0 \exp\left\{-c_2 \left(\frac{\alpha^2}{kTN(\mu)}\right)^{1/3}\right\}$) obtained in several papers: [1] (MOTT, 1968, 1969), [2] (AMBEGAOKAR et al., 1971), [3] (POLLAK, 1972), [4] (SEAGER and PIKE, 1974), [5] (SKAL and SHKLOVSKII, 1976)

ref.	[1]	[2]	[3]	[4]	[5]
c_3	2.05	2.00	1.84	2.28	2.15
c_2	2.05	2.00	1.74	2.42	2.40

Similar values for T_c result from other estimations (POLLAK, 1972, SHKLOVSKII, 1972a)

The problem under consideration may also be investigated by considering hops between nearest neighbours in a four-dimensional space (BRENIG et al., 1971, 1973a, APSLEY and HUGHES, 1974). The distance between two points in this space is given by $2\alpha R + \beta E = \mathcal{R}$, i.e. it is determined by four variables: three spatial coordinates and the energy. Assuming $\sigma \sim \exp(-\bar{\mathcal{R}})$ (cf. (5.1.11)), where $\bar{\mathcal{R}}$ is the average nearest-neighbour distance in the four-dimensional space ($\bar{\mathcal{R}} = \int d\mathcal{R} P(\mathcal{R})\, \mathcal{R}$, $P(\mathcal{R})$ — probability distribution of nearest neighbours), in the low-temperature regime σ agrees with (5.1.19). However, the number c_3 differs from that found by Mott (cf. Table 5.1). At high temperatures the form of the exponent thus obtained agrees with (5.1.9).

Analogously, for two-dimensional systems, one finds that $\sigma = \sigma_0 \exp[-(T_0/T)^{1/3}]$ (T_0 is given in Table 5.1) (BRENIG et al., 1973). In one-dimensional systems, however, hard hops cannot be by-passed by the current. The resistance of such a system is governed by the hardest hop (and not by average ones), i.e. Mott's approach to variable range hopping does not apply to this case (BRENIG et al., 1973a, BRENIG et al., 1973b, BLOCH et al., 1973, KURKIJÄRVI, 1973). For variable-range hopping in one-dimensional systems, one finds a simple $1/T$ activated behaviour. According to ALEXANDER (1982), this result is valid only for sufficiently long chains and long times of observations, while for intermediate lengths and times, anomalous variable-range hopping with a time-dependent diffusion constant should occur (cf. Section 11.4.2.).

The expression (5.1.19) was obtained by assuming the density of states $N(E)$ near the Fermi level μ to be constant. Such an assumption is also sufficient for a not too strong energy dependence of $N(E)$ near μ. However, if $N(E)$ goes to zero at μ, the law (5.1.19) strongly changes. Assuming $N(E) = c\,|E - \mu|^n$ (c is a constant), with the aid of (5.1.17), we find instead of (5.1.19) the following expression (POLLAK, 1972, HAMILTON, 1972)

$$\sigma = \sigma_0 \exp[-(T_0/T)^{(n+1/n+4)}]. \tag{5.1.21}$$

Note that the low-temperature hopping behaviour can be strongly affected by various effects, such as electron-electron correlation or long-range potential fluctuations. Effects of the electron-electron correlation on hopping will be considered in Section 5.4. For effects of long-range potential fluctuations, which occur, for example, in heavily doped and amorphous semiconductors, the reader is referred to the review by BÖTTGER and BRYKSIN (1976a) and the book of SHKLOVSKII and EFROS (1979) and references therein.

In Section 5.3.3. Mott's law for variable-range hopping will be rederived by using the percolation theory. Furthermore, in this section we will discuss some experimental findings.

5.1.5. *Polaron effects in disordered systems*

As noted in Section 5.1.2., at low temperatures the hopping probability for strong interaction with phonons has the same form as that for weak interaction. At high and intermediate temperatures, however, the polaron effect may play an important role giving rise to a strong change in the temperature dependence of σ. For $T > T_D/2$ (T_D — Debye temperature) the integral in (5.1.4) may be evaluated by the saddle-point method, which leads to (SCHNAKENBERG, 1968)

$$Z_{mm'}^{-1} = \frac{e^2}{kT} f_{m'}(1 - f_m) \, e^{\varepsilon_{m'm}/2kT} J_0^2 \, e^{-2\alpha|\boldsymbol{R}_m - \boldsymbol{R}_{m'}|} \, e^{-E_a/kT} \times e^{-(\varepsilon_{m'm})^2/16E_a kT} \frac{\sqrt{\pi}}{\hbar\sqrt{4kTE_a}}. \quad (5.1.22)$$

The energy E_a is defined in (3.2.8). The parameter $\Gamma^2/16E_a kT$ is usually small, so that the last exponential on the right-hand side of (5.1.22) may be replaced by unity. At high temperatures ($kT \leqq \Gamma, \varepsilon_3$) the basic temperature dependence is described by the factor $\exp(-E_a/kT)$ as in the usual small-polaron theory. For strong interaction with phonons, an activated temperature dependence may therefore be expected, where in the low temperature regime the activation energy is given by ε_3 and in the high temperature one by E_a (SCHNAKENBERG, 1968) (cf. Fig. 5.4). At very low temperatures the temperature dependence of σ is described by Mott's law (5.1.19).

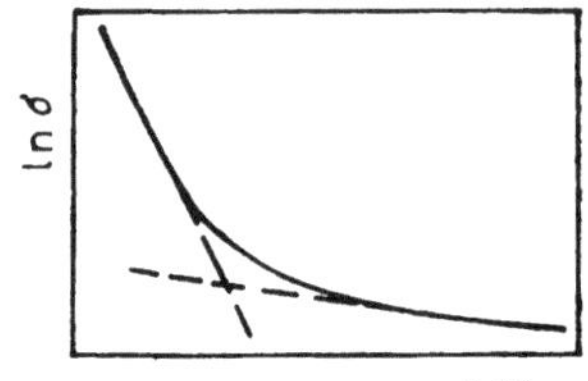

Fig. 5.4. Schematic representation of the temperature dependence of the logarithm of the conductivity in disordered systems in the presence of strong coupling with phonons. At low temperatures the activation energy is equal to ε_3, while at high temperatures to E_a

If kT is smaller than Γ and ε_3 (but, as previously, larger than $kT_D/2$), in (5.1.22) the factor $f_{m'}(1 - f_m) \exp(\varepsilon_{m'm}/2kT)$ may be replaced by $\exp(-|\varepsilon_{m'm}|/2kT)$, and we obtain

$$Z_{mm'}^{-1} \sim \exp\left\{-2\alpha|\boldsymbol{R}_m - \boldsymbol{R}_{m'}| - \left(E_a + \frac{1}{2}|\varepsilon_{mm'}|\right)\Big/ kT\right\}. \quad (5.1.23)$$

At high temperatures, (5.1.23) leads to the activation energy $E_a + \frac{1}{2}\Gamma$ instead of E_a (cf. MOTT and DAVIS, 1979).

For strong interaction with phonons, EMIN (1974) found by a numerical inspection of $Z_{mm'}^{-1}$ that this quantity exhibits in a large temperature range a temperature dependence which resembles the analytical law (5.1.19)[1]). Since for a small spread of the atomic energies, in the resistor network of Miller and Abrahams the temperature dependence of the individual resistors governs that of the overall resistance, the temperature dependence of the conductivity in the situation in question may thus resemble Mott's law (5.1.19).

As can be seen from (3.2.8), the activation energy E_a depends on the hopping distance $|\boldsymbol{R}_m - \boldsymbol{R}_{m'}|$ (in the general case, in (3.2.8) $\boldsymbol{g}$ has to be replaced by $(\boldsymbol{R}_m - \boldsymbol{R}_{m'})$). According to KILLIAS (1966), $E_a(R)$ may be approximately written as $E_a(R) = E_a(0) + aR$ $(a > 0)$. The second term in $E_a(R)$ leads in (5.1.22) to a renormalization (enhancement) of α as $\alpha \to \alpha + a/2kT$. This renormalization leads to a decrease of the temperature at which hops between nearest neighbours become important, i.e. the activated dependence (5.1.9) holds down to lower temperatures. Consequently, in materials with a strong coupling with phonons, Mott's law is valid at lower temperatures than in materials with weak coupling.

Recently, TRIBERIS and FRIEDMAN (1981) studied dc hopping conduction of small polarons in disordered systems in the high-temperature regime ($kT \gg \hbar\omega_{op}$, ω_{op} — optical phonon frequency) by a percolation treatment. Assuming that disorder mainly affects the small-polaron binding energy, for high T they predicted $\sigma \sim \exp\left(-(T_0'/T)^{2/5}\right)$ (T_0' of order 10^6 K), while for the low T the $T^{-1/4}$ law (5.1.19) was obtained. On account of correlation between successive hops due to the energy of the common site Mott's $T^{-1/4}$ law is obtained for high T as well (TRIBERIS and FRIEDMAN, 1985). For a recent discussion of small-polaron formation in disordered solids see, e.g., EMIN (1983).

5.2. Percolation theory

The results (5.1.11) and (5.1.19) for the dc conductivity were obtained by using the following two assumptions: (i) the current through the resistor network of Miller and Abrahams is governed by certain one-dimensional paths and (ii) the major contribution to the resistance of such a path comes from resistors between neighbouring sites with a spacing of the order of the average one (at low temperatures in a four-dimensional space, at high temperatures in the usual three-dimensional one). It is now necessary firstly to verify such an approach and to find the limits of its validity; secondly to find the values of the numbers η (cf. (5.1.11)) and c_3 (cf. (5.1.19)) in the exponentials; and thirdly to determine the pre-exponential factors in (5.1.11) and (5.1.19). The percolation theory has proved to be an effective method for studying these problems. In this section we give a concise introduction into the percolation theory, while its application to hopping is the topic of the next section.

---

[1]) This $T^{1/4}$-like dependence was obtained with a model in which only acoustic phonons are involved. AHMED and BLACKMAN (1980a, b) found that local defect phonon modes tend to destroy such a dependence, i.e. local modes tend to stabilize an activated form of the hopping rate. A similar effect was predicted for optical phonons, too (GORHAM-BERGERON and EMIN, 1977).

5.2.1. *Basic ideas of the percolation theory*

The problem of percolation was first raised and studied by BROADBENT and HAMMERSLEY (1957), who introduced lattice models for the flow of a fluid through a porous random medium. It was rigorously shown that no fluid will flow if the concentration of the active medium (pores) is smaller than some non-zero threshold value. Ideas of percolation theory were first applied to conductivity in disordered systems by ZIMAN (1968), and by AMBEGAOKAR et al. (1971), SHKLOVSKII and EFROS (1971) and POLLAK (1972) to the dc hopping transport in such systems. In the recent decade there has been a rapidly increasing interest in the percolation theory and its application to various physical problems, so that there is now a vast number of research papers published on this subject. The reader interested in a detailed representation of the percolation theory is referred, for example, to the reviews by ESSAM (1972, 1980) and STAUFFER (1979). Here we describe only some basic ideas and methods of the percolation theory.

The classical percolation problem consists in investigating the flow of a fluid through "bonds" connecting "sites" in a regular lattice. The most common statistical assumptions are known as the "site percolation" and "bond percolation" problems. In the site problem some fraction p of lattice sites is randomly occupied and a bond is assumed to exist between any pair of occupied neighbour sites (Fig. 5.5). In the bond problem some fraction p of bonds is randomly distributed on a lattice between neighbour sites (Fig. 5.6).[1])

It is interesting to note that every bond percolation problem can be converted to a site one on another lattice (the covering lattice) so that there is no principal differ-

---

[1]) The conventional site (bond) percolation problem is defined on an isotropic regular lattice and the likelihood that a site (bond) is present is independent of that of its neighbours. In recent years the percolation problem has been generalized in many respects. For instance, in the correlated percolation (see, e. g., CONIGLIO et al., 1977, CONIGLIO et al., 1979, TURBAN and GUILMIN, 1979, MURATA, 1979, TUTHILL and KLEIN, 1982) the probability of occupancy of an element (site, bond) may depend on the presence or absence of neighbouring elements. Such a correlated percolation problem may arise, for example, in a magnetic alloy if a magnetic atom carries a moment only when a sufficient fraction of its nearest neighbours is magnetic. In the case of percolation with restricted valence (see, e.g., GAUNT et al., 1979, KERTESZ et al., 1982) lattice sites (bonds) are occupied at random, but the occupation is prohibited if it would lead to an occupied site having valence (number of bonds) which exceeds some prescribed maximum value. Another problem is percolation in anisotropic systems (see, e.g., SHKLOVSKII, 1978, TURBAN, 1979, and MENDELSON and KARIORIS, 1980). Furthermore, in the continuum percolation (see below) the sites are distributed in a continuum instead of on a lattice. Such continuum percolation may arise, for example, in disordered systems conducting through a hopping mechanism. Further percolation problems are surface percolation (see, e.g., DE BELL and ESSAM, 1981), quantal percolation (see, e.g., PANT and TONG, 1980, ODAGAKI et al., 1980) and first-passage percolation (SMYTHE and WIERMAN, 1978), which account for the effect of surfaces, of the quantum character of the carriers and the time necessary for a carrier to run through a bond, respectively. Finally, a very active field of research is at present the directed percolation, where the bonds are diode-like (see Sections 7.2.2. and 11.3.1.).

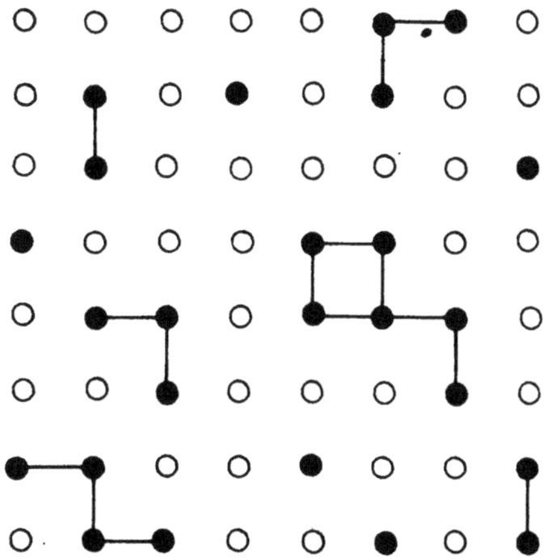

Fig. 5.5. Typical configuration of a site percolation problem. Occupied (empty) sites are denoted by solid (open) circles. Nearest neighbour occupied sites are connected by bonds

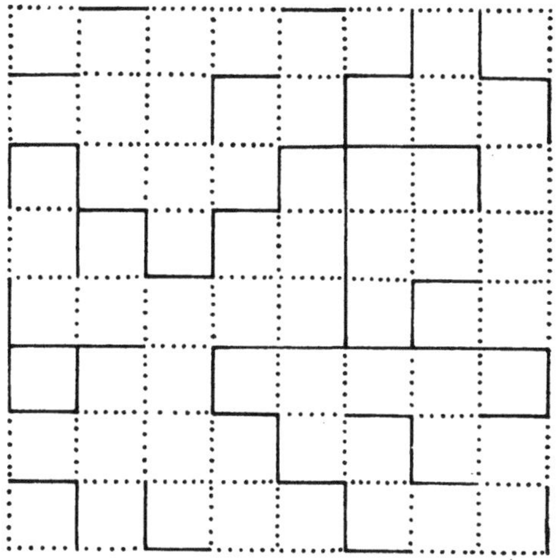

Fig. 5.6. Typical configuration of a bond percolation problem. Free (occupied) and blocked (unoccupied) bonds are denoted by solid and dotted lines, respectively

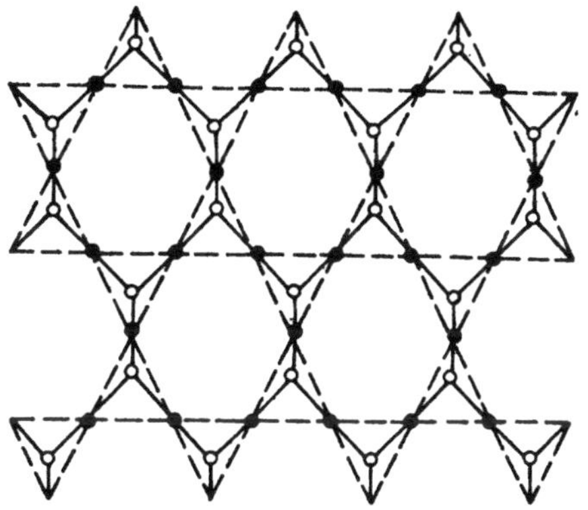

Fig. 5.7. The honeycomb lattice (open circles and solid lines) and its covering lattice, the Kagomé lattice (solid circles and broken lines) (after FISCHER, 1964)

ence between both the problems. The covering lattice is found by putting a site on each bond of the original lattice and drawing a new bond between any such sites in which the corresponding old bonds are neighbours (Fig. 5.7). Not every site problem can be converted to a bond one, so that the former problem is the more fundamental one.

In a percolation model any two sites are said to be connected, if between them there exists at least one sequence of sites which are pairwise linked by bonds. The sites in a distribution may partitioned into clusters such that in a given cluster all sites are connected, but between different clusters there is not any pair of connected sites.

The most important feature of a percolation model is the occurrence of an infinite cluster of sites (assuming an infinite number of lattice sites in an unbounded volume) for $p > p_c$, where p_c denotes a certain threshold value (critical probability). In the fluid context such a cluster renders possibly a macroscopic flow through the system. Its occurrence is characterized by the percolation probability $P(p)$, which is the probability that a given site belongs to an infinite cluster. For the site percolation problem, $P(p)$ is the fraction of occupied sites belonging to an infinite cluster (analogous definition of $P(p)$ for bond percolation). The quantity $P(p)$ vanishes below p_c and is nonzero above p_c (Fig. 5.8). For $p = p_c$ only a marginal fraction of occupied sites contributes to the percolating cluster and the dominant fraction forms isolated clusters. With increasing p the number of isolated clusters decreases rapidly leading (near p_c) to the power law

$$P(p) \sim (p - p_c)^\beta, \qquad p \geqq p_c, \tag{5.2.1}$$

which reminds of a phase transition, although the transition to the percolating state is a purely geometric phenomenon in which the clusters are static objects.

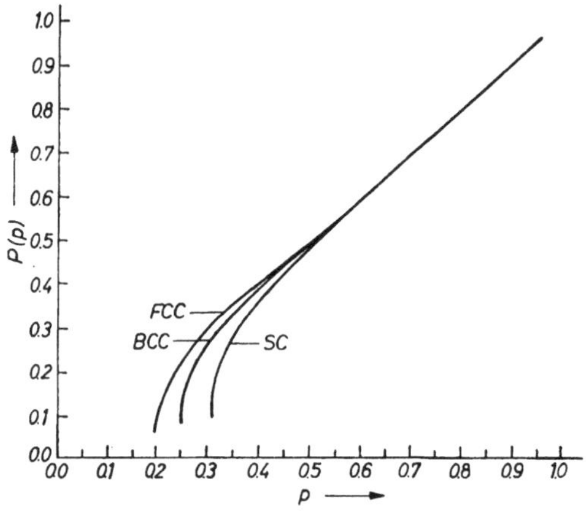

Fig. 5.8. Site percolation probability $P(p)$ for three different three-dimensional lattices (SC — simple cubic, FCC — face centered cubic, BCC — body centered cubic) plotted as a function of the fraction p of sites present (after KIRKPATRICK, 1973)

There are no general rigorous methods of solving the percolation problem. Exact solutions may be obtained for some highly symmetric lattices (see, e.g., the review by ESSAM, 1972) — for the Bethe lattice and for certain two-dimensional structures. In all other cases approximate methods are used (see, e.g., the review by ESSAM 1980): by forming a series expansion of the mean cluster size, by utilizing Monte Carlo methods in conjunction with a large computer and by making use of the renormalization-group (RG) theory (cf. Section 5.2.3.), which becomes also numerical in nature if one is interested in concrete results. By all these methods, $P(p)$ has been shown to behave as described above.

It is difficult to determine precisely the critical exponent β. For different lattice structures, one finds β between 0.13 and 0.15 for dimension $d = 2$, and between 0.4 and 0.5 for $d = 3$, depending on the method used for estimating it (see, e.g., ESSAM,

1980). These results suggest that β is a dimensional invariant. From general considerations, β is estimated to be in the interval $0 \leqq \beta \leqq (d-1)/d$ (SHANTE and KIRKPATRICK, 1971). For a Bethe lattice $\beta = 1$, i.e. for such a lattice formally $d = \infty$ holds (cf. SHANTE and KIRKPATRICK, 1971).

Another invariant of the site percolation problem is the critical-volume-fraction v_c (SCHER and ZALLEN, 1970). Inscribing spheres (or circles) about each site of the lattice, with radius of half the nearest-neighbour distance, the available volume fraction v is then defined as the ratio of the volume of the spheres centered on occupied sites to that of the whole system. At a critical value $v = v_c$ an infinite cluster occurs. To good accuracy, v_c depends only on the dimensionality: $v_c = 0.15 \pm 0.01$ for $d = 3$ and $v_c = 0.45 \pm 0.02$ for $d = 2$ (SHANTE and KIRKPATRICK, 1971, SCHER and ZALLEN, 1970).

For the bond percolation problem, the mean number B_c of allowed bonds per site at the threshold value p_c turns out to be a dimensional invariant. The relation $B_c = zp_c \approx d/(d-1)$ (z — coordination number) holds to good accuracy for all common two- and three-dimensional lattices (SHANTE and KIRKPATRICK, 1971, SCHER and ZALLEN, 1970).

Note that a (threshold) behaviour of type (5.2.1) is not restricted to percolation on lattices with nearest-neighbour bonds or even to regular lattices at all. It also occurs in lattices with bonds between more remote sites, and even in systems with a random site distribution[1]). A situation of the latter type is encountered on applying percolation theory to the current flow in the random resistor network of Miller and Abrahams. Such a situation requires that we first define "bonds". To do this, let $\zeta_{mm'}$ be a function of $\boldsymbol{R}_{mm'} = \boldsymbol{R}_m - \boldsymbol{R}_{m'}$, where $\boldsymbol{R}_m$ and $\boldsymbol{R}_{m'}$, respectively, are the position vectors of sites m and m'. As bonding criterion the condition

$$\zeta_{mm'} \leqq \zeta \qquad (5.2.2)$$

may be used, where ζ is a given number. If two sites are directly connected by (5.2.2) or via a sequence of sites connected pairwise by (5.2.2), these two sites belong to a cluster. The quantity of particular interest is ζ_c, the minimal value of ζ for which an infinite cluster occurs.

The simplest bonding criterion of type (5.2.2) reads

$$|\boldsymbol{R}_{mm'}| \leqq R, \qquad (5.2.3)$$

---

[1]) An interesting question is whether the critical exponents (such as β (5.2.1), γ (5.2.9), ν (5.2.15)) are the same if the sites are distributed randomly in a continuum instead of in a regular lattice. So far, only relatively few papers (e.g. PIKE and SEAGER, 1974, FREMLIN, 1976, HAAN and ZWANZIG, 1977, WINTLE and PUHACH, 1978, POWELL, 1979, 1980, VICSEK and KERTESZ, 1981, SHALITIN, 1981, GAWLINSKI and STANLEY, 1981. KLEIN, 1982; see also Section 11.3.1.) on continuum percolation have been published. The percolation exponents found for continuum percolation are in close agreement with the values for ordinary lattice percolation. Consequently, continuum percolation appears to be in the same universality class as lattice percolation. The conductivity exponents (cf. (5.2.17)) are also close to that for lattice percolation (WEBMAN et al., 1977). The problem at hand is as yet not completely solved.

which is fulfilled if the site m' is inside a sphere (circle) with radius R drawn about the site m (Fig. 5.9). The smallest radius R_c for which an infinite cluster occurs is termed the percolation radius. Sometimes spheres (circles) are constructed with radius $R/2$ and circumscribed about each site. In this construction, two sites are bonded if their associated spheres overlap (Fig. 5.9). According to PIKE and SEAGER (1974), the former figures are called the inclusive figures, the latter ones the overlapping figures.

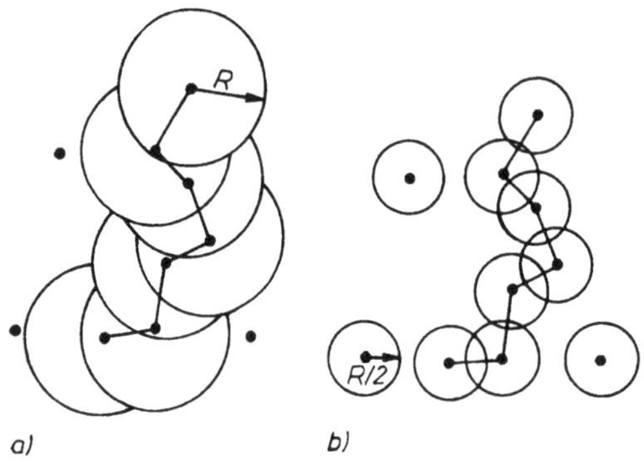

Fig. 5.9. Construction of inclusive (a)) and overlapping (b)) circles. Bonds are denoted by straight lines

Introducing the dimensionless parameter $B = \frac{4\pi}{3} \mathcal{N} R^3$ for $d = 3$ and $B = \pi \mathcal{N} R^2$ for $d = 2$ ($\mathcal{N}$ — concentration of sites, R — radius of the inclusive figures) which has the meaning of the average number of bonds per site, then close to

$$B_c = \frac{4\pi}{3} \mathcal{N} R_c^3 \qquad (d = 3), \tag{5.2.4}$$

$$B_c = \pi \mathcal{N} R_c^2 \qquad (d = 2), \tag{5.2.5}$$

the percolation probability $P(B)$ behaves as (cf. (5.2.1))

$$P(B) \sim (B - B_c)^\beta, \qquad B \geqq B_c. \tag{5.2.6}$$

PIKE and SEAGER (1974) found that to good approximation (for spheres or circles) B_c depends on the dimensionality d as

$$(B_c)^d = \text{const} \approx 20 \tag{5.2.7}$$

(see also SHKLOVSKII and EFROS, 1979, section 18).

Let us now turn to the problem of the critical-volume fraction v_c for the case of overlapping spheres (circles). As before, v_c is defined as the ratio of the volume lying inside the spheres of radius $R_c/2$ centered on the sites to the volume of the whole system. Thus, v_c equals the probability that a point X chosen at random, lies inside the volume enclosed by the spheres with radius $R_c/2$. To calculate this volume requires careful treatment of the regions in which several spheres overlap. Note that the probability that the point X is not inside one of the spheres with radius $R_c/2$

agrees with the probability $\exp\left\{-\frac{4\pi}{3}\mathcal{N}(R_c/2)^3\right\}$ that inside a sphere with radius $R_c/2$ centered about the point X there is not any random site. Therefore, the probability that the point X lies in one of the spheres gives v_c and equals

$$v_c = 1 - \exp\left\{-\frac{4\pi}{3}\mathcal{N}\left(\frac{R_c}{2}\right)^3\right\} = 1 - \exp\left(\frac{B_c}{8}\right) = 0.29 \tag{5.2.8}$$

for $d = 3$ (with $B_c = 2.7$[1]), while for $d = 2$ one finds $v_c = 0.67$ (SHANTE and KIRKPATRICK, 1971).

It is interesting to note that such values v_c are also obtained for percolation on lattices if bonds between sufficiently distant neighbours are accounted for. In this case the periodicity of the underlying structure becomes meaningless for the value v_c to be obtained, i.e. v_c found in this way also applies to irregular structures. An account of more remote neighbours leads to an enhancement of v_c, which rapidly converges, if more and more remote neighbours are accounted for. The dimensional invariants v_c and B_c for irregular structures are critically discussed by PIKE and SEAGER (1974), who in particular analyzed the dependence of these quantities on the shape of the characteristic figures.

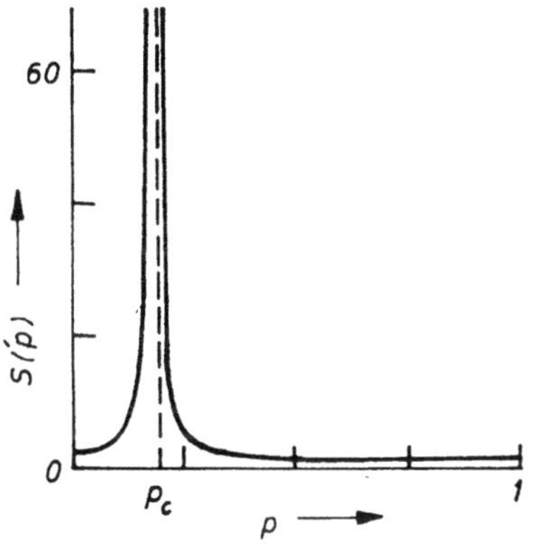

Fig. 5.10. The mean size $S(p)$ of finite clusters for site percolation on the FCC lattice (after ESSAM, 1980)

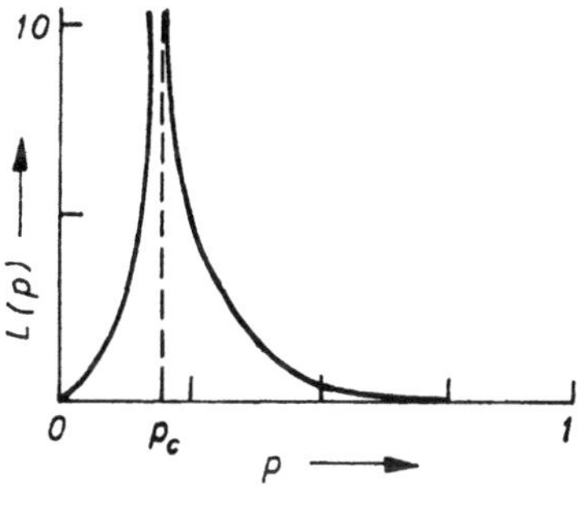

Fig. 5.11. The correlation length $L(p)$ for site percolation on the FCC lattice. For $p > p_c$, $L(p)$ is sketched in (after ESSAM, 1980)

[1]) The value $B_c = 2.7$ was found by PIKE and SEAGER (1974) by studying percolation through finite systems by Monte Carlo technique. For systems with $N = 1000$ and $N = 8000$ randomly distributed sites, they obtained $B_c^{(1000)} = 2.92 \pm 0.03$ and $B_c^{(8000)} = 2.80 \pm 0.02$, respectively. Extrapolation by the formula

$$B_c^{(N)} = B_c + \text{const.}\, N^{-1/3}$$

yields $B_c = 2.7$. Such an interpolation formula agrees with the conception that a finite system cannot contain all information on critical behaviour. The error is smaller the larger the system.

Now let us consider a second important percolation function, the mean size $S(p)$ of the finite cluster to which a given occupied site belongs. $S(p)$ is finite at $p < p_c$, since below p_c there is zero probability of an infinite cluster. It is also finite at $p < p_c$, since, by definition, the infinite cluster is not included. But at $p \to p_c \pm 0$ $S(p)$ diverges (Fig. 5.10). Close to p_c, the behaviour of $S(p)$ can be described by a power law (SYKES and ESSAM, 1964)

$$S(p) \sim |p - p_c|^{-\gamma}, \tag{5.2.9}$$

where, as usual, the critical exponent γ has been assumed to be the same above and below p_c. By series expansion and numerical methods γ is found between 2.4 and 2.5 for $d = 2$, and between 1.6 and 1.7 for $d = 3$ (see, e.g., ESSAM, 1980, Appendix 1.), i.e. γ is a dimensional invariant.

A third important percolation function is the correlation length L (also termed the connectivity length or coherence length), which is given by the root mean-square distance between the cluster sites and a given site of the cluster.[1]) L is finite below and above p_c and diverges at $p \to p_c \pm 0$ (Fig. 5.11). Close to p_c, $L(p)$ can be described by a power law

$$L(p) \sim |p - p_c|^{-\nu}. \tag{5.2.10}$$

Here again the critical exponent ν is assumed to be the same above and below p_c. ν is a dimensional invariant, too. Current estimates of ν range from 1.3 to 1.4 for $d = 2$ (cf., e.g., KARIORIS and MENDELSON, 1981) and from 0.8 to 0.9 for $d = 3$ (see, e.g., ESSAM, 1980, Appendix 1.) At $p < p_c$ the finite clusters are being placed in the "pores" of the infinite cluster (Fig. 5.12). Thus, above p_c, a small probability of finding finite clusters with a linear dimension larger than $L(p)$ implies also a small probability of finding pores with a linear dimension larger than $L(p)$, so that at $p > p_c$, $L(p)$ may be viewed as characterizing the linear dimension of the pores.

The correlation length L may be defined (LEVINSHTEIN et al., 1975) as the characteristic length satisfying the condition that at $p > p_c$ in a cube with the length $l \gg L$ (l and L in units of the lattice constant) percolation with a probability close to unity occurs, whereas for $l \ll L$ a percolating realization is very unlikely. To the contrary, at $p < p_c$, for $l \gg L$ percolation is very unlikely, whereas for $l \ll L$ percolation with probability close to unity occurs. In accordance with this definition

---

[1]) The pair-connectedness function (correlation function) $G(\boldsymbol{r}, p)$ is the probability that a given site (origin) is occupied, belongs to a finite cluster, and is connected to the site at $\boldsymbol{r}$. Taking the sum over $\boldsymbol{r}$ we obtain

$$\sum_{\boldsymbol{r}} G(\boldsymbol{r}, p) = S(p)\, p\big(1 - P(p)\big),$$

where the last two factors on the right-hand side describe the probability that the origin is occupied and belongs to a finite cluster. The mean-square distance from the origin of a particle which is connected to the origin is given by

$$L^2(p) = \sum_{\boldsymbol{r}} \boldsymbol{r}^2 G(\boldsymbol{r}, p) \Big/ \sum_{\boldsymbol{r}} G(\boldsymbol{r}, p).$$

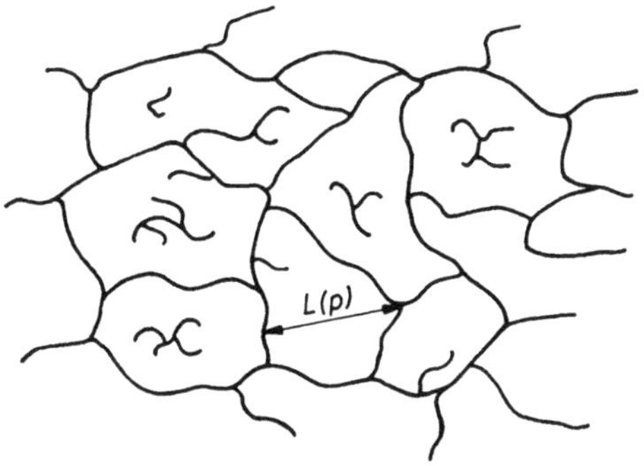

Fig. 5.12. Schematic representation of the infinite cluster and of large finite clusters for $0 < p - p_c \ll 1$. $L(p)$ denotes the correlation length

the exponent ν may be determined by calculating the spread w_l of the percolation threshold p_{cl} for a cube with length l of an edge as follows (SKAL et al., 1975).

Monte Carlo calculations yield for p_{cl} a nearly Gaussian distribution $f_l(p_{cl})$ (Fig. 5.13), with the width $w_l^2 = \langle (p_{cl} - \langle p_{cl} \rangle)^2 \rangle$, which decreases with increasing l as

$$w_l = Bl^{-1/\nu} \tag{5.2.11}$$

(B is a constant). To show that ν from (5.2.11) equals the critical exponent of the correlation length, we introduce the quantity

$$\mathcal{P}_l(p) = \int_0^p dp' f_l(p'), \tag{5.2.12}$$

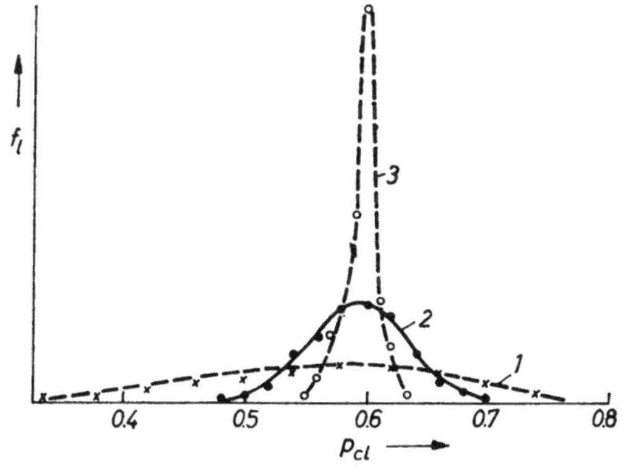

Fig. 5.13. Distribution function for the threshold values p_{cl} of the site percolation problem on a quadratic lattice with length l of an edge. The length l in units of the lattice constant is 8(×), 32(•), 128(○) (after LEVINSHTEIN et al., 1975)

which has the meaning of the probability that at given p the cube with length l percolates, i.e. $f_l(p)$ is the corresponding probability density. From the above definition of L, it follows that, at $p < p_c$ and $l \gg L(p)$, we have

$$1 - \mathcal{P}_l(p) \ll 1, \tag{5.2.13}$$

i.e. the cube with the length l percolates with probability close to unity, unlike the case $l \ll L(p)$, where percolation is unlikely. On the other hand at $p < p_c$ and

$l \gg L(p)$ we have

$$\mathcal{P}_l(p) \ll 1, \tag{5.2.14}$$

in contrast to the case where $l \ll L(p)$.

Let us fix p at $p > p_c$ and study, by diminishing l, at which value l the inequality (5.2.13) breaks down. Evidently, this occurs when the width w_l of f_l approaches $p - p_c$. Thus, using $w_l = BL^{-1/\nu} \approx p - p_c$ we find

$$L \sim (p - p_c)^{-\nu}, \tag{5.2.15}$$

i.e. the exponent ν is indeed the critical exponent of the correlation length.[1])

Now let us turn to the question: How does an electrical current flow through a random system, for example, through a simple cubic lattice which is composed of randomly distributed electrically well conducting (metallic) pieces with probability p and non-conducting (insulating) pieces with probability $1 - p$? Historically, in a percolating system the relation between $P(p)$ and the conductivity $\sigma(p)$ was first investigated by LAST and THOULESS (1971) on a conducting layer containing randomly punched holes and by KIRKPATRICK (1971) by a Monte Carlo procedure on resistor networks. According to these investigations, the threshold value p_c for $\sigma(p)$ agrees with that for $P(p)$, but the dependence on p, at $p > p_c$, is quite different for both these quantities: $\sigma(p)$ approaches zero for $p \to p_c + 0$ with a zero slope, whereas $P(p)$ approaches zero for $p \to p_c + 0$ with an infinite slope (see, Fig. 5.14). LAST and THOULESS (1971) suggested that the behaviour is different, since for p just exceeding p_c the most probable infinite clusters are ineffective conductors because

---

[1]) Size dependence of percolation properties is the basis of the finite-size scaling method (e.g. NIGHTINGALE, 1982) for determining critical percolation exponents (SUR et al., 1976, LOBB and FRANK, 1979, MITESCU et al., 1982), which is based on the fact that so long as the correlation length L is small compared with the linear dimension l of the system, the system will display the behaviour of an infinite system. However, on approaching the critical point, L eventually becomes comparable to l and the finite system exhibits a smeared-out transition instead of a sharp transition as an infinite system. Thus the operative universal variable becomes $l/L \sim l(p - p_c)^\nu$. The finite-size scaling method is based on the assumption that $\mathcal{P}_l(p)$ (5.2.12) (or another function exhibiting critical behaviour) can be described as (cf., e.g., VANNIMENUS and NADAL, 1984)

$$\mathcal{P}_l(p) \sim l^{-x} X(z)$$

where X is a homogeneous function (scaling function) of the modified universal variable $z = l^{1/\nu}(p - p_c)$, which describes the smearing-out of the transition for finite l. Since for $l \to \infty$, $\mathcal{P}_\infty(p) \equiv P(p) \sim (p - p_c)^\beta$, it must hold that $X(z) \sim z^\beta$ for $z \to \infty$, so that $x = \beta/\gamma$, in order to guarantee the independence of l, for large l. Accordingly, $\mathcal{P}_l(p) \sim l^{-\beta/\nu} X(l^{1/\nu} \times (p - p_c))$ (cf. MARGOLINA and HERRMANN, 1984). Note that there is a rather intimate relationship between finite-size scaling and the renormalization-group method (see below) (cf. REYNOLDS et al., 1978, NAKANISHI, 1984). For extensive studies of the statistical distribution of the percolation threshold in finite systems see, e.g., LEVINSHTEIN et al., 1975, ROUSSENQ et al., 1976, REYNOLDS et al., 1980, KARIORIS and MENDELSON, 1981).

of their tortuous and constricted nature. Thus, writing the conductivity as (LAST and THOULESS, 1971)

$$\sigma(p) = u(p)\, P(p), \tag{5.2.16}$$

the investigations mentioned before suggest that the effective mobility $u(p)$ vanishes at p_c.[1])

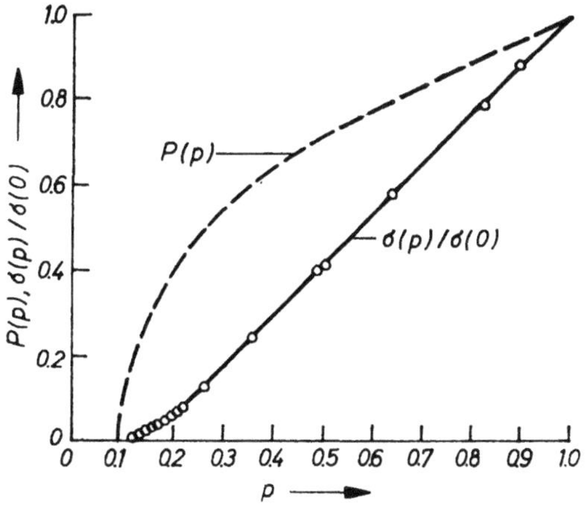

Fig. 5.14. Numerical results of the percolation probability P and the conductance σ for site percolation on a simple cubic lattice (after KIRKPATRICK, 1973). The site percolation data are plotted against p, the fraction of bonds present, rather than the fraction of sites present. Note that in this model the fraction of bonds present is the square of the fraction of sites present, since a bond is present only if the sites at both ends are present

Close to the percolation threshold p_c an exponent t can be defined through (KIRKPATRICK, 1973)

$$\sigma(p) \sim (p - p_c)^t\,. \tag{5.2.17}$$

Below p_c the conductivity is zero. The power law (5.2.17) has been studied by analog experiments (e.g. LAST and THOULES, 1971, WATSON and LEATH, 1974, FITZPATRICK et al., 1974, LEVINSHTEIN, 1977, OTTAVI et al., 1978, CLARKE et al., 1978), by numerical calculations (e.g. KIRKPATRICK, 1973, 1976, ONIZUKA, 1975, STRALEY, 1977a) and by other methods (see below). The critical exponent t is approximately a dimensional invariant. However, estimates for this exponent are less accurate than those for the other exponents. t is about 1.2 (between 1.0 and 1.5) for $d = 2$ and about 1.7 (between 1.5 and 2.0) for $d = 3$. A power of the type (5.2.17) holds also for

[1]) It is worth noticing that the mobility $u(p)$ is immediately connected with the stiffness of spin waves, $D(p)$, in disordered Heisenberg ferromagnets with a concentration p of magnetic sites (KIRKPATRICK, 1973). In the limit of long wavelengths, the spin-wave spectrum of the disordered ferromagnet is given by $\omega(k) = 2JSD(p)\, k^2$, where $D(p)$ and $u(p)$ are connected by the relation $u(p) = D(p)\, \sigma(1)$ (J — exchange integral, S — total spin on a site, k — wave number, $\sigma(1) = \sigma(p = 1)$).

the exactly soluble Bethe lattice (STINCHCOMBE, 1973, 1974, STRALEY, 1976), where $t = 3$ was found (STRALEY, 1977b)[1]).

Now we ask whether the exponent t can be related to our earlier introduced exponents. This problem can be attacked by the simple but not well established node-link model of SKAL and SHKLOVSKII (1974) and DE GENNES (1976) (cf. STRALEY, 1982). According to this model, slightly above p_c the infinite percolation cluster mainly consists of one-dimensional chains with crosslinks (Fig. 5.15). The distance between two nodes is of the order of the correlation length $L \sim (p - p_c)^{-\nu}$ (cf. Fig. 5.12). As the chain segments (macrobonds) between two nodes can be twisted, their length $\mathcal{L}$ in general exceeds L. $\mathcal{L}$ is assumed to diverge as

$$\mathcal{L} \sim (p - p_c)^{-\zeta}, \tag{5.2.18}$$

where ζ is a new exponent. An actual percolation cluster contains additional elements: "dead" ends and loops (cf. Fig. 5.16) (above all, loops occur in two-dimensional systems), which are not accounted for in Fig. 5.15, since these elements contribute to $P(p)$, but (at least for weak electrical field) not to $\sigma(p)$, so that they may be ignored here, which yields the so-called "backbone" of the infinite cluster.

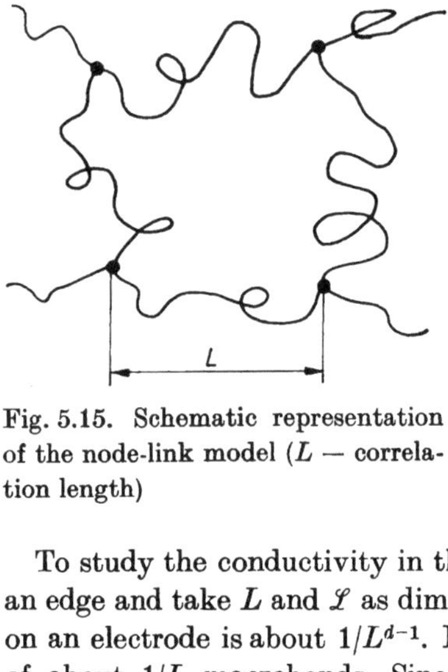

Fig. 5.15. Schematic representation of the node-link model (L — correlation length)

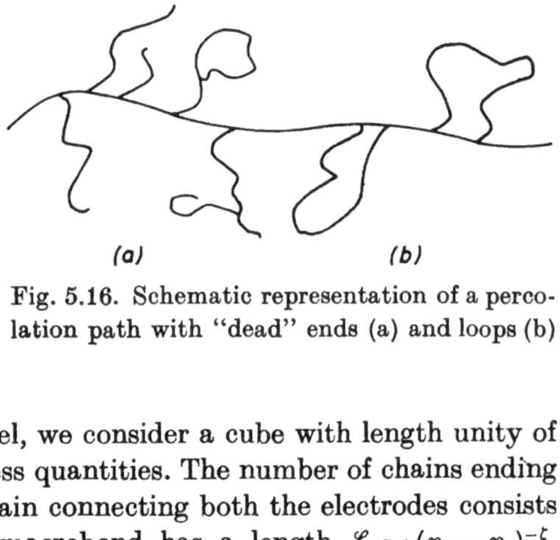

Fig. 5.16. Schematic representation of a percolation path with "dead" ends (a) and loops (b)

To study the conductivity in this model, we consider a cube with length unity of an edge and take L and $\mathcal{L}$ as dimensionless quantities. The number of chains ending on an electrode is about $1/L^{d-1}$. Each chain connecting both the electrodes consists of about $1/L$ macrobonds. Since each macrobond has a length $\mathcal{L} \sim (p - p_c)^{-\zeta}$, such a chain between the electrodes has a length $\sim (p - p_c)^{\nu-\zeta}$ and therefore a conductivity $\sim (p - p_c)^{\zeta-\nu}$. If we account for all $L^{1-d} \sim (p - p_c)^{(d-1)\nu}$ "parallel" chains we find that the conductivity of the cube varies as $\sigma \sim (p - p_c)^{(d-1)\nu+\zeta-\nu}$,

[1]) An alternative resistor network model consists of a mixture of two kinds of resistors, with a concentration p of resistors with infinite conductivity and with concentration $(1 - p)$ of finite conductivity (see, e.g., STRALEY, 1976, 1977b, 1979, 1980). For this model, one finds that the conductivity σ is infinite for $p > p_c$, and for $p < p_c$, $\sigma \sim (p_c - p)^{-s}$. For $d = 2$, $s = 1.1$; $d = 3$, $s = 0.70$; and for the Bethe lattice, $s = 0$ (with possible logarithmic corrections; see also BUNDE et al., 1985, and references therein).

which yields for the critical exponent t the following expression (SKAL and SHKLOVSKII, 1974; cf. Section 11.3.2.)

$$t = \zeta + (d - 2)\,\nu\,. \tag{5.2.19}$$

SKAL and SHKLOVSKII (1974) indicated that $\zeta = 1$. Since $\nu > 1$ for $d = 2$, such a value ζ contradicts the requirement that $\mathscr{L} \geqq L$, i.e. the node-link model may not be applied to $d = 2$. On the other hand, for $d = 3$, we have $\nu = 0.9$, so that there is no contradiction. For $d = 3$, (5.2.19) yields $t = 1 + \nu \approx 1.9$, a value somewhat larger than the value 1.7 given above.

The node-link model has also been used for studying spin-waves in dilute magnets (ZIMAN, 1979). Shortcomings of this model have been discussed by several authors (DASGUPTA et al., 1978, GEFEN et al., 1981, and references therein). A more sophisticated node-link model was proposed by CONIGLIO (1981, 1982, 1983), who replaced the links by more complicated structures (by a "necklace" of blobs of all sizes; HERRMANN and STANLEY, 1984).

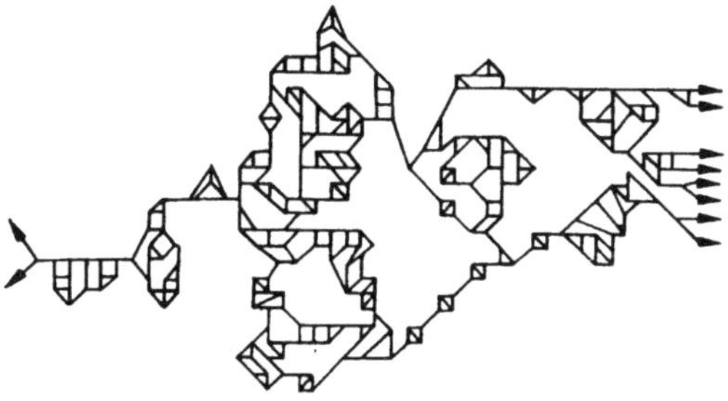

Fig. 5.17. Example of the infinite cluster graph for a two-dimensional system with 100 × 100 sites, obtained by the Monte Carlo method at the percolation threshold (after SARYCHEV and VINOGRADOFF, 1981)

An alternative model for the infinite cluster is suggested by Monte Carlo simulation: the self-similar drop (fractal)-model (KIRKPATRICK, 1978, GEFEN et al., 1981, SARYCHEV and VINOGRADOFF, 1981, GEFEN et al., 1982, 1983). In this model the infinite cluster at the percolation threshold has the form of a drop on a thread for any length scale (cf. Fig. 5.17). For a given scale (below the correlation length L and well above the microscopic lattice constant a) each drop (cluster of sites) consists of similar drops on a smaller scale, etc. Two different adjacent drops are connected by a macrobond. Each drop is connected with two or more other drops. Thus, in this model the infinite cluster is a network consisting of drops with a diameter of the order of the correlation length, which are connected by threads (macrobonds).

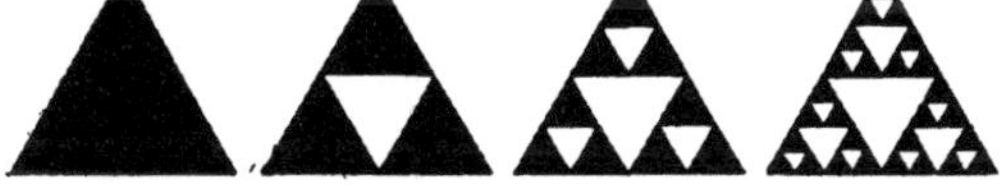

Fig. 5.18. Illustration of self-similar fractals. The Sierpinski gasket's initial triangle, and the first construction stages

Note that on length scales l with $a \ll l \ll L$ the physical behaviour of such a model (e.g. the diffusion for short times; GEFEN et al., 1983) reflects the self-similarity and is independent of L. Bulk, properties are in this case obtained by juxtaposition of pieces of size L (GEFEN, 1981; cf. Section 11.3.2.).

An important characteristic of such a drop model is the order of ramification R of a given point P, defined as the number of threads which must be cut to isolate the subset of drops surrounding P. When the smallest value of R, $R_{\min}$, equals two, then the structure is termed quasi one-dimensional (node-link-blob model of CONIGLIO, 1981). When the largest value of R, $R_{\max}$, is finite, then the structure is termed finitely ramified. There is strong indication that the percolation cluster is finitely ramified but not quasi one-dimensional (cf. GEFEN et al., 1981).

As a model for such a percolation cluster, GEFEN et al. (1981) studied a d-dimensional non-random scale invariant fractal lattice. In this model d-dimensional hypertetrahedra are the self-similar fractals (Fig. 5.18). This model allows the percolation and the conductivity problems to be exactly solved at the percolation threshold. It has the property $R_{\min} = d + 1$ and $R_{\max} = 2R_{\min} - 2 = 2d$.

The conductivity exponent of the model under consideration may be obtained as follows (GEFEN et al., 1981). Suppose we construct the fractal lattice iteratively by connecting the midpoints of the edges of hypertetrahedra and erasing the volume at the center bounded by the faces of the new tetrahedra (cf. Fig. 5.18). This procedure is repeated down to the microscopic length scale a. Let l be the length of an edge (bond) of the initial tetrahedron and $Z(l)$ the resistance of such a bond. After the first iteration step, the resistance of a bond is $Z(l/b)$, where $b = 2$. When a current I is sent into one corner of the initial tetrahedron, then a current I/d flows out of the other d corners. The voltages at these corners remain unchanged after iteration, provided that we assume $Z(l) = [(d + 3)/(d + 1)]\, Z(l/b) \equiv b^{\bar{\zeta}} Z(l/b)$. After n iteration steps, where $b^n = l/a$, we have $Z(l) = b^{n\bar{\zeta}} Z(a) \sim l^{\bar{\zeta}}$, and the conductivity becomes

$$\sigma(l) \sim l^{2-d} Z(l)^{-1} \sim l^{-\bar{t}} \tag{5.2.20}$$

with

$$\bar{t} = d - 2 + \bar{\zeta}. \tag{5.2.21}$$

When l exceeds the correlation length L, the conductivity does not further change with l and we have

$$\sigma \sim L^{-\bar{t}}. \tag{5.2.22}$$

Apart from the specific value of $\bar{\zeta}$, the relations (5.2.20) and (5.2.21) hold for any model which, up to a length scale L, is composed of self-similar fractals with finite order of ramification.

GEFEN et al. (1981) found that $\bar{t} = 0.73$ for $d = 2$ and $\bar{t} = 1.58$ for $d = 3$, in rather good agreement with Monte Carlo studies ($\bar{t} \approx 0.82$ for $d = 2$, $\bar{t} \approx 1.9$ for $d = 3$). Note that $\bar{t} = t/\nu$, which results from (5.2.22) with the aid of $L \sim (p - p_c)^{-\nu}$.

5.2.2. *The percolation problem in the context of critical phenomena*

The transition at a definite critical probability p_c from a non-percolating to a percolating state in a percolation model has many features in common with the transition at a critical temperature T_c from, say, a paramagnetic (disordered) to a ferromagnetic (ordered) phase in a magnetic system.

Let us first consider the latter type of transition. When the temperature T is much larger than T_c, the system is in a completely disordered phase. As T decreases, thermodynamic fluctuations give rise to droplets of the ordered phase, the average diameter of which (characterized by the correlation or coherence length) increases with decreasing T and becomes infinitely large close to T_c. Below T_c, in the ordered phase there exist droplets of the disordered phase. The average diameter of these droplets decreases with decreasing temperature.

Close to the critical temperature T_c, the so-called critical behaviour of the system (including the critical exponents characterizing the singular behaviour of thermodynamic quantities near T_c) is governed by droplets with a typical size (scaling size), which is large compared to the characteristic microscopic length scale and which becomes infinitely large as $T \to T_c$. Therefore, the critical behaviour does not depend on characteristics of the system such as the type of interaction between the particles, the structure of the lattice etc. But it depends only on the dimensionality of the space (and the order parameters). This is the essence of the principle of universality (KADANOFF, 1971).

Hence, one can define equivalence classes of systems such that the systems of each class are characterized by the same value of the corresponding critical exponents. Each such class can include very distinct systems. A consequence of the universality principle is the existence of simple relations between the critical exponents, i.e. the existence of so-called scaling laws (scaling relations). These laws can be obtained by assuming that the thermodynamic potentials and the correlation functions are homogeneous functions of their relevant variables (i.e. of the variables characterizing the critical behaviour, such as $T - T_c$, the magnetic field etc.). This is the essence of the scaling hypothesis.

Now let us turn to percolation. First, we note that percolation is a purely geometric problem, a problem of non-interacting particles. Nevertheless, it exhibits critical behaviour analogous to thermodynamic systems. However, in the percolation problem, singularities occur in physical properties of clusters, rather than in thermodynamic properties.

At $p < p_c$, with increasing p the average size (and diameter) of the finite clusters increases, so that close to p_c very large clusters dominate and govern the critical behaviour of the system. Therefore, the critical behaviour may be expected to be independent of characteristics of the system such as the structure, but to depend decisively on the dimensionality of the system. The latter dependence is suggested by the fact that, at $p > p_c$, in two-dimensional systems only a single infinite cluster (of occupied sites) exists, whereas in three-dimensional systems two infinite clusters (one of occupied and one of unoccupied sites) coexist. The idea of universality and scaling applies to percolation, too.

Particular limits of systems described by a Hamiltonian reduce to percolation systems. This provides a bridge between critical phenomena in thermodynamic and percolation systems. A correspondence between percolation and magnetic properties was first obtained by KASTELEYN and FORTUIN (1969) (see, e.g., also GINZBURG, 1976, HARRIS and LUBENSKY, 1981, WU, 1982) by considering the $Q \to 1$ limit of a Q-state Potts model (see Appendix 3.)[1]). A similar correspondence was found by DUNN et al. (1975) by considering the percolation limit $J \gg kT$ (J — exchange energy) in a dilute ferromagnet. The latter approach will be described in the present section in some more detail. Note that a Hamiltonian is not in any case necessary for studying percolation by methods of the theory of phase transitions. As will be shown in the following section, the existence of a diverging length scale is sufficient to justify application of the efficient real-space renormalization-group (RG) method (see, e.g., BURKHARD and VAN LEEUWEN, 1982) to critical behaviour, including the critical behaviour of percolation models.

Now consider a nearest-neighbour ferromagnetic spin 1/2-Ising model in which magnetic atoms, with probability $(1 - p)$, are independently replaced by non-magnetic atoms. In the limit $J \gg kT$, the thermal energy is too low to break up the exchange bonds. Therefore, the spins fall into non-interacting clusters, the spins within a cluster all being parallel to one another. The clusters behave as non-interacting superspins with magnetic moment proportional to the size of the cluster. Let μ be the magnetic moment of a single spin, then a cluster containing s spins has magnetic moment $s\mu$. In a magnetic field H such a cluster has the probability $p_+ (p_- = 1 - p_+)$,

$$p_+ = \frac{\exp(s\mu H/kT)}{\exp(-s\mu H/kT) + \exp(s\mu H/kT)}, \tag{5.2.23}$$

of being parallel (antiparallel) to the field. The average magnetic moment of a spin belonging to a cluster of size s is $p_+ - p_- = \tanh(s\mu H/kT)$ (in units of the saturation magnetization). The contribution of all finite clusters to the total magnetization (in units of the saturation magnetization) is $\left(\sum_s sn_s\right)^{-1} \sum_s sn_s \tanh(s\mu H/kT)$. Here n_s denotes the average number of clusters of size s devided by the total number of lattice sites. For an infinite cluster, we have $p_+ = 1$ ($s \to \infty$ in (5.2.23)) independently of H. Thus, the contribution of the infinite cluster to the total magnetizaiion (in units of the saturation magnetization) is given by the fraction $P(p)$ of spins tn the infinite cluster. The total magnetization M (in units of the saturation magnetization) of the system becomes

$$M = P(p) + (1/p) \sum_s sn_s \tanh(s\mu H/kT), \tag{5.2.24}$$

where, for the saturation magnetization per site, we have used the equation $pP(p) + \sum_s sn_s = p$, which results from

$$1 - p + pP(p) + \sum_s sn_s = 1. \tag{5.2.25}$$

---

[1]) On the basis of the $Q \to 1$ limit of the Potts model, GINZBURG (1976) calculated by means of field theoretical methods the critical indices of the bond percolation problem. For space dimensions $3 < d < 6$, he found $\gamma = 8/(2 + d)$, $\nu = 4/(2 + d)$, $\beta = 2(d - 2)/(d + 2)$, $\Delta = \beta + \gamma = 2$. These results are in agreement with computer calculations which yield for $d = 3$: $\beta = 0.35 \pm 0.05$, $\gamma = 1.69 \pm 0.05$, $\nu = 0.9 \pm 0.05$, $\Delta = 2.2 \pm 0.3$.

Equation (5.2.25) implies that the sum of the following three probabilities is equal to unity: $(1 - p)$ — the probability that a given site is not occupied by a magnetic atom, $pP(p)$ — that this site belongs to an infinite magnetic cluster, and $\sum_s sn_s$ — that the given site belongs to a finite cluster of magnetic sites. With the aid of (5.2.25), equation (5.2.24) can be rewritten as

$$pM = \left(p - \sum_s sn_s\right) + \sum_s sn_s \tanh\,(s\mu H/kT). \tag{5.2.26}$$

Accordingly, for $J \gg kT$ the magnetization of a dilute ferromagnet is directly related to the cluster number n_s. At $p > p_c$, the zero-field magnetization (the "spontaneous" magnetization) M_0 is proportional to the percolation probability (cf. (5.2.24)), that is

$$M_0(p) \sim (p - p_c)^\beta. \tag{5.2.27}$$

At $p < p_c$ and $H = 0$, the finite clusters give zero magnetization, corresponding to the paramagnetic state. From (5.2.24) or (5.2.26), we see that the susceptibility $\chi \equiv (\partial M/\partial H)_{H=0}$ is proportional to $\sum_s s^2 n_s$, i.e. proportional to the mean size $S(p)$ of the finite clusters, so that

$$\chi \sim (p - p_c)^{-\gamma}. \tag{5.2.28}$$

Calculating also the free energy of the system under consideration and introducing further the pair correlation function (and thus the correlation length) one finally finds (Dunn at al., 1975, cf. Kasteleyn and Fortuin, 1969) the following correspondence between the percolation transition and the magnetic phase transition in a dilute Ising model:

mean number of clusters	↔	zero-field free energy	(5.2.29)
percolation probability	↔	spontaneous magnetization	
mean cluster size	↔	initial susceptibility	
pair connectedness	↔	correlation function	

This correspondence suggests that the principle of universality and the scaling laws will also apply to percolation (Essam and Gwilym, 1971). The scaling hypothesis in percolation theory formulated by Dunn et al. (1975) and Levinshtein et al. (1975) is completely analogous to Kadanoff's classical formulation in the phenomenological theory of phase transitions.

Thus, a ferromagnetic system may exhibit two types of transition: (i) a percolative transition at some definite probability p_c and (ii) a thermodynamic phase transition at some critical temperature T_c. For the latter type of transition, close to T_c, the temperature dependence of the magnetization and the initial susceptibility, respectively, can be described by formulae (5.2.27) and (5.2.28) after replacing in these formulae $(p - p_c)$ by $(T_c - T)$. Analogous relations hold for the other quantities (5.2.29). Accordingly, the percolation behaviour at $p > p_c (p < p_c)$ corresponds to the thermodynamic behaviour at $T < T_c (T > T_c)$. However, for both types of transition, the corresponding critical exponents somewhat differ (see, e.g., Böttger and Bryksin, 1976a, Table 2).

Let us now turn to the problem of the scaling laws for percolation. We restrict ourselves to the site percolation problem. Following DUNN et al. (1975) and LEVINSHTEIN et al. (1975) (see also the review by ESSAM, 1980), we introduce the generating function

$$F(p, h) = \sum_{s=1}^{\infty} n_s(p) \exp(-sh). \tag{5.2.30}$$

Here $n_s(p)$ is the mean number of clusters per site of size s, and h is a parameter which plays role of a dimensionless magnetic field ($h = H\mu/kT$). The function $F(p, h)$ is the analogue of the free energy of a dilute ferromagnet: at $h = 0$ it reduces to the mean cluster number (cf. (5.2.29)) and its first and second derivatives with respect to h determine the percolation probability and the mean cluster size, respectively.

Before explicitely pointing out this, we study the following problem. Consider clusters within a cube of volume Ω with N sites which is centred on a given lattice site. Let $\gamma_i(s) = 1$, if site i belongs to a cluster of size s, and $\gamma_i(s) = 0$ otherwise. The number N_s of sites of Ω belonging to clusters of size s is then given by

$$N_s = \sum_{i \in \Omega} \gamma_i(s), \tag{5.2.31}$$

and N_s/sN is the number of clusters per site of size s. Averaging (5.2.31) over all possible configurations in Ω, in the thermodynamic limit ($\Omega \to \infty$, $N \to \infty$) (assuming translational invariance) we find that

$$sn_s = p_s(p), \tag{5.2.32}$$

where $p_s(p) = \langle \gamma_i(s) \rangle$ ($\langle \cdots \rangle$ indicates the configuration average) is the probability that a given site i (say, the origin) belongs to a cluster of size s and $n_s = \langle N_s/sN \rangle$ is the mean number of clusters per site of size s. Let h be the average number per site of additional independently distributed "ghost" particles (allowing multi-occupation of sites), then (according to the Poisson distribution) $\exp(-h)$ is the probability that any site has no ghost particle on it[1]). Accordingly, the function $F(p, h)$ defined in (5.2.30) represents the mean number of clusters per site which are free from ghost particles. $F(p, 0)$ is the mean number of clusters per site.

The probability $P(p)$ that a given occupied site (origin) belongs to the infinite cluster may be expressed as (cf. (5.2.25))

$$P(p) = 1 - \frac{1}{p} \sum_s p_s(p) = 1 - \frac{1}{p} \sum_s sn_s(p) = 1 + \frac{1}{p} \left. \frac{\partial F(p, h)}{\partial h} \right|_{h=0}, \tag{5.2.33}$$

which corresponds to the thermodynamic relation between the spontaneous magnetization and the first derivative of the free energy with respect to the magnetic field (cf. (5.2.29)).

[1]) LEVINSHTEIN et al. (1975) interpreted $1 - e^{-h}$ as the probability that a site is connected via a ghost bond with a single ghost site (and hence with an infinite cluster), which is equivalent to the above use of h. Note that REYNOLDS et al. (1980) use h, instead of $1 - e^{-h}$, for the probability of such a ghost bond. To the leading order, there is no discrepancy between the two different utilizations of h.

Furthermore, the mean size $S(p)$ of the cluster belonging to the occupied origin is given by

$$S(p) = \sum_s s p_s(p) \Big/ \sum_s p_s(p) = \sum_s s^2 n_s(p) \Big/ \sum_s s n_s(p)$$
$$= \frac{1}{p - pP(p)} \left. \frac{\partial^2 F(p, h)}{\partial h^2} \right|_{h=0}, \tag{5.2.34}$$

which is the analogue to the thermodynamic relation between the susceptibility and the second derivative of the free energy with respect to the magnetic field (cf. (5.2.29)).

For percolation (as for thermodynamics) the scaling hypothesis implies that the singular part[1]) of the "free energy" $F(p, h)$ is a generalized homogeneous function satisfying

$$F_{\text{sing}}(\lambda^{a_\varepsilon} \varepsilon, \lambda^{a_h} h) = \lambda F_{\text{sing}}(\varepsilon, h), \tag{5.2.35}$$

where

$$\varepsilon = (p_c - p)/p_c. \tag{5.2.36}$$

Note that the relation (5.2.35) was confirmed by Monte Carlo calculations (NAKANISHI and STANLEY, 1978).

A scaling relation of the type (5.2.35) is based on the following idea. Close to the percolation threshold, the properties of a percolation model are governed by large clusters (up to a nonsingular part). By changing the length scale, say by a factor $b \equiv \lambda^{1/d} > 1$, the large clusters and thus the singularities of the percolation problem should be not affected. The functions of interest, such as the free energy $F(p, h)$, are to maintain the original singular structure. The variables ε and h, however, must now have new values, to compensate for the change in the length scale (for the change in the cluster size): $\varepsilon \to \lambda^{a_\varepsilon} \varepsilon$ and $h \to \lambda^{a_h} h$, with definite values of the scaling powers a_ε and a_h. Equation (5.2.35) relates the singular part of the mean number of clusters (per site) before and after the change of the length scale. The factor $\lambda = b^d$ on the right-hand side of this equation is due to the fact that the new lattice contains a factor b^d fewer sites.

The scaling relation (5.2.35) can be obtained as follows (DUNN et al., 1975). Expanding the exponential in (5.2.30) in powers of h, we get

$$F(p, h) = \sum_{l=0}^{\infty} \frac{F_l}{l!} h^l, \tag{5.2.37}$$

where

$$F_l = (-1)^l \sum_{s=1}^{\infty} s^l n_s(p). \tag{5.2.38}$$

For p close to p_c, in (5.2.38) the sum over s is assumed to be dominated by clusters of a typical size $\mathscr{S}(p)$ (i.e. $n_s(p)$ is assumed to be peaked at $s \approx \mathscr{S}(p)$) of which there

[1]) Here and in the following the term "singular" denotes the leading nonanalytic part of a quantity. It does not necessarily mean that the corresponding quantity diverges.

are $\mathcal{N}(p)$ per site. It may be expected that $\mathcal{S}(p) \to \infty$ as $p \to p_c$, which suggests that as $\varepsilon \to 0$

$$\mathcal{S}(p) \sim |\varepsilon|^{-\Delta}. \tag{5.2.39}$$

Assuming further that as $\varepsilon \to 0$

$$\mathcal{N}(p) \sim |\varepsilon|^{2-\alpha}, \tag{5.2.40}$$

we obtain from (5.2.38) to (5.2.40)

$$F_l \sim \mathcal{N}\mathcal{S}^l \sim |\varepsilon|^{2-\alpha-l\Delta}. \tag{5.2.41}$$

Inserting (5.2.41) into (5.2.37), we find the following expression for the singular part of the free energy[1])

$$F_{\text{sing}}(p, h) = |\varepsilon|^{2-\alpha} \sum_{l=0}^{\infty} c_l^{\pm} (h/|\varepsilon|^{\Delta})^l = |\varepsilon|^{2-\alpha} \mathcal{F}_{\pm}(h/|\varepsilon|^{\Delta}), \tag{5.2.42}$$

where $+(-)$ refers to $\varepsilon > 0$ $(\varepsilon < 0)$.

Putting

$$1/(2 - \alpha) = a_\varepsilon, \tag{5.2.43}$$

$$\Delta = a_h/a_\varepsilon, \tag{5.2.44}$$

then (5.2.42) becomes

$$F_{\text{sing}}(p, h) = |\varepsilon|^{1/a_\varepsilon} \mathcal{F}_{\pm}(h/|\varepsilon|^{a_h/a_\varepsilon}). \tag{5.2.45}$$

As can readily be seen, $F_{\text{sing}}(p, h)$ from (5.2.45) satisfies the scaling property (5.2.35). Setting in (5.2.35) $\lambda = |\varepsilon|^{-1/a_\varepsilon}$, we obtain

$$F_{\text{sing}}(\pm 1, |\varepsilon|^{-a_h/a_\varepsilon} h) = |\varepsilon|^{-1/a_\varepsilon} F_{\text{sing}}(\varepsilon, h), \tag{5.2.46}$$

and by (5.2.45), (5.2.46) we find that

$$F_{\text{sing}}(\pm 1, x) = \mathcal{F}^{\pm}(x). \tag{5.2.47}$$

Now let us turn to the problem of the scaling laws. Differentiating $F_{\text{sing}}(p, h)$ from (5.2.42) with respect to h, we get

$$\frac{\partial^l F_{\text{sing}}}{\partial h^l} = |\varepsilon|^{2-\alpha-l\Delta} \frac{\partial^l \mathcal{F}_{\pm}(h/|\varepsilon|^{\Delta})}{\partial (h/|\varepsilon^{\Delta}|)^l}. \tag{5.2.48}$$

---

[1]) Note that, strictly speaking, (5.2.42) reads $F_{\text{sing}}(p, h) = q_0 \, |q_1\varepsilon|^{2-\alpha} \, \mathcal{F}_{\pm}(h/|q_1\varepsilon|^{\Delta})$, where q_0 and q_1 are nonuniversal amplitudes and $\mathcal{F}_{\pm}$ are universal functions. Note further that the scaling hypothesis (5.2.42) is equivalent to the assumption that the function $n_s(p)$ has the following scaling form (see, e.g., the reviews by Stauffer, 1979 and Essam, 1980)

$$n_s(p) = q_0 s^{-\tau} f(q_1 \, |\varepsilon| \, s^{\sigma}), \qquad s \gg 1,$$

where q_0 and q_1 are nonuniversal amplitudes, f is a universal function and $\tau = 2 + 1/\delta$, $\sigma = 1/\beta\delta$ (cf. (5.2.52)). This is due to the fact that $n_s(p)$ is the inverse Laplace transform of $F(p, h)$.

By (5.2.33), (5.2.48) and (5.2.1), we find for β (the critical exponent of the singular part of $P(p)$)

$$\beta = 2 - \alpha - \Delta, . \tag{5.2.49}$$

and by (5.2.34), (5.2.48) and (5.2.9) for γ (the critical exponent of the singular part of $S(p)$)

$$\gamma = 2\Delta - 2 + \alpha . \tag{5.2.50}$$

We may define the extension of the percolation probability to non-zero h as (cf. (5.2.33))

$$P(p, h) = 1 - \frac{1}{p} \sum_s e^{-hs} p_s(p) = 1 - \frac{1}{p} \sum_s e^{-hs} s n_s(p) = 1 + \frac{1}{p} \frac{\partial F(p, h)}{\partial h}, \tag{5.2.51}$$

which is the probability that a given site (origin) belongs to an infinite cluster or to a cluster which is occupied by at least one ghost particle ($p_s(p) \exp(-sh)$ is the probability that the origin is occupied and belongs to a cluster with s sites all of them are free from ghost particles). For $h \ll 1$, a new critical exponent δ may be defined by

$$P(p = p_c, h) \sim h^{1/\delta}. \tag{5.2.52}$$

By means of (5.2.51) and (5.2.48), we obtain

$$P(p = p_c, h) \sim h^{\left(\frac{2-\alpha}{\Delta} - 1\right)} \lim_{z \to \infty} z^{-\left(\frac{2-\alpha}{\Delta} - 1\right)} \mathcal{F}'_{\pm}(z), \tag{5.2.53}$$

and from (5.2.52) and (5.2.53) we find that

$$1/\delta = \frac{2 - \alpha}{\Delta} - 1 . \tag{5.2.54}$$

Finally, from (5.2.42), we see that close to the percolation threshold, the asymptotic form of the singular part of the mean number of clusters per site is given by

$$F_{\text{sing}}(p, 0) \sim (p_c - p)^{2-\alpha}, \tag{5.2.55}$$

i.e. $(2 - \alpha)$ is the corresponding critical exponent. Accordingly, all the critical exponents introduced here, with the exception of ν, can be expressed by Δ and α, which renders it possible to write down relations (scaling laws) between these exponents.

A scaling relation (analogous to (5.2.35)) may also be formulated for the correlation function (pair-connectedness), which leads to a new independent critical exponent η. One finds that (see, e.g., Essam, 1980)

$$\gamma = (2 - \eta)\, \nu . \tag{5.2.56}$$

The number of independent exponents can be reduced from three to two, which can be shown by the following simple argument (Dunn et al., 1975). In the region just below p_c a typical cluster occupies a volume L^d (L — correlation length). Thus, the number $\mathcal{N}$ of such clusters is of order L^{-d}. Assuming $F_{\text{sing}}(p, 0) \sim \mathcal{N}$, then

$$F_{\text{sing}}(p, 0) \sim L^{-d} \sim \varepsilon^{d\nu}, \tag{5.2.57}$$

and by means of (5.2.55) we obtain the hyperscaling relation (Dunn et al., 1975, Levinshtein et al., 1975)

$$d\nu = 2 - \alpha = 2\beta + \gamma. \tag{5.2.58}$$

For more details on the problem of scaling and universality in the percolation theory the reader is referred to the reviews by Stauffer (1979) and Essam (1980).

5.2.3. *The real-space renormalization-group method*

Wilson's renormalization-group (RG) technique (see, e.g., Wilson and Kogut, 1974) for studying the critical behaviour of thermodynamic systems is based on the idea that at the critical point the correlation length diverges, such that close to this point the system looks similar, irrespective of the length scale used. The RG technique is to remove successively the number of degrees of freedom of the system leaving only those related to the long-wavelength fluctuations, which govern critical behaviour. It is a method for calculating critical exponents. It allows also to see, why the principle of universality holds.

The two basic realizations of the RG technique is the real (position-)space approach and the ε-expansion. In the real-space approach the lattice points are successively thinned out by replacing local properties of sites and bonds by smaller numbers of sites and bonds with renormalized parameters in such a manner that local connectedness and translational symmetry are preserved. In the ε-expansion the number of degrees of freedom is reduced by removing high wavenumber fluctuations. On applying the ε-expansion to percolation, the equivalence between the percolation problem and the $Q \to 1$ limit of the Q-state Potts model is used. Application of the real-space RG approach to percolation does not require a Hamiltonian (but it can also be set up by using a Hamiltonian of the Potts type). In this approach the original lattice is rescaled by a decimation procedure and the renormalized parameters are calculated directly. Only the latter approach, which was proposed by Young and Stinchcombe (1975), Kirkpatrick (1977), and Reynolds et al. (1977, 1980), is discussed in the present section.

In the real-space RG approach the "site" problem is mapped onto a "block" problem by partitioning the lattice into equal blocks (cells) covering the lattice. The cells become renormalized sites in a new lattice of the same symmetry, but with a lattice constant which is by a factor $b = N^{1/d}$ (rescaling factor) times larger than that of the original lattice[1]), where N is the number of sites in a cell. On the renormal-

---

[1]) Although here above all, we have in view, site percolation, the above representation also holds analogously for bond percolation.

In the site problem, a block is considered to be occupied if it is possible to traverse it (in any direction). The probability of traversing it is then identified with the rescaled concentration.

In the bond problem, lattice rescaling is realized by removing sites and connecting the remaining sites with new bonds. The new lattice has a new bond concentration in order to leave the same probability of getting between the remaining sites via the new bonds as via the old ones.

ized lattice the correlation length L (measured in units of the new lattice constant) is smaller by a factor b and we are thus further from the percolation threshold, where L becomes indefinitely large. Therefore, in order to compensate for the change in the length scale, the set of parameters $\{p_i\}$ (occupation probabilities of various kinds) must take on new renormalized values $\{p_i'\}$. We may write

$$L(\{p_i'\}) = \frac{1}{b} L(\{p_i\}). \tag{5.2.59}$$

The set $\{p_i'\}$ is related to $\{p_i\}$ by the renormalization-group transformation (recursion relations)

$$\{p_i'\} = R(\{p_i\}), \tag{5.2.60}$$

which has a "fixed point" $\{p_i^*\}$ at which R leaves $\{p_i\}$ unchanged

$$\{p_i^*\} = R(\{p_i^*\}). \tag{5.2.61}$$

From (5.2.59) we see that a fixed point corresponds to either $L = 0$ or $L = \infty$. Accordingly, a critical point (at which L becomes indefinitely large) is associated with a fixed point.

In the simplest case the only quantities in the set $\{p_i\}$ are p, the site (bond) occupation probability, and the ghost field h (cf. Section 5.2.2.). In this case the recursion relations (5.2.60) become

$$\begin{aligned} p' &= R_p(p, h), \\ h' &= R_h(p, h). \end{aligned} \tag{5.2.62}$$

The percolation threshold corresponds to the fixed point at $p = p^*$ and $h^* = 0$ Note that the transformation (5.2.62) is approximate, since it neglects bonds between distant sites and correlations which are introduced by the scaling procedure (see below).

As discussed in the preceding subsection, by a change of length scale the mean total number of clusters is left constant up to a nonsingular term. Thus, the singular part of the free energy $F(p, h)$ (the mean number of clusters per site which are free from ghost particles) before and after renormalization is related by

$$F_{\text{sing}}\big((p' - p^*), h'\big) = b^d F_{\text{sing}}\big((p - p^*), h\big), \tag{5.2.63}$$

where the factor b^d accounts for the fact the new lattice contains b^d fewer sites than the original one.

Linearizing the renormalization group equations (5.2.62) close to the fixed point, we obtain

$$(p' - p^*) = \lambda_p(p - p^*), \tag{5.2.64}$$

$$h' = \lambda_h h, \tag{5.2.65}$$

where

$$\lambda_p = \mathrm{d}p'/\mathrm{d}p|_{\substack{p=p^* \\ h=0}}, \tag{5.2.66}$$

$$\lambda_h = \mathrm{d}h'/\mathrm{d}h|_{\substack{p=p^* \\ h=0}}. \tag{5.2.67}$$

Defining scaling powers y_x by

$$\lambda_x \equiv b^{y_x}, \qquad x = (p, h), \tag{5.2.68}$$

and using (5.2.64) and (5.2.65), then (5.2.63) becomes

$$F_{\text{sing}}\big(b^{y_p}(p - p^*), b^{y_h}h\big) = b^d F_{\text{sing}}(p - p^*, h). \tag{5.2.69}$$

With

$$b^d = \lambda, \qquad y_p/d = a_\varepsilon, \qquad y_h/d = a_h, \text{ and } p - p^* = \varepsilon,$$

equation (5.2.69) takes the form of the relation (5.2.35), which was discussed in detail in the preceding section. In particular, from the fact that F_{sing} is a generalized homogeneous function follows that the critical exponents α, β, δ, and γ can be expressed by y_p and y_h as (cf. (5.2.49), (5.2.50), (5.2.54))

$$2 - \alpha = d/y_p, \tag{5.2.70}$$

$$\beta = (d - y_h)/y_p, \tag{5.2.71}$$

$$\delta^{-1} = (d - y_h)/y_h, \tag{5.2.72}$$

and

$$\gamma = (2y_h - d)/y_p. \tag{5.2.73}$$

The critical exponent ν of the correlation length may also be expressed by y_p and y_h. To do this, by means of (5.2.68), we write (5.2.64) as

$$(p' - p^*) = b^{y_p}(p - p^*), \tag{5.2.74}$$

and raise (5.2.74) to the $(-\nu)$th power

$$(p' - p^*)^{-\nu} = b^{-\nu y_p}(p - p^*)^{-\nu}. \tag{5.2.75}$$

Since $L \sim (p - p^*)^{-\nu}$, by means of (5.2.75) and (5.2.59) we find (cf. (5.2.58)) that

$$\nu = y_p^{-1} = d/(2 - \alpha). \tag{5.2.76}$$

To illustrate thq basic scaling procedure, let us consider examples of bond and site percolation on a lattice, and of percolation on a continuum. At this point, we deal with renormalization of only the occupation probability p.

First, as example of bond percolation, we scale the square lattice of Fig. 5.19 to a square lattice with a lattice constant larger by a factor $b = \sqrt{2}$ (cf. Young and Stinchcombe, 1975). The remaining sites after scaling are indicated by crosses in this figure. To relate the probability p of a nearest-neighbour bond in the original lattice to the probability p' of a nearest-neighbour bond in the new lattice, consider the four-site cluster $A1B2$ in Fig. 5.19. Combining the probabilities of the possible nearest-neighbour bond configurations, we find the renormalization-group transformation

$$p' = 2p^2(1-p)^2 + 4p^3(1-p) + p^4$$
$$= 2p^2 - p^4. \tag{5.2.77}$$

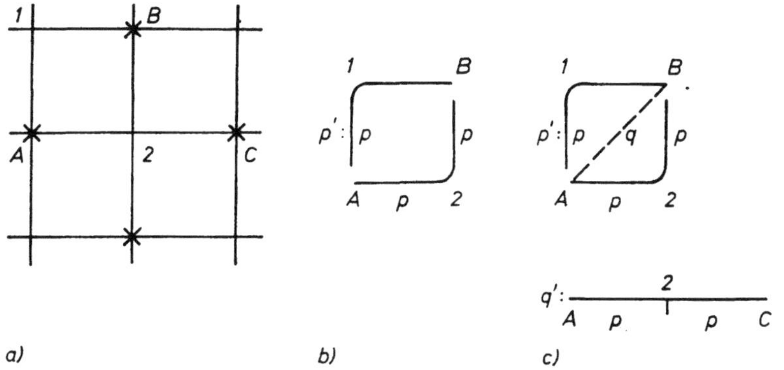

Fig. 5.19. a) Scaling of a square lattice by $b = \sqrt{2}$; b) combination of paths for the scaled nearest-neighbour probability p'; c) combination of paths for scaling of nearest-neighbour and next-nearest-neighbour probabilities; (see text) (after YOUNG and STINCHCOMBE, 1975)

Note that actually, after one stage of the scaling procedure, in the new lattice a next-nearest-neighbour bond (e.g. between sites A and C via 2) occurs (with probability $q' = p^2$). Consistency of the renormalization procedure requires that we also account for next-nearest-neighbour bonds in the original lattice (probability q, cf. Fig. 5.19c)). On account of paths of no more than two steps on the original lattice, the renormalization-group transformation now becomes

$$p' = q + 2p^2 - p^4 - 2qp^2 + p^4q,$$
$$q' = p^2. \tag{5.2.78}$$

Continuing the scaling procedure, more distant bonds must be incorporated. An exact treatment in this manner would involve an infinite number of bonds, such that (5.2.60) would then become an infinite set of coupled equations, which in practice must be truncated somewhere.

Now let us turn to the site percolation problem. As mentioned above, the original lattice is partitioned into equal blocks, which are taken as new renormalized sites. A block is said to be percolative, when it contains a cluster of occupied sites which renders it possible to traverse the block between opposite edges (Fig. 5.20). This definition leads to the renormalization-group transformation

$$p' = \sum_{\mathcal{G} \in P} W_{\mathcal{G}}(p) \equiv R(b; p), \tag{5.2.79}$$

where p and p', respectively, are the site occupation probabilities before and after scaling, $W_{\mathscr{C}}(p)$ is the probability of a configuration $\mathscr{C}$, and the summation is over the set P of percolating configurations.

$R(b; p)$ can be calculated by Monte Carlo simulation by using (5.1.12), which (with $l \equiv b = N^{1/d}$, $\mathscr{P}_b(p) \equiv R(b; p)$, N — number of sites in a block), may be written as

$$R(b; p) = \int_0^p \mathrm{d}p' f_b(p'), \tag{5.2.80}$$

where $f_b(p)$ (cf. Section 5.2.1.) is the distribution function (normalized to unity) obtained by determining the percolation threshold for finite blocks of length $b = N^{1/d}$ of an edge.

By means of (5.2.76), (5.2.68), and (5.2.66) the correlation length critical exponent ν is given by

$$\nu = \ln b/\ln \lambda_p, \tag{5.2.81}$$

where

$$\lambda_p = \mathrm{d}R(b; p)/\mathrm{d}p|_{p=p^*}, \tag{5.2.82}$$

and p^* denotes the fixed point of (5.2.79). Note that (5.2.81) and (5.2.82) agree with (cf. Section 5.2.1.) $f_b(p^*) \sim 1/w_b$, $w_b \sim b^{-1/\nu}$ (w_b — half width of $f_b(p)$, which is nearly a Gaussian).

The transformation (5.2.79) is only approximate, since it neglects the correlations between blocks arising from the necessity that there must exist paths of occupied

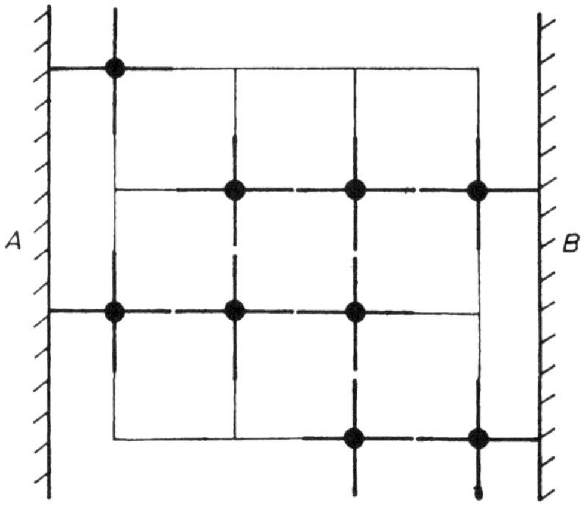

Fig. 5.20. Illustration of the renormalization-group procedure for site percolation. An occupied site is represented by a diagram with legs (half-bonds) directed towards nearest-neighbour sites. A pair of touching legs form a nearest-neighbour bond. The horizontal direction is singled out as a direction of percolation. A block (the figure shows an example for $d = 2$ and $b = 4$) is said to be percolative (i.e. the corresponding renormalized site is occupied) if it renders possible percolation between the boundaries (electrodes) A and B

sites linking the blocks (for more details see, e.g., REYNOLDS et al., 1980, PAYANDEH, 1980, JAN et al., 1984).

Now let us apply real-space RG technique to percolation on a continuum. Following VICSEK and KERTESZ (1981), here we study scaling behaviour of overlapping circles (cf. Section 5.2.1.), i.e. of circles of equal radii ($R/2$) centred on sites which are randomly distributed on a plane. The sites whose circles intersect are considered as belonging to the same cluster. The critical radius ($R_c/2$) at which the inifinite cluster occurs depends on the concentration $\mathcal{N}$ of sites. It is convenient to introduce the dimensionless variable

$$s = 1 - \exp\left(-\pi(R/2)^2 \mathcal{N}\right), \tag{5.2.83}$$

which is the so-called area fraction, the $d = 2$ analogue of the volume fraction v introduced in Section 5.2.1. (cf. (5.2.8)). s plays the role of the occupation probability.

We subdivide the irregular lattice into square-shaped cells (blocks) with length of an edge characterized by the quantity $b = N^{1/2}$, where N is the average number of sites per cell. The actual number of sites in the ith such cell, N_i, is distributed according to the Poisson distribution: $P_{N_i} = N^{N_i}\,\mathrm{e}^{-N}/N_i!$. Analogous to the site percolation problem, where cells become renormalized sites, here the decimation procedure consists in replacing cells with the average number $N = b^2$ of sites by such with smaller average number $N' = b'^2$ of sites, i.e. the rescaling factor is b/b'. The renormalization group is defined by

$$R'(b'; s') = R(b; s), \tag{5.2.84}$$

where $R(b; s)$ denotes the probability that a cell of size b percolates and $R'(b'; s')$ is the corresponding probability for the renormalized problem with area fraction s'. $R(b; s)$ is given by

$$R(b; s) = \sum_{k=1}^{\infty} P_k(N)\, R[b; s(k)], \tag{5.2.85}$$

where $P_k(N)$ is the probability of finding k circles in a cell for a given N and $R[b; s(k)]$ is the probability of finding a percolating cluster (connecting two opposite edges) in a cell containing k circles. Here $s(k)$ is the corresponding area fraction given by

$$s(k) \approx 1 - \left(1 - \pi\mathcal{N}(R/2)^2/N\right)^k. \tag{5.2.86}$$

Analogous to the site percolation problem, $R(b; s)$ can be obtained by Monte Carlo simulation by (cf. (5.2.80))

$$R(b; s) = \int_0^s \mathrm{d}s' f_b(s'), \tag{5.2.87}$$

where $f_b(s)$ is the distribution of the critical area fraction for finite cells of size $b = N^{1/2}$. Similarly as for site percolation, the correlation length critical exponent for the problem at hand can be found from

$$\nu = \ln (b/b')/\ln \lambda_{b,b'} \tag{5.2.88}$$

with

$$\lambda_{b,b'} = \left(\frac{\mathrm{d}R(b;s)}{\mathrm{d}s} \Big/ \frac{\mathrm{d}R'(b';s)}{\mathrm{d}s} \right)\Bigg|_{s=s^*}, \tag{5.2.89}$$

where s^* is the fixed point of the transformation (5.2.84). By the method under consideration VICSEK and KERTESZ (1981) found for overlapping circles $\nu = 1.33 \pm 0.07$, which well agrees with values obtained on triangular and square lattices $\nu = 1.32^{+0.02}_{-0.07}$ (COX and ESSAM, 1976), $\nu = 1.354 \pm 0.015$ (REYNOLDS et al., 1978) and $\nu = 4/3$ (DEN NIJS, 1979), which indicates regular-irregular lattice universality.

We close this section by applying the real-space RG technique to percolation conductivity of dilute resistor networks, in which the conductance between two adjacent lattice sites may assume either some finite value σ (with probability p) or zero (with probability $1 - p$). As mentioned above, the total conductance G of the lattice is zero for p below the percolation threshold p_c; and at p slightly above p_c it increases as $G \sim (p - p_c)^t$, where t is the critical conductivity exponent (cf. (5.2.17)).

The conductance $G(p)$ is a more complex quantity than the percolation quantities such as the percolation probability $P(p)$ and the mean cluster size $S(p)$, since it also depends on the topology of the infinite cluster. However, at p very close to p_c, where the correlation length L becomes very large, the underlying length scale disappears and the distribution of the cluster size and the topology of the infinite cluster become independent of the length scale. Accordingly, analogous to the critical percolation, the critical percolation conductivity may be studied by constructing a renormalization-group transformation. Following ROSMAN and SHAPIRO (1977), let us demonstrate this procedure for the square lattice model shown in Fig. 5.19. We require that the average conductance between sites A and B (cf. Fig. 5.19) remains the same after removing sites 1 and 2. Then, in addition to (5.2.77) we have

$$p'\sigma' = \frac{\sigma}{2} \left(2p^2(1-p)^2 + 4p^3(1-p)\right) + \sigma p^4 = p^2\sigma. \tag{5.2.90}$$

Here σ' is the conductance of a nearest-neighbour bond in the rescaled lattice. Close to p^* (fixed point) the conductance may be written as

$$G = \sigma l^{d-2}(p - p^*)^t, \tag{5.2.91}$$

where l is the linear dimension of the lattice measured in units of the lattice constant, and d is the dimensionality. The requirement, that a change of the length scale must left unchanged the conductance, leads to

$$\sigma'(l/b)^{d-2} (p' - p^*)^t = \sigma l^{d-2}(p - p^*)^t, \tag{5.2.92}$$

where b is the rescaling factor. Using $(p' - p^*) = \lambda(p - p^*)$ (cf. (5.2.64)), from (5.2.92), we find for the conductivity critical exponent the expression

$$t = [\ln(b^{d-2}\sigma/\sigma')]/\ln \lambda, \tag{5.2.93}$$

which, by $\nu = \ln b/\ln \lambda$ (cf. (5.2.81)), may be written as

$$t = [d - 2 + \ln (\sigma/\sigma')/\ln b]\,\nu. \tag{5.2.94}$$

Here ν is the correlation length critical exponent, and, by (5.2.90), it holds that $\sigma/\sigma' = 1/p$. In this way, the reasonable value $t = 1.134$ was found. On account of next-nearest-neighbour bonds (see above) one gets $t = 1.359$ (Rosman and Shapiro, 1977).

Note that in a more refined treatment of the percolation conductivity, one renormalizes the conductance-distribution function rather than the average conductance (Stinchcombe and Watson, 1976). Note further, the real-space RG approach can also be used for studying site-dilute resistor networks (Yeomans and Stinchcombe, 1978). Different real-space RG studies lead to appreciable different values for the conductivity critical exponent (Bernasconi, 1978). Certain "pathological" conductivity distribution functions give rise to non-universal conductivity behaviour (Kogut and Straley, 1979, Bergman and Kantor, 1981, Straley, 1982).

5.2.4. *Relationship between percolation and effective-medium treatments of random-resistor networks*

As noted above, theoretical discussions of random-resistor networks have been given by means of numerical solutions of finite networks, the Bethe lattice, the node-link model, the fractal model and the real-space renormalization group transformation. Moreover, random-resistor networks have been studied by a scaling theory (Straley, 1976), by relating the resistor network to the zero-state Potts model (Dasgupta et al., 1978, cf. Wu, 1982), by a series expansion (Fisch and Harris, 1978), by a mean-field theory (in close analogy to the problem of thermal phase transitions, after properly defining an order parameter) (Stephen, 1978) and by the effective-medium approximation (EMA) (cf. Section 6.4.) whose relation to the percolation approach will be examined in the present section.

The EMA is a standard approximation for transport in disordered systems (cf., e.g., Yonezawa et al., 1982). It has its origin in the work of Bruggeman (1935) and is formally equivalent to the CPA (coherent potential approximation) (Kirkpatrick, 1973), a familiar approximation for compositionally disordered systems. The EMA was first applied to resistor networks by Kirkpatrick (1971, 1973), who pointed out that this approximation is useful for some uniform resistor distributions and for binary resistor distributions, except near the conduction threshold, while there appears a discrepancy for broad log-uniform distributions of the kind considered in Section 5.3.1. (cf. Koplik, 1981, and references therein). Let us consider results obtained for binary resistor distributions, where two kinds of resistors are distributed on a regular lattice, with a concentration p of resistors with a conductivity σ_1 and with a concentration $(1 - p)$ with a conductivity σ_2. If σ_1 and σ_2 are comparable, then the EMA well agrees with computer results. For $\sigma_2 \ll \sigma_1$, the EMA gives a linear decrease with decreasing p (i.e. for the critical exponent t, we have $t = 1$[1]), whereby the conductivity goes to σ_2 at $p_c = 1/3$ for a simple cubic network, and at $p_c = 1/2$ for a square network (for comparison, the generally accepted "exact" values are $p_c = 0.500$ for a square and $p_c = 0.247$ for a simple cubic lattice; and $1.0 < t < 1.5$ for $d = 2$, $1.5 < t < 2.0$ for $d = 3$). Computer results show this linear dependence, except near the threshold. Note that the EMA is exact in two limits of coordination number z, for $z = 2$ and $z \to \infty$ (KOPLIK, 1981).

In the EMA, all resistors are associated with a single conductivity value which is chosen such that the effect of changing any one conductance back to its true value will vanish on the average (cf. Section 6.4.). The fact that near the percolation threshold the EMA gives rather poor agreement with Monte Carlo results is not surprising, because the (single-bond) EMA excludes correlations between the values of the bonds. Analogously, the single-site CPA cannot produce, for example, structure in a density of states and describe correctly the behaviour near a band edge, because it excludes spatial correlations. A description of such phenomena requires account of cluster effects (see, e.g., ELLIOTT et al., 1974, BÖTTGER, 1983).

Making use of a perturbation method for the effective conductivity, a cluster theory of bond percolation was developed by BLACKMAN (1976) and later extended by AHMED and BLACKMAN (1979) and NAGATANI (1981, 1982). It turns out that a cluster theory yields an improvement over the single-bond EMA, but the convergence to the exact result (achieved by such a theory with increasing cluster size) appears to be very slow. For example, by a two-bond EMA NAGATANI (1981) obtained for a simple cubic lattice the threshold value $p_c = 0.315$ ("exact" value $p_c = 0.247$). Compared with the single-bond EMA value $p_c = 0.333$, this result shows a slight improvement.

The EMA has also been applied to site percolation (e.g., WATSON and LEATH, 1974, BERNASCONI and WIESMANN, 1976, PIKE and SEAGER, 1976, NAKAMURA, 1984). In this case, one considers the cluster of bonds immediately around a lattice site, i.e. one accounts for correlations between the values of bonds with a common site. An alternative technique of the type of the EMA is the cumulant theory proposed by HORI and YONEZAWA (1974, 1975, 1977), who applied this technique to conduction in a continuum. Its application to bond percolation was studied by AHMED and BLACKMAN (1979). The cumulant technique also allows to account for cluster effects. It appears to be a good approximation even near the percolation threshold, at least for a simple cubic lattice. At present, it is not clear, whether this technique is generally good in three dimensions. For the square lattice, the EMA gives better results than the cumulant technique. A combination of real-space RG and EMA techniques has been proposed by SAHIMI et al. (1983, 1984). — We conclude this section with some remarks on the upper critical dimension d_c above which percolation exponents are independent of the dimensionality d of the system. According to the present-day theories, for thermal phase transitions, a critical dimension exists above that the

[1]) NAKAMURA (1982) proposed a method to improve the EMA such that near the percolation threshold the effective conductivity increases less than linearly. The results obtained by this method well agree with computer data, both for $d = 2$ and $d = 3$.

mean field critical exponents are exact. Therefore, owing to the intimate relation between percolation and thermal phase transitions (which results from the Potts model (cf. Appendix 3)), one may expect that for percolation an analogous critical dimension d_c exists.

In this context note that the dimensionality of a Bethe lattice (Cayley tree) (cf. Section 6.6.) is infinite, because the number of distinct sites which can be reached in n steps from the origin of the lattice grows exponentially in n, while in a lattice with finite dimension d it grows as n^d. This led TOULOUSE (1974) to the idea that, for percolation, the Bethe lattice is the analog of the mean field approximation for thermal phase transitions, so that the Bethe lattice, which can be solved exactly, yields the corrent critical percolation exponents for $d \geqq d_c$. Thus, using the values of the critical exponents of the Bethe lattice ($\alpha = -1, \nu = 1/2, \beta = 1, \gamma = 1$) in the hyperscaling relation (cf. (5.2.58))

$$d\nu = 2 - \alpha = 2\beta + \gamma, \tag{5.2.95}$$

Toulouse finds that $d_c = 6$ (note that for thermal phase transitions, $d_c = 4$, and for directed percolation, $d_c = 5$ (OBUKHOV, 1980)).

This result was later verified by Monte Carlo calculations (see, e.g., HARRIS and FISCH, 1977, FISCH and HARRIS, 1978). It was also confirmed by GINZBURG (1976) and STEPHEN (1977), who gave field theories for the percolation problem in which the vertex renormalization becomes irrelevant above $d_c = 6$. For further discussions of the critical dimension $d_c = 6$, see, e.g., KIRKPATRICK (1977), STEPHEN (1978), STRALEY (1982) and references therein.

Note that the value $d_c = 6$ suggests that renormalization-group methods basing on an expansion with respect to the "small" parameter $\varepsilon = 6 - d$ do not yield reliable results for $d = 2$ or 3, so that real-space renormalization group methods are necessary. According to FUCITO and PARISI (1981), an ε-expansion gives good results for $5 \leqq d \leqq 3$, but not for $d = 2$.

5.3. The percolation approach to hopping conduction

According to AMBEGAOKAR et al. (1971) and SHKLOVSKII and EFROS (1971), the random-resistor network problem of Miller and Abrahams (cf. Section 5.1.3.) may be related to the classical percolation problem as follows: The resistors $Z_{mm'}$ are all removed from the network and after that replaced one by one, the smallest first. The value of $Z_{mm'}$ at which the first infinite cluster occurs is Z_c, the threshold value in the language of percolation theory. It may be argued that for a very broad distribution of $Z_{mm'}$ the current flow is not strongly affected by replacing further resistors with $Z_{mm'} > Z_c$, since the magnitude of these resistors rapidly increases. The average conductivity σ of the network may be expressed as

$$\sigma = \frac{1}{l} Z_c^{-1}, \tag{5.3.1}$$

where l is some characteristic length weakly (as powers) depending on the concentration of sites and the temperature. Percolation theory allows the exponential dependence (the values of the numbers η (5.1.11) and ε_3 (5.1.13) in the exponentials) and, up to a numerical factor, also the pre-exponential factor of the hopping conductivity σ (5.3.1) to be found.

In the present section, first we study conductivity in a simple cubic lattice with resistors joining nearest neighbour sites (nodes), by assuming an exponentially large spread of the magnitude of the individual resistors. Such a spread is characteristic of actual hopping systems. Accordingly, some basic problems (applicability of the percolation approach, exponential dependence of the conductivity, pre-exponential factor) encountered on investigating the hopping conductivity can be studied by means of this simple lattice model.

In a hopping system, at high temperatures the spread of the resistors $Z_{mm'}$ is determined by the dispersion of the mutual spatial distances between the sites (situation of the so-called R-percolation or R-hopping) and at low temperatures by the dispersion of the mutual spatial and energetical distances (situation of the so-called R-ε-percolation or R-ε-hopping; regime of the variable-range hopping). The exponential dependence of the hopping conductivity in the case of R-percolation is discussed in Section 5.3.2. and that of R-ε-percolation in Section 5.3.3. Section 5.3.4. is devoted to the pre-exponential factor for R- and R-ε-percolation.

5.3.1. *Resistor network on a simple cubic lattice*

Following Kirkpatrick (1971) and Seager and Pike (1974), let us study the conductivity of a simple cubic lattice of sites (nodes) with randomly distributed resistors Z between nearest neighbours, by assuming

$$Z = Z_0 \exp \zeta' \tag{5.3.2}$$

where ζ' is uniformly distributed between $-\zeta_0$ and ζ_0, with $\zeta_0 \gg 1$, and Z_0 is a constant. As mentioned above, application of the percolation concept to the problem in question requires that first we remove all resistors from the lattice and then reincorporate them in the order of their magnitudes, starting with the smallest resistor and stopping with the critical one, Z_c, at which the first path through the infinite system (the first infinite cluster) opens.

The critical value ζ_c, defined as

$$Z_c = Z_0 \exp \zeta_c \tag{5.3.3}$$

can be related to the critical probability p_c of the nearest-neighbour bond percolation problem on the lattice under consideration. To see this, consider the quantity $p(\zeta)$ defined as

$$p(\zeta) = \int_{-\zeta_0}^{\zeta} d\zeta' f(\zeta'), \tag{5.3.4}$$

where

$$f(\zeta') = \begin{cases} (2\zeta_0)^{-1} & \text{if} \quad |\zeta| \leqq \zeta_0, \\ 0 & \text{if} \quad |\zeta| > \zeta_0 \end{cases} \tag{5.3.5}$$

is the distribution function of the values ζ'. For a lattice where all resistors with $\zeta' > \zeta$ are removed (cut), $p(\zeta)$ denotes the probability that any resistor picked out is not cut. Thus, $p_c = p(\zeta_c)$. By means of (5.3.4) and (5.3.5), we find that

$$p(\zeta) = (\zeta_0 + \zeta)/2\zeta_0 \tag{5.3.6}$$

and, in particular

$$p_c = (\zeta_0 + \zeta_c)/2\zeta_0. \tag{5.3.7}$$

Fig. 5.21 shows the conductivity of a simple cubic network (numerically computed for samples with 1000 sites each) in dependence of p (5.3.6) for several values ζ_0. We see that for large ζ_0, at p above the critical bond fraction $p_c \approx 0.25$, the conductivity steeply increases. For $\zeta_0 \geqq 9$ this increase is sharply peaked near p_c.

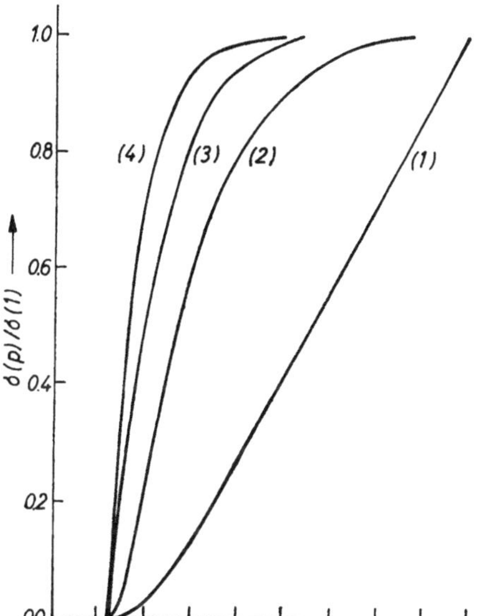

Fig. 5.21. Computed conductivity σ of simple cubic networks in dependence of the fraction p of conductances present in the networks. The individual conductances $Z_{mm'}^{-1}$ are ranging from $\exp(1 - 2p)\ \zeta_0$ to $\exp \zeta_0$, where in the curve (1) $\zeta_0 = 0$, (2) 4.605, (3) 9.21, (4) 13.85 (after Seager and Pike, 1974)

Accordingly, enhancing ζ from ζ_c (where the first percolation path opens) up to about $\zeta_c + 1$, the number of parallel percolation paths increases to such a degree that, for sufficiently large ζ_0, a further enhancement of ζ only slightly affects the conductivity. That is, for sufficiently large ζ_0, the conductivity of the network under consideration agrees to a good approximation with that of the subnetwork (termed the critical network in the following) formed by the resistors with ζ' between $-\zeta_0$ and $\zeta_c + 1$.

Replacing in the critical network all resistors by the largest one, by Z_c (5.3.3) (approximating $\exp(\zeta_c + 1) \approx \exp(\zeta_c)$, on account of $\zeta_0 \gg 1$), the critical network provides as lower bound for the conductivity

$$\sigma = \sigma_0 \exp(-\zeta_c). \tag{5.3.8}$$

The pre-exponential factor occuring in (5.3.8) will be discussed later. Note that for the system under consideration, from (5.3.7) with $p_c \approx 0.25$, we obtain $\zeta_c = -\zeta_0/2$.

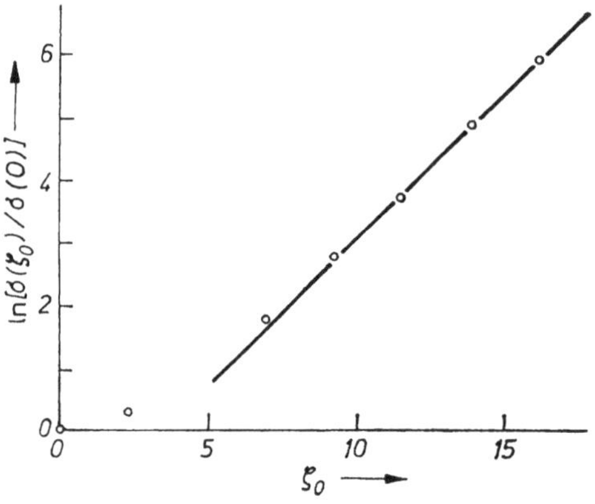

Fig. 5.22. Conductivity of a simple cubic network as a function of the parameter ζ_0 characterizing the conductance spread (after SEAGER and PIKE, 1974). The circles denote the computed data. The solid straight line has a slope close to the theoretical value 1/2

In Fig. 5.22, results of a numerical computation (using cubic samples of 1000 sites each) of the conductivity as function of the resistor spread are compared with the prediction of formula (5.3.8). We see that above $\zeta_0 \approx 9$ the computer results well agree with the percolation formula (5.3.8), but for $\zeta_0 < 9$, the predictions of the percolation model become unsatisfactory. Consequently, a percolation path analysis yields accurate prediction for the exponential dependence of the conductivity if the resistor spread ($e^{2\zeta_0}$) is at least seven to eight decades.

Now let us turn to the pre-exponential factor occurring in (5.3.8). Following KURKIJÄRVI (1974) and SHKLOVSKII and EFROS (1975) (see also SHKLOVSKII and EFROS, 1979, § 21), we imagine that the infinite resistor lattice consists of small cubes each of length l of an edge. We pick out such a cube and determine by the procedure described above the corresponding critical exponent ζ_{cl}. For sufficiently small cubes, the critical resistor is much smaller than the subsequent one. In this case the critical resistor governs the total conductivity of the cube, which becomes

$$\sigma_l = (Z_0 l)^{-1} \exp(-\zeta_{cl}). \tag{5.3.9}$$

For sufficiently small cubes, the critical exponent ζ_{cl} strongly fluctuates from cube to cube. By means of (5.2.11) and (5.3.7) we find for the root mean-square fluctuation

$$\delta\zeta_{cl} = B'\zeta_0 \left(\frac{a}{l}\right)^{1/\nu}, \tag{5.3.10}$$

where ν is the correlation-length critical exponent, a is the lattice constant, and B' is a number of order unity. When, with increasing l, $\delta\zeta_{cl}$ becomes small compared to

ζ_c, (5.3.9) yields the macroscopic conductivity. This happens when $\delta\zeta_{cl} \approx 1$, i.e. when

$$l \approx a\zeta_0{}^\nu \equiv L_0. \tag{5.3.11}$$

Accordingly, up to a numerical factor, the conductivity of the resistor lattice becomes

$$\sigma = (Z_0 a)^{-1} \zeta_0{}^{-\nu} \exp(-\zeta_c). \tag{5.3.12}$$

Note that L_0 defined in (5.3.11) is the correlation length of the critical network, since, by means of (5.2.15), (5.3.6), and (5.3.7), the correlation length L becomes

$$L \sim a(p - p_c)^{-\nu} \sim a(\zeta - \zeta_c)^{-\nu} \zeta_0{}^\nu, \tag{5.3.13}$$

which for the critical network, where $\zeta = \zeta_c + 1$, agrees with L_0 from (5.3.11). This is in accordance with the above assertion that the critical network governs the current flow through the system.

According to AMBEGAOKAR et al. (1973) (cf. KURKIJÄRVI, 1974), equation (5.3.12) implies that

$$\zeta_{cl} = \zeta_c + \zeta_0 (l^3 \mathcal{N})^{-1/3} \tag{5.3.14}$$

(and vice versa). Here $\mathcal{N}$ is the density of sites and $l^3\mathcal{N} = N$ is the number of sites in a cube of length l of an edge. To see this, we insert (5.3.14) into (5.3.9) and take the thermodynamic limit requiring N and l^3 to increase proportionally

$$\begin{aligned}\sigma &= \frac{\mathcal{N}^{1/3}}{Z_0} \lim_{N\to\infty} \frac{1}{N^{1/3}} \exp\left[-(\zeta_c + \zeta_0 N^{-1/3\nu})\right] \\ &= \frac{\mathcal{N}^{1/3}}{Z_0} \zeta_0{}^{-\nu} \mathrm{e}^{-\zeta_c} \lim_{x\to\infty} x^\nu \mathrm{e}^{-x}.\end{aligned} \tag{5.3.15}$$

The latter expression agrees with (5.3.12).

5.3.2. *Concentration dependence of the hopping conductivity (R-percolation)*

As discussed in Section 5.1.3., according to MILLER and ABRAHAMS (1960), the dc hopping conductivity leads to the problem of calculating the current flow through a random-resistor network, where the resistance $Z_{mm'}$ between sites (nodes) m and m' is given by the expression (cf. (5.1.7))

$$Z_{mm'}^{-1} = Z_0{}^{-1} \mathrm{e}^{-2\alpha|\boldsymbol{R}_{mm'}| - \frac{\beta}{2}(|\varepsilon_{m'} - \mu| + |\varepsilon_m - \mu| + |\varepsilon_{m'} - \varepsilon_m|)}, \tag{5.3.16}$$

which was obtained by assuming s-like "impurity" wave functions with Bohr radius α^{-1}. At sufficiently high temperatures (for $\Gamma/kT \ll \alpha/\mathcal{N}^{1/3}$; Γ — width of the spread of the site energies ε_m, $\mathcal{N}$ – concentration of sites), the intersite seperation dominates the individual resistors. In this case of the so-called R-percolation (POLLAK, 1972) the temperature exponent in (5.3.16) is unimportant for finding the current path across the system, and the resistance depends only on the distance $|\boldsymbol{R}_{mm'}|$ between the sites, i.e.

$$Z_{mm'}^{-1} = Z_0^{-1}\, e^{-2\alpha|\boldsymbol{R}_{mm'}|}. \tag{5.3.17}$$

Then, the average conductivity becomes (cf. (5.1.9), (5.1.11))

$$\sigma = \varrho_3^{-1}\, e^{-\varepsilon_3/kT},\ \varrho_3 = \varrho_0\, e^{f(\mathcal{N})},\ f(\mathcal{N}) = \eta\alpha/\mathcal{N}^{1/3}. \tag{5.3.18}$$

The problem is to find the value of the number η in (5.3.18).

On using the percolation theory for determining η, owing to (5.3.17), the bonding criterion (5.2.2) takes the simple form

$$\zeta_{mm'} \equiv 2\alpha|\boldsymbol{R}_{mm'}| \leqq \zeta, \tag{5.3.19}$$

i.e. for given parameter ζ, any two sites are connected by a bond, when their distance is less than or equal to $\zeta/2\alpha$. Thus, the characteristic figures of the R-percolation problem are spheres (circles) of radius $R = \zeta/2\alpha$ (inclusive figures) or of radius $R/2 = \zeta/4\alpha$ (overlapping figures) circumscribed about each site, and any two sites are viewed as connected by a bond, if $|\boldsymbol{R}_{mm'}| \leqq R$.

To obtain the exponential dependence of the conductivity for R-percolation, we now proceed as described above: We switch on all bonds (resistors) with $2\alpha|\,\boldsymbol{R}_{mm'}| \leqq \zeta$ and increase ζ. At some threshold value

$$\zeta_c = 2\alpha R_c \tag{5.3.20}$$

a percolation path opens across the infinite system. Increasing ζ further, from ζ_c to $\zeta_c + 1$, the critical network is formed, which is characterized by the critical resistor

$$Z_c = Z_0 \exp(\zeta_c), \tag{5.3.21}$$

and which yields (as lower bound) for the conductivity of the hopping system the following expression

$$\sigma = \sigma_0\, e^{-\zeta_c}\, e^{-\varepsilon_3/kT}. \tag{5.3.22}$$

The pre-exponential factor will be considered in Section 5.3.4.

According to PIKE and SEAGER (1974), for the $3d$ system under consideration, the average number B_c of bonds per site (cf. Section 5.2.1.) is given by

$$B_c \equiv \frac{4\pi}{3}\, \mathcal{N} R_c^3 = 2.7 \pm 0.1, \tag{5.3.23}$$

i.e.

$$R_c = (0.865 \pm 0.015)\, \mathcal{N}^{-1/3}. \tag{5.3.24}$$

By means of (5.3.18), (5.3.20) and (5.3.22), we obtain

$$\eta = \zeta_c \mathcal{N}^{1/3}/\alpha = 2R_c \mathcal{N}^{1/3}, \tag{5.3.25}$$

and by (5.3.25) and (5.3.24), finally we find for the number η in (5.3.18) the value

$$\eta = 1.73 \pm 0.03. \tag{5.3.26}$$

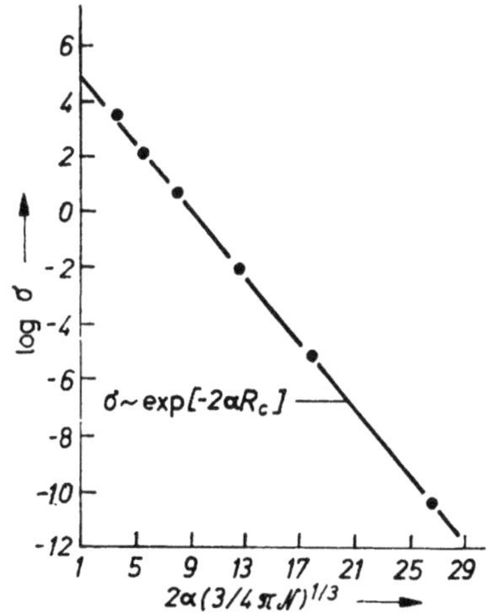

Fig. 5.23. Concentration dependence of the conductivity σ in the R-percolation limit. The circles denote the computed conductivity of a random network. The straight line is the percolation theoretical prediction (α — reciprocal Bohr radius, $\mathcal{N}$ — concentration of sites) (after SEAGER and PIKE, 1974)

To test the exponential concentration dependence of the conductivity predicted by (5.3.18) and (5.3.26), SEAGER und PIKE (1974) calculated numerically the conductivity of computer generated random networks with 1000 sites each. As shown in Fig. 5.23, their results are in excellent agreement with the prediction of (5.3.18) and (5.3.26), except for the highest density employed. Thus, the percolation approach holds for $\alpha\mathcal{N}^{-1/3} > 4$ to 5. Note that from a computer printout of the 10 smallest

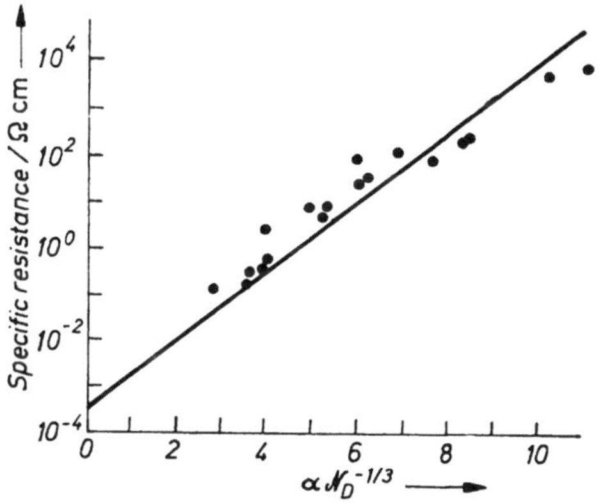

Fig. 5.24. The specific resistance of weakly doped n- GaAs as function of the donor concentration $\mathcal{N}_D$ (α-reciprocal Bohr radius) (experimental data from EMEL'JANENKO et al., 1973, OLIVER, 1962, ASAI et al., 1970, and EDDOLS, 1966)

resistors from each site, SEAGER and PIKE (1974) found (except for the highest density employed) a resistor spread of at least 10^8, in accordance with the situation in the random-resistor lattice considered in the preceding section. Furthermore, it is interesting to note that the conductivity was found to be rather independent of the number n of the smallest resistors originating from each site, for n beyond 5 or 6.

The dependence (5.3.18) was experimentally studied on several doped semiconductors. Fig. 5.24 shows the concentration dependence of the specific resistance of weakly doped n-GaAs, which can be well fitted by (5.3.18) with the theoretical value

(5.3.26) for the parameter η. Among other experimental findigs for η are: 1.9 ± 0.2 for n-InP (EMELJANENKO et al., 1975), 1.9 ± 0.1 for p-Ge(Ga) (FRITZSCHE and CUEVAS, 1960), 1.75 ± 0.25 for p-Ge(In) (FRITZSCHE, 1955), 2.0 ± 0.3 for p-Ge(Cu) (ROLLIN and RUSSEL, 1963) and 1.8 ± 0.2 for p-Si (RAY and FAN, 1961). The origin of the discrepancy between the theoretical and experimental values of η is not quite clear at present (cf. SHKLOVSKII and EFROS, 1979, § 22).

Percolation theory has also been used to study the hopping conductivity in doped semiconductors with anisotropic impurity states. Such states are encountered, for instance, in n-type germanium and silicon and also in the presence of external uniaxial pressure or of a magnetic field. In this case, the transition probability does not simply depend on $|\boldsymbol{R}_{mm'}|$ (as in the case of the R-percolation), but it also depends on the angles between $\boldsymbol{R}_{mm'}$ and certain selected directions (SKAL and SHKLOVSKII, 1973, SHKLOVSKII, 1977, cf. Section 9.3.).

5.3.3. *Variable-range hopping (R-ε-percolation)*

As discussed in Section 5.1.4., Mott's law of variable-range hopping (cf. (5.1.19))[1])

$$\sigma = \sigma_0 \exp\left[-(T_0/T)^{1/4}\right], \; T_0 = c_3{}^4\alpha^3/kN(\mu) \tag{5.3.27}$$

$N(\mu)$ — density of states at the Fermi level μ, c_3 — numerical factor) describes dc hopping conductivity at low temperatures, where the spread of the resistors in the network of Miller and Abrahams depends on the dispersion of the spatial as well as the energetical separation between the sites. By means of the percolation theory, the law (5.3.27) was first derived by AMBEGAOKAR et al. (1971) and POLLAK (1972). The percolation approach to this law is the subject of the present section.

According to (5.3.16), the bonding criterion (cf. (5.2.2.)) now becomes

$$\zeta_{mm'} \equiv 2\alpha\,|\boldsymbol{R}_{mm'}| - \frac{\beta}{2}\left(|\varepsilon_{m'} - \mu| + |\varepsilon_m - \mu| + |\varepsilon_{m'} - \varepsilon_m|\right) \leqq \zeta\,. \tag{5.3.28}$$

The problem is to find the threshold value $\zeta = \zeta_c$ at which the first percolation path across the system opens. Adopting the above conception of successively switching on resistors with increasing magnitude, the conductivity of the system is determined by the critical resistor Z_c, which is related to ζ_c by (5.3.21). According to (5.3.28), in the case of R-ε-percolation, the existence of a bond depends on five parameters: three coordinates, $\boldsymbol{R}_{mm'}$, and the energies $\varepsilon_m - \mu$ and $\varepsilon_{m'} - \mu$, i.e. the characteristic figure has in general to be constructed in a five-dimensional space, while in the case of R-percolation, the corresponding space is three-dimensional.

---

[1]) Note that in addition to variable-range hopping, a $T^{-1/4}$ law of the type (5.3.27) can, at least in principle, also arise from strong interaction with acoustic phonons (cf. Section 5.1.5.) or from electron-electron interaction (cf. Section 5.4.). For instance, SZPILKA and VISĆOR (1982) argue that the $T^{-1/4}$ dependence they have found on evaporated amorphous germa nium is rather due to strong interaction with phonons than due to variable-range hopping.

An approximate treatment of the R-ε-problem in three-dimensional space was given by AMBEGAOKAR et al. (1971) as follows. Let us rewrite (5.3.28) in the dimensionless form

$$\frac{|\boldsymbol{R}_{mm'}|}{R_{\max}} + \frac{|\varepsilon_m - \mu| + |\varepsilon_{m'} - \mu| + |\varepsilon_m - \varepsilon_{m'}|}{2E_{\max}} \leqq 1, \tag{5.3.29}$$

where

$$R_{\max} = \zeta/2\alpha, \; E_{\max} = kT\zeta. \tag{5.3.30}$$

We see that in any case $|\boldsymbol{R}_{mm'}| > R_{\max}$ and $|\varepsilon_m - \mu| > E_{\max}$ or $|\varepsilon_{m'} - \mu| > E_{\max}$ will violate the condition (5.3.29), i.e. $R_{\max}$ and $E_{\max}$ are the maximal admitted spatial distance and energetical separations from the Fermi level, respectively.

Assuming a constant density of states in the energy interval from $\mu - E_{\max}$ to $\mu + E_{\max}$, the density n of sites satisfying (5.3.29) is given by

$$n = 2N(\mu)\, E_{\max}. \tag{5.3.31}$$

Adopting the conception of overlapping figures, Ambegaokar et al. inscribe each site into a sphere whose radius is energy dependent (see below), since in the present situation $Z_{mm'}$ also depends on the energies of the sites m and m'. By averaging the volume of such a sphere with respect to its dependence on energy, the problem is reduced to a three-dimensional one. Making use of the conception of critical-volume fraction v_c (cf. Section 5.2.1., see also WINTLE, 1975), the percolation condition reads

$$\frac{4\pi}{3} \bar{R}^3 n = \Phi_c, \tag{5.3.32}$$

where $\bar{R}$ denotes the average radius of a sphere, and Φ_c is related to v_c by $v_c = 1 - \exp(-\Phi_c)$ (cf. (5.2.8)). One gets (see below)

$$\bar{R} = \gamma R_{\max}, \tag{5.3.33}$$

where γ is a number depending on assumptions on the energy dependence of the radius of the spheres.

Setting $\zeta = \zeta_c$, by means of (5.3.30) to (5.3.33), we find

$$\zeta_c = (T_0/T)^{1/4}, \tag{5.3.34}$$

where T_0 is given in (5.3.27) with $c_3{}^4 = 3\Phi_c/\pi\gamma^3$ (cf. Table 5.1). Accordingly, by means of (5.3.21) and (5.3.34), we obtain Mott's law (5.3.27).

To obtain the spheres with energy dependent radius mentioned above, AMBEGAOKAR et al. (1971) argue that there will certainly be not a bond between sites m and m', unless $|\boldsymbol{R}_{mm'}|/R_{\max}$ satisfies

$$|\boldsymbol{R}_{mm'}|/R_{\max} < 1 - \frac{1}{2}\,|\epsilon_m| - \frac{1}{2}\,|\epsilon_{m'}|, \tag{5.3.35}$$

where $\epsilon_m = (\varepsilon_m - \mu)/E_{\max}$. The condition (5.3.35) can be fulfilled by associating each site m with a sphere of radius R_m centered at $\boldsymbol{R}_m$ with

$$R_m = \begin{cases} R_{\max}\left(\frac{1}{2} - |\epsilon_m|\right) & \text{if } |\epsilon_m| \leqq 1/2, \\ 0 & \text{if } 1/2 < |\epsilon_m| \leqq 1, \end{cases} \tag{5.3.36}$$

and assuming a bond between sites m and m' only if their spheres intersect or touch each other. The average volume of such a sphere is

$$\frac{4\pi}{3} R_{\max}^3 \int\limits_{-1/2}^{1/2} d\epsilon \left(\frac{1}{2} - |\epsilon|\right)^3 \Big/ \int\limits_{-1}^{1} d\epsilon = \frac{1}{48} \pi R_{\max}^3. \tag{5.3.37}$$

Analogously, for variable-range hopping in a two-dimensional system, one finds the law (cf. Section 5.1.4.)

$$\sigma = \sigma_0 \exp\left[-(T_0/T)^{1/3}\right], \; T_0 = c_2{}^3\alpha^2/kN(\mu). \tag{5.3.38}$$

To obtain this law by the procedure described above, $N(\mu)$ in (5.3.31) must be viewed as the density of states at the Fermi level in a two-dimensional system, as characteristic figures circles must be used instead of spheres, and (5.3.32) must be replaced by

$$\pi \overline{R} n = \Phi_c{}', \tag{5.3.39}$$

where $\Phi_c{}'$ is related to the critical-area fraction s_c by $s_c = 1 - \exp(-\Phi_c{}')$.

Another approach to the percolation problem under consideration was proposed by SEAGER and PIKE (1974) and MASCHKE et al. (1974) by using a description in a four-dimensional space (for a $3d$ system). Let us briefly outline this approach for $2d$ systems.

The bonding condition (5.3.29) implies the following restrictions on the energies $E_m \equiv \varepsilon_m - \mu$ and $E_{m'} \equiv \varepsilon_{m'} - \mu$ and the distance $|\boldsymbol{R}_{mm'}|$ between the sites:

$$|E_m|, |E_{m'}| \leqq \zeta kT,$$

$$2\alpha |\boldsymbol{R}_{mm'}| \leqq \begin{cases} \zeta - |E_m|/kT, \; |E_m| > |E_{m'}|, \text{sign}(E_m) = \text{sign}(E_{m'}), \\ \zeta - |E_{m'}|/kT, \; |E_m| < |E_{m'}|, \text{sign}(E_m) = \text{sign}(E_{m'}), \\ \zeta - |E_m - E_{m'}|/kT, \; \text{sign}(E_m) \neq \text{sign}(E_{m'}). \end{cases} \tag{5.3.40}$$

According to (5.3.40), a pair of sites m and m' is connected by a bond if site m' is within the following figure of two space coordinates around site m and the energy coordinate E_m: A cylinder of height E_m and radius $(1/2\alpha)\,(\zeta - |E_m|/kT)$ at both ends capped by a cone of the same radius and of height $(kT\zeta - |E_m|)$ (see Fig. 5.25).

The volume enclosed by this figure about site m is

$$V_m = \frac{\pi}{12\alpha^2} (\zeta - |E_m|/kT)^2 (E_m - 2kT\zeta). \tag{5.3.41}$$

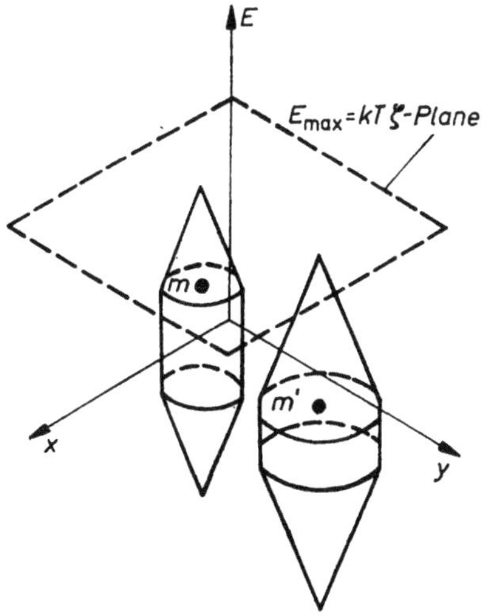

Fig. 5.25. Surfaces of constant conductance associated with two arbitrary, unconnected sites m and m' in an inclusive figure construction simulating variable-range hopping in two spatial dimensions (x, y) and one energetic dimension (E) (see text) (after SEAGER and PIKE, 1974)

Multiplying this volume by the density of states $N(\mu)$, one obtains the average number of bonds for site m (cf. Section 5.2.1.). By averaging the latter quantity over the energy E_m lying between $\pm\zeta kT$, one finds the average number B of bonds per site:

$$B = \delta\pi \frac{N(\mu)\, kT}{16\alpha^2} \zeta^3, \tag{5.3.42}$$

where $\delta = 1$ (SEAGER and PIKE, 1974) or $\delta = 176/105$ (MASCHKE et al., 1974), depending on performing the average over E_m without or with a weighting factor proportional to the number bonds (sites with a small number of bonds are expected to form dead ends).

Inverting (5.3.42), we find the critical exponent

$$\zeta_c = (T_0/T)^{1/3}, \tag{5.3.43}$$

where T_0 is given by (5.3.38) with

$$c_2{}^3 = 16B_c/\pi\delta. \tag{5.3.44}$$

The critical average number of bonds per site, B_c, turns out to be independent of $N(\mu)$ and T (SEAGER and PIKE, 1974, MASCHKE et al., 1974). Therefore, by means of (5.3.21) and (5.3.43), we rediscover (5.3.38). Analogously, the $T^{-1/4}$ law can be obtained for $3d$ systems.

Note that WINTLE and PUHACH (1978) derived the $T^{-1/4}$ law by reducing the percolation problem in question to probabilistic bond percolation in random arrays of sites. In this treatment explicit energy dependence of the characteristic percolation figures is ignored, but some fraction of bonds is assumed to be removed, because they are inhibited for energy reasons.

Mott's law (5.3.27) and (5.3.38) was also confirmed by computer studies (MASCHKE et al., 1974, SKAL and SHKLOVSKII, 1976, cf. also SEAGER and PIKE, 1974).

The values c_3 and c_2 (cf. (5.3.27), (5.3.38)) obtained by the various methods are given in Table 5.1.

Note that so far, here we have considered only the case of a constant density of states at the Fermi level. For a rectangular such state density with width 2Γ, Mott's law holds for $E_{max} \leqq \Gamma$, which, by means of (5.3.30), (5.3.34), and (5.3.27), defines the following temperatur T_c up to which Mott's law applies:

$$T_c = \frac{1}{k}\left(\frac{\Gamma}{c_3}\right)^{4/3} \mathcal{N}^{1/3}\alpha^{-1}. \tag{5.3.45}$$

As mentioned in Section 5.1.4., for a density of states $N(E) = c\,|E - \mu|^n$ (c is the constant), the dependence (5.3.27) must be replaced by (5.1.21).

A number of experimental observations of variable-range hopping in amorphous semiconductors and doped crystalline semiconductors have been reported in the literature. A large amount of the existing experimental data were critically analyzed by Hill (1976a, b) (see also Zabrodskii, 1977, Ahmed and Blackman, 1980b, and references therein). According to Hill, the conventional plot of $\ln \sigma$ versus $T^{-1/4}$ is unsuitable for a comparison of experimental data with (5.3.27) (cf. Fig. 5.26). He proposed to perform such a comparison in terms of the local activation energy ΔE defined as the gradient of an Arrhenius plot of the conductivity, i.e.

$$\Delta E = -\frac{d(\ln \sigma)}{d(1/kT)}. \tag{5.3.46}$$

For a conductivity of the form

$$\sigma \sim T^b \exp\left(-BT^{-p}\right), \tag{5.3.47}$$

then we have

$$\Delta E = bkT + BpkT^{(1-p)}. \tag{5.3.48}$$

Analysis of the dependence $\lg \Delta E$ on $\lg (1/kT)$ yields for amorphous semiconductors values p between 0.18 and 0.7, which are grouped round $p = 1/4$ (Hill, 1976b, Zabrodskii, 1977a) and for doped semiconductors values between 0.25 and 0.7, which are grouped rather round $p = 1/2$ then round $p = 1/4$ (Zabrodskii, 1977).

As will be discussed in Section 5.4.2., a value $p = 1/2$ may be due to Coulomb-type electron-electron correlation, which was predicted to lead to a weak gap, the so-called Coulomb gap, near the Fermi level due to the exciton effect. The resulting density of states near the Fermi level has the form $N(E) \sim |E - \mu|^2$. With regard to the effect of the Coulomb gap on dc hopping conductivity there exist contradictory results in the literature (cf. Section 5.4.2.).

Note that the parameter T_0 in Mott's law (5.3.27) is inversely proportional to $N(\mu)$, the density of states at the Fermi level. Abkowitz et al. (1976) determined $N(\mu)$ from the values of T_0 for amorphous germanium, obtained experimentally by different groups of authors, and showed that the values found in this way do not agree with $N(\mu)$ obtained from ac conductivity measurements. Furthermore, all values of T_0 are close to 10^8 K, which is surprising, since the ac data show a systematic variation with preparation conditions. This suggests that, at least for amorphous germanium, $N(\mu)$ deduced from T_0 does not represent the correct density of states at the Fermi level (cf. GILBERT and ADKINS, 1976). In this context note that according to HILL and JONSCHER (1979), in hopping systems the question of the relation between the dc and ac data cannot be answered unambiguously at present, because of the lack of reliable data and because of insufficient theoretical understanding of the underlying transport mechanisms.

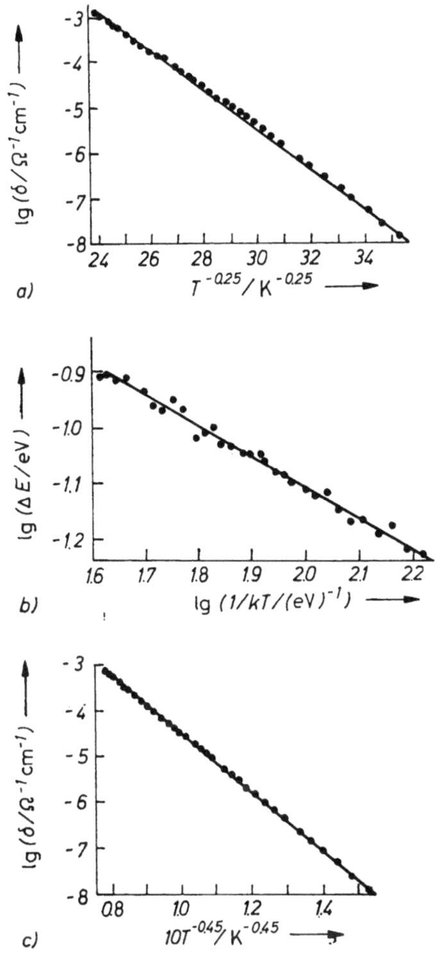

Fig. 5.26. Analysis of variable-range hopping data obtained on implanted amorphous silicon (after PFEILSTICKER et al., 1978). a) Standard plot lg σ vs. $T^{0.25}$. b) Data analyzed according to Hill's method (cf. (5.3.46) to (5.3.48)). The slope of the fitted line yields $p = 0.45$. c) The data of a) replotted for lg σ vs. $T^{-0.45}$. For the parameters σ_0 and T_0 in $\sigma = \sigma_0 \exp (T_0/T)^p$ one finds a) $T_0/\mathrm{K} = 9.4 \times 10^7$ ($\pm$ 35%), $\sigma_0/\Omega^{-1}\ \mathrm{cm}^{-1} = 2.2 \times 10^7$ (+150%, −75%), c) $T_0/\mathrm{K} = 6.4 \times 10^4$ ($\pm$ 0.2%), $\sigma_0/\Omega^{-1}\ \mathrm{cm}^{-1} = 6.6 \times 10^1$ ($\pm$ 10%). The reduction of the errors by about one order of magnitude reflects the correctness of Hill's method

Concluding this section, let us draw attention to thickness dependency of hopping conductivity in amorphous films. With increasing thickness, in germanium, silicon and carbon films transition from $3d$ hopping behaviour (5.3.27) to $2d$ behaviour (5.3.38) has been observed (KNOTEK et al., 1973, KNOTEK, 1975, HAUSER, 1975, 1977). This transition occurs when the thickness becomes smaller than the mean hopping length. The case where the thickness is larger than the average hopping distance but smaller than the correlation length was studied by SHKLOVSKII (1975, 1977b).

Note that in the quasi-one dimensional system NMP-TCNQ, with increasing temperature, a transition from $T^{-1/4}$ to $T^{-1/3}$ to T^{-1} hopping behaviour was observed indicating a transition from 3d to 2d to 1d behaviour due to decrease of the average hopping length with increasing temperature (SHANTE, 1977).

5.3.4. *The pre-exponential factor of the hopping conductivity*

By arguments analogous to that used in Section 5.3.2. for a resistor lattice, the coductivity of a hopping system may be written (KURKIJÄRVI, 1974, SHKLOVSKII and EFROS, 1975)

$$\sigma = \sigma_0 \exp(\zeta_c), \; \sigma_0 \approx (L_0 Z_0)^{-1}, \tag{5.3.49}$$

where L_0 denotes the correlation length of the critical network and Z_0 is the pre-exponential factor in the expression for the resistor $Z_{mm'}$ (cf. (5.1.6)).

Close to the percolation threshold ζ_c, the correlation length L behaves as (cf. (5.2.10))

$$L(\zeta) \approx \mathscr{L} \left| \frac{\zeta - \zeta_c}{\zeta_c} \right|^{-\nu}, \tag{5.3.50}$$

where $\mathscr{L}$ denotes the natural length scale of the system. For R-percolation, $\mathscr{L}$ is given by the average distance between the sites

$$\mathscr{L} \approx \mathscr{N}^{-1/3}, \tag{5.3.51}$$

and for R-ε-percolation, by the average hopping distance $R_{\min}$ (5.1.18), i.e.

$$\mathscr{L} \approx [9\beta/8\pi\alpha N(\mu)]^{1/4}. \tag{5.3.52}$$

As discussed in Section 5.3.2., the critical network is characterized by $\zeta = \zeta_c + 1$. Accordingly, the correlation length L_0 becomes

$$L_0 \equiv L(\zeta + 1) \approx \mathscr{L}\zeta^{\nu}, \tag{5.3.53}$$

where, according to (5.3.25) and (5.3.34), for R-percolation, ζ_c is given by

$$\zeta_c = \eta\alpha\mathscr{N}^{-1/3}, \tag{5.3.54}$$

and for R-ε-percolation, by

$$\zeta_c = (T_0/T)^{1/4}, \; T_0 = c_3{}^4\alpha^3/kN(\mu). \tag{5.3.55}$$

According to (5.1.6), the quantity Z_0 involved in (5.3.49) depends algebraically on $|\boldsymbol{R}_{mm'}|$ and $|\varepsilon_m - \varepsilon_{m'}|$. Hence, any investigation of the pre-exponential factor requires a detailed study of this dependence, in particular, for a given situation it must be investigated whether or not the quantity J_0 (in $J_{mm'} = J_0 \exp(-\alpha |\boldsymbol{R}_{mm'}|)$) depends on $|\boldsymbol{R}_{mm'}|$ (see, e.g., WUERTZ and THOMAS, 1978).

In the spirit of the present approach, in Z_0 from (5.3.49) $\zeta_{mm'}$ must be replaced by ζ_c. Accordingly, for R-percolation, in Z_0 we must put $|\varepsilon_m - \varepsilon_{m'}| \approx \varepsilon_3$ and $|\boldsymbol{R}_{mm'}| \approx R_c \approx \mathscr{N}^{-1/3}$, and for R-ε-percolation, $|\varepsilon_m - \varepsilon_{m'}| \approx E_{\max} = kT\zeta_c$, $|\boldsymbol{R}_{mm'}| \approx R_{\max}$

$= \zeta_c/2\alpha$, with ζ_c from (5.3.55). Denoting symbolically the corresponding value of Z_0 by $(Z_0)_{\zeta_c}$, then, by means of (5.3.49), (5.3.53), and (5.3.54), we find for R-percolation that

$$\sigma \approx (\alpha \mathcal{N}^{-1/3})^{-(1+\nu)} \left(\alpha^{-1}(Z_0)_{\zeta_c}\right)^{-1} \exp\left(-\eta\alpha\mathcal{N}^{-1/3}\right), \tag{5.3.56}$$

and, by means of (5.3.49), (5.3.53), and (5.3.55), for R-ε-percolation that

$$\sigma \approx \left(\frac{T_0}{T}\right)^{-(1+\nu)/4} \left(\alpha^{-1}(Z_0)_{\zeta_c}\right)^{-1} \exp\left(-(T_0/T)^{1/4}\right). \tag{5.3.57}$$

Consequently up to a numerical factor, the percolation theory allows exponential and pre-exponential behaviour of dc hopping conductivity to be determined.

The pre-exponential factor has also been studied by Ambegaokar et al. (1973), Pollak (1972), Brening et al. (1973a), Mott (1972) and Butcher et al. (1977) (cf. Butcher, 1980) by means of different methods. The results obtained differ somewhat from each other.

5.3.5. *Results of other methods*

In addition to the percolation approach, a number of different methods has been proposed for studying the dc hopping conductivity $\sigma(0)$. Among these methods are intuitive approaches (cf. Sections 5.1.3. and 5.1.4.), computer studies (e.g., Ambegaokar et al., 1973, Maschke et al., 1974, Kurkijärvi, 1974, Seager and Pike, 1974, Butcher et al., 1977, Butcher and Mc Innes, 1978), the continous-time random walk (CTRW) model (Scher and Lax, 1973; cf. Section 6.8.3.), effective-medium approximations (EMA) (e.g., Bryksin, 1980, Movaghar et al., 1980, Odagaki and Lax, 1981a, cf. Sections 6.4. and 6.6.), and the analytical treatment of Butcher et al. (1977) (cf. Butcher, 1980).

First, consider results of the CTRW model. For R-percolation, one finds that $\ln \sigma(0) \sim (\alpha\mathcal{N}^{-1/3})^{3/2}$ (cf. Scher and Lax, 1973), and for R-ε-percolation, $\ln \sigma(0) \sim (T_0/T)^{1/3}$ (cf. Kivelson, 1980), in contrast to the results of the percolation approach $\ln \sigma(0) \sim \alpha\mathcal{N}^{-1/3}$ and $\ln \sigma(0) \sim (T_0/T)^{1/4}$, respectively ($\alpha^{-1}$ — Bohr radius, $\mathcal{N}$ — concentration). Consequently, the CTRW underestimates the dc conductivity in the simple percolation problems to which it has been applied. This is due to the single-site approximation scheme involved in this model (cf. Kivelson, 1980).

Now let us turn to the EMA. The basic idea of this method consists in replacing the disordered system by some effective ordered medium the parameters of which are subsequently chosen such that it describes the properties of the actual medium as well as possible. In the theory of charge transport in disordered systems this method has been widely used to study random-resistor networks (cf. Section 5.2.4.). With regard to ac and dc hopping transport, the EMA was employed by Bryksin (1980) (cf. Section 6.4.). Another variant of the EMA, which uses the exact solution of the Bethe lattice, was given by Movaghar et al. (1980) (cf. Section 6.6.). Furthermore, Odagaki and Lax (1981a) employed a cluster EMA, which is identical with the homomorphic cluster CPA (coherent potential approximation) elaborated by

Odagaki and Yonezawa (1978). It turns out that the EMA-results well agree with computer results obtained on the basis of the percolation approach. Moreover, the effective medium approximations also provide expressions for the pre-exponential factors.

Starting point of the treatment of Butcher et al. (1977) is the following formula for the real part of the conductivity

$$\sigma = \frac{1}{2\Omega E^2} \sum_{mm'} |U_{mm'}|^2/Z_{mm'}, \tag{5.3.58}$$

where Ω is the volume of the system, E the amplitude of the electrical field directed in x-direction, and $U_{mm'} = U_m - U_{m'}$. This formula is obtained by considering the resistor network of Miller and Abrahams (cf. Section 5.1.) as a real resistor network, with resistors $Z_{mm'}$ between sites (nodes) m and m' and potential U_m on site m, and by equating the Joule heating in the volume Ω to the sum of powers dissipated in the individual resistors in Ω (cf. Butcher and Summerfield, 1981).

The configuration average of (5.3.58) is performed by using some ansatz for $\langle U_{mm'}^2 \rangle'$, the mean-square voltage between sites m and m' averaged over all stochastic variables, except those associated with sites m and m' (cf. (8.2.27), (8.2.28)). To illustrate the procedure, consider, for simplicity, the case of R-percolation, where (cf. (5.1.7))

$$Z_{mm'} = Z_0\, e^{2\alpha|\boldsymbol{R}_{m'm}|}. \tag{5.3.59}$$

Here Z_0 is a constant and $|\boldsymbol{R}_{m'm}|$ the distance between sites m and m'. Then the configuration average of (5.3.58) reduces to

$$\sigma = \frac{2\pi\mathcal{N}^2}{E^2} \int_0^\infty dR\, R^2 P(R), \tag{5.3.60}$$

where

$$\begin{aligned} P(R) &= W^2(R)/Z(R) \\ &= I^2(R)\, Z(R). \end{aligned} \tag{5.3.61}$$

Here $W(R) \equiv \langle U_{mm'}^2 \rangle'$, $P(R)$ is the mean power dissipated in $Z(R)$, and $I(R) = W(R)/Z(R)$ is the mean current running through $Z(R)$.

When R is a macroscopic distance, i.e. $R \gg \mathcal{N}^{-1/3} \gg \alpha^{-1}$, then $U_{mm'} = E(X_m - X_{m'})$ and, hence, $W^2(R) = E^2R^2/3$. On the other hand, when $R \ll \mathcal{N}^{-1/3}$, then $Z(R)$ behaves like a short-circuit fed by the currents entering from the surrounding resistors, so that $I(R)$ is approximately constant in this "well-conducting island". Hence, by (5.3.61) and (5.3.59), we see that for $R \gg \mathcal{N}^{-1/3}$, $P(R) \sim e^{-2\alpha R}$, and for $R \ll \mathcal{N}^{-1/3}$, $P(R) \sim e^{2\alpha R}$. This suggests that $P(R)$ has a single maximum at some value R_c, which, according to physical intuition and computer calculations (Butcher et al., 1979), agrees with the critical percolation radius.

Using $W^2(R) = E^2R^2/3$ for $R > R_c$ and $I^2(R) = E^2R_c^2/3Z^2(R_c)$ for $R < R_c$, then, by means of (5.3.61), (5.3.59), and (5.3.60), we find that

$$\sigma(0) = 2\pi\mathcal{N}^2(R_c^4/3\alpha)\, Z_0^{-1} \exp(-2\alpha R_c). \tag{5.3.62}$$

It turns out that the formula (5.3.62) is in excellent agreement with computer data (Butcher and Mc Innes, 1978) not only for $\alpha \mathcal{N}^{-1/3} \gg 1$, the region of applicability of percolation theory, but also for small values of $\alpha \mathcal{N}^{-1/3}$. For more details on this approach, see Butcher (1980) and references therein.

5.4. Effects of electron-electron interaction

5.4.1. *On the interplay of randomness and electron-electron interaction*

As described in Section 4.1., according to the present concept of electron states in disordered systems, any kind of randomness can give rise to localized states. The localization picture is strongly supported by the hopping transport, in particular, by the variable-range hopping observed in certain materials at low temperatures.

The familiar localization and hopping pictures (as described in the preceding sections) rest upon the one-electron approximation. However, electron-electron interaction may also be important in disordered systems. For instance, in impurity bands the electron-electron interaction energy is of the order of the disorder energy, except for very large and very small compensation. In general, electron-electron interaction may be neglected only if it is much smaller than the characteristic energies of the problem such as the disorder energy, the resonance energy, and in the case of hopping also kT. The question now arises how the localization and the hopping pictures are affected by electron-electron interaction. This is a rather difficult question, because localization and hopping are complex problems even in the absence of interaction. In the following we cast a look on some problems of the interplay of randomness and interaction which have recently been studied in the literature.

The importance of electron-electron interaction on hopping transport was first emphasized by Pollak (1970, 1971) and Srinivasan (1971), who pointed out that the interaction leads to a gap in the one-electron excitation spectrum near the Fermi energy μ. This gap has been termed the Coulomb gap (Efros and Shklovskii, 1975) Pollak (1970, 1971) also showed that the need to overcome the Coulomb gap can be alleviated by correlated successive or simultaneous hops of several electrons, i.e. by the so-called multi-electron hopping.

In recent years the Coulomb gap and the multi-electron hopping have been studied by a number of authors (see, for example, Knotek and Pollak, 1977, Mott, 1975, 1976, Efros, 1976, Bonch-Bruevich, 1977, Baranovskii et al., 1979, Pollak and Knotek, 1979, Pollak, 1980, 1981, Chroboczek et al., 1981, Davies et al., 1982, and references therein). Although at present a rigorous analytical description of the problem is still lacking and some of the results obtained are contradictory, it has become quite clear that the Coulomb gap and the multi-electron hopping may be very important at low temperatures (for more details on these phenomena, see Sections 5.4.2., 5.4.3., 11.2.1., and 11.4.2.).

On studying the Coulomb gap and multi-electron hopping, in general only the effect of inter-site electron-electron interaction is considered. The intra-site interaction (i.e. the Hubbard energy U, c.f. Section 4.1.) is assumed to be very large so that each site may be occupied only by a single electron. In recent years metal-non-metal transitions in doped semiconductors have also required considerable attention (see, for example, YAMAGUCHI et al., 1979, MOTT and DAVIES, 1980, KAMIMURA, 1980, KAMIMURA et al., 1982, and references therein). In this case the positions and properties of the upper Hubbard band (for instance, of the band of the D^- centres in Si:P) are of great interest, which requires account of intersite and intra-site interaction in an Anderson-localized system, since both these interactions are in general of the same order.

Furthermore, electron-electron interaction may play a role in delocalization or localization of electrons in disordered systems by modifying the randomness felt by the electrons. According to POLLAK and KNOTEK (1979) (cf. POLLAK, 1980) inter-site interaction may be important in the delocalization of electron states. On the other hand, for a model with attractive intra-site interaction ("negative-U" model) RITALA and KURKIJÄRVI (1982) showed that the randomness is increased by the interaction, which facilitates localization.

Note that such an attractive intra-site interaction (cf. Section 2.2.), which tends to pair free spins, was introduced by ANDERSON (1975) to explain the peculiar behaviour of chalcogenid glasses (i.e. the failure to detect paramagnetism for intrinsic defects and the absence of variable-range hopping in these materials). Subsequently this idea was picked up and studied in more detail by a number of authors (see, for example, MOTT et al., 1975, LICCIARDELLO et al., 1981, ROBASKIEWICZ et al., 1981, 1982, VANDERBILT and JOANNOPOULOS, 1982, and references therein).

Another interesting problem is the stability of excited states in an Anderson-localized system in the presence of electron-electron interaction. It was shown (FLEISHMAN et al., 1978, FLEISHMAN and ANDERSON, 1980) that the range of interaction (which is not known at present) and the dimensionality of the system are crucial in determining the stability. If the interaction is long-range, the low-energy excitations will decay with a rate proportional to $\exp\left(-(T_0/T)^{1/4}\right)$ at low temperature. This rate is related to the diffusion constant which in turn is related to the conductivity. Accordingly, a $T^{-1/4}$ — law of the conductivity does not necessarily arise from thermally activated hopping. It may also arise from electron-electron interaction (cf. BUTCHER and SWIERKOWSKI, 1980). In the latter case, the electron-electron interaction represents the energy reservoir for the localized excitations. Unlike the case of long-range interaction, for short-range interaction the excited states are non-conducting.

A very active field of research are at present two-dimensional disordered systems. As mentioned in Section 4.3., theoretical (scaling) arguments have been advanced (ABRAHAMS et al., 1979, GORKOV et al., 1979, KAVEH and MOTT, 1981a, see also THOULESS, 1982, and references therein) suggesting that in such systems all states are localized with the mobility edge being a localization edge seperating strongly (exponentially) localized states, which give variable-range hopping, and weakly (algebraically) localized states, which do not. The scaling theory of localization predicts that for weak disorder in two dimensions the conductivity decreases as $\ln l$, where at zero temperature, l is the length of the sample, but at finite temperature,

l is the inelastic diffusion length[1]). In the latter case one finds a logarithmic decrease of the conductivity as the temperature is lowered.

An alternative theory which predicts the same logarithmic temperature dependence of the conductivity was proposed by ALTSHULER et al. (1980) (see also ABRAHAMS et al., 1981, LEE, 1982). In this theory (which does not assume localization) this temperature dependence results from a logarithmic singularity in the density of states at the Fermi level which arises from interference between impurity and electron-electron scattering (cf. GIRVIN et al., 1982). According to KAVEH and MOTT (1981b), both the mechanisms should be present in actual two-dimensional systems, the localization affecting the diffusivity and the electron-electron interaction affecting the density of states. It should be noted that these two mechanisms can be distinguished by studying the response to a magnetic field (UREN et al., 1980, cf. LEE and RAMAKRISHNAN, 1982).

Two typs of two-dimensional systems have been used for studying localization effects: inversion layers of semiconductor surfaces[2]) and thin metallic films. On both these systems the above logarithmic corrections to the conductivity have been found (BISHOP et al., 1980, DOLAN and OSHEROFF, 1979). Note that there are experimental findings (temperature dependence of the conductivity of metallic doped semiconductors such as Si:P and Ge:Sb) which indicate that effects of localization and Coulomb interaction may also be important in three-dimensianal systems (THOMAS et al., 1982, and references therein).

ADKINS (1978) suggested that in inversion layers a single-particle approach is not valid. He proposed a model in which the electrons from a disordered version of a Wigner crystal (see also AOKI, 1979, PEPPER and UREN, 1982), which is pinned by disorder. At finite temperatures, thermal motions will provide the energy to overcome the pinning barrier, so that an activated collective motion of the electrons results.

Interacting disordered systems are still a very active field of research (cf. Section 11.2.1.).

---

[1]) More precisely, this l dependence of the conductivity σ is given by (cf. KAVEH, 1982, and references therein)

$$\sigma = \sigma_B[1 - \gamma \ln (l/\lambda)] \qquad \text{for } l \ll l_0,$$

$$\sigma = \sigma_B(\lambda/l)^\gamma \qquad \text{for } l \geqq l_0,$$

where σ_B is the Boltzmann term, $\sigma_B = e^2k_F\lambda/2\pi\hbar$. Here k_F is the Fermi wavevector, λ the elastic mean-free path, $l_0 = \lambda \exp(1/\gamma)$, and $\gamma = 2/\pi k_F l \approx 2/\pi$.

[2]) An inversion layer is formed at a semiconductor surface when a sufficiently strong electrical field is applied perpendicular to the surface. The field is such as to repel the majority carriers (holes in a p-semiconductor) so that, below the surface, there occurs a region depleted of holes, which causes a band bending near the surface. When the band bending is sufficiently strong, the bottom of the conduction band passes through the Fermi level (which is constant through the semiconductor) and electrons (minority carriers) enter the region below the surface. The inversion layer formed in this way represents a pseudo two-dimensional system, in which the electron density may be controlled over several orders of magnitude.

5.4.2. The Coulomb gap

In a Anderson-localized system the spatial distribution of electrons is affected by long-range Coulomb interaction. As mentioned in the preceding section, this causes a depletion in the one-electron density of states near the Fermi level μ, which is called the Coulomb gap.

Let us demonstrate the existence of a Coulomb gap by using the argumentation of POLLAK and KNOTEK (1979). To do so, we consider a system with random site energies. We assume for simplicity that in the system the number of electrons N is one-half the number of sites N_s and that in the absence of electron-electron interaction the system has a constant density of states near μ. Spin and tunneling between the sites are neglected.

First let us consider a system with $N - 1$ electrons. The ground state energy of this system is given by

$$E(N - 1) = \sum_m \varepsilon_m + \sum_{m<n} e^2/R_{mn} \equiv \varepsilon + \sum_{m<n} e^2/R_{mn}, \tag{5.4.1}$$

where ε_m is the energy of an electron at site m and R_{mn} is the distance between the sites m and n. The second term in (5.4.1) describes the electron-electron interaction.

Now, leaving the $(N - 1)$ electrons in their positions, we add the Nth electron and put it at a site with an energy as low as possible. We may argue that in a large enough system there must be a number of sites with equally small energies, otherwise the system would already exhibit a gap. Let 1 and 2 be such sites with equally low energy, then we have

$$\begin{aligned} E_1(N) &= \varepsilon + \sum_{m<n} e^2/R_{mn} + \varepsilon_1 + \sum_m e^2/R_{m1} \\ &= \varepsilon + \sum_{m<n} e^2/R_{mn} + \varepsilon_2 + \sum_m e^2/R_{m2} = E_2(N), \end{aligned} \tag{5.4.2}$$

where $E_1(N)$ and $E_2(N)$ are the energies of N-electron systems with $(N - 1)$ electrons in the ground state and an additional electron at site 1 and 2, respectively.

Now we keep the Nth electron at site 1 and admit relaxation of the other $(N - 1)$ electrons so as to lower the energy of the system. The new energy, $E_1'(N)$, becomes

$$E_1'(N) = \varepsilon' + \sum_{m'<n'} e^2/R_{m'n'} + \varepsilon_1 + \sum_{m'} e^2/R_{m'1} \leqq E_1(N), \tag{5.4.3}$$

where $\varepsilon' = \sum_{m'} \varepsilon_{m'}$ is the energy of the $(N - 1)$ electrons in the relaxed (primed) positions.

Finally, we transfer the electron from site 1 to site 2, while leaving the remaining $(N - 1)$ electrons fixed. The energy of the system becomes

$$E_2'(N) = \varepsilon' + \sum_{m'<n'} e^2/R_{m'n'} + \sum_{m'} e^2/R_{m'2} \geqq E_2(N). \tag{5.4.4}$$

Here the last equation was obtained by assuming

$$\sum_{m'} e^2/R_{m'2} = \sum_m e^2/R_{m2}, \tag{5.4.5}$$

and by accounting for

$$\varepsilon' + \sum_{m'<n'} e^2/R_{m'n'} \geqq \varepsilon + \sum_{m<n} e^2/R_{mn}. \tag{5.4.6}$$

The assumption (5.4.5) may be expected to be true when the charge redistribution due to an electron at site 1 is of no relevance to an electron at site 2. This is the case, when the distance between site 1 and 2 is of the order or longer than the correlation length of the electron-electron interaction, which, according to POLLAK (1971), is equal to the average separation between the electrons. The relation (5.4.6) holds, because its right hand side represents the ground state energy of the $(N-1)$ electron system.

From (5.4.2), (5.4.3) and (5.4.4) we find

$$E_1'(N) \leqq E_1(N) = E_2(N) \leqq E_2'(N). \tag{5.4.7}$$

Accordingly, in a disordered interacting system a transition from the ground state to another state requires a positive energy $E_2'(N) - E_1'(N)$, which is equivalent to the existence of a gap in the one-electron density of states at the Fermi level.

So far, in the literature no full consensus has been achieved on the characteristic features of the Coulomb gap and on its role in hopping conduction, though a number of analytical (see, for example, EFROS and SHKLOVSKII, 1975, MOTT, 1975, 1976, EFROS, 1976, POLLAK and KNOTEK, 1979) and numerical studies (see, for example, EFROS, 1976, BARANOVSKII et al., 1979, EFROS et al., 1979, DAVIES et al., 1982) have been devoted to this problem. Obviously, many of these features are model dependent.

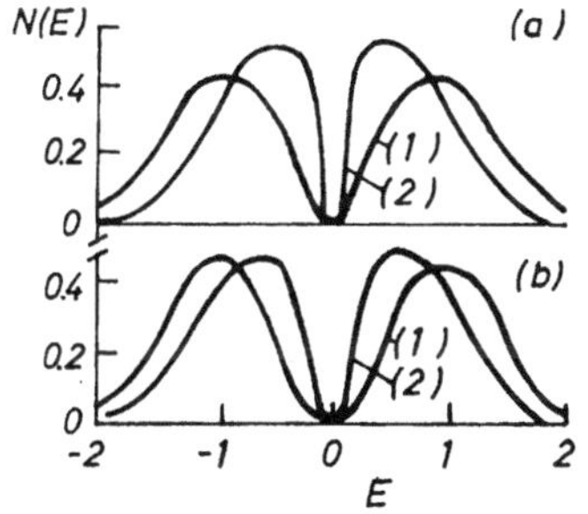

Fig. 5.27. Bare (1) and polaron (2) single-particle densities of states $N(E)$ for two- (a) and three-dimensional (b) systems of localized interacting electrons. E is in units of half the width of the distribution of the site energies (after DAVIES et al., 1982) (see text)

Let us consider some results obtained for the Coulomb gap. Fig. 5.27 shows single-particle densities of states calculated numerically by DAVIES et al. (1982) by using a model of a disordered interacting system which was introduced by EFROS (1976). This model is of the type as described above. The sites are put on a square or simple cubic lattice. The number of electrons is assumed to be one-half the number of sites. Charge neutrality is realized by placing positive charges on the unoccupied sites. The site energies are selected from a rectangular distribution. Starting point for the calculation of the density of states is a random distribution of the electrons over the sites. Successive one-electron hops which lower the total energy are admitted.

The state with the lowest energy is believed to be the ground state. Note that the model in question exhibits are large number of metastable states, but the corresponding one-electron densities of states (distribution functions of the site energies on account of the Coulomb interaction) do not signifficantly differ from each other (cf. GRÜNEWALD et al., 1982).

As illustrated in Fig. 5.27, the one-electron density of states exhibits a well-defined Coulomb gap near the Fermi level. In three dimensions, the gap may be fitted by the exponential form

$$N(E) \sim \exp\left(-(E_0/(E - \mu))^{1/2}\right), \tag{5.4.8}$$

which was first proposed by EFROS (1976). In two dimensions ($d = 2$) the gap is stronger than the "soft" gap given by the law

$$N(E) \sim |E - \mu|^{d-1}, \tag{5.4.9}$$

which was first predicted by SHKLOVSKII and EFROS (1975). Note that the computer results of BARANOVSKIJ et al. (1979), which were also obtained by the model at hand, can be well fitted for $d = 2$ and $d = 3$ by the law (5.4.9). According to DAVIES et al. (1982) the discrepancy between the numerical results of the two groups may be due to the different boundary conditions used.

It has been suggested (POLLAK, 1970, 1971, MOTT, 1975, 1976, EFROS, 1976) that in interacting Anderson-localized systems at low temperatures many-particle excitations ("electronic polarons") may be important. This polaron formation is due to the fact that around a given occupied site the electron-electron interaction raises the level of the surrounding sites so that some of the sites will be empty, while in the ground state of the system they would be occupied if the given sites were empty.

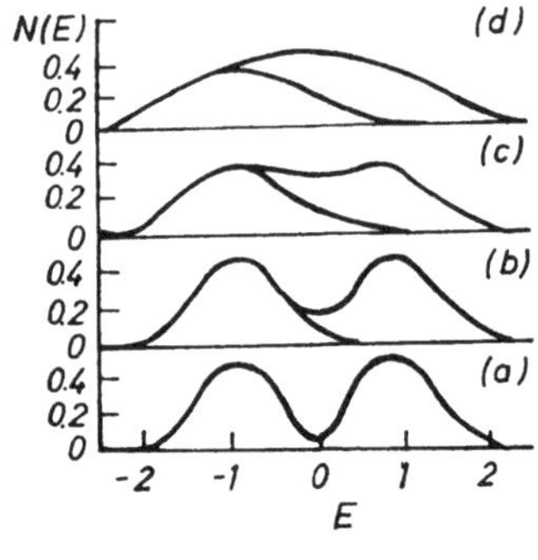

Fig. 5.28. Bare single-particle density of states $N(E)$ for a two-dimensional system of localized interacting electrons at various temperatures: $kT = 0.05$(a), 0.1(b), 0.3(c), 0.5(d). The lower curve in each case shows the density of occupied states. The energies are in units of half the width of the distribution of the site energies (after DAVIES et al., 1982) (see text)

According to MOTT (1975, 1976) there is no gap in the polaron density of states, while according to EFROS (1976) a certain gap exists. The numerical results of DAVIES et al. (1982) exhibit a polaron gap, which is much narrower than that for bare electrons (Fig. 5.27). In the latter paper the polaron density of states was calculated by putting an additional electron on a site and allowing then one-electron transitions of the surrounding electrons in order to lower the energy of the system until the system was stabel against such transitions.

Davies et al. (1982) extended their calculations to finite temperature T by associating a hop, which requires an energy E, with unit probability if $E < 0$, and with probability $\exp(-E/kT)$ if $E > 0$ $(k = 1)$ (algorithm of Metropolis et al., 1953). As shown in Fig. 5.28, the Coulomb gap fills with increasing temperature.

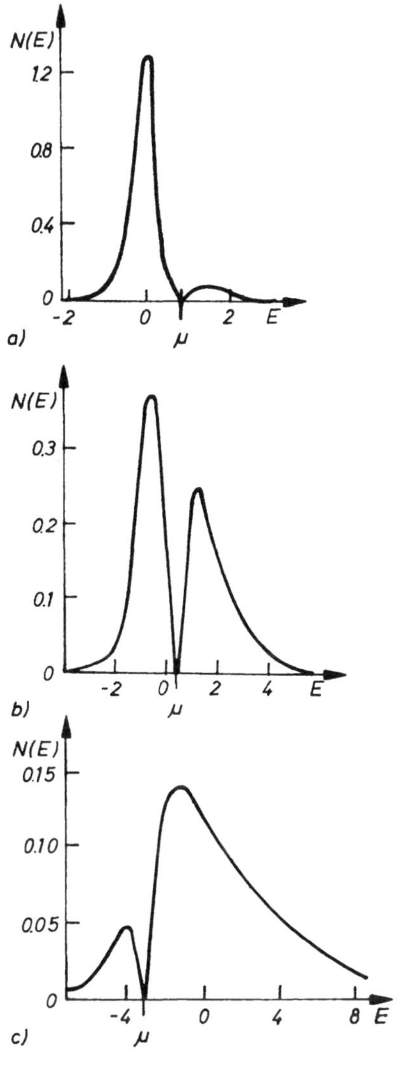

Fig. 5.29. Numerical results for the density of states $N(E)$ and the position of the Fermi level μ in an impurity band at zero temperature for various degrees of compensation K: 0.1(a)), 0.5(b)), 0.9(c)). $N(E)$ is in units of $\mathcal{N}_D(e^2\mathcal{N}^{1/3}/\varkappa)^{-1}$ and E in units of $e^2\mathcal{N}_D{}^{1/3}/\varkappa$ ($\mathcal{N}_D$ — donor concentration, $\varkappa$ — dielectric constant) (after Efros et al., 1979)

Fig. 5.29 shows computer results of Efros et al. (1979) for the density of states and the position of the Fermi level in an impurity in a three-dimensional system at zero temperature for different degrees of compensation. The density of states exhibits a soft Coulomb gap in the vicinity of the Fermi level. Close to the Fermi level, the gap does not depend on the degree of compensation.

The role of the Coulomb gap in hopping conduction is discussed in the ne xt section

5.4.3. *Correlation effects in hopping transport*

Let us discuss in a qualitative way the temperature dependence of the dc hopping conductivity in an Anderson-localized system in the presence of electron-electron interaction. To this end, consider for simplicity a system with a square distribution of site energies, say between $-\Gamma/2$ and $\Gamma/2$, a half-filled band and a sharp Coulomb gap Δ at the Fermi level (see Fig. 5.30).

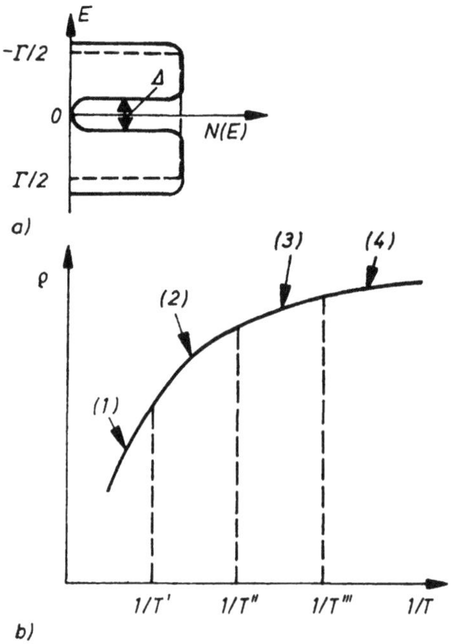

Fig. 5.30. a) Schematic representation of the density of states $N(E)$ with (solid line) and without (broken line) intersite Coulomb interaction in a system of localized electrons; b) Qualitative picture of the temperature dependence of the hopping resistivity ϱ to be expected for such a density of states: (1) nearest-neighbour hopping (activation energy Γ), (2) Mott $T^{-1/4}$ law, (3) activation across the Coulomb gap (activation energy Δ), (4) correlated hopping; (after POLLAK and KNOTEK, 1979)

At high enough temperatures we get activated hopping between nearest neighbours, with an activation energy $\Gamma/2$. At lower temperatures, the electrons tend to hop longer distances with less activation energy, i.e. we get variable-range hopping with the $\ln \sigma \sim -T^{-1/4}$ behaviour. When the activation energy becomes comparable to, or less than the Coulomb gap, then the activation energy can no longer be reduced by enlarging the hopping distance. Accordingly, at low temperatures we except hopping behaviour with an activation energy Δ, resulting from the need to activate across the Coulomb gap (cf. ORTUNO and POLLAK, 1983, and references therein).

In the case of a weak Coulomb gap, however, variable-range hopping is still possible near the Fermi level, but the temperature dependence of the conductivity differs from the $T^{-1/4}$ behaviour, which requires a constant density of states. For a Coulomb gap of the form (5.4.9), a $\ln \sigma \sim -T^{-1/2}$ dependence results, according to (5.1.21) (for $d = n = 2$). Note that for doped semiconductors, experimental findings (ZABRODSKII, 1977) are in favour of a $T^{-1/2}$ rather than a $T^{-1/4}$ dependence, which suggests the importance of correlation effects in these materials (see also, for example, BARANOVSKII et al., 1979, ZABROOLSKII, 1980)

The temperature regime, where the hopping is determined by one-electron excitations across the Coulomb gap, does not extend to the lowest temperatures. At low enough temperatures the need for such excitations can be alleviated by correlated motion of several electrons. Such correlation can exist in simultaneous hops of several electrons (polaron-like multi-electron hops, see Fig. 5.31) or in successive hops (cascade-like multi-electron hops, see Fig. 5.32). Estimations of the temperatures separating the above four temperature regimes can be found in the paper of POLLAK and KNOTEK (1979).

Fig. 5.31. Schematic representation of a polaron-like multi-electron hop

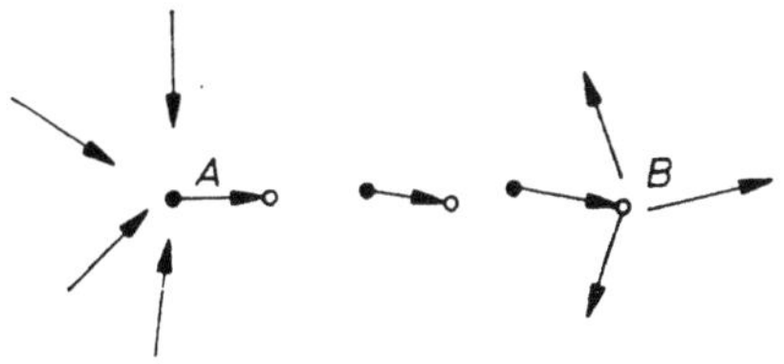

Fig. 5.32. Schematic representation of a cascade-like multi-electron hop

Let us now discuss some aspects of multi-electron hopping. In the absence of electron-electron interaction, only one-electron hopping transitions have non-vanishing transition rate. Expressions for many-electron transition rates have been derived with (KNOTEK and POLLAK, 1974, POLLAK, 1980) and without (GOSAR, 1980, 1983, POLLAK, 1981) using the Hartree-Fock approximation. Both these approaches yield the same exponential dependences on temperature and hopping distances. For n-electron hopping (in the case of strongly localized states), the dc conductivity is given by

$$\sigma(n) \sim \exp\left(-2\alpha \sum_{i=1}^{n} R_i\right) \exp\left(-E(n)/kT\right), \tag{5.4.10}$$

where the summation is over the hopping distances of the n electrons. The current paths are such that the activation energy $E(n)$ becomes minimum. Since for many-electron hopping, the number of possible excitations increases rapidly with the number n, $E(n)$ is a decreasing function of n. On the other hand, $\sum_i R_i$ is an increasing function of n. Accordingly, there must exist a number n which is optimal for the current flow. Because of this property, many-electron hopping is also called the "variable-number hopping" (cf. POLLAK, 1980).

According to (5.4.10), with decreasing temperature it becomes advantageous for more and more electrons to hop simultaneously. Since $E(n)$ decreases with increasing n, we expect the activation energy to decrease with decreasing T. Note that such behaviour has been reported for doped Ge below 1K by ALLEN and ADKINS (1972) and GORDY (1972). For high enough temperature, n becomes unity, which is the usual one-electron hopping.

As displayed in (5.4.10), an expansion of the Bohr radius α^{-1} decreases the quantity $\alpha \sum_i R_i$ and thus increases the optimal number n of simultaneous excitations. This dependence is of special interest because α can be controlled experimentally: by large uniaxial stress α is reduced (see, for example, MIKOSHIBA, 1972) and by large magnetic field α is enhanced (see, for example, FRITZSCHE, 1962). Since σ depends on the product of α and R (cf. (5.4.10)), the shrinking (extension) of α should affect σ similarly as an increase (decrease) of the impurity concentration $\mathcal{N}$.

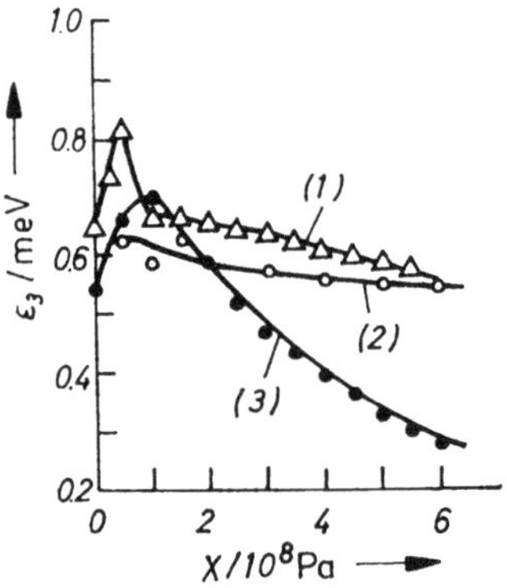

Fig. 5.33. The dependence of the activation energy ε_3 upon compressive stress X for hopping transport in p-Ge with different acceptor concentrations $\mathcal{N}_A$: 4.7×10^{15} cm^{-3} (1), 1.2×10^{15} cm^{-3} (2), 4.0×10^{16} cm^{-3} (3) (after CHROBOCZEK et al., 1981). The decrease of ε_3 with X is attributed to the increase of the Bohr radius of the acceptor function, which increases the optimal number of simultaneous excitations in correlated hopping. The effect increases with increasing concentration, as expected

The effect of large uniaxial stresses and high magnetic fields on hopping conduction was studied by CHROBOCZEK (1980) and CHROBOCZEK et al. (1981) in p-Ge. It was found strong evidence for the importance of correlated multi-electron hopping in this material at acceptor concentrations above about $\mathcal{N}_A = 3 \times 10^{15}$ cm^{-3} (cf. Fig. 5.33). Similarly, investigations of hopping conductivity in strong magnetic fields also give evidence for correlation effects in n-InSb (MANSFIELD and KUSZTELAN, 1978, TOKUMOTO et al., 1982). It was shown that in this material, with increasing magnetic field the activation energy ε_3 and the constant T_0 (occuring in the variable-range hopping law (5.1.19)) gradually approach their one-electron values. Note that the decrease of the activation energy at the onset of many-electron hopping (with decreasing temperature) is accompanied with a decrease of the pre-exponential factor of the conductivity (see, for example, CHROBOCZEK et al., 1981). For a survey of the experimental work which is relevant to the effect of correlation on hopping transport the reader is referred, for example, to POLLAK (1980).

At present, there is no quantitative theory of multi-electron hopping which is valid for an arbitrary number n of simultaneous hops. There are attempts to calculate the conductivity at not too low temperatures where n is small (KNOTEK and POLLAK, 1972, 1974) and others which approach the problem from the low temperature end (MOTT, 1975, 1976, EFROS and SHKLOVSKII, 1975, EFROS, 1976). The former approach yields for two-electron hops a lowering of the activation energy which resembles qualitatively the $T^{-1/4}$ dependence characteristic of one-electron variable-range hopping.

For the very low-temperature regime, contradictory results have been obtained.

MOTT (1975, 1976) predicts $\ln \sigma \sim -T^{-1/4}$, while EFROS and SHKLOVSKII (1975) and EFROS (1976) obtain $\ln \sigma \sim -T^{-1/2}$. This difference arises from a disagreement about the density of states of the "electronic polarons" near the Fermi level. According to Efros and Shklovskii the density of states exhibits a weak gap of the form (5.4.9) (with $d = 3$), while according to Mott no gap exists.

POLLAK (1980) adapted Mott's procedure for variable-range hopping (minimization of the conductivity exponent, see Section 5.1.4.) to correlated hopping. He found that at sufficiently low temperatures the conductivity is proportional to a power of the temperature

$$\sigma \sim T^s, \tag{5.4.11}$$

where s is roughly equal to the ratio of a characteristic hopping length to the Bohr radius. According to POLLAK (1981) there is some experimental evidence for such a law (with $s \approx 11$).

Concluding this section note that despite of the considerable experimental evidence for the existence of a finite intrasite Coulomb interaction, there exist only few papers dealing with hopping transport on account of such interaction (e.g. ZVYAGIN, 1977, BANYAI et al., 1981, KAMIMURA et al., 1982, 1983, KUROBE and KAMIMURA, 1982, GRÜNEWALD et al., 1983, DERSCH et al., 1983, see also the monograph by BONCH-BRUEVICH et al., 1981, and cf. Section 11.4.2.).

6. AC hopping conductivity

6.1. Introductory remarks

As pointed out in Section 2.5., in disordered systems the hopping current is additively composed of an Ohmic and a diffusive contribution. At not too high frequencies (in the Markovian limit, where $\omega\delta t \ll 1$), the Ohmic contribution does not depend on frequency, while the diffusive one decreases with increasing parameter $\omega\tau$, where τ is a characteristic time between consecutive hops. Owing to this, in the high-frequency limit $\omega\tau \gg 1$ (but $\omega\delta t \ll 1$), the frequency dependence of the conductivity passes into a plateau, which is described by (2.5.19). In this limit, the diffusive contribution completely vanishes. At frequencies $\omega\tau < 1$, the diffusive contribution partly compensates the Ohmic one (the lower ω the stronger the compensation) This produces a strong decrease of the conductivity with decreasing ω, so that for a large spread of the parameters characterizing the system, the dc conductivity turns out to be many orders of magnitude smaller than the conductivity in the range of the plateau. Note that the strong increase of the real part of the conductivity, σ', with increasing ω is one of the major features of the hopping conduction in disordered systems.

For $\omega\tau < 1$, the investigation of the dependence $\sigma(\omega)$ by means of analytical methods is a very intricate problem, because it requires the solution of the problem of a resistor network with capacitors (cf. Fig. 2.3). Above all, evaluation of $\sigma'(\omega)$ in the low frequency limit (close to dc limit) turns out to be a very difficult task.

The complexity and the topicality of these problems have stimulated the elaboration of new methods for studying charge transport in disordered systems (cluster approaches, effective-medium approximations, use of exact solutions of the Bethe lattice, a diagram approach, etc.). These methods allow the hopping conductivity to be calculated in a large frequency range, from the dc limit up to the high-frequency plateau. Moreover, they may be used also to examine other kinetic phenomena in disordered solids such as the hopping Hall effect (cf. Section 8).

Historically, the first step towards a theoretical understanding of $\sigma(\omega)$ was taken by Pollak and Geballe (1961), who introduced a two-site model for studying the ac hopping conductivity. This model is valid only at sufficiently high frequencies, where during half a period of the oscillation of the external field, electrons can hop solely between nearest-neighbour sites. At lower frequencies, however, during half a period of the oscillation of the external field, electrons may perform many hops (multiple hopping). The current then arises from transitions within highly conducting clusters of sites, the size of which increases with decreasing frequency, so that it becomes macroscopically large for $\omega \to 0$. Up to the end of seventies, the two-site model was the only tool for interpreting experimental ac data.

In view of the latter fact, here we first describe the two-site model, and only after that we will study more complex approaches, which contain the two-site model as special case (for reviews of the ac hopping conductivity see, e.g., Böttger and Bryksin, 1976b, 1982, Long 1982).

6.2. The two-site model

6.2.1. *Basic formulae for the two-site model*

One possible approach to the two-site model is to neglect in the equation for the density matrix (2.5.16) all sites except those of the selected pair (with site indices $m = 1, 2$). However, in this section we use another approach, which, in particular, is not based on the assumption of a small resonance integral J and which also takes into account the nondiagonal elements of the density matrix (cf. Böttger and Bryksin, 1976c). The latter approach allows the repulsion of the atomic levels (due to their interaction with each other) and "phononless" transitions (photon-assisted hopping; see below) to be accounted for.

On account of two sites only, and assuming weak interaction with phonons, then the Hamiltonian (2.1.5) becomes

$$\tilde{\mathscr{H}} = \mathscr{H}_0 + \mathscr{H}_{\text{ph}} + \mathscr{H}_1, \tag{6.2.1}$$

$$\mathscr{H}_0 = V_1 a_1^+ a_1 + V_2 a_2^+ a_2 + J_{12} a_1^+ a_2 + J_{21} a_2^+ a_1, \tag{6.2.2}$$

$$\mathscr{H}_1 = (S_1 - S_2) [J_{21} a_2^+ a_1 - J_{12} a_1^+ a_2], \tag{6.2.3}$$

where

$$V_m = \varepsilon_m - e\boldsymbol{E}_m\boldsymbol{R}_m \ (m = 1, 2)$$

and

$$S_m = \frac{1}{\sqrt{2N}} \sum_{\boldsymbol{q}} \{b_{\boldsymbol{q}}\gamma(\boldsymbol{q})\, \mathrm{e}^{\mathrm{i}\boldsymbol{q}\boldsymbol{R}_m} - b_{\boldsymbol{q}}^{+}\gamma^*(\boldsymbol{q})\, \mathrm{e}^{-\mathrm{i}\boldsymbol{q}\boldsymbol{R}_m}\}. \tag{6.2.4}$$

The operator S_m occurs in the second order on expanding $\hat{\Phi}$ in powers of γ. The corresponding first order terms are incorporated into $\mathscr{H}_0$.

It is convenient to introduce a new representation (α-representation) diagonalizing $\mathscr{H}_0$:

$$a_1 = c_{11}\alpha_1 + c_{12}\alpha_2; \qquad a_2 = c_{21}\alpha_1 + c_{22}\alpha_2, \tag{6.2.5}$$

where

$$c_{11} = c_{22}^* = \frac{1}{\sqrt{2\widetilde{W}}} \sqrt{\widetilde{W} - V_{12}}\, \mathrm{e}^{\mathrm{i}\varphi/2}, \qquad c_{12} = -c_{21}^* = \frac{1}{\sqrt{2\widetilde{W}}} \sqrt{\widetilde{W} + V_{12}}\, \mathrm{e}^{\mathrm{i}\varphi/2}.$$

Here

$$V_{mm'} = V_m - V_{m'}, \qquad \widetilde{W} = \sqrt{V_{12}^2 + 4J^2}, \qquad J_{21} = J_{12}^* \equiv J\, \mathrm{e}^{-\mathrm{i}\varphi}.$$

In this representation the operators $\mathscr{H}_0$ and $\mathscr{H}_1$ read

$$\left.\begin{aligned} \mathscr{H}_0 &= W_1\alpha_1^{+}\alpha_1 + W_2\alpha_2^{+}\alpha_2, \\ \mathscr{H}_1 &= J(\hat{\Phi}_{12} - \langle\hat{\Phi}_{12}\rangle)\, \alpha_1^{+}\alpha_2 + J(\hat{\Phi}_{21} - \langle\hat{\Phi}_{21}\rangle)\, \alpha_2^{+}\alpha_1, \end{aligned}\right\} \tag{6.2.6}$$

where

$$W_{1,2} = \frac{1}{2}\left\{V_1 + V_2 \mp \sqrt{V_{12}^2 + 4J^2}\right\} \tag{6.2.7}$$

may be viewed as the energy levels of a two-atomic molecule on account of the level-repulsion effect, which is in that the energetic spacing between the levels is the larger the more the two atoms are close together (i.e. the larger J is). The smallest possible level spacing (for $\varepsilon_1 = \varepsilon_2$, $\boldsymbol{E} = 0$) is equal to J.

In the operator $\mathscr{H}_1$ we have replaced $S_1 - S_2$ by $\exp(S_1 - S_2) - \langle\exp(S_1 - S_2)\rangle = \hat{\Phi}_{12} - \langle\hat{\Phi}_{12}\rangle$, in order to may use results from Section 2., which were derived without expanding $\hat{\Phi}$ in powers of $\gamma(\boldsymbol{q})$. Such a procedure requires that in the final results, γ is assumed to be weak, and the transition probabilities are expanded in powers of $\gamma(\boldsymbol{q})$ whereby only the terms of the leading order are accounted for. $\langle\hat{\Phi}\rangle$ denotes the phonon average of $\hat{\Phi}$. It is equal to $\exp(-S_{\mathrm{T}})$ (cf. Appendix 1.).

For the two-site model, let us now write down the expression for the current (2.3.3) in the α-representation

$$\boldsymbol{j}(t) = \boldsymbol{j}_1(t) + \boldsymbol{j}_2(t), \tag{6.2.8}$$

$$\boldsymbol{j}_1(t) = \frac{e}{2\Omega} (\boldsymbol{R}_1 - \boldsymbol{R}_2) \frac{V_{12}}{\widetilde{W}} \frac{\mathrm{d}}{\mathrm{d}t} [\varrho_2^2(t) - \varrho_1^1(t)], \tag{6.2.9}$$

$$\boldsymbol{j}_2(t) = \frac{e}{\Omega} (\boldsymbol{R}_1 - \boldsymbol{R}_2) \frac{J}{\widetilde{W}} \frac{\mathrm{d}}{\mathrm{d}t} [\varrho_2^1(t) - \varrho_1^2(t)], \tag{6.2.10}$$

where $\varrho^i_{i'}$ is the density matrix in the α-representation (cf. (2.4.4)):

$$\varrho^i_{i'}(t) = \langle U^+(t)\, \alpha^+_{i'} \alpha_i U(t) \rangle \qquad (i = 1, 2),$$

and $U(t)$ is the time-evolution operator associated with the Hamiltonian $\tilde{\mathscr{H}}$.

Within the framework of the two-site model, the thermodynamic average is performed by means of the canonical distribution (i.e. by means of $\exp(-\beta\tilde{\mathscr{H}}^{(0)})$, where the superscript (0) indicates the absence of an electrical field), and not by means of the Gibbs distribution as in (2.3.6). Evidently, pairs of sites without electrons or with two electrons do not contribute to the current. We must account only for pairs occupied with a single electron, for which it holds that $\alpha_1{}^+\alpha_1 + \alpha_2{}^+\alpha_2 = 1$.

The Hamiltonian in the α-representation agrees with that in the site representation if, in the latter Hamiltonian, we replace V_m by W_i. Therefore, we may use the equation for the density matrix (2.4.6), derived in Section 2., after replacing in this equation the site indices m by the quantum numbers of the two-site molecule i. Since we may restrict ourselves to pairs occupied with a single electron, (2.4.6) is exact in the sense that we must not decouple many-particle density matrices. Moreover, because of $\langle \mathscr{H}_1 \rangle = 0$, diagrams with an isolated point do not occur, so that $\overset{(1)}{\mathscr{W}} = 0$. Note that the transformation (6.2.5), which from the Hamiltonian removes the term $J_{12}a_1{}^+a_2 + J_{21}a_2{}^+a_1$, is equivalent to a summing-up of all diagrams with an arbitrary number of isolated points.

Since we are interested only in the leading order with respect to the coupling constant, $\mathscr{W}$ is given by the second-order diagrams illustrated in Fig. 2.2. Thus, we have

$$\frac{\mathrm{d}\varrho^i_{i'}(t)}{\mathrm{d}t} = \frac{\mathrm{i}}{\hbar}\,(W_{i'} - W_i)\,\varrho^i_{i'}(t) + \sum_{i_1 i_2} \int\limits_0^\infty \mathrm{d}t'\varrho^{i_1}_{i_2}(t - t')\,\mathscr{W}^{i_1 i}_{i_2 i'}(t - t', t). \qquad (6.2.11)$$

On deriving (6.2.11), we have adopted the limit of large times, at which the transition processes occurring after turning on the electrical field have run out (cf. the transition from (2.4.6) to (2.4.16)).

Using for $\mathscr{W}$ the diagram rules from Appendix 1., (6.2.11) may be rewritten as

$$\frac{\mathrm{d}\varrho^1_1(t)}{\mathrm{d}t} = 2\int\limits_0^\infty \mathrm{d}t'\{\varrho^2_2(t - t')\,\mathrm{Re}\,[I(t, t')\,\mathscr{W}(t')] - \varrho^1_1(t - t')\,\mathrm{Re}\,[I^*(t, t')\,\mathscr{W}(t')]\}, \qquad (6.2.12)$$

$$\frac{\mathrm{d}\varrho^1_2(t)}{\mathrm{d}t} = \frac{\mathrm{i}}{\hbar}\,\widetilde{W}(t)\,\varrho^1_2(t) + 2\int\limits_0^\infty \mathrm{d}t'[\varrho^2_1(t - t') - \varrho^1_2(t - t')]\,\mathrm{Re}\,\mathscr{W}(t'), \qquad (6.2.13)$$

where

$$I(t, t') = \exp\left\{-\frac{\mathrm{i}}{\hbar}\int\limits_0^{-t'} \mathrm{d}\tau\widetilde{W}(\tau + t)\right\}, \qquad (6.2.14)$$

$$\mathscr{W}(t) = \frac{J^2}{\hbar^2}\sum_{\boldsymbol{q}} \frac{|\gamma(\boldsymbol{q})|^2\,[1 - \cos \boldsymbol{q}(\boldsymbol{R}_1 - \boldsymbol{R}_2)]}{N \sinh(\hbar\omega_{\boldsymbol{q}}\beta/2)} \cos \omega_{\boldsymbol{q}}\left(t + \frac{\mathrm{i}\hbar\beta}{2}\right).$$

Here $\mathscr{W}$ is already given in the lowest nonvanishing order with respect to $\gamma(\boldsymbol{q})$.

Equations (6.2.12) and (6.2.13) must be completed by the conditions $\varrho_1^1 + \varrho_2^2 = 1$ and $\varrho_1^2 = (\varrho_2^1)^*$. Note that the form of these equations implies the use of a real quantity for the electrical field (cf. the remarks concerning (2.4.9)).

By means of (6.2.12), (6.2.13), and the expression for the current (6.2.8) to (6.2.10), the charge transport can be described in the presence of an electrical field of arbitrary strength (including, for instance, absorption of laser radiation). Note that the problem of charge transport in disordered systems in the presence of strong electrical fields was studied by Böttger and Bryksin (1975c). Here, however, we restrict ourselves to an investigation of the Ohmic behaviour of the system under consideration. To linearize the above equations with respect to $\boldsymbol{E}$, we write

$$\varrho_{i'}^{i}(t) = \overset{(0)}{\varrho_{i'}^{i}} + \overset{(1)}{\varrho_{i'}^{i}}\, e^{i\omega t}, \tag{6.2.15}$$

where the superscripts (0) and (1), respectively, indicate the zeroth and first order with respect to $\boldsymbol{E}$. On neglection of electron-phonon interaction, $\overset{(0)}{\varrho_{i'}^{i}}$ becomes

$$\overset{(0)}{\varrho_{i'}^{i}} = \overset{(0)}{\varrho_{i}}\delta_{ii'}, \qquad \overset{(0)}{\varrho_{1,2}} = 2\exp\{\pm\beta\tilde{\varepsilon}/2\}/\cosh(\beta\tilde{\varepsilon}/2), \tag{6.2.16}$$

where $\tilde{\varepsilon} = \sqrt{\varepsilon_{12}^2 + 4J^2}$, $\varepsilon_{12} = \varepsilon_1 - \varepsilon_2$.

Evaluating $\overset{(1)}{\varrho}$ by means of (6.2.12) and (6.2.13), and inserting the result into the formula for the current, then we find the following expression for the real part of the conductivity $\sigma'(\omega)$:

$$\sigma'(\omega) = \sigma_1(\omega) + \sigma_2(\omega), \tag{6.2.17}$$

$$\sigma_1(\omega) = \frac{e^2(\boldsymbol{R}_1 - \boldsymbol{R}_2)_x^2}{2\Omega\cosh^2(\tilde{\varepsilon}\beta/2)}\,\frac{\varepsilon_{12}^2}{\tilde{\varepsilon}^2}\,\frac{\omega^2/\tau_1'(\omega)}{\omega^2 + 1/\tau_1^2(\omega)}, \tag{6.2.18}$$

$$\sigma_2(\omega) = \frac{e^2(\boldsymbol{R}_1 - \boldsymbol{R}_2)_x^2}{\Omega\tilde{\varepsilon}}\,\frac{J^2}{\tilde{\varepsilon}^2}\tanh(\hbar\omega\beta/2)\,\frac{\omega^2}{\tau_2(\omega)} \times\left\{\left[\left(\omega - \frac{\tilde{\varepsilon}}{\hbar}\right)^2 + 1/\tau_2^2(\omega)\right]^{-1} + \left[\left(\omega + \frac{\tilde{\varepsilon}}{\hbar}\right)^2 + 1/\tau_2^2(\omega)\right]^{-1}\right\}. \tag{6.2.19}$$

Here we have introduced the following relaxation times

$$\frac{1}{\tau_1(\omega)} = \pi\frac{J^2}{\hbar^2}\sum_{\boldsymbol{q}} g(\boldsymbol{q})\cosh(\hbar\omega_{\boldsymbol{q}}\beta/2)\sum_{\eta,\eta'=\pm1}\delta\left(\omega + \eta\omega_{\boldsymbol{q}} + \eta'\frac{\tilde{\varepsilon}}{\hbar}\right), \tag{6.2.20}$$

$$\frac{1}{\tau_1'(\omega)} = \pi\frac{J^2}{\hbar^2}\frac{\sinh(\hbar\omega\beta/2)}{\hbar\omega\beta/2}\cosh(\tilde{\varepsilon}\beta/2)\sum_{\boldsymbol{q}} g(\boldsymbol{q})\sum_{\eta,\eta'=\pm1}\delta\left(\omega + \eta\omega_{\boldsymbol{q}} + \eta'\frac{\tilde{\varepsilon}}{\hbar}\right), \tag{6.2.21}$$

$$\frac{1}{\tau_2(\omega)} = \pi\frac{J^2}{\hbar^2}\sum_{\boldsymbol{q}} g(\boldsymbol{q})\cosh(\hbar\omega_{\boldsymbol{q}}\beta/2)\sum_{\eta=\pm1}\delta(\omega + \eta\omega_{\boldsymbol{q}}), \tag{6.2.22}$$

where

$$g(\boldsymbol{q}) = |\gamma(\boldsymbol{q})|^2 \,[1 - \cos \boldsymbol{q}(\boldsymbol{R}_1 - \boldsymbol{R}_2)]/N \sinh (\hbar\omega_{\boldsymbol{q}}\beta/2).$$

The contribution σ_1 (6.2.18) to the conductivity originates from phonon-induced nonresonance transitions of electrons. Note that for small J, where $\tilde{\varepsilon} \approx |\varepsilon_{12}|$, the expression (6.2.18) can be obtained by the formalism described in Section 2, on account of only the diagonal (with respect to the site indices) elements of the density matrix. In the limit of small ω (Markovian limit) we have

$$\frac{1}{\tau_1(0)} = \frac{1}{\tau_1'(0)} \equiv \frac{1}{\tau} = 2\pi \frac{J^2}{\hbar^2} \sum_{\boldsymbol{q}} g(\boldsymbol{q}) \cosh (\hbar\omega_{\boldsymbol{q}}\beta/2)\, \delta\left(\omega_{\boldsymbol{q}} - \frac{\tilde{\varepsilon}}{\hbar}\right). \tag{6.2.23}$$

An expression of the type (6.2.18) (in the Markovian limit and for $J \to 0$, i.e. $\tilde{\varepsilon} = |\varepsilon_{12}|$) was first obtained by LYO and HOLSTEIN (1973), while σ_1 and σ_2 in the form given above was first obtained by BÖTTGER and BRYKSIN (1976c).

The contribution σ_2 (6.2.19) to the conductivity does not vanish in the limit as $\gamma(\boldsymbol{q}) \to 0$ (i.e. as $\tau_2 \to \infty$). In this limit (i.e. in the absence of electron-phonon interaction) we have

$$\sigma_2(\omega) = \pi \frac{e^2(\boldsymbol{R}_1 - \boldsymbol{R}_2)_x^2}{\Omega\tilde{\varepsilon}} \frac{J^2}{\tilde{\varepsilon}^2} \tanh (\hbar\omega\beta/2)\, \omega^2\delta\left(\omega - \frac{\tilde{\varepsilon}}{\hbar}\right). \tag{6.2.24}$$

This formula indicates that σ_2 originates from photon-induced transitions between the levels of the two-site molecule under consideration. Accordingly, σ_2 may be termed the phononless or photon-assisted hopping contribution.

To get $\sigma'(\omega)$, the expression obtained for some pair must be averaged over all possible configurations of the pair. However, before performing this average, let us now discuss the validity of the two-site approximation.

According to (6.2.24), for small $1/\tau_2$, the phononless contribution to the conductivity arises from pairs, for which $\tilde{\varepsilon} = \sqrt{\varepsilon_{12}^2 + 4J^2} \approx \hbar\omega$, i.e. $2J \leqq \hbar\omega$ holds. Hence, it follows that the spacing between the sites in a pair, R, must satisfy the inequality $R > R_{\min} = (1/\alpha) \ln (2J_0/\hbar\omega)$ (here we have accounted for $J = J_0 \exp(-\alpha R)$; see above). Consequently, for $\omega \to 0$, σ_2 goes to zero. It should be noted, however, that this fact does not indicate the existence of localized states, since for $\omega \to 0$, $R_{\min}$ exceeds the average distance between the sites. In this case, it is not reasonable to account for the repulsion effect, because the position of the energy levels of the selected pair of sites is not determined by "intramolecular" interaction, but by interaction with the surrounding sites, i.e. the two-site model fails for low frequencies.

As pointed out in Section 3., in the small-polaron model, the current is additively composed of a hopping and a tunnelling contribution. Note that these contributions are the analoga to σ_1 and σ_2, respectively. It is interesting to ask what happens with the contribution σ_2 in disordered systems in the low frequency limit, where the two-site mode breaks down. Evidently, in the case of localized charge carriers, for $\gamma \to 0$, σ_2 must vanish, while for $\gamma \neq 0$, it may remain finite, for instance, due to tunnelling processes. If the carriers are not localized, then σ_2 must describe the usual band transport with scattering on structural irregularities.

Let us now turn to the averaging of the expression for σ'. To this end, we must multiply (6.2.17) to (6.2.19) by the probability of finding one electron on the pair selected out. This probability is given by $f(E_1)\,[1 - f(E_2)] + f(E_2)\,[1 - f(E_1)]$, where $E_{1,2} = \frac{1}{2}\left\{\varepsilon_1 + \varepsilon_2 \mp \sqrt{\varepsilon_{12}^2 + 4J^2}\right\}$ are the molecular levels (6.2.7) in the absence of a field, and $f(E_i)$ $(i = 1, 2)$ is the Fermi function

$$f(E_i) = \left[\exp\left(\beta(E_i - \mu) + 1\right)\right]^{-1}. \tag{6.2.25}$$

On account of the fact that in a system with N sites the total number of pairs is equal to $\frac{1}{2} N(N - 1) \approx \frac{1}{2} N^2$, the configuration average of the real part of the conductivity becomes

$$\sigma'(\omega) = \frac{1}{2} N^2 \int \mathrm{d}\boldsymbol{R}\, \mathrm{d}\varepsilon_1\, \mathrm{d}\varepsilon_2 \Pi(\boldsymbol{R})\, P(\varepsilon_1)\, P(\varepsilon_2)\, \sigma'(\varepsilon_1 - \varepsilon_2, \boldsymbol{R})$$
$$\times \{f(E_1)\,[1 - f(E_2)] + f(E_2)\,[1 - f(E_1)]\}, \tag{6.2.26}$$

where $\sigma'(\varepsilon_1 - \varepsilon_2, \boldsymbol{R})$ denotes the unaveraged conductivity of a pair (6.2.17) to (6.2.19), $\Pi(\boldsymbol{R})$ the probability of finding a pair with spacing R between the sites, and $P(\varepsilon)$ the distribution function for the site energies. Such an averaging includes the assumption of uncorrelated energy levels of the (isolated) sites, and that there is no correlation between energy and spatial position of a site. Assuming that the sites are distributed in the space at random, then we have $\Pi(\boldsymbol{R}) = 1/\Omega$, where Ω is the total volume of the system.

6.2.2. *The phonon contribution to the conductivity*

Inserting now (6.2.18) into (6.2.26), we obtain for the phonon contribution to the conductivity the following expression

$$\sigma_1(\omega) = \frac{\pi}{3} e^2\omega^2\beta \int_0^\infty \mathrm{d}R R^4 \frac{2\nu_{\mathrm{ph}}^2 \exp(-2\alpha R)}{\omega^2 + 4\nu_{\mathrm{ph}}^2 \exp(-4\alpha R)} \int_{-\infty}^{\infty} \mathrm{d}\varepsilon \int_{2J_0\exp(-\alpha R)}^{\infty} \mathrm{d}\tilde{\varepsilon}$$
$$\times \frac{\sqrt{\tilde{\varepsilon}^2 - 4J_0{}^2 \exp(-2\alpha R)}}{\tilde{\varepsilon}}\, N\left(\varepsilon + \frac{1}{2}\sqrt{\tilde{\varepsilon}^2 - 4J_0{}^2 \exp(-2\alpha R)}\right)$$
$$\times N\left(\varepsilon - \frac{1}{2}\sqrt{\tilde{\varepsilon}^2 - 4J_0{}^2 \exp(-2\alpha R)}\right) [\sinh(\beta\tilde{\varepsilon})]^{-1}$$
$$\times \left\{[\mathrm{e}^{\beta(\varepsilon - (\tilde{\varepsilon}/2) - \mu)} + 1]^{-1} - [\mathrm{e}^{\beta(\varepsilon + (\tilde{\varepsilon}/2) - \mu)} + 1]^{-1}\right\}. \tag{6.2.27}$$

Here we have adopted the Markovian limit (6.2.23) and τ approximated by $\tau^{-1} = 2\nu_{\mathrm{ph}} \exp(-2\alpha R)$. The characteristic frequency ν_{ph} can be shown to depend only algebraically on the random variables R and ε. Since the quantity (6.2.27) is above all determined by the exponential dependences on the random variables, on evaluating (6.2.27), ν_{ph} may be considered as a non-random quantity. The function $N(\varepsilon) = \mathcal{N}P(\varepsilon)$ is the density of states of the site levels ($\mathcal{N} = N/\Omega$).

For not too large ω, in the integral (6.2.27), the characteristic value of R is equal to $(1/2\alpha) \ln (2\nu_{\mathrm{ph}}/\omega)$. Consequently, the repulsion effect may be neglected if the characteristic values of $\tilde{\varepsilon}$, which are of order kT, are larger than $2J_0 \sqrt{\omega/2\nu_{\mathrm{ph}}}$. Hence, it follows that in the integrand in (6.2.27) the term $2J_0\, \mathrm{e}^{-\alpha R}$ may be made equal to zero, provided that

$$\omega < \nu_{\mathrm{ph}} \frac{1}{2} \left(\frac{kT}{J_0}\right)^2. \tag{6.2.28}$$

Here we restrict ourselves to this limiting case (the different limiting cases were studied in detail by Böttger and Bryksin (1976d)).

Thus, on neglection of the level repulsion, we find that

$$\sigma_1(\omega) = \frac{\pi}{3}\, e^2\omega^2\beta(2\alpha)^{-5} \int\limits_0^\infty \mathrm{d}x x^4\, \frac{2\nu_{\mathrm{ph}}\, \mathrm{e}^{-x}}{\omega^2 + 4\nu_{\mathrm{ph}}^2\, \mathrm{e}^{-2x}}\, \Phi_{\mathrm{T}}, \tag{6.2.29}$$

where

$$\Phi_{\mathrm{T}} = \int\limits_{-\infty}^{\infty} \mathrm{d}\varepsilon \int\limits_0^\infty \mathrm{d}\tilde{\varepsilon} N\left(\varepsilon + \frac{1}{2}\,\tilde{\varepsilon}\right) N\left(\varepsilon - \frac{1}{2}\,\tilde{\varepsilon}\right) [\sinh (\beta\tilde{\varepsilon})]^{-1}$$
$$\times \left\{[\mathrm{e}^{\beta(\varepsilon-(\tilde{\varepsilon}/2)-\mu)} + 1]^{-1} - [\mathrm{e}^{\beta(\varepsilon+(\tilde{\varepsilon}/2)-\mu)} + 1]^{-1}\right\}. \tag{6.2.30}$$

For $\omega \ll \nu_{\mathrm{ph}}$, we get

$$\sigma_1(\omega) \approx \frac{\pi^2}{6} \frac{e^2\omega}{kT} (2\alpha)^{-5}\, \Phi_{\mathrm{T}} \ln^4 (2\nu_{\mathrm{ph}}/\omega), \tag{6.2.31}$$

while for $\omega \geqq 2\nu_{\mathrm{ph}}$,

$$\sigma_1(\omega) \approx 16\pi\, \frac{e^2\nu_{\mathrm{ph}}}{kT} (2\alpha)^{-5}\, \Phi_{\mathrm{T}}. \tag{6.2.32}$$

Accordingly, within the framework of the two-site model, at low frequencies, σ' increases as $\omega \ln^4 (2\nu_{\mathrm{ph}}/\omega)$, while at high frequencies σ' passes over into a plateau.

In the frequency range $\omega \ll \nu_{\mathrm{ph}}$, where the characteristic value of the spacing R is equal to $(1/2\alpha) \ln (2\nu_{\mathrm{ph}}/\omega)$, the lower limiting frequency results from the requirement that this spacing must be smaller than $\varkappa\mathcal{N}^{-1/3}$ ($\varkappa$ — numerical factor of order unity), i.e. the spacing R may not exceed a value of order the average spacing between the sites. Thus, for the phonon contribution to the conductivity, we obtain the following criterion for the applicability of the two-site model:

$$\omega > 2\nu_{\mathrm{ph}} \exp (-2\varkappa\alpha\mathcal{N}^{-1/3}). \tag{6.2.33}$$

Note that (6.2.33) is a rather rough criterion, because the value of the factor $\varkappa$ is uncertain and in general $\alpha\mathcal{N}^{-1/3} \gg 1$ holds. This problem will be discussed in more detail below, in the context with solutions which are valid for arbitrary ω.

With increasing frequency, the characteristic spacing between the sites of a pair decreases, so that at the high-frequency plateau, the main contribution to σ_1 arises from the small concentration of pairs having a spacing of order $1/2\alpha$.

In the integrand in (6.2.30), the characteristic value of $\tilde{\varepsilon}$ is of order kT, while with regard to ε, there are in general two maxima, at $\varepsilon = \varepsilon_0$ and $\varepsilon = \mu$, where ε_0 denotes the position of the maximum of the density of states $N(\varepsilon)$. In the case that the main contribution to the integral arises from the region about the point $\varepsilon = \varepsilon_0$, hopping occurs near the maximum of the density of states, owing to thermal excitation of charge carriers from the Fermi level. In Section 5., transport of this type was termed R-percolation (R-hopping). For $\Gamma < kT$ (Γ — width of the spread of the energy levels), in this case $N(\varepsilon)$ may be replaced by $N(\varepsilon) = \mathcal{N}\delta(\varepsilon - \varepsilon_0)$, so that we get

$$\Phi_{\mathrm{T}} = \mathcal{N}^2 \exp\left(-\varepsilon_3/kT\right), \qquad \varepsilon_3 > kT, \tag{6.2.34}$$

where $\varepsilon_3 = |\mu - \varepsilon_0|$ is the activation energy of the dc conductivity (cf. (5.1.9) (5.1.12)).

Note that within the framework of the two-site model, σ_1 does not contain the exponentially small factor $\exp\left(-\eta\alpha\mathcal{N}^{-1/3}\right)$, which appears in the expression for the dc conductivity in the case of R-percolation (cf. (5.1.9), (5.1.11)).

At low temperatures, in (6.2.30) the main contribution to the integral over ε arises from the region about the point $\varepsilon = \mu$, i.e. hopping is realized close to the Fermi level (R-ε-percolation). Assuming that $N(\varepsilon)$ depends only weakly on ε in the region from $\mu - kT$ to $\mu + kT$, we take $N\left(\varepsilon + \frac{1}{2}\tilde{\varepsilon}\right)$ and $N\left(\varepsilon - \frac{1}{2}\tilde{\varepsilon}\right)$, at $\varepsilon = \mu$ and $\tilde{\varepsilon} = 0$, out of the integral. Then, we obtain

$$\Phi_{\mathrm{T}} = \frac{\pi^2}{4} (kT)^2 N^2(\mu). \tag{6.2.35}$$

Accordingly, in contrast to dc conductivity in the case of R-ε-percolation, described by the Mott $T^{-1/4}$ law (5.1.19), in the two-site model σ_1 depends only weakly (algebraically) on temperature. Note that the temperature separating the R- and R-ε-regimes, obtained by equating (6.2.34) and (6.2.35), possibly differs from the temperature separating the regions of activated dc conductivity and the Mott $T^{-1/4}$ law.

6.2.3. *The phononless contribution to the conductivity*

Let us now turn to the phononless contribution to σ'. According to (6.2.26) and (6.2.24), this quantity is given by

$$\begin{aligned}\sigma_2(\omega) = {} & \frac{\pi^2}{6} e^2\omega(2\alpha)^{-5} \int_0^{\xi} \mathrm{d}z(1-z)^{-1/2} \ln^4\left[\hbar\omega z/2J_0\right] \\ & \times \int_{-\infty}^{\infty} \mathrm{d}\varepsilon N\left(\varepsilon + \frac{1}{2}\hbar\omega\sqrt{1-z}\right) N\left(\varepsilon - \frac{1}{2}\hbar\omega\sqrt{1-z}\right) \\ & \times \left\{\left[\mathrm{e}^{\beta\left(\varepsilon - \frac{1}{2}\hbar\omega - \mu\right)} + 1\right]^{-1} - \left[\mathrm{e}^{\beta\left(\varepsilon + \frac{1}{2}\hbar\omega - \mu\right)} + 1\right]^{-1}\right\},\end{aligned} \tag{6.2.36}$$

where

$$\xi = 1 \text{ if } 2J_0 > \hbar\omega, \quad \text{and} \quad \xi = (2J_0/\hbar\omega)^2 \text{ if } 2J_0 < \hbar\omega .$$

In the case of the phononless contribution to the current, the level repulsion may be neglected (i.e. in (6.2.36) we may assume $z \ll 1$) at very high frequencies, $\hbar\omega \gg 2J_0$, whereas for σ_1, the level repulsion may be neglected at low frequencies. We restrict ourselves to a discussion of the low frequency limit $\hbar\omega < \Gamma$, kT, $2J_0$ (for a more detailed investigation, see Böttger and Bryksin, 1976d). In this frequency range, (6.2.36) becomes

$$\sigma_2(\omega) = \frac{16}{3}\,\pi^2\,\frac{e^2\omega^2\hbar}{kT}\,(2\alpha)^{-5}\ln^4(2J_0/\hbar\omega)\,\Psi_T,$$

$$\Psi_T = \frac{1}{4}\int\limits_{-\infty}^{\infty} d\varepsilon N^2(\varepsilon)/\cosh^2[(\varepsilon-\mu)/2kT]. \tag{6.2.37}$$

For hops near the maximum of the density of states, we have $\Psi_T \approx \mathcal{N}^2(\lambda/\Gamma) \times \exp(-\varepsilon_3/kT)$, where λ is a number defined as $\lambda = \Gamma \int d\varepsilon P^2(\varepsilon)$. On the other hand, for hops near the Fermi level, $\Psi_T \approx N^2(\mu)\,kT$ (for this case, σ_2 was given by Mott, 1970).

Hence, within the framework of the two-site model, the phononless contribution $\sigma_2(\omega)$ behaves as $\omega^2 \ln^4(2J_0/\hbar\omega)$ up to frequencies of order $2J_0/\hbar$, and at higher frequencies, analogous to the phonon contribution, $\sigma_2(\omega)$ passes over into a plateau, as can be readily seen from (6.2.36). Note that for actual values of J_0 ($J_0 \geqq 0.1$ eV), the plateau is in the infrared region. Accordingly, owing to strong lattice absorption in this region, the plateau can hardly be observed.

From an inspection of the expression for σ_2, one finds that the characteristic spacing between the sites in a pair is equal to $(1/\alpha)\ln(2J_0/\hbar\omega)$. Since this spacing must be smaller than the mean distance between the sites, the lower frequency limit for the applicability of the two-site model becomes

$$\hbar\omega > 2J_0 \exp(-\varkappa'\alpha\mathcal{N}^{-1/3}), \tag{6.2.38}$$

where $\varkappa'$ is a number of order unity. Analogous to (6.2.33), the relation (6.2.38) is a rather rough criterion.

On comparing (6.2.31) with (6.2.37), we see that the phonon contribution dominates at low frequencies, while the phononless one at higher frequencies. This fact indicates that it is not reasonable to study σ_2 by means of a cluster approach. If the transition from the regime dominated by the phonon contribution to that dominated by the phononless one occurs at frequencies smaller than $2\nu_{ph}$ ($\nu_{ph} \approx 10^{12}\ s^{-1}$), the corresponding characteristic frequency ω_0 can be found by equating (6.2.31) and (6.2.37). For hops near the Fermi level, this yields $\hbar\omega_0 = (\pi^2/128)\,kT\,[\ln(2\nu_{ph}/\omega_0)/\ln(2J_0/\hbar\omega_0)]^4$ (T in Kelvin). On deriving this expression, we have assumed that J_0 exceeds $\hbar\nu_{ph}$ at least by one order of magnitude.

A characteristic feature of the transition from the regime dominated by the phonon contribution to that dominated by the phononless one is that the experimental dependence $\sigma'(\omega) \sim \omega^s$ changes from $s \leqq 1$ to $s \leqq 2$. Such a change has been reported, for example, for disordered polymers. For a more detailed discussion of the frequency dependence of σ', see Section 6.6.

On examining the two-site model, we have assumed that there is no correlation between the energy and the spatial position of the sites. A consequence of this assumption is the multiplicative dependence of σ_1 on frequency and temperature (cf. (6.2.29)). However, experimental findings (cf. Section 6.6.) indicate that the frequency and temperature dependence of σ_1 is often more complicated. Therefore, several authors have recently studied two-site models on account of correlations between energy and spatial position. For instance, Hill (1977) considered the case of a spatially inhomogeneous distribution of energy levels, and Elliott (1977) studied over-barrier hops by taking into account that the height of a barrier depends on the spacing between the sites as a result of the Coulomb interaction. Efros (1981) investigated the effect of Coulomb correlation on hops near the Fermi level. For further references on this topic see, e.g., the review by Böttger and Bryksin (1982).

6.3. Cluster approximation

6.3.1. *The basic idea underlying the cluster approximation*

The two-site approximation has its major deficiency in that it is only valid for sufficiently high frequencies. If one formally extends the formulae obtained by this approximation to the low frequency limit, the resulting expressions do not join with the dc formulae found by means of the percolation theory (cf. Section 5). In this section we attack the intricate problem of low-frequency ac hopping conductivity by using ideas contained in the percolation theory.

Recall that the ac hopping conductivity can be described by an equivalent circuit consisting of a random-resistor network with capacitors (cf. Section 2.5.). At sufficiently high frequencies the impedance of the capacitors becomes very small so that the potential at a site is determined solely by the external field. The current is then governed by transitions of electrons between pairs of sites, and the two-site approximation is adequate. At lower frequencies, the current arises from transitions within highly conducting clusters of sites (multiple hopping), the size of which increases with decreasing frequency so that it becomes macroscopically large for $\omega \to 0$. At zero frequency the capacitors may be neglected, and the network reduces to one with resistors only. As pointed out in Section 5., in the dc limit the current is governed by the infinite critical cluster of bonds (resistors).

In the present section, we describe a cluster approximation for the low-frequency hopping conductivity (Böttger et al., 1979a, b, c). It is based on the idea that the problem of the random resistor network with capacitors may be related to an analogous network containing two types of resistors only. Resistances of the first type are assumed to be of finite size and equal in magnitude to one another, those of the second type are assumed to be infinite. This will be referred to as the disconnected network (broken-bond) model. In one dimension ($d = 1$) it can be solved rigorously. For $d = 3$, scaling arguments may be used to obtain a solution. The original resistor network problem is related to the disconnected-network model through methods reminiscent of those used to relate the random-resistor network describing dc hopping conductivity to a percolation problem. Note that the cluster approximation has first of all guiding character. It is very suitable for visualizing the physical situation. Its results are in qualitative agreement with the results of other methods, which will be described below.

For simplicity, we restrict ourselves to the case of R-percolation (cf. Section 5.), where in the equivalent circuit in question all capacitors C_m are equal to each other, and according to (2.5.1) it holds that

$$C_m \equiv C = \frac{e^2}{kT} \exp\left(-\varepsilon_3/kT\right). \tag{6.3.1}$$

Here ε_3 is the activation energy of the dc conductivity (cf. Section 5.1.3.).

In this model, the only random quantities are the resistances $Z_{m'm}$, which depend on the random parameter — the distance between the sites, $\boldsymbol{R}_{m'} - \boldsymbol{R}_m$, as (cf. Section 5.1.3.)

$$Z_{m'm} = Z_0 \exp\left\{\frac{\varepsilon_3}{kT} + 2\alpha\left|\boldsymbol{R}_{m'} - \boldsymbol{R}_m\right|\right\}. \tag{6.3.2}$$

Analogous to the procedure for determining the dc conductivity by methods of percolation theory, (cf. Section 5.), let us now remove resistors from the system, (cutting bonds), the largest first, down to some optimal one, Z_{opt}, which is a function of the portion ε of the resistors removed. By definition, all remaining resistors are smaller than Z_{opt}. We now replace all remaining resistors by Z_{opt}, i.e. we enlarge them, which yields the disconnected-network model mentioned above. Evidently, for this model, $|\sigma(\omega)|$ is smaller than for the original system. If we know the conductivity of the disconnected-network model for given values ε and $Z_{\text{opt}}(\varepsilon)$, then the optimal value $\varepsilon = \varepsilon_{\text{opt}}$ is that for which $|\sigma(\omega, \varepsilon)|$ becomes maximum:

$$\mathrm{d}|\sigma(\omega, \varepsilon)|/\mathrm{d}\varepsilon|_{\varepsilon=\varepsilon_{\text{opt}}} = 0. \tag{6.3.3}$$

It is evident that ε_{opt} depends on frequency. The existence of such a maximum results from the fact that in the two limiting cases $\varepsilon = 1$ (all bonds are cut) and $\varepsilon = 0$ (no bond is cut), σ vanishes, since in both limiting cases all resistors are replaced by resistors of infinite size. For $\omega = 0$, Z_{opt} is simply the critical resistor Z_c (cf. Section 5.) and the existing bonds form the infinite critical cluster. For $\omega \neq 0$, we have $Z_{\text{opt}} < Z_c$, and $\varepsilon_{\text{opt}} > \varepsilon_c$ (ε_c — critical portion of lacking bonds at the percolation threshold). In this case the bonds form finite clusters only. With increasing frequency, ε_{opt} increases and Z_{opt} decreases.

6.3.2. *One-dimensional systems*

In this section the cluster approximation is applied to one-dimensional systems with nearest-neighbour hopping (Böttger et al., 1979a). In this case, the calculations can be readily performed and the results obtained can be compared with some rigorous expressions.

According to (2.5.17), for the disconnected-network model with resistors Z_{opt}, at $d = 1$ (d — dimensionality), the current becomes

$$j(\omega) = -\frac{1}{NZ_{\text{opt}}} \sum_s \delta U_s, \tag{6.3.4}$$

where δU_s is the potential difference between the ends of a cluster that consists of s (existing) bonds, $N = L/a$ is the total number of bonds in a chain of length L and nearest-neighbour distance a (which is assumed to be non-random). The summation is over all clusters. Replacing this summation by a configuration average, we obtain

$$j(\omega) = -\frac{1}{Z_{\text{opt}}} \sum_{s=1}^{\infty} n_s \delta U_s. \tag{6.3.5}$$

Here Nn_s is the number of clusters with s bonds in the chain. For an infinite chain, n_s is given by

$$n_s = \varepsilon^2(1 - \varepsilon)^s. \tag{6.3.6}$$

Thus the problem is to calculate δU_s.

According (2.5.16), in the disconnected-network model, the equation for the potential U_m at site m reads

$$\mathrm{i}\omega\tau(U_m - u_m) = U_{m+1} + U_{m-1} - 2U_m. \tag{6.3.7}$$

Here $\tau = Z_{\text{opt}}C$, where C is the capacity (6.3.1), and $u_m = -Eam$ is the potential of the external field at site m.

Equation (6.3.7) is valid for all sites of a cluster, except the boundary sites, for which it holds that

$$\left.\begin{aligned} \mathrm{i}\omega\tau(U_{m_0} - u_{m_0}) &= U_{m_0+1} - U_{m_0}, \\ \mathrm{i}\omega\tau(U_{m_0+s} - u_{m_0+s}) &= U_{m_0+s-1} - U_{m_0}, \end{aligned}\right\} \tag{6.3.8}$$

where m_0 and $m_0 + s$ denote the boundary sites.

Solution of the set of equations (6.3.7) and (6.3.8) gives

$$U_m = u_m + Ea \frac{\sinh \varphi(m - m_0 - s/2)}{2 \sinh (\varphi/2) \cosh [\varphi(s + 1)/2]}, \tag{6.3.9}$$

where φ is defined as

$$\cosh \varphi = 1 + \frac{1}{2} \mathrm{i}\omega\tau. \tag{6.3.10}$$

Accounting for $\delta U_s = U_{m_0+s} - U_{m_0}$, then from (6.3.5) and (6.3.9), with the aid of (6.3.6), we obtain the following expression for the conductivity of the disconnected network:

$$\sigma(\omega) = \frac{a}{Z_{\text{opt}}} \frac{\varepsilon^2}{1 - \varepsilon} \sum_{s=1}^{\infty} (1 - \varepsilon)^s [s - \coth (\varphi/2) \tanh (s\varphi/2)]. \tag{6.3.11}$$

At low frequencies $\omega\tau \ll 1$ and low concentration of cut bonds $\varepsilon \ll 1$, $\varphi \approx (\mathrm{i}\omega\tau)^{1/2}$ and $|\varphi| \ll 1$ (the parameter $(\omega\tau)^{1/2}/2\varepsilon$ remains arbitrary). The summation in (6.3.11) may then be replaced by an integral to give

$$\sigma(\omega, \varepsilon) = \frac{a}{Z_{\text{opt}}} \int_0^\infty \mathrm{d}s\, \mathrm{e}^{-s} s \left\{1 - \frac{\tanh\left(s\sqrt{\mathrm{i}\omega\tau/\varepsilon^2}\right)}{s\sqrt{\mathrm{i}\omega\tau/\varepsilon^2}}\right\}. \tag{6.3.12}$$

Note that (6.3.12) is a rigorous result (BÖTTGER et al., 1979a, cf. HEINRICHS, 1980, 1982). Therefore, it may be used to test the validity of approximate methods. Here we need (6.3.12) as an interim result on formulating the cluster approximation.

Let us now determine the relation between Z_{opt} and ε, the concentration of cut bonds. According to (6.3.2), the resistance between two sites depends on their spacing R. Thus removal of resistors larger than Z_{opt} requires bonds to be cut between neighbouring sites with a spacing larger than R_{opt} given by

$$Z_{\text{opt}} = Z_0 \exp(2\alpha R_{\text{opt}}) \exp(\varepsilon_3/kT).$$

For a random site distribution, the distance between nearest neighbours is described by the Poisson distribution. Accordingly, the relation between ε and R_{opt} is given by

$$\left.\begin{aligned} &\frac{1}{a}\int_{R_{\text{opt}}}^\infty \mathrm{d}R \exp(-R/a) = \varepsilon, \\ &R_{\text{opt}} = a \ln(1/\varepsilon). \end{aligned}\right\} \tag{6.3.13}$$

Hence, we find that

$$Z_{\text{opt}}^{-1} = \frac{e^2}{kT} \nu_{\text{ph}} \exp(-\varepsilon_3/kT)\, \varepsilon^{-2\alpha a}. \tag{6.3.14}$$

Here, for convenience, a new parameter ν_{ph} has been introduced, which has the dimension of a frequency and which is connected to Z_0 through $Z_0^{-1} = (e^2/kT)\,\nu_{\text{ph}}$. For R-percolation, this parameter is related to the hopping probability as (cf. the corresponding quantity introduced in Section 3.2.)

$$W_{m'm} = \nu_{\text{ph}} \exp\{-2\alpha\, |\boldsymbol{R}_{m'} - \boldsymbol{R}_m|\}. \tag{6.3.15}$$

Insertion of (6.3.14) into (6.3.12) gives

$$\sigma(\omega, \varepsilon) = \frac{e^2 a}{kT} \nu_{\text{ph}} \left(\frac{\omega}{\nu_{\text{ph}}}\right)^{\alpha a/(\alpha a+1)} \mathrm{e}^{-\varepsilon_3/kT} f[\lambda(\omega, \varepsilon)], \tag{6.3.16}$$

where

$$\left.\begin{aligned} \lambda(\omega, \varepsilon) &= \frac{1}{2}\left(\frac{\omega}{\nu_{\text{ph}}}\right)^{1/2} \varepsilon^{-(2\alpha a+1)}, \\ f(\lambda) &= \lambda^{2\alpha a/(\alpha a+1)} \int_0^\infty \mathrm{d}s\, \mathrm{e}^{-s} s \left\{1 + \frac{\tanh\left(\lambda s\sqrt{i}\right)}{\lambda s\sqrt{i}}\right\}. \end{aligned}\right. \tag{6.3.17}$$

Now from (6.3.3) and (6.3.16), we may determine $\varepsilon_{\text{opt}}(\omega)$. Since $\sigma(\omega, \varepsilon)$ depends on ε only through the parameter λ, the solution of (6.3.3) may be written as follows

$$\lambda(\varepsilon_{\text{opt}}) = \frac{1}{2}\left(\frac{\omega}{\nu_{\text{ph}}}\right)^{1/2} \varepsilon_{\text{opt}}^{-(\alpha a+1)} = c',$$

where the number c' is only weakly dependent on the parameter αa. Since we are not interested in numerical factors, we may insert the above expression for ε_{opt} into (6.3.16). Thus, we finally obtain

$$\sigma(\omega) = c\,\frac{e^2 a}{kT}\,\nu_{\text{ph}}\left(\frac{\omega}{\nu_{\text{ph}}}\right)^{\alpha a/(\alpha a+1)} \exp\left(-\varepsilon_3/kT\right), \tag{6.3.18}$$

where $c \equiv f(c')$ is a number, which only weakly depends on αa. Hence, for a one-dimensional system, the cluster approximation in question gives a dependence of the type $\sigma(\omega) \sim \omega^{s'}$ with $s' = \alpha a/(\alpha a + 1)$. The fact that $\sigma(\omega) \to 0$ for $\omega \to 0$, is in agreement with the results obtained in Section 5.

By means of the above expression for ε_{opt}, the frequency dependence of the characteristic cluster size s_{opt} can be found (s_{opt} is defined as the value s at which $sn_s(\varepsilon)$ becomes maximum). According to (6.3.6), for $\varepsilon \ll 1$, we have $s_{\text{opt}} = 1/\varepsilon_{\text{opt}}$, and thus we find that

$$s_{\text{opt}} = (\omega/\nu_{\text{ph}})^{-1/(2\alpha a+2)}, \tag{6.3.19}$$

i.e. the characteristic cluster size decreases with increasing frequency.

This yields an upper frequency limit for the range of application of the cluster approach. Since the cluster size s_{opt} may not be less than a few lattice spacings, we find from (6.3.19) that

$$\omega < \nu_{\text{ph}} \exp\left(-\varkappa' 2\alpha a\right). \tag{6.3.20}$$

Here $\varkappa' = \ln \varkappa > 0$, where $\varkappa$ is the admitted minimal number of spacings (bonds) in the characteristic cluster. Owing to the uncertainty in choosing the value $\varkappa$, the criterion (6.3.20) has guiding character only, but it shows that for multiple hopping, the upper limiting frequency decreases with increasing αa. It should be emphasized that the above cluster approach is suitable only if the spread of the random parameters is large, i.e. if methods of the percolation theory may be applied. Hence, in the situation under consideration the parameter αa must be sufficiently large (cf. Section 5.3.2.). The result (6.3.18) will be discussed in more detail below.

6.3.3. *Three-dimensional systems*

For $d \neq 1$, it is not possible to evaluate the frequency dependence of the disconnected network model rigorously[1]). Therefore, in the vicinity of the percolation threshold, we use for $\sigma(\omega, \varepsilon)$ a scaling expression. As discussed in Section 5., close to the perco-

---

[1]) For the ac hopping conductivity in the disconnected-network model, for $d \neq 1$, below the percolation threshold, some exact results were obtained by ODAGAKI and LAX (1982), in particular Re σ and Im σ were shown to vanish linearly and quadratically, respectively, with frequency. Such a frequency dependence was also obtained by a Monte Carlo simulation by VICSEK (1981).

lation threshold, the behaviour of a percolation system is analogous to that of a thermodynamic system close to the point of a phase transition. In particular, near the percolation threshold, the dc conductivity of the disconnected network model goes to zero according to the power law (5.2.17). On the other hand for $\omega \neq 0$, i.e. in the presence of capacitors in the resistor network (cf. Fig. 2.3), $\sigma(\omega)$ does not vanish at $\varepsilon = \varepsilon_c$. This suggests that for the disconnected network with capacitors the frequency of the external field plays a role analogous to that of a magnetic field in the theory of phase transitions. Therefore, near the percolation threshold $|\varepsilon - \varepsilon_c|/\varepsilon_c \ll 1$, it seems reasonable to write $\sigma(\omega, \varepsilon)$ in a scaling form conventional for critical phenomena:[1])

$$\sigma(\omega, \varepsilon) \sim \frac{1}{Z_{\text{opt}}} (\mathrm{i}\omega\tau)^2 \varphi\left(\frac{\varepsilon - \varepsilon_c}{(\mathrm{i}\omega\tau)^{r/t}}\right), \tag{6.3.21}$$

where $\varphi(x)$ is an unknown function. Note that a similar scaling expression was used by Efros and Shklovskii (1976) to describe a random-resistor network with two types of resistors Z_D and Z_M, where $Z_M/Z_D \ll 1$.

With regard to φ, we must distinguish between two regimes. The first regime is characterized by $|\varepsilon - \varepsilon_c|/(\omega\tau)^{r/t} \ll 1$ (regime of "strong fields"). In this regime, for any sign of x, $\varphi(x)$ may be expanded in powers of x as

$$\varphi(x) = a_0 + a_1 x + a_2 x^2 + \cdots. \tag{6.3.22}$$

Accordingly, (6.3.21) becomes

$$\sigma(\omega, \varepsilon) \sim \frac{1}{Z_{\text{opt}}} (\mathrm{i}\omega\tau)^r \left\{a_0 + a_1 \frac{\varepsilon - \varepsilon_c}{(\mathrm{i}\omega\tau)^{r/t}} + a_2 \frac{(\varepsilon - \varepsilon_c)^2}{(\mathrm{i}\omega\tau)^{2r/t}} + \cdots\right\}. \tag{6.3.23}$$

The coefficients a_i are real quantities, since in the case of analytic continuation of ω to the imaginary axis, φ is a real function. The expression (6.3.23) indicates that for $\varepsilon = \varepsilon_c$, the critical exponent r describes the frequency dependence of σ.

The second regime is characterized by $|\varepsilon - \varepsilon_c|/(\omega\tau)^{r/t} \gg 1$ (regime of "weak, fields"). In this regime, the conductivity can be represented as a power series of ω, i.e. it holds that

$$\left.\begin{aligned} \varphi(x) &= |x|^t \{b_0 + b_1 |x|^{-t/r} + b_2 |x|^{-2t/r} + \cdots\}, \\ \sigma(\omega, \varepsilon) &\sim \frac{1}{Z_{\text{opt}}} |\varepsilon - \varepsilon_c|^t \left\{b_0 + b \frac{\mathrm{i}\omega\tau}{|\varepsilon - \varepsilon_c|^{t/r}} + b_2 \frac{(\omega\tau)^2}{|\varepsilon - \varepsilon_c|^{2t/r}} + \cdots\right\}. \end{aligned}\right\} \tag{6.3.24}$$

Accordingly, t is the critical exponent of the dc conductivity, in accordance with (5.2.17). Note that for $\varepsilon > \varepsilon_c$, the form of the expansion (6.3.24) differs from that for $\varepsilon < \varepsilon_c$. In particular, for $\varepsilon > \varepsilon_c$ (below the percolation threshold) $b_0 = 0$, since in

---

[1]) A scaling relation of this type was considered by Stephens (1978), who pointed out that the critical exponents t and r may be related to the critical index ν of the correlation length as $t = \Phi + (d - 2)\nu$, $t/r = \Phi + \Phi_1$, where Φ and Φ_1 are the so-called cross-over exponents. Using the estimated values $\Phi \approx 1$, $\Phi_1 \approx 2$, then, for $d = 3$, $t = 1 + \nu \approx 1.9$, $t/r \approx 3$.

this regime $\sigma = 0$ for $\omega = 0$. The remaining coefficients b_i may differ for $\varepsilon > \varepsilon_c$ and $\varepsilon < \varepsilon_c$, but they are real numbers.

Let us determine the relation between Z_{opt} and $\varepsilon - \varepsilon_c$. In what follows, for conveniency we use the number of existing bonds per site p instead of ε. Accordingly, in formulae (6.3.21) to (6.3.24) $\varepsilon - \varepsilon_c$ must be replaced by $p_c - p$. To get p, we need $P(\varkappa)$, the probability of finding $\varkappa$ sites within a sphere of radius R_{opt} circumscribed about a given site. For a random distribution of sites, we find that $P(\varkappa) = a^\varkappa \, e^{-a}/\varkappa!$, where $a = (4\pi/d) \mathcal{N} R_{opt}^d$, and $\mathcal{N}$ is the concentration of sites. Thus, we obtain $p = \sum_\varkappa \varkappa P(\varkappa) = a$. For $|p_c - p|/p_c \ll 1$, we have

$$R_{opt} = \mathcal{N}^{-1/d} \left(\frac{p_c d}{4\pi}\right)^{1/d} \left[1 - \frac{p_c - p}{p_c d}\right]. \tag{6.3.25}$$

Hence, the relation between Z_{opt} and p reads

$$Z_{opt} = Z_c \exp\{-\gamma\alpha\mathcal{N}^{-1/d}(p_c - p)\}. \tag{6.3.26}$$

where

$$Z_c = \frac{kT}{e^2} \tau_c \exp(\varepsilon_3/kT), \quad \tau_c = \frac{1}{\nu_{ph}} \exp(\eta\alpha\mathcal{N}^{1/d}),$$

$$\eta = 2(p_c d/4\pi)^{1/d}, \quad \gamma = \frac{1}{2\pi}\left(\frac{2}{\eta}\right)^{d-1}.$$

Z_c is the critical resistance in the resistor network of Miller and Abrahams, which describes the dc conductivity (cf. Section 5.). Formula (6.3.26) relates the number η (cf. (5.1.11)) to p_c. For $d = 3$, $\eta \approx 1.7$ so that according to (6.3.26) $p_c \approx 2.6$ and $\gamma \approx 0.2$.

Inserting now the expression found for Z_{opt} into (6.3.21), we obtain

$$\sigma(\omega, p) \sim \frac{1}{Z_c} (i\omega\tau_c)^r \exp\{\gamma\alpha\mathcal{N}^{-1/d}(p_c - p)(1 - r)\} \, \varphi\left\{\frac{\lambda(p_c - p)}{(i\omega\tau_c)^{r/t}}\right\}, \tag{6.3.27}$$

where

$$\lambda(p) = p \exp\{\gamma\alpha\mathcal{N}^{-1/d} pr/t\}.$$

From an inspection of (6.3.24) for $p < p_c$, it follows that r must be less than unity, since it is physically clear that σ increases with decreasing $p_c - p$, which requires that $r < 1$. Hence, it follows that the right-hand side of (6.3.27), as function of p, has a maximum, since $\varphi(x)$ increases monotonically with x. Thus, by means of (6.3.3), we may determine $p_{opt}(\omega)$.

Let us first study the static limit of $p_{opt}(\omega)$. Setting $\omega = 0$ in (6.3.27), and using (6.3.24), then we obtain

$$\sigma(0, p) \sim \frac{1}{Z_c} \exp\{\gamma\alpha\mathcal{N}^{-1/d}(p_c - p)\} (p - p_c)^t.$$

Hence, it follows that $p_{\text{opt}} - p_c = t/(\gamma\alpha\mathcal{N}^{-1/d})$ and

$$\sigma(\omega)|_{\omega=0} \sim (\alpha\mathcal{N}^{-1/d})^{-t} \exp\left\{-\eta\alpha\mathcal{N}^{-1/d} - \frac{\varepsilon_3}{kT}\right\}. \tag{6.3.28}$$

The result (6.3.28) is in agreement with (5.3.56) if we replace in (6.3.28) the exponent t by $1 + \nu$. Such a relation between t and ν is suggested by physical arguments and numerical calculations (Shklovskii and Efros, 1975).

Let us now turn to non-zero frequencies. At sufficiently low ω, for $\varphi(x)$ in (6.3.27), we may use the expansion (6.3.24). Thus, for

$$\omega\tau_c(\alpha\mathcal{N}^{-1/d})^{t/r} \ll 1, \tag{6.3.29}$$

we find that

$$p_{\text{opt}} - p_c \simeq \frac{t}{\gamma\alpha\mathcal{N}^{-1/d}} \{1 - a(\omega\tau_c)^2 (\alpha\mathcal{N}^{-1/d})^{2t/r}\}, \tag{6.3.30}$$

where $a > 0$ is a number. Accordingly, with increasing frequency, p_{opt} approaches the percolation threshold (from above), i.e. in the frequency range (6.3.29), the charge transport continues to arise mainly from the infinite cluster. Inserting (6.3.30) into (6.3.27) and making use of the expansion (6.3.24), in the frequency range (6.3.29), the conductivity becomes

$$\sigma(\omega) = \sigma(0) \{1 + i\omega\tau_c(\alpha\mathcal{N}^{-1/d})^{t/r} + (\omega\tau_c)^2 (\alpha\mathcal{N}^{-1/d})^{2t/r}\}. \tag{6.3.31}$$

Note that in one-dimensional systems this frequency range does not exist, because in such systems an infinite cluster is lacking, so that $\sigma \to 0$ for $\omega \to 0$.

With increasing frequency, p_{opt} continues approaching p_c, and at some $\omega_0 \sim 1/[\tau_c(\alpha\mathcal{N}^{-1/d})^{t/r}]$, p_{opt} becomes equal to p_c. This can be seen by using the expansion (6.3.22). At frequencies above ω_0, the infinite cluster begins to decay and the charge transport is realized within clusters of finite size. At not too high frequencies, the size of the clusters is very large. The corresponding transport process is called the multiple hopping. Note that in the frequency range from zero-frequency up to ω_0, the quantity $\text{Re}\,\sigma(\omega)$ does not very strongly depend on ω. It holds that

$$[\text{Re}\,\sigma(\omega) - \sigma(0)]/\sigma(0) \lesssim 1.$$

At sufficiently high frequencies (the concrete criterion will be given below) the parameter $(p_c - p_{\text{opt}})\,\alpha\mathcal{N}^{-1/d}$ becomes larger than unity ($p_c > p_{\text{opt}}$). Then the dependence σ on $(p_c - p)$ is governed by exponentials of the type $\exp\{\alpha\mathcal{N}^{-1/d}(p_c - p)\}$ so that in an algebraical dependence $p_c - p$ may be replaced by $p_c - p_{\text{opt}}$. After performing such a replacement, it can be readily seen that (6.3.3) is solved by

$$(p_c - p_{\text{opt}})^{t/r} \exp\{\gamma\alpha\mathcal{N}^{-1/d}(p_c - p_{\text{opt}})\} = c\omega\tau_c,$$

where c is a number. This solution implies that the argument of the function φ in (6.3.27) is of order unity. This solution is suitable in the frequency range $(\alpha\mathcal{N}^{-1/d})^{t/r} \times \omega\tau_c \gg 1$. Thus, we find that

$$p_c - p_{\text{opt}} \approx \ln\{\omega\tau_c(\alpha\mathcal{N}^{-1/d})^{t/r}\}/(\alpha\mathcal{N}^{-1/d}) \tag{6.3.32}$$

and

$$\sigma(\omega) \sim e^{-\varepsilon_3/kT}\, i\omega \ln \{i\omega\tau_c(\alpha\mathcal{N}^{-1/d})^{t/r}\}^{-t\left(\frac{1}{r}-1\right)}. \tag{6.3.33}$$

Note that, according to (6.3.32) and (6.3.25), in the multiple-hopping regime the characteristic hopping distance R_{opt} only weakly differs from $\mathcal{N}^{-1/d}$. Owing to this fact, in algebraical dependences in the original expression for the current, $\boldsymbol{R}_m - \boldsymbol{R}_{m'}$ may be replaced by $\mathcal{N}^{-1/3}$, and in the disconnected-network model, fluctuations of the distances between nearest neighbours may be neglected, since the major random quantities are the resistors, which exponentially depend on $\boldsymbol{R}_m - \boldsymbol{R}_{m'}$.

According to (6.3.33), in the multiple hopping regime, $\sigma(\omega)$ behaves as $i\omega \times \{\ln(i\omega\tau_c)\}^{-\varrho}$, where $\varrho = t\left(\frac{1}{r} - 1\right) > 0$. This means that, using the approximation $\sigma(\omega) \sim \omega^s$, the exponent s increases with increasing frequency and approaches asymptotically unity.

Let us now investigate the frequency dependence of the characteristic cluster size. According to the scaling hypothesis (STAUFFER, 1976), close to the percolation threshold ($p < p_c$), for the portion of the clusters with s bonds, n_s, it holds that

$$n_s(p_c - p) = s^{-(2+1/\delta)} g[(p_c - p)\, s^{1/\delta\beta}], \tag{6.3.34}$$

where β and δ are the critical exponents describing the dependence of the size of the infinite cluster of the concentration of bonds and the "magnetic field", respectively (cf. Section 5). For $d = 3$, one finds that $\beta \approx 0.4$ and $\delta \approx 5$. The function $g(x)$ exhibits a maximum at $p_c - p = cs^{-1/\beta\delta}$ (c is a number), which follows from a dimensional analysis. Hence, the characteristic cluster size s_{opt} at $p = p_{\text{opt}}$ is of order $(p_c - p_{\text{opt}})^{-\beta\delta}$. Consequently, according to (6.3.32), in the multiple-hopping regime, the characteristic cluster size depends on ω as

$$s_{\text{opt}} \sim \{\ln[\omega\tau_c(\alpha\mathcal{N}^{-1/d})^{t/r}]/\alpha\mathcal{N}^{-1/d}\}^{-\beta\delta}, \tag{6.3.35}$$

i.e. it decreases logarithmically with increasing ω. For the multiple-hopping regime, the restriction $s_{\text{opt}} > 1$ leads to an upper frequency limit. One finds that, up to a pre-exponential factor, the corresponding limiting frequency obeys the relation (6.3.20).

In summary, one may say that the cluster approximation yields a dependence $\sigma(\omega)$, which for $\omega \to 0$, smoothly changes into the dc limit known from Section 5. One of the major results obtained by this approximation is in that the frequency dependence of σ is governed by the parameter $\omega\tau_c = (\omega/\nu_{\text{ph}}) \exp(\eta\alpha\mathcal{N}^{-1/d})$ and not, as in the case of the two-site model, by ω/ν_{ph}. Since in general it holds that $\alpha\mathcal{N}^{-1/d} \gg 1$, both parameters differ by many orders of magnitude, and the frequency dependence of σ sets in at very low frequencies, at ω of order $\nu_{\text{ph}} \exp(-\eta\alpha\mathcal{N}^{-1/d})$ (the probability of the critical hop). The frequency dependence of σ is caused by a reconstruction of the infinite cluster and by transport across clusters of finite size.

Note that the cluster approximation does not yield $\sigma(\omega)$ within a large frequency range. In particular, it does not include the results of the two-site approximation studied in Section 6.2. In the following sections, we will describe methods which allow the conductivity to be obtained within a large frequency range. By means of these methods, the validity of the estimations given in the present section can be verified.

6.4. Effective-medium approximation

6.4.1. *Equation for self-consistent evaluation of the effective hopping probability*

The basic idea of the effective-medium approximation (EMA) consists in replacing a disordered medium by some effective ordered medium, whose parameters are subsequently chosen such that it describes the properties of the actual medium as well as possible. Kirkpatrick (1973) was the first who used the EMA to study the conductivity of a disconnected network. On comparing results of the EMA with those of computer investigations, in general one finds good agreement, except in the vicinity of the percolation threshold. The latter fact can readily be understood by taking into account that the EMA is related to the self-consistent-field method of the theory of phase transitions, which, as well known, yields unsatisfactory results in the fluctuation regime near the point of a phase transition. Accordingly, in applying the EMA to the hopping transport, one may expect this approximation to fail in the limit as $\alpha\mathcal{N}^{-1/3}$ becomes very large, i.e. in the region where the percolation approach is appropriate (cf. Kertész and Metzger, 1984).

In what follows, we restrict ourselves to the case of the R-percolation, where the spread of the energy levels may be neglected. Furthermore, in this section we illustrate the EMA by considering one-dimensional systems. In the following section we will discuss aspects of applying the EMA to three-dimensions and present results for the latter case. As starting point we use the conductivity in terms of the conditional-probability function (2.6.14):

$$\sigma = \frac{e^2 n}{2kT} s^2 \left\langle \sum_{m'} (X_{m'} - X_m)^2 P_{m'm}(s) \right\rangle. \tag{6.4.1}$$

Here $n = \mathcal{N}f = \mathcal{N}e^{-\varepsilon_3/kT}$ is the concentration of electrons, X_m is the Cartesian x-component of the position vector of the site m, $\langle\ldots\rangle$ indicates the configuration average, and the parameter s, which occurs through the Laplace transform with respect to time, must be replaced in the final results by $i\omega$.

The conditional-probability function P satisfies equation (2.6.5), which, after separating the incoming and outgoing contributions, reads

$$sP_{m'm} = \delta_{m'm} + \sum_g W_{m+g,m}(P_{m',m+g} - P_{m'm}). \tag{6.4.2}$$

Here, for simplicity, we have neglected the superscript (0) at the probabilities $W_{m+g,m}$, which are defined in the absence of an external electrical field and which are given (for R-percolation) by (6.3.15), whereby $W_{m'm} = W_{mm'}$. The index $m + g$ indicates a nearest-neighbour of site m.

We solve the set of equations (6.4.2), by using the EMA. We consecutively number all sites in the one-dimensional chain under consideration from left to right and take into account only nearest-neighbour hops. Furthermore, we write $W_{m,m+g} = W + u_{m,m+g}$, where W is some effective hopping probability between nearest neighbours, which will be determined below. Furthermore, we represent $P_{m'm}$ as an infinite power series of the t-matrix (cf., e.g., KIRKPATRICK, 1973):

$$P_{m'm} = P(m' - m) + \sum_{m_1 g} \{P(m' - m_1)\, P(m_1 + g - m) - P(m' - m_1)\, P(m_1 - m)\}\, t_{m_1+g,m_1} + \cdots, \tag{6.4.3}$$

where

$$t_{m+g,m} = u_{m+g,m}\{1 + 2[P(0) - P(g)]\, u_{m+g,m}\}^{-1}, \tag{6.4.4}$$

$$P(m) = (1/2\pi) \int_{-\pi}^{\pi} \mathrm{d}k\, \mathrm{e}^{\mathrm{i}km} \left\{s + W \sum_g (1 - \cos kg)\right\}^{-1}. \tag{6.4.5}$$

Here $P(m)$ is the conditional-probability function of the ordered effective medium.

The right-hand side of (6.4.3) represents an infinite power series of the t-matrix. On examining problems of the disconnected-network model by means of the EMA, the usual condition for determining the probability W reads $\langle t \rangle = 0$. This procedure allows, e.g., the configuration average of $P_{m'm}$ to be calculated. However, according to (6.4.1), here we are interested in another quantity, the configuration average of $(X_{m'} - X_m)^2 P_{m'm}$. Since t and $X_m - X_{m'}$, involved in (6.4.1), are correlated, these quantities may not be averaged independently. This fact is especially important in the region of not too low frequencies. Indeed, at small s, in the sum (6.4.1), terms with large $X_m - X_{m'}$ play the leading role. Consequently, the correlation between t (depending on the nearest-neighbour distance only) and $X_m - X_{m'}$ is weak. With increasing s the "effective" difference $X_m - X_{m'}$ decreases (since at sufficiently high frequencies the main contribution to the sum (6.4.1) arises from nearest-neighbour hops, i.e. from two-site hops) and the correlation between t and $X_m - X_{m'}$ becomes important. To perform the configuration average on account of such correlation, we introduce the quantity

$$P(q) = \sum_{m'} \langle P_{m'm} \exp [\mathrm{i}q(X_m - X_{m'})]\rangle. \tag{6.4.6}$$

Then the expression (6.4.1) becomes

$$\sigma = -\sigma_0 \frac{\mathcal{N}^2}{2\nu_{\mathrm{ph}}} s^2 \left.\frac{\partial^2 P(q)}{\partial q^2}\right|_{q=0}, \qquad \sigma_0 = \frac{e^2 n}{kT} \mathcal{N}^{-2} \nu_{\mathrm{ph}}. \tag{6.4.7}$$

In one-dimension, $P(q)$ can be evaluated rigorously on the basis of (6.4.6) and (6.4.4), with the aid of the Poisson distribution

$$\Pi(x) = a^{-1} \exp(-x/a) \tag{6.4.8}$$

for the nearest-neighbour distance x. To do this, first we evaluate $\langle \exp [iq(X_m - X_{m'})] \rangle$ and $\langle \exp [iq(X_m - X_{m'})]\, t_{m_1,m_1+g} \rangle$. Consider for definiteness the case $m' \geqq m$ (for $m' < m$, it is sufficient to replace q by $-q$). Then, using (6.4.8), we obtain

$$\langle \exp [iq(X_m - X_{m'})] \rangle = \left\{ \int_0^\infty \mathrm{d}x\, \mathrm{e}^{-(1+\mathrm{i}qa)x} \right\}^{m'-m} = (1 + \mathrm{i}qa)^{m-m'}.$$

Analogously, we evaluate $\langle \exp [iq(X_m - X_{m'})]\, t_{m_1,m_1+g} \rangle$. If m_1 and $m_1 + g$ are inside the interval between m and m', then we obtain

$$\langle \exp [iq(X_m - X_{m'})]\, t_{m_1,m_1+g} \rangle = (1 + \mathrm{i}qa)^{m-m'+1}\, t(q),$$

where

$$t(q) = \int_0^\infty \mathrm{d}x \exp [-(1 + qa\,\mathrm{i})\, x]\, \frac{\nu_{\mathrm{ph}}\, \mathrm{e}^{-2\alpha a x} - W}{\nu_{\mathrm{ph}}\, \mathrm{e}^{-2\alpha a x} + \frac{1}{4}\, s(y + 1)} \tag{6.4.9}$$

with $y = (1 + 4W/s)^{1/2}$.

If m_1 and $m_1 + g$ are outside the interval between m and m', then this average becomes a product of two averages:

$$\langle \exp [iq(X_m - X_{m'})]\, t_{m_1,m_1+g} \rangle = (1 + \mathrm{i}qa)^{m-m'}\, t(0).$$

This formula holds, of course, for $m = m'$, too.

Now, by means of the relations obtained, in (6.4.6) the summation over m' can readily be performed. We find that

$$\begin{aligned} P(q) = (sy)^{-1} \operatorname{Re} \Bigg\{ 1 &+ \frac{2b}{1 - b + \mathrm{i}qa} - \frac{1}{2}\, t(0)\, \frac{y + 1}{y} \left(1 + \frac{2b}{1 - b + \mathrm{i}qa}\right) \\ &+ 2t(q)\, \frac{1}{y + 1} \left(\frac{1 + \mathrm{i}qa}{1 - b + \mathrm{i}qa}\right)^2 \Bigg\}, \end{aligned} \tag{6.4.10}$$

where $b = (y - 1)/(y + 1)$. Note that neglection of the correlation between t and $X_m - X_{m'}$ corresponds to the replacement $t(q) \to t(0)\, (1 + \mathrm{i}qa)^{-1}$.

Inserting the expression for $P(q)$ (6.4.10) into (6.4.7), we obtain

$$\sigma = \sigma_0\, \frac{W}{\nu_{\mathrm{ph}}} \left\{ 1 + \frac{1}{y} + \left[\lambda_0 \left(1 - \frac{y + 1}{2y^2}\right) + \frac{2\lambda_1}{y} + \frac{\lambda_2}{y(y - 1)} \right] \right\}, \tag{6.4.11}$$

where

$$\lambda_n = (2\alpha a)^{-n-1} \int_0^\infty \mathrm{d}x x^n\, \mathrm{e}^{-x/2\alpha a}\, \frac{\nu_{\mathrm{ph}}\, \mathrm{e}^{-x} - W}{\nu_{\mathrm{ph}}\, \mathrm{e}^{-x} + \frac{1}{4}\, s(y + 1)}. \tag{6.4.12}$$

Now we require that the contribution to σ from the term linear in the t-matrix must vanish. This requirement leads to the following self-consistency equation for determining W:

$$\lambda_2 + 2(y-1)\,\lambda_1 + (y-1)^2\left(1 + \frac{1}{2y}\right)\lambda_0 = 0. \tag{6.4.13}$$

Then, the expression for σ becomes

$$\sigma = \sigma_0\,\frac{W}{\nu_{\mathrm{ph}}}\left(1 + \frac{1}{y}\right). \tag{6.4.14}$$

Before discussing the frequency dependence of σ for a model with a random site distribution, let us study $\sigma(\omega)$ for the disconnected-network model, where the probability of nearest-neighbour hops can assume two values — either 0 with probability ε, or ν_{ph} with probability $1-\varepsilon$. The investigation of this model allows the above formulae to be compared with the exact results (6.3.12). It is easy to show that, owing to the absence of fluctuations of the nearest-neighbour distances, for this model, the self-consistency equation (6.4.13) reads $\lambda_0 = 0$. On account of the given form for the distribution function of the probabilities, the latter equation becomes

$$\left\langle (u - \tilde{W}) \Big/ \left(u + \frac{1}{2}\,\tilde{s}(1+y)\right)\right\rangle = 0, \tag{6.4.15}$$

where $\tilde{W} = W/\nu_{\mathrm{ph}}$, $\tilde{s} = s/\nu_{\mathrm{ph}}$, $\sigma = \sigma_0\tilde{W}$ and the average is over the random quantity u, which assumes either the value 0 or 1. After simple manipulations, from (6.4.15), we find that

$$\operatorname{Re}\sigma \equiv \sigma' = \sigma_0\left\{1 - \sqrt{2}\left[\sqrt{\left(1 - \frac{\tilde{\omega}^2}{4\varepsilon^2}\right)^2 + \frac{\tilde{\omega}^2}{\varepsilon^4}} + 1 - \frac{\tilde{\omega}^2}{4\varepsilon^2}\right]^{-1/2}\right\},$$

$$\operatorname{Im}\sigma \equiv \sigma'' = \sigma_0\,\frac{2\varepsilon^2}{\tilde{\omega}}\,\frac{\sigma'}{\sigma_0 - \sigma'}. \tag{6.4.16}$$

For $\tilde{\omega} \ll 1$, $\varepsilon \ll 1$, $\varepsilon^2/\tilde{\omega} \ll 1$, (6.4.16) reduces to

$$\sigma' \approx \sigma_0\left(1 - \sqrt{2}\,\varepsilon/\sqrt{\tilde{\omega}}\right), \qquad \sigma'' \approx \sigma_0\sqrt{2}\,\varepsilon/\sqrt{\tilde{\omega}}.$$

This result completely agrees with the prediction of (6.3.12) for the same limiting case. In the other limiting case $\tilde{\omega} \ll 1$, $\varepsilon \ll 1$, but $\varepsilon^2/\tilde{\omega} \gg 1$, from (6.4.16) it follows that $\sigma' \approx \sigma_0\tilde{\omega}^2/8\varepsilon^4$, $\sigma'' \approx \sigma_0\tilde{\omega}/4\varepsilon^2$, while from (6.3.12) $\sigma' \approx \sigma_0\tilde{\omega}^2/\varepsilon^4$, $\sigma'' \approx \sigma_0\tilde{\omega}/2\varepsilon^2$, i.e. both results differ only by a numerical factor.

Let us now study $\sigma(\omega)$ for randomly distributed sites. First, consider the self-consistency equation (6.4.13) for low frequencies, where $2W/s \gg 1$. In this case, we have $y \approx (4W/s)^{1/2} \gg 1$, and (6.4.13) takes the form $\lambda_0 = 0$, or explicitely

$$\int\limits_0^\infty \mathrm{d}x\;\mathrm{e}^{-x/2\alpha a}\,\frac{\mathrm{e}^{-x} - \tilde{W}}{\mathrm{e}^{-x} + \frac{1}{2}\sqrt{\tilde{W}\tilde{s}}} = 0. \tag{6.4.17}$$

On account of the fact that in this frequency range $\tilde{W}\tilde{s} \ll 1$ holds, we obtain

$$\sigma \approx 4\sigma_0 \, \mathrm{i}\tilde{\omega}^{\frac{2\alpha a-1}{2\alpha+1}} \left(\frac{\sin(\pi/2\alpha a)}{2\pi/\alpha a}\right)^{\frac{4\alpha a}{2\alpha a+1}} \exp\left(-\frac{\mathrm{i}\pi}{2\alpha a+1}\right). \tag{6.4.18}$$

Let us now compare this result with formula (6.3.18), which we have obtained by means of the cluster approximation. According to both approaches, it holds that $\sigma \sim \omega^s$, where the cluster method gives $s = 1 - 1/(\alpha a + 1)$, and the EMA $s = 1 - 2/(2\alpha a + 1)$. Since the cluster method may be used only for $\alpha a \gg 1$, in this limiting case, for both approaches, the exponent s is equal to $1 - 1/\alpha a$. Note that during the past few years a number of papers have been published which are devoted to the low-frequency limit of disordered one-dimensional systems. For instance, this problem has been attacked by Bernasconi et al. (1980) by using the method of Dyson (1953) and by Prigodin (1984) by means of the method of Berezinskii (1973). Up to a numerical factor, the results of both these papers are in complete agreement with (6.4.18).

Let us now discuss the range of application of (6.4.18). The restriction $2W/s \gg 1$ gives an upper limiting frequency. Using W from (6.4.14), this restriction becomes $\tilde{\omega}^{1/2\alpha a} \ll 1$, which can readily be written as (6.3.20). Accordingly, the expression (6.4.18) describes transport of the multiple-hopping type. It is interesting to note that in one dimension, there is no "hydrodynamic" regime, in which $\sigma(\omega)$ may be represented by an expansion of the type (6.3.31), i.e. by an analytical expansion with respect to powers of ω.

Let us now turn to high frequencies $2W/s \ll 1$, where $y \approx 1$, the self-consistency equation (6.4.13) reads $\lambda_2 = 0$, i.e.

$$\int_0^\infty \mathrm{d}x \, x^2 \, \mathrm{e}^{x/2\alpha a} \frac{\mathrm{e}^{-x} - \tilde{W}}{\mathrm{e}^{-x} + \frac{1}{2}\tilde{s}} = 0, \tag{6.4.19}$$

and $\sigma = 2\sigma_0\tilde{W}$. For $2W/s \ll 1$ and $2\alpha a \gg 1$, the solution of the self-consistency equation reads

$$\mathrm{Re}\,\tilde{W} = \frac{1}{2}(2\alpha a)^{-3}\,\mathrm{e}^{\eta} \int_0^\infty \mathrm{d}x\, x^2/\cosh(x+\eta),$$

$$\mathrm{Im}\,\tilde{W} = \frac{1}{16}(2\alpha a)^{-3}\,\mathrm{e}^{\eta} \int_0^\infty \mathrm{d}x\, x^2\{1 + \mathrm{e}^{x+2\eta}\}^{-1}, \tag{6.4.20}$$

where $\eta = \ln(\tilde{\omega}/2)$. It is easy to show that σ' from (6.4.20) agrees with the result of the two-site model (6.2.29), provided that all levels are at $\varepsilon = \varepsilon_3$ and in the integrand in (6.2.29) x^4 is replaced by x^2 (due to consideration of a one-dimensional system).

After evaluating the integrals in (6.4.20), for $\tilde{\omega} \ll 1$, we obtain

$$\left.\begin{aligned} \sigma' &\approx \frac{\pi}{4}\,\sigma_0(2\alpha a)^{-3}\,\tilde{\omega}\,\ln^2(2/\tilde{\omega}), \\ \sigma'' &\approx \frac{1}{6}\,\sigma_0(2\alpha a)^{-3}\,\tilde{\omega}\,\ln^3(2/\tilde{\omega}), \\ &\text{and for } \tilde{\omega} \gtrsim 1, \\ \sigma' &\approx 2\sigma_0(2\alpha a)^{-3}\left(1 - \frac{4}{27}\,\tilde{\omega}^{-2}\right), \\ \sigma'' &\approx \frac{1}{2}\,\sigma_0(2\alpha a)^{-3}\,\tilde{\omega}^{-1}. \end{aligned}\right\} \tag{6.4.21}$$

Accordingly, for $\omega \ll \nu_{ph}$, we have $\sigma' \sim \omega \ln^2 2\nu_{ph}/\omega$, and for $\omega \gtrsim \nu_{ph}$, σ' exhibits a plateau.

The lower limiting frequency for the range of application of (6.4.20) results from the restriction $2W/s \ll 1$, which leads to the inequality (6.2.33). In this context note that, for instance, for $\alpha a = 5$, the upper limiting frequency of the multiple-hopping regime becomes about 10^2 to $10^3\,s^{-1}$, while the lower limiting frequency of the two-site hopping regime is given by about 10^7 to $10^8\,s^{-1}$. This indicates that between both regimes there is a large frequency range characterized by hopping across medium-sized clusters. It should be noticed that with increasing parameter αa, both these limiting frequencies decrease, so that at very large αa, the two-site model dominates, beginning at very low frequencies.

Note that the above analytical results for the ac hopping conductivity in one-dimensional systems well agree with computer data of Moore (1974). For a detailed comparison of analytical and numerical results, see Bryksin (1980).

6.4.2. *Results for the* ac *conductivity in three-dimensional systems*

In applying the EMA to hopping conductivity in three-dimensional systems, the basic conception of the calculation is the same as in the one-dimensional case. The only additional difficulty arises from the uncertainty in the definition of nearest-neighbour sites. Such a definition is needed for mapping the irregular system onto a regular lattice, which allows the EMA to be applied. The concept of bonds between nearest neighbours can be formulated only rather arbitrarily. However, it turns out that the resulting expressions for the conductivity depend only weakly on the choice of the underlying lattice model.

Let us define a nearest-neighbour bond as follows. The full solid angle around some site is subdivided into z segments of equal size, and we put $z = 2d$ (cubic lattice). After that in each such segment we determine the nearest neighbour of the site under consideration. The z sites found in this way are then used as the nearest-neighbour sites quoted. Of course, such a procedure for constructing a lattice is not rigorous, since for random positions of sites, the bonds constructed do not merge into a network. Moreover, unlike the percolation approach (cf. Section 5.), this procedure ignores the fact that the number of nearest neighbours fluctuates from site to site. However, it turns out that the choice of the number of nearest neighbours affects only some numerical coefficients in the final results. The most important coefficients (occurring in exponents) are found to be in good agreement with computer data.

According to our choice of neighbouring sites, the probability of finding a nearest neighbour at a distance between r and $r + \mathrm{d}r$ is given by the Poisson distribution

$$\mathrm{d}p(r) = \frac{\mathcal{N}\gamma_d{}^d}{z}\, r^{d-1}\, \mathrm{d}r \exp\left\{-\frac{\gamma_d \mathcal{N} r^d}{z}\right\}, \tag{6.4.22}$$

where $\gamma_2 = \pi$, $\gamma_3 = 4\pi/3$.

Recall that in one dimension, the procedure of evaluating the conductivity by the EMA is to insert the expression for P (6.4.3) into the formula for σ (6.4.1) and to perform the configuration average. The self-consistency condition is in that the linear-in-t contribution to σ must vanish, which is equivalent to the approximation $\sigma \sim \sum_m \langle X_m^2 \rangle\, P(m)$. For three-dimensional systems, the EMA may be formulated analogously. Since for the latter systems, an evaluation of $\langle X_m^2 \rangle$ is a very intricate problem, we use in the following the approximate expression $\mathcal{N}^{-2/d} m^2$, which holds to within a numerical factor. Then we get

$$\sigma \approx \sigma_0 W/\nu_{\mathrm{ph}}\,. \tag{6.4.23}$$

Note that this expression for σ differs from the rigorous expression (6.4.14) (obtained for $d = 1$) only by a factor which weakly depends on frequency (it changes from 1 to 2 by rising ω from 0 to ∞).

Let us now turn to the self-consistency equation for determining W. First consider the low frequency regime. As mentioned in the preceding section, for $s/W \ll 1$, the correlation between X_m and P is unimportant. In this case the self-consistency equation reads $\langle P_{m'm} \rangle = P(m' - m)$ or $\langle t \rangle = 0$, i.e.

$$\left\langle \frac{\mathcal{W} - W}{\mathcal{W} + \chi W} \right\rangle = 0, \tag{6.4.24}$$

where

$$\chi = \frac{z}{2}\left\{1 - \frac{z}{(2\pi)^d} \int \frac{dk}{s + W \sum\limits_g (1 - \cos kg)}\right\}^{-1} - 1$$

$$= \frac{z}{2}\left\{1 - \frac{z}{W} \int\limits_0^\infty \mathrm{d}t I_0{}^d(2t) \exp\left[-\left(\frac{s}{W} + z\right) t\right]\right\}^{-1} - 1. \tag{6.4.25}$$

Here $\mathcal{W} = \nu_{\mathrm{ph}} \exp(-2\alpha r)$ is the random quantity, I_0 is the Bessel function, and in (6.4.24) it must be averaged over r by means of the Poisson distribution (6.4.22).

For high frequencies $s/W \gg 1$, the major contribution to the conductivity comes from two-site hops. Accordingly, the correlation between P and X_m must be accounted for. In this case, the self-consistency equation becomes

$$\left\langle r^2 \frac{\mathcal{W} - W}{\mathcal{W} + \chi W} \right\rangle = 0. \tag{6.4.26}$$

It turns out that the expressions obtained for W at low frequencies (6.4.24) and high frequencies (6.4.26) well agree in the regime of moderate frequencies, which suggests that the results of the EMA allow σ to be described in the whole frequency range.

Consider now the low frequency regime $s/W \ll 1$. On account of the inequality $(2\alpha)^d \gg \mathcal{N} \, |\ln (\chi W/\nu_{\text{ph}})|^2$, which holds up to the regime of the two-site approximation, the self-consistency equation (6.4.24) may be written as

$$1 + \frac{1}{\chi} = \exp\left\{\frac{\mathcal{N}\gamma_d}{z} \left[\ln\left(\frac{\nu_{\text{ph}}}{\chi W}\right) \Big/ 2\alpha\right]^d\right\}. \tag{6.4.27}$$

In what follows, for χ we use the expansion

$$\chi \approx d - 1 + \mathrm{i} \frac{\omega}{W} d\xi_d. \tag{6.4.28}$$

For $d = 3$, the quantity ξ_d is a number (cf. (6.4.25))

$$\xi_3 = \frac{1}{2} \int\limits_0^\infty \mathrm{d}t \, \mathrm{e}^{-3t} I_0{}^3(t) \approx 0.253.$$

For $d = 2$, χ exhibits a logarithmic singularity (being proportional to $\ln (\omega/W)$. Inserting (6.4.28) into (6.4.27), we obtain

$$W = W_{\mathrm{c}} \exp\left(\mathrm{i}\tilde{\Omega} \frac{W_{\mathrm{c}}}{W}\right), \tag{6.4.29}$$

where we have introduced the dimensionless frequency

$$\tilde{\Omega} = \frac{\omega}{W_{\mathrm{c}}} \alpha \mathcal{N}^{-1/d} \eta_d \xi_d \left[d(d-1) \ln \frac{d}{d-1}\right]^{-1}, \tag{6.4.30}$$

and W_{c} has the meaning of the probability of a critical hop

$$W_{\mathrm{c}} \equiv W|_{\omega=0} = \frac{\nu_{\text{ph}}}{d-1} \exp\left(-\eta_d \alpha \mathcal{N}^{-1/d}\right). \tag{6.4.31}$$

Here the number

$$\eta_d = 2 \left\{\frac{2d}{\gamma_d} \ln \frac{d}{d-1}\right\}^{1/d} \tag{6.4.32}$$

corresponds to the number η in the exponent of the dc conductivity (cf. (5.1.11)), which may be evaluated by using methods of the percolation theory (cf. Section 5.)

According to (6.4.32), for $d = 3$, $\eta_3 = 1.669$, and for $d = 2$, $\eta_2 = 1.879$. By computer calculations, one has obtained for η values between 1.62 and 1.78 for $d = 3$, and a value 1.86 for $d = 2$. Accordingly, the η's passably well coincide.

From (6.4.29), closed analytical expressions can be obtained for W in the limiting cases $\tilde{\Omega} \ll 1$, and $\tilde{\Omega} \gg 1$. For $\tilde{\Omega} \ll 1$, we find that (but cf. (11.4.1))

$$\sigma \approx \sigma(0)\,(1 + \mathrm{i}\tilde{\Omega} + 1/2\tilde{\Omega}^2), \tag{6.4.33}$$

where $\sigma(0) = \sigma_0 W_c/\nu_{\rm ph}$ is the dc conductivity.

According to (6.4.33), for $d = 3$ and for sufficiently low frequencies, $\sigma(\omega)$ may be represented by a regular expansion, which is in agreement with the result of the cluster approximation (6.3.31). They differ only in respect of the power of $\alpha\mathcal{N}^{-1/d}$, which according to (6.3.31) is equal to t/r, but in the EMA (6.4.30)) equal to unity. For the range of application of (6.4.33), the upper limiting frequency is of order $W_c/\alpha\mathcal{N}^{-1/d}$. The change of σ' is small in this frequency range, $\sigma' - \sigma(0) \approx \sigma(0)$. These facts are also in agreement with the cluster approximation.

In the other limiting case $\tilde{\Omega} \gg 1$, from (6.4.29) and (6.4.23), we get

$$\sigma = \sigma_0\tilde{\Omega}\left\{\frac{\pi}{2}\left[\ln\left(\frac{\pi}{2}\,\tilde{\Omega}\right)\right]^{-2} + \mathrm{i}\left[\ln\left(\frac{\pi}{2}\,\tilde{\Omega}\right)\right]^{-1}\right\}, \tag{6.4.34}$$

which is also in qualitative agreement with the result of the cluster approximation for multiple hopping (6.3.33). The results (6.4.34) and (6.3.33) differ only in respect of the power of the logarithm. In consideration of the crude character of the estimations, we may state rather good agreement between the results of both approximations.

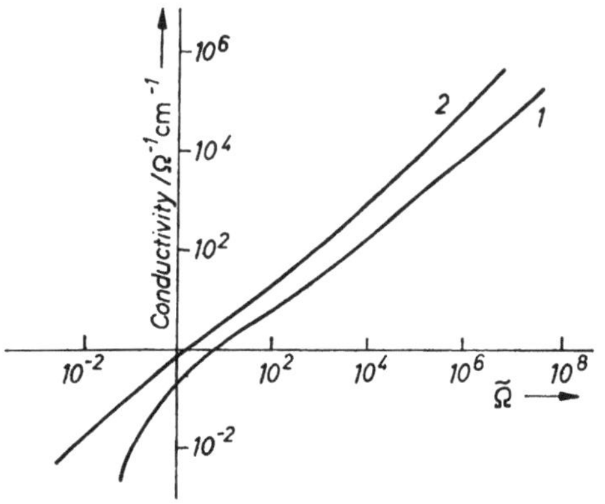

Fig. 6.1. Frequency dependence of Re $\sigma - \sigma(0)/\sigma(0)$ (curve 1) and Im $\sigma/\sigma(0)$ (curve 2) evaluated by means of formula (6.4.29). The dimensionless frequency $\tilde{\Omega}$ is defined in (6.4.30). $\tilde{\Omega} > 1$ corresponds to the multiple-hopping regime, while $\tilde{\Omega} < 1$ to the hydrodynamic one

Results of a numerical calculation of $\sigma(\omega)$ performed on the basis of (6.4.23) and (6.4.29) are shown in Fig. 6.1. As can be seen from this figure, if we use the approximation $\sigma' - \sigma(0) \sim \omega^{s'}$ and $\sigma'' \sim \omega^{s''}$, then for $\tilde{\Omega} < 1$, $s' = 2$ and $s'' = 1$. At $\tilde{\Omega} \approx 1$, s' as well as s'' begin to decrease, and for $\tilde{\Omega} > 1$, they again increase, as $s' = 1 - 2/\ln\tilde{\Omega}$ and $s'' = 1 - 1/\ln\tilde{\Omega}$.

Let us now investigate the range of application of equation (6.4.29) and with it also the range of application of the dependence $\sigma(\omega)$ represented in Fig. 6.1. The strongest restriction is $\omega/|W| < 1$, which becomes effective at $\tilde{\Omega} > 1$. Using for W the expression holding for $\tilde{\Omega} > 1$, $|W| \sim W_c\tilde{\Omega}/\ln\tilde{\Omega}$, then we find the restriction

$$\ln(\alpha\mathcal{N}^{-1/d}\omega/W_c) \ll \alpha\mathcal{N}^{-1/d}, \tag{6.4.35}$$

which to within non-exponential factors agrees with (6.3.20).

Let us now turn to the high-frequency limit, which is described by the self-consistency equation (6.4.26). On account of the restriction $s/W \gg 1$, the latter equation becomes

$$\int_0^\infty dr r^{d+1} \exp\left\{-\frac{r^d\mathcal{N}\gamma_d}{z(2\alpha)^d}\right\} \frac{\nu_{ph}\,e^{-r} - W}{\nu_{ph}\,e^{-r} + s/2} = 0, \tag{6.4.36}$$

which is equivalent to (6.4.1). Analogous to the procedure of changing from (6.4.19) to (6.4.20), we find for σ the following expressions, which are characteristic of the two-site approximation (cf. (6.2.29)):

$$\begin{aligned} \sigma' &= \sigma_p \frac{1}{48} e^\eta \int_0^\infty dx x^4[\cosh(x+\eta)]^{-1}, \\ \sigma'' &= \sigma_p \frac{1}{768} e^\eta \int_0^\infty dx\, x^4[1 + e^{x+2\eta}]^{-1}, \end{aligned} \tag{6.4.37}$$

where $\eta = \ln(\omega/2\nu_{ph})$, and σ_p is a quantity having the dimension of a conductivity and the meaning of the conductivity on the high-frequency plateau

$$\sigma_p = \sigma_0\, 16\pi\mathcal{N}^{5/3}(2\alpha)^{-5} \tag{6.4.38}$$

(the above formulae hold for $d = 3$).

For very high frequencies, $\omega \gtrsim \nu_{ph}$, the dependence $\sigma(\omega)$ passes into a plateau

$$\sigma = \sigma_p\left\{1 + i\frac{1}{16}\frac{\nu_{ph}}{\omega} - \frac{4}{243}\frac{\nu_{ph}^2}{\omega^2}\right\}. \tag{6.4.39}$$

On the other hand, for $\omega \ll \nu_{ph}$, we obtain

$$\omega = \sigma_p \frac{\omega}{2\nu_{ph}}\left\{\frac{\pi}{48}\ln^4\frac{2\nu_{ph}}{\omega} + i\frac{1}{80}\ln^5\frac{2\nu_{ph}}{\omega}\right\}, \tag{6.4.40}$$

i.e., as usual in the low frequency range of the two-site approximation, we find that $\sigma' \sim \omega \ln^{d+1}(\nu_{ph}/\omega)$. Results of numerical studies of $\sigma(\omega)$ on the basis of (6.4.37) are represented in Fig. 6.2.

For the range of application of (6.4.37), the lower limiting frequency results from the restriction $\omega/W \gg 1$. This restriction leads to an inequality of the type (6.2.33).

Hence, in the low-frequency regime, the results of the EMA thoroughly agree with those obtained by the cluster approximation, while in the high frequency regime, they agree with those of the two-site approximation. The most serious deficiency of the EMA is that (for $d \neq 1$) a regular lattice is required, which cannot be found without arbitrariness. Therefore, before discussing $\sigma(\omega)$ in detail, we yet present another method of calculating $\sigma(\omega)$, which is, no doubt, rather formal, but which does not require additional uncontrollable assumptions. Furthermore, for $d \neq 1$, this method allows the conductivity to be determined in the whole frequency range, whereas in applying the EMA, in this case explicit analytical expressions can be obtained only in certain limiting cases.

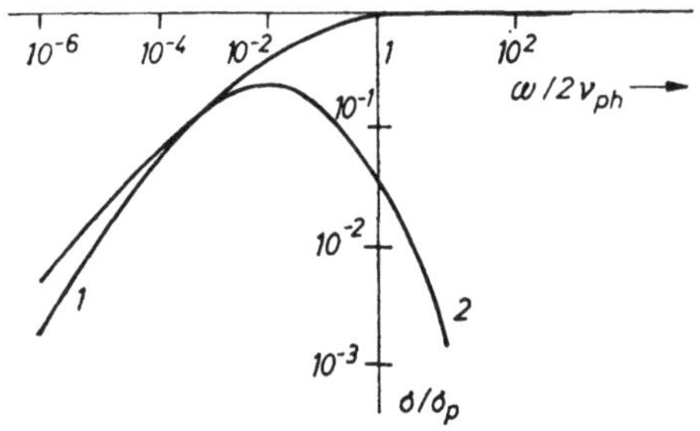

Fig. 6.2. Frequency dependence of the real part of the conductivity (curve 1) and the imaginary part (curve 2) in the regime of the two-site approximation, evaluated by means of formula (6.4.37). σ_p is defined in (6.4.38)

6.5. Diagram technique and some problems of diffusion in disordered systems

The major difficulty in applying the EMA to the conductivity is to construct an appropriate lattice model. This difficulty is intimately related to the fact that the underlying formalism is based on the use of site numbers m, while the random quantities are the position vectors. The necessity of constructing a lattice model arises, strictly speaking, from the impossibility of numbering the sites in a disordered system in such a manner that nearest neighbours in the space of the numbers m are also nearest neighbours in the configuration space.

On the other hand, it is intuitively clear, that in configuration averaged systems, the diffusion (and consequently also the conductivity, cf. (2.6.15)) should be adequately described in the space of the continuous coordinates $\boldsymbol{r}$. Accordingly, restricting ourselves to the case of R-hopping (where the hopping probability $W_{m'm}$ depends only on the spatial distance between the sites), we use as starting point for a consequent theory of diffusion in disordered systems the conditional-probability function $P(\boldsymbol{r}', \boldsymbol{r})$ defined as

$$P(\boldsymbol{r}', \boldsymbol{r}) = \sum_{m'm} P_{m'm} \delta(\boldsymbol{r}' - \boldsymbol{R}_{m'})\, \delta(\boldsymbol{r} - \boldsymbol{R}_m), \tag{6.5.1}$$

where $P_{m'm}$ is the conditional-probability function introduced above.

Multiplying the equation for $P_{m'm}$ (6.4.2) by $\delta(\boldsymbol{r}' - \boldsymbol{R}_{m'})\, \delta(\boldsymbol{r} - \boldsymbol{R}_m)$, performing the summations over m and m', and taking into account that $W_{m'm} = W(\boldsymbol{R}_{m'} - \boldsymbol{R}_m)$, then we obtain the following equation for $P(\boldsymbol{r}', \boldsymbol{r})$:

$$sP(\boldsymbol{r}', \boldsymbol{r}) = \eta(\boldsymbol{r})\,\delta(\boldsymbol{r}' - \boldsymbol{r}) + \int \mathrm{d}\boldsymbol{r}_1 W(\boldsymbol{r}_1 - \boldsymbol{r})\,\{P(\boldsymbol{r}', \boldsymbol{r}_1)\,\eta(\boldsymbol{r}) - P(\boldsymbol{r}', \boldsymbol{r})\,\eta(\boldsymbol{r}_1)\}, \tag{6.5.2}$$

where we have accounted for that $W(\boldsymbol{r}) = W(-\boldsymbol{r})$. The function

$$\eta(\boldsymbol{r}) = \sum_m \delta(\boldsymbol{r} - \boldsymbol{R}_m) \tag{6.5.3}$$

represents the electron density fluctuating in space at random.

The conductivity (6.4.1) may be expressed by $P(\boldsymbol{r}', \boldsymbol{r})$ as

$$\sigma = \frac{e^2 n}{2kT}\, s^2 \int \mathrm{d}\boldsymbol{r} x^2 P(\boldsymbol{r}). \tag{6.5.4}$$

Here $P(\boldsymbol{r}' - \boldsymbol{r})$ denotes the configuration average (over all possible realizations of the random quantity $\eta(\boldsymbol{r})$) of $P(\boldsymbol{r}', \boldsymbol{r})$:

$$\mathcal{N} P(\boldsymbol{r}' - \boldsymbol{r}) = \langle P(\boldsymbol{r}', \boldsymbol{r})\rangle, \tag{6.5.5}$$

where $\langle\ldots\rangle$ indicates the configuration average.

It is convenient to use the momentum representation

$$P(\boldsymbol{q}', \boldsymbol{q}) = \int \mathrm{d}\boldsymbol{r}'\,\mathrm{d}\boldsymbol{r} P(\boldsymbol{r}', \boldsymbol{r}) \exp(\mathrm{i}\boldsymbol{q}'\boldsymbol{r}' - \mathrm{i}\boldsymbol{q}\boldsymbol{r}), \tag{6.5.6}$$

$$\eta(\boldsymbol{q}) = \int \mathrm{d}\boldsymbol{r}\eta(\boldsymbol{r}) \exp(\mathrm{i}\boldsymbol{q}\boldsymbol{r}) = \sum_m \exp(\mathrm{i}\boldsymbol{q}\boldsymbol{R}_m). \tag{6.5.7}$$

In this representation, equation (6.5.2) reads

$$sP(\boldsymbol{q}', \boldsymbol{q}) = \eta(\boldsymbol{q}' - \boldsymbol{q}) + \int \frac{\mathrm{d}\boldsymbol{q}_1}{(2\pi)^3} P(\boldsymbol{q}', \boldsymbol{q}_1)\,\mathscr{W}(\boldsymbol{q}_1, \boldsymbol{q})\,\eta(\boldsymbol{q}_1 - \boldsymbol{q}), \tag{6.5.8}$$

where

$$\mathscr{W}(\boldsymbol{q}', \boldsymbol{q}) = W(\boldsymbol{q}') - W(\boldsymbol{q}' - \boldsymbol{q}). \tag{6.5.9}$$

Here $W(\boldsymbol{q})$ is the Fourier transform of $W(\boldsymbol{r})$.

In what follows, we iterate equation (6.5.8) with respect to the free term (i.e. with respect to $1/s$) and perform the configuration average of the series obtained. Assuming a random distribution of the positions of the sites (no short-range order), then the configuration average is defined as

$$\langle A(\{\boldsymbol{R}_m\})\rangle = \Omega^{-N} \int \prod_m \mathrm{d}\boldsymbol{R}_m A(\{\boldsymbol{R}_m\}), \tag{6.5.10}$$

where N is the total number of sites in the volume Ω. Upon configuration averaging the iteration series, there occur averages of products of the type $\eta(\boldsymbol{q}_1)\,\eta(\boldsymbol{q}_2)\ldots\eta(\boldsymbol{q}_n)$:

$$\langle\eta(\boldsymbol{q})\rangle = \delta(\boldsymbol{q})\,(2\pi)^3\,\mathcal{N},$$

$$\langle \eta(\boldsymbol{q}_1)\, \eta(\boldsymbol{q}_2)\rangle = \Omega^{-2} \sum_{m_1 \neq m_2} \int \mathrm{d}\boldsymbol{R}_{m_1}\, \mathrm{d}\boldsymbol{R}_{m_2} \exp\left[\mathrm{i}(\boldsymbol{q}_1\boldsymbol{R}_{m_1}) + \mathrm{i}(\boldsymbol{q}_2\boldsymbol{R}_{m_2})\right]$$

$$+ \Omega^{-1} \sum_{m_1} \int \mathrm{d}\boldsymbol{R}_{m_1} \exp\left[\mathrm{i}(\boldsymbol{q}_1 + \boldsymbol{q}_2)\, \boldsymbol{R}_{m_1}\right]$$

$$= (2\pi)^6\, \delta(\boldsymbol{q}_1)\delta(\boldsymbol{q}_2)\, \mathcal{N}^2 \left(1 - \frac{1}{N}\right) + (2\pi)^3\, \delta(\boldsymbol{q}_1 + \boldsymbol{q}_2)\, \mathcal{N}.$$

...

In large systems, the factor $(1 - 1/N)$ may be replaced by unity. Analogous replacements may be performed in correlators of higher order. Compared to the averaging procedure for disordered alloys (see, e.g., ELLIOTT et al., 1974), this implies that cumulant coefficients are lacking here. This is due to the fact that in the system under consideration the number of impurity sites (between which hopping occurs) is much less than the number of sites of the host lattice, which allows the continuous coordinate $\boldsymbol{r}$ to be introduced instead of the discrete set of coordinates of the host lattice.

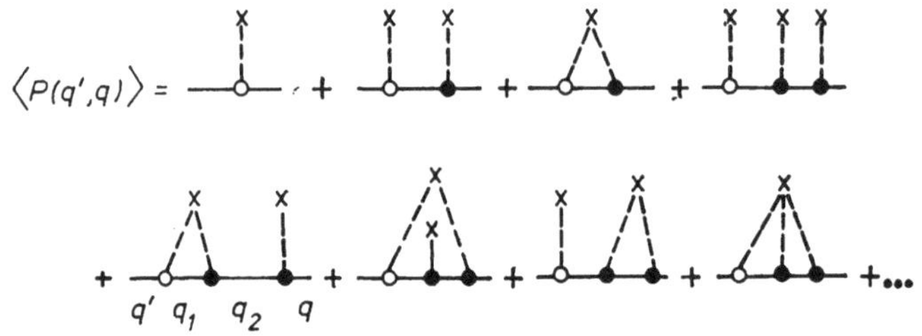

Fig. 6.3. Diagram series for the averaged conditional-probability function $\langle P(\boldsymbol{q}', \boldsymbol{q})\rangle$ (see text)

An inspection of the terms of the iteration series shows that these terms may be represented by diagrams, which are constructed according to the following rules (cf. the diagram technique for lattice systems in the paper by ELLIOTT at al., 1974): On a straight line n points are set up, whereby all points are "bold", except the utmost left one, which is empty (cf. Fig. 6.3). Each segment of the line between two neighbouring points is associated with an arrow directed from the left to the right and is associated with the momentum $\boldsymbol{q}_i$. The utmost left line is associated with the momentum $\boldsymbol{q}'$ and the utmost right one with $\boldsymbol{q}$. Each bold point is associated with a factor $\mathcal{W}(\boldsymbol{q}_i', \boldsymbol{q}_i)$, where $\boldsymbol{q}_i'$ and $\boldsymbol{q}_i$ are the indices of the incoming and outgoing lines, respectively. The empty point is associated with the factor unity. All lines, except the utmost left one, are associated with a factor $1/s$. In every possible way, all points are connected with vertices by means of dashed lines, whereby the number of lines running into a vertex may be one, two, etc. Each vertex is associated with a factor $\mathcal{N}$ and a factor $(2\pi)^3\, \delta\left(\sum_i \boldsymbol{q}_i' - \sum_i \boldsymbol{q}_i\right)$ describing the conservation of the momentum, where $\boldsymbol{q}_i'$ and $\boldsymbol{q}_i$, respectively, are the momenta of the lines running into and away from the points connected with the given vertex. It must be integrated over all intermediate momenta. The resulting expression must be multiplied by a factor $(2\pi)^{-3\varkappa}$, where $\varkappa$ denotes the number of integrations involved. As example, let us write down the analytical expression associated with the fifth graph in Fig. 6.3:

$$\frac{\mathcal{N}^2}{s^3} \int \frac{d\boldsymbol{q}_1}{(2\pi)^3} \frac{d\boldsymbol{q}_2}{(2\pi)^3} \mathcal{W}(\boldsymbol{q}_1, \boldsymbol{q}_2) \mathcal{W}(\boldsymbol{q}_1, \boldsymbol{q}) (2\pi)^6 \delta(\boldsymbol{q}_2 - \boldsymbol{q}') \delta(\boldsymbol{q}_2 - \boldsymbol{q})$$

$$= \mathcal{N}(2\pi)^3 \delta(\boldsymbol{q}' - \boldsymbol{q}) \frac{\mathcal{N}}{s} \int \frac{d\boldsymbol{q}_1}{(2\pi)^3} \mathcal{W}(\boldsymbol{q}_1, \boldsymbol{q}) \frac{1}{s} \mathcal{W}(\boldsymbol{q}, \boldsymbol{q}) \frac{1}{s}.$$

This example indicates that, as usual, it is convenient to introduce the concept of irreducible blocks which do not include elements that are not mutually connected through dashed lines. This concept allows the Dyson equation for the diagram series to be written down. To do this, we introduce the function $P(\boldsymbol{q})$ defined as

$$\langle P(\boldsymbol{q}', \boldsymbol{q})\rangle = (2\pi)^3 \delta(\boldsymbol{q}' - \boldsymbol{q}) \mathcal{N} P(\boldsymbol{q}). \qquad (6.5.11)$$

The diagram technique for $P(\boldsymbol{q})$ differ from that for $\langle P(\boldsymbol{q}', \boldsymbol{q})\rangle$ only in that the factor $\mathcal{N}$ for a node connected with an empty point is lacking, the incoming and outgoing momenta are equal to each other, and the corresponding δ-function is lacking. For $P(\boldsymbol{q})$, it holds that

$$P(\boldsymbol{q}) = S(\boldsymbol{q}) D(\boldsymbol{q}). \qquad (6.5.12)$$

Here $S(\boldsymbol{q})$ is the irreducible block with an empty left point, and $D(\boldsymbol{q})$ describes the set of reducible blocks without an empty point. $D(\boldsymbol{q})$ satisfies the relation (Dyson equation)

$$D(\boldsymbol{q}) = [s - \Pi(\boldsymbol{q})]^{-1}, \qquad (6.5.13)$$

where $\Pi(\boldsymbol{q})$ is an irreducible block without an empty site. For $S(\boldsymbol{q})$ and $\Pi(\boldsymbol{q})$, the diagram series are illustrated in Fig. 6.4. It is easy to see that $S(\boldsymbol{q})$ and $\Pi(\boldsymbol{q})$ represent power series of $1/s$, i.e. for low frequencies, it is impossible to restrict oneself

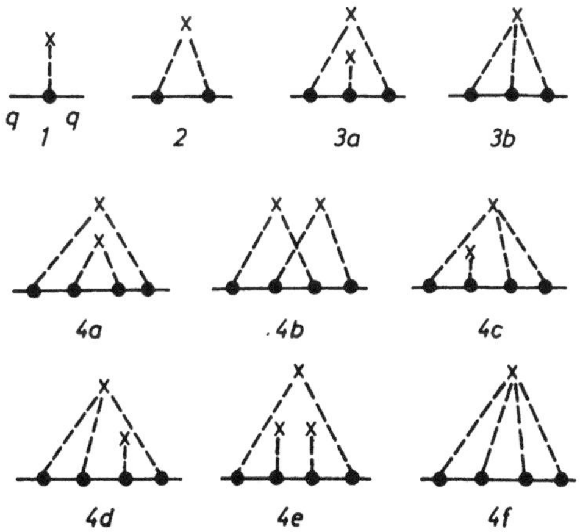

Fig. 6.4. Diagrams for the irreducible part $\Pi(\boldsymbol{q})$ up to the fourth order. The diagrams for $S(\boldsymbol{q})$ can be obtained by replacing the utmost left point in each diagram by an empty point

to a finite number of diagrams. To remove the corresponding divergencies, we perform a partial summation by replacing the internal lines between neighbouring points by bold lines, whereby each bold line is associated with a factor $D(\boldsymbol{q}_i)$ (instead of $1/s$, as in the case of the original lines). Owing to this procedure, diagrams of the type 3a, 4a, 4c, 4d, and 4e in Fig. 6.4 and analogous higher order diagrams (where the lines are replaced by bold ones) are excluded. The set of modified diagrams is illustrated in Fig. 6.5. The basic property of the modified diagrams is that they remain finite in the limit as $s \to 0$.

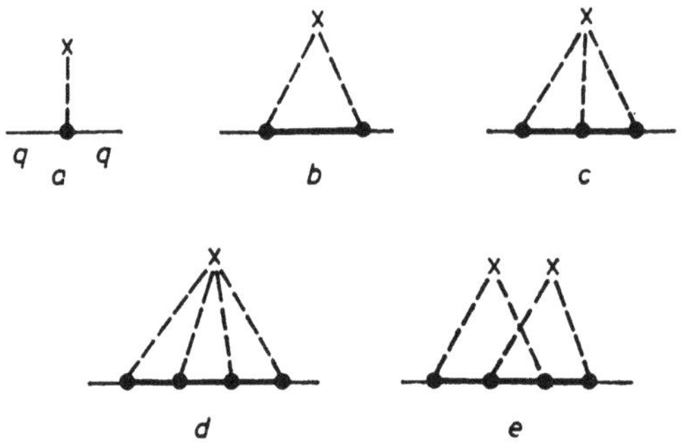

Fig. 6.5. Modified diagrams for $\Pi(\boldsymbol{q})$ up to the fourth order. The diagrams for $S(\boldsymbol{q})$ can be obtained by replacing in the diagrams the utmost left point by an empty point

To make the diagram summation feasible, in what follows we neglect all interference diagrams (diagrams of the type e) in Fig. 6.5). This approximation gives the correct expression for σ in the high-frequency regime (in the regime of the two-site approximation). In the low-frequency regime, it yields reasonable results, which are in agreement with the results of the cluster approximation and the effective-medium approximation to within numerical factors.

The summation of diagrams with a single vertex leads to the following integral equation for $S(\boldsymbol{q})$:

$$S(\boldsymbol{q}) = 1 + \int \frac{\mathrm{d}\boldsymbol{q}'}{(2\pi)^3} S(\boldsymbol{q}') D(\boldsymbol{q}') \mathscr{W}(\boldsymbol{q}', \boldsymbol{q}), \tag{6.5.14}$$

which can readily be verified by iterating this equation.

Somewhat more complicated is the summation of the diagrams for $\Pi(\boldsymbol{q})$. To perform this summation, we introduce the quantity $\Pi(\boldsymbol{q}', \boldsymbol{q})$ which differs from $\Pi(\boldsymbol{q})$ in that the incoming and outgoing momenta do not coincide. As can readily be verified by iteration, $\Pi(\boldsymbol{q}', \boldsymbol{q})$ satisfies the integral equation

$$\Pi(\boldsymbol{q}', \boldsymbol{q}) = \mathcal{N}\mathscr{W}(\boldsymbol{q}', \boldsymbol{q}) + \int \frac{\mathrm{d}\boldsymbol{q}_1}{(2\pi)^3} \Pi(\boldsymbol{q}', \boldsymbol{q}_1) D(\boldsymbol{q}_1) \mathscr{W}(\boldsymbol{q}_1, \boldsymbol{q}), \tag{6.5.15}$$

and, by definition,

$$\Pi(\boldsymbol{q}) \equiv \Pi(\boldsymbol{q}', \boldsymbol{q})|_{\boldsymbol{q}'=\boldsymbol{q}}. \tag{6.5.16}$$

The procedure of finding $\Pi(\boldsymbol{q})$ (or $D(\boldsymbol{q})$) is to solve (6.5.15), which yields a relation between $\Pi(\boldsymbol{q}', \boldsymbol{q})$ and $D(\boldsymbol{q})$. Then, with the aid of (6.5.16) and (6.5.13), we can obtain an equation for $D(\boldsymbol{q})$. To make this procedure feasible, we take into account that for $\boldsymbol{q} = 0$, the diffusion function $D(\boldsymbol{q})$ is equal to $1/s$, and that at sufficiently large values $\boldsymbol{q}$ (of order the reciprocal linear dimension of the spreading electron wave packet which was located at time $t = 0$ at the origin), $D(\boldsymbol{q})$ reaches the limiting value D_∞. At not too small times (i.e. not too large s), this characteristic momentum is less than that for $W(\boldsymbol{q})$, which is of order α (reciprocal Bohr radius). Therefore, the simplest approximation is to replace $D(\boldsymbol{q})$ by D_∞ in (6.5.15), and to take D_∞ out of the sign of integration. This procedure requires that the point $\boldsymbol{q} = 0$ is not singular, which is the case for $d \neq 1$. Hence, (6.5.15) becomes

$$\Pi(\boldsymbol{q}', \boldsymbol{q}) = \mathcal{N}\mathcal{W}(\boldsymbol{q}', \boldsymbol{q}) + D_\infty \int \mathrm{d}\boldsymbol{q}_1 (2\pi)^{-3}\, \Pi(\boldsymbol{q}', \boldsymbol{q}_1)\, \mathcal{W}(\boldsymbol{q}_1, \boldsymbol{q}). \tag{6.5.17}$$

Using the expression for $\mathcal{W}(\boldsymbol{q}', \boldsymbol{q})$ (6.5.9), and writing $\Pi(\boldsymbol{q}', \boldsymbol{q}) = \varphi(\boldsymbol{q}) + \tilde{\varphi}(\boldsymbol{q}' - \boldsymbol{q})$, then on account of $\int \mathrm{d}\boldsymbol{q}' \mathcal{W}(\boldsymbol{q}', \boldsymbol{q}) = 0$, we find that

$$\left.\begin{aligned} \tilde{\varphi}(\boldsymbol{q}) &= -\mathcal{N} W(\boldsymbol{q}) - D_\infty \int \frac{\mathrm{d}\boldsymbol{q}'}{(2\pi)^3}\, \tilde{\varphi}(\boldsymbol{q}')\, W(\boldsymbol{q}' - \boldsymbol{q}), \\ \varphi(\boldsymbol{q}) &= \mathcal{N} W(\boldsymbol{q}) + D_\infty \int \frac{\mathrm{d}\boldsymbol{q}'}{(2\pi)^3}\, \tilde{\varphi}(\boldsymbol{q})\, W(\boldsymbol{q}' - \boldsymbol{q}). \end{aligned}\right\} \tag{6.5.18}$$

Hence, in particular, we have $\tilde{\varphi}(\boldsymbol{q}) = -\varphi(\boldsymbol{q})$. Using the $\boldsymbol{r}$-representation, then from (6.5.18), we obtain

$$\tilde{\varphi}(\boldsymbol{r}) = -\frac{\mathcal{N} W(\boldsymbol{r})}{1 + D_\infty W(\boldsymbol{r})}, \tag{6.5.19}$$

which yields

$$\Pi(\boldsymbol{q}) = -\mathcal{N} \int \mathrm{d}\boldsymbol{r}\, \frac{W(\boldsymbol{r})}{1 + D_\infty W(\boldsymbol{r})}\, (1 - \mathrm{e}^{-\mathrm{i}\boldsymbol{q}\boldsymbol{r}}). \tag{6.5.20}$$

To get $D_\infty = (s - \Pi_\infty)^{-1}$, we need Π_∞. Setting in (6.5.20) $\boldsymbol{q} \to \infty$, we obtain the following equation for Π_∞:

$$\Pi_\infty = \mathcal{N}(\Pi_\infty - s) \int \mathrm{d}\boldsymbol{r}\, \frac{W(\boldsymbol{r})}{s - \Pi_\infty + W(\boldsymbol{r})}. \tag{6.5.21}$$

For $D(\boldsymbol{q}) \to D_\infty$, from (6.5.14) (on account of $\int \mathrm{d}\boldsymbol{q}' \mathcal{W}(\boldsymbol{q}', \boldsymbol{q}) = 0$) we get $S(\boldsymbol{q}) = 1$. Hence, the conductivity (6.5.4.) may be expressed by $\Pi(\boldsymbol{q})$ as

$$\sigma = -\frac{e^2 n}{2kT} \left.\frac{\partial^2 \Pi(\boldsymbol{q})}{\partial q_x^2}\right|_{\boldsymbol{q}=0} = \frac{e^2 n}{6kT}\, \mathcal{N} \int_0^\infty \mathrm{d}\boldsymbol{r} r^2\, \frac{W(\boldsymbol{r})}{1 + D_\infty W(\boldsymbol{r})}. \tag{6.5.22}$$

On deriving (6.5.22), we have accounted for that $\Pi(\boldsymbol{q}) = 0$ and $\nabla_{\boldsymbol{q}} \Pi(\boldsymbol{q}) = 0$ for $\boldsymbol{q} = 0$.

Using the explicit expression for the probability $W(\boldsymbol{r}) = \nu_{\text{ph}} \exp(-2\alpha r)$, formula (6.5.22) may be written as

$$\sigma = \frac{2\pi}{3kT} \frac{e^2 n}{(2\alpha)^5} \mathcal{N}(s - \Pi_\infty) \, 4! \, F_4\left(\ln \frac{\nu_{\text{ph}}}{s - \Pi_\infty}\right), \tag{6.5.23}$$

where $F_j(x)$ denotes the well-known Fermi-Dirac function

$$F_j(x) = \frac{1}{j!} \int\limits_0^\infty \mathrm{d}z \, z^j (1 + \mathrm{e}^{z-x})^{-1}. \tag{6.5.24}$$

Accordingly, the equation for Π_∞ (6.5.21) becomes

$$\frac{\Pi_\infty}{\Pi_\infty - s} = (2\alpha R_c)^{-3} \, 2F_2\left(\ln \frac{\nu_{\text{ph}}}{s - \Pi_\infty}\right), \tag{6.5.25}$$

where, as will be shown below, $R_c = \mathcal{N}^{-1/3}(3/4\pi)^{1/3} \approx 0.62 \, \mathcal{N}^{-1/3}$ has the meaning of the characteristic hopping distance in the dc limit (cf. Section 5.). For not too high frequencies, where $|\nu_{\text{ph}}/(\mathrm{i}\omega - \Pi_\infty)| \gg 1$, we may use the following asymptotic expression for the Fermi-Dirac integrals:

$$F_j(x) \approx x^{j+1}/(j+1)! \tag{6.5.26}$$

Then, (6.5.23) and (6.5.25) take the form

$$\sigma(\omega) = \frac{2\pi}{15} \frac{e^2 n}{(2\alpha)^5} \mathcal{N} \frac{\mathrm{i}\omega - \Pi_\infty}{kT} \ln^5 \left(\frac{\nu_{\text{ph}}}{\mathrm{i}\omega - \Pi_\infty}\right), \tag{6.5.27}$$

$$\mathrm{i}\omega - \Pi_\infty = \nu_{\text{ph}} \exp\left\{-2\alpha R_c \left(\frac{\Pi_\infty}{\Pi_\infty - \mathrm{i}\omega}\right)^{1/3}\right\}. \tag{6.5.28}$$

From an inspection of these formulae, it follows that up to frequencies of the order of ν_{ph}, i.e. up to frequencies where $\operatorname{Re} \sigma(\omega)$ passes into the plateau, the resulting dependence $\sigma(\omega)$ agrees qualitatively with that obtained by means of the effective-medium approximation. Indeed, according to (6.5.28), in the dc limit it holds that $\Pi_\infty = -\nu_{\text{ph}} \exp(-2\alpha R_c)$ and $\sigma(0) \sim \exp(-2\alpha R_c)$ (i.e. R_c is the critical hopping distance in the dc limit). For low frequencies, $\omega \ll |\Pi_\infty|$, the self-consistency equation (6.5.28) agrees with the corresponding equation (6.4.29) to within a numerical factor. This frequency range covers the multiple-hopping regime and the hydrodynamic low-frequency limit. For $\omega \gg |\Pi_\infty|$, the expression for σ (6.5.27) represents the result obtained by means of the two-site approximation for frequencies below the high-frequency plateau (cf. (6.4.40)).

For very high frequencies, $\omega \gtrsim \nu_{\text{ph}}$, the approximation (6.5.26) does not hold, but we have $F_j(x) \approx e^x(1 + O(1/x))$. With the aid of the latter expression, (6.5.23) and (6.5.25) lead to a dependence $\sigma(\omega)$ of the type (6.4.39), which describes the frequency regime where $\operatorname{Re} \sigma(\omega)$ changes into the plateau (while $\operatorname{Im} \sigma \sim 1/\omega$).

The major difference between the results obtained by the diagram technique and the effective-medium approximation (and the percolation approach) is that the diagram technique yields for the parameter η in the critical hopping probability $W_c \simeq \nu_{ph} \times \exp(-\eta\alpha\mathcal{N}^{-1/3})$ the value $\eta = 0.62$, which is somewhat less than the correct value $\eta \approx 0.85$. This difference is due to neglection of the interference diagrams. An account of such diagrams with two intersecting bundles of dashed lines leads to an increase of η. However, so far this problem has not yet been completely solved.

The major advantage of the diagram technique is in its "automatism", i.e. it does not require additional assumptions (e.g. an underlying regular lattice). In principle, this approach may also be extended to describe R-ε-hopping. In this case, instead of the function $P(\boldsymbol{r}', \boldsymbol{r})$, a function $P(\boldsymbol{r}', \varepsilon'; \boldsymbol{r}, \varepsilon)$ must be introduced, where ε' and ε, respectively, denote the energies of the initial and final states. However, to do this, at first the equation for $P_{m'm}$ must be generalized to the case of the Fermi statistics, since in Section 2.6. the equation derived for the conditional-probability function holds only in the one-electron approximation (i.e. for the Boltzmann statistics).

Note that the diagram technique may also be used to study other phenomena such as transport in strong electrical field, the Hall effect etc. However, this requires certain modifications of the technique, which have not been performed so far.

We close this section by noting that the diagram technique in question may also be used to obtain the diffusion function $P(\boldsymbol{q})$ (cf. (6.5.12) and (6.5.13)). For $S(\boldsymbol{q}) = 1$ (see above), we have $P(\boldsymbol{q}) = [s - \Pi(\boldsymbol{q})]^{-1}$. If we are not interested in too small distances, $\Pi(\boldsymbol{q})$ may be expanded in a power series of $\boldsymbol{q}$. Then, according to (6.5.22), it holds that

$$\Pi(\boldsymbol{q}) \approx \frac{1}{2}\, \boldsymbol{q}^2 \Pi''(\boldsymbol{q})|_{\boldsymbol{q}=0} = -\boldsymbol{q}^2 \frac{\sigma kT}{e^2 n} = -\boldsymbol{q}^2 \frac{ukT}{e} = -\boldsymbol{q}^2 D,$$

where u is the drift mobility related to the diffusion coefficient D through the Einstein relation.

In this approximation, the diffusion is described by

$$P(\boldsymbol{q}, s) = [s + \boldsymbol{q}^2 D(s)]^{-1}. \tag{6.5.29}$$

Accordingly, in the t-$\boldsymbol{r}$-representation the diffusion function $P(\boldsymbol{r}, t)$ satisfies the diffusion equation, whereby the frequency dependence of D indicates non-Markovian behaviour (2.6.26). Hence, it follows that non-Gaussian diffusion behaviour in disordered systems is intimately related to the frequency dependence of σ (cf. Section 2.6. and 6.8.).

Let us draw attention to the fact that (6.5.29) was obtained by using the approximation $S(\boldsymbol{q}) = 1$. It can be shown that a function $S(\boldsymbol{q})$ that differs from unity is closely related to the initial condition for the function $P(\boldsymbol{r}', \boldsymbol{r})$. If in (6.5.2) the random quantity $\eta(\boldsymbol{r})$ were lacking in the free term (initial condition), then $S(\boldsymbol{q})$ would be equal to unity exactly. However, so far the conditions of obtaining (6.5.29) have not been studied in detail. In this context note that, according to BERNASCONI et al. (1980) and STEPHENS and KARIOTIS (1982), in one dimension $P(\boldsymbol{r}, s)$ satisfies the scaling relation $P(\boldsymbol{r}, s) = f\left(\boldsymbol{r}\sqrt{s/D}\right)$, in agreement with the expression (6.5.29).

In the case that (6.5.29) holds, the diffusion process is described by (2.6.27), which, after performing the integration over τ and changing over from the Laplace representation to the time representation, becomes

$$P(\boldsymbol{r}, t) = \frac{1}{2\pi i} \int_{-i\infty}^{i\infty} \frac{ds}{4\pi D(s)\, r} \exp\left(st - r\sqrt{\frac{s}{D(s)}}\right). \tag{6.5.30}$$

In the regime of not too small times, for $D(s)$, a relation of the type (6.4.29) may be used. To see this, note that, analogous to σ, $D(s)$ is proportional to the effective transition probability W, which can be obtained from (6.5.28) for $\omega \ll |\Pi_\infty|$). Accordingly, we have

$$D(s) = D_0 \exp\left\{\frac{D_0}{D(s)} \frac{s}{\omega_{0D}}\right\}. \tag{6.5.31}$$

Here, according to (6.4.30), up to a numerical factor, it holds that $\omega_{0D} = W_c/\alpha\mathcal{N}^{-1/3}$, and D_0 is the diffusion coefficient in the dc limit.

A detailed analysis of the diffusion function (6.5.30) has not been performed so far. However, for very large times and not too large distances, the main contribution to the integral in (6.5.30) comes from the vicinity of the point $s = 0$ in the complex s-plane. In this case, in the integrand of (6.5.30), $D(s)$ may be replaced by D_0, which yields the familiar Gaussian behaviour

$$P(\boldsymbol{r}, t) = (4\pi D_0 t)^{-3/2} \exp\left(-r^2/4D_0 t\right).$$

Note that a similar diagram technique was proposed by Gochanour et al. (1979) (see also Section 11.4.1.).

6.6. Use of the topology of the Bethe lattice (Cayley tree) in the theory of the ac conductivity

6.6.1. *General expression for the* ac *conductivity of a Bethe lattice*

A further method of evaluating $\sigma(\omega)$ was proposed by Movaghar et al. (1980) and Movaghar and Schirmacher (1981) (cf. Movaghar et al., 1983). In this method, the linearized rate equation is solved by the Green-function technique. The Green function is evaluated by using the renormalized perturbation expansion (cf. Section 4.2.), and the configuration average is performed with the aid of an effective-medium approximation which becomes exact for a Bethe lattice. A Bethe lattice is illustrated in Fig. 6.6. A characteristic of it is the absence of closed loops of bonds, i.e. any two points are connected only through a single path of bonds tied together. For the case of R-hopping, the results obtained by this method agree with those of the diagram technique described above. For R-ε-hopping, this method allows a closed expression for $\sigma(\omega)$ to be obtained, while the cluster approximation (Böttger et al., 1979b) yields only qualitative results for this case.

Since the approach of Movaghar and coworkers "under way" contains some approximations which cannot be easily understood, we do not follow this approach, but start directly from the equivalent scheme (cf. Fig. 2.3) for a Bethe lattice. This allows the results of the above papers to be obtained and the underlying assumptions to be understood.

The first problem encountered on evaluating σ is the procedure of decoupling averages of the kind $\langle X^2P\rangle$ contained in the expression for σ (2.6.11). This decoupling cannot be performed without arbitrariness, since the numbering of the sites of a

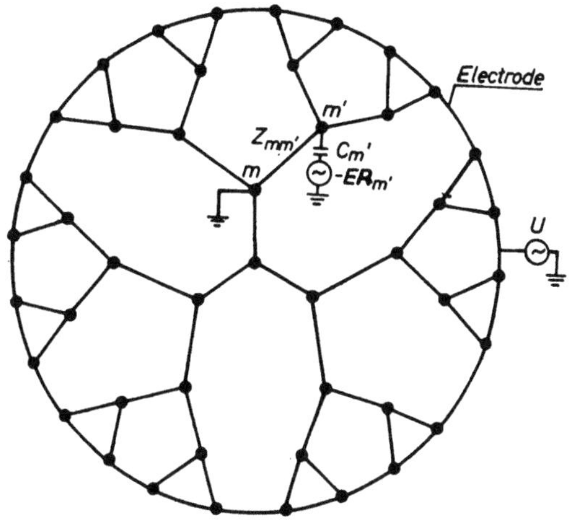

Fig. 6.6. Bethe lattice with the coordination number $z = 3$. The extrinsic sites of the lattice are placed on a metallic electrode on which the potential U (the potential of the external electrical field) is imposed. Each bond between sites m and m' is associated with a resistor $Z_{mm'}$, and each site m' with a capacitor $C_{m'}$ and the voltage $-\boldsymbol{ER}_{m'}$. The second "electrode", the site m, is grounded

lattice is not directly related to the positions of the sites (cf. the discussion in the preceding section). Moreover, in the case of a Bethe lattice the sites cannot be disposed in a space with a finite number of dimensions (for a coordination number z larger than two). Nevertheless, MOVAGHAR et al. (1980) proposed that for a Bethe lattice σ may be approximately written as

$$\sigma = \frac{\langle X^2\rangle}{2} \mathcal{N}\mathcal{G}, \tag{6.6.1}$$

where $\mathcal{G}$ is the conductance of a resistor network which is topologically related to the Bethe lattice (see below), and $\langle X^2\rangle = \frac{1}{3}\langle R^2\rangle$, where $\langle R^2\rangle$ is some average square distance between nearest neighbours. The concrete expression for $\langle R^2\rangle$ remains open for the present (see below).

The second problem is to correctly define the quantity $\mathcal{G}$. According to Section 2.5., for $\omega = 0$, $\mathcal{G}$ is determined by the total current running through an electrode (cf. STINCHCOMBE, 1974). For $\omega \neq 0$, the situation is more complicated. Here, owing to the voltage generators at the capacitors, the current, and thus $\mathcal{G}$, is not simply given by the current through an electrode (cf. Section 2.5). However, on studying the case $\omega \neq 0$, MOVAGHAR et al. (1980) ignore the voltage generators at the capacitors, i.e. they define $\mathcal{S}$ analogously as for $\omega = 0$. Note that this prescription fails at least at high frequencies, where the two-site approximation is appropriate, because in this case the equivalent scheme reduces to a single resistor with two capacitors and two voltage generators, and the external circuit does not play any role. Because of this, one imposes a frequency dependence on $\langle R^2 \rangle$ such that the two-site regime can also be covered.

To evaluate $\mathcal{S}$, we introduce $\mathcal{S}^{(m)}$, the conductance between site m and the electrode. By definition, it holds that $\mathcal{S} = \langle \mathcal{S}^{(m)} \rangle$, where $\langle \ldots \rangle$ indicates the configuration average. Owing to the fact that all z branches connected with site m represent parallel circuits, we have

$$\mathcal{S}^{(m)} = \sum_{m'} \mathcal{S}^{(m)}_{m'}, \tag{6.6.2}$$

where $\mathcal{S}^{(m)}_{m'}$ is the conductance of the branch through the m'th nearest-neighbour site; m' may take z values. $(\mathcal{S}^{(m)}_{m'})^{-1}$ represents the resistance of the following two resistors connected in series: (i) the resistor (with resistance $Z_{mm'}$) between sites m and m' and (ii) the parallel connection of the capacitor at site m' (with capacity $C_{m'}$) and the resistor describing the resistance between site m' and the electrode in the absence of $Z_{mm'}$. The conductance of the latter resistor is denoted by $\mathcal{S}_{mm'}$.

Hence, we have

$$\mathcal{S}^{(m)}_{m'} = \{Z_{mm'} + [\mathrm{i}\omega C_{m'} + \mathcal{S}_{mm'}]^{-1}\}^{-1}. \tag{6.6.3}$$

It is easy to see that $\mathcal{S}_{mm'} = \sum_{m''(\neq m)} \mathcal{S}^{(m')}_{m''}$, where $\mathcal{S}^{(m')}_{m''}$ is the conductance associated with site m'. Accordingly, (6.6.3) can be written as

$$\mathcal{S}^{(m)}_{m'} = \left\{Z_{mm'} + \left[\mathrm{i}\omega C_{m'} + \sum_{m''(\neq m)} \mathcal{S}^{(m')}_{m''}\right]^{-1}\right\}^{-1}. \tag{6.6.4}$$

This formula may be viewed as the equation for determining the quantity $\mathcal{S}^{(m)}_{m'}$.

Let us now turn to the configuration average. Using (6.6.2), we obtain

$$\mathcal{S} = \langle \mathcal{S}^{(m)} \rangle = z \langle \mathcal{S}^{(m)}_{m'} \rangle \equiv \frac{e^2}{kT} W, \tag{6.6.5}$$

i.e. by means of (6.6.4), we must evaluate the quantity $W = (kTz/e^2) \langle \mathcal{S}^{(m)}_{m'} \rangle$. STINCHCOMBE (1974) pointed out that the configuration average of an equation of the type (6.6.4) leads to a nonlinear integral equation (for $\omega \neq 0$, this equation was given by MOVAGHAR and SCHIRMACHER, 1981). Since the solution of such an equation is a very intricate problem, here we use an approximation in the spirit of the self-consistent field: On the right-hand side of the configuration averaged equation (6.6.4) we replace $\mathcal{S}^{(m')}_{m''}$ by $e^2 W/zkT$, which yields the following closed equation for W:

$$W = z \left\langle \left\{ \Gamma^{-1}_{mm'} + \left[\mathrm{i}\omega f_{m'}(1 - f_{m'}) + \frac{z-1}{z} W \right]^{-1} \right\}^{-1} \right\rangle, \tag{6.6.6}$$

where we have used the expressions for $Z_{mm'}$ and $C_{m'}$ from (2.5.15).

The expression for σ (6.6.1) becomes

$$\sigma = \frac{\langle R^2 \rangle}{6kT} e^2 \mathcal{N} W . \tag{6.6.7}$$

At this stage we dispose of $\langle R^2 \rangle$. Above all, note that W has the meaning of a characteristic hopping probability, since the expression for σ (6.6.7) has a form characteristic of an ordered system. In order that this expression has really such form, $\sqrt{\langle R^2 \rangle}$ must be the hopping distance. Accordingly, we must require that

$$\langle R^2 \rangle = N \langle (R_m - R_{m'})^2 \{ \Gamma_{mm'}^{-1} + [\mathrm{i}\omega f_{m'}(1 - f_{m'}) + W]^{-1} \}^{-1} \rangle / W , \tag{6.6.8}$$

where we have already assumed that the coordination number z is equal to the total number of sites in the system, N, and $N \to \infty$.

6.6.2. *Results for R-percolation*

Consider the R-(hopping) percolation model, where $f_m = f$, $Z_{mm'}^{-1} = e^2 \beta f(1 - f) W_{mm'}$, and $W_{mm'}$ is given in (6.3.15). The configuration average must be performed only with respect to the distance r between pairs of sites. For a chaotic distribution of site positions, the corresponding distribution function is given by $4\pi r^2/\Omega$. Then, if we put $W = -f(1 - f) \Pi_\infty$, the self-consistency equation (6.6.6) agrees exactly with equation (6.5.21), which was obtained by means of the diagram technique developed in Section 6.5. Furthermore, on account of (6.6.8), the formula for the conductivity (6.6.7) takes the form of (6.5.23). The only difference is that the factor $(1 - f)$ is lacking in (6.5.23), which is due to the use of the Boltzmann statistics (i.e. $f \ll 1$) in Section 6.5. Hence, the use of a Bethe lattice is obviously equivalent to the neglection of the interference diagrams in the diagram technique.

Note that a renormalization of $\langle R^2 \rangle$ in the sense of (6.6.8) is very important. With increasing frequency, $\langle R^2 \rangle$ increases from $\mathcal{N}^{-2/3}$ (at $\omega = 0$) to $(2\alpha)^{-2}$ (on the high-frequency plateau). By neglecting this frequency dependence, for instance, in the regime of the two-site approximation, we would obtain $\omega \ln^2 \omega$, instead of the correct expression $\omega \ln^4 \omega$, which is also obtained on account of (6.6.8).

As mentioned in Section 6.5., the major deficiency of (6.5.21) is in that the value of the parameter η (occurring in the dc conductivity $\sigma(0) \sim \exp(-\eta \alpha \mathcal{N}^{-1/d})$) is too small compared with computer results. MOVAGHAR et al. (1980) noted that in actual systems, between any two sites there exists a multitude of possible paths, and not only a single one, which leads to a reduction of the right-hand side of (6.5.21). In a semi-empirical manner, these authors pointed out that this effect leads to a multiplication of the right-hand side of (6.5.21) by a number $1/e$. Accordingly, the self-consistency equation takes the form

$$\Pi_\infty = \frac{4\pi}{e} \frac{\mathcal{N}}{(2\alpha)^5} (\Pi_\infty - \mathrm{i}\omega) \int_0^\infty \mathrm{d}r\, r^2 \frac{e^{-r}}{\mathrm{i}\omega - \Pi_\infty + e^{-r}} . \tag{6.6.9}$$

Owing to this renormalization, the number η becomes equal to $(2de/\pi)^{1/d}$, where d is the dimensionality of the space (equation (6.6.9) can readily be generalized to $d = 2$).

According to this formula, for $d = 3$, we have $\eta = 1.732$, and for $d = 2$, $\eta = 1.860$, in good agreement with computer data (cf. Section 5.). This agreement suggests that the expression found for η is possibly an exact one. Correspondingly, for the critical-volume fraction v_c (cf. Section 5.), which is related to η by $\eta = [v_c 16(2d - 3)/\pi]^{1/d}$ we find that $v_c = ed/[8(2d - 3)]$, which for $d = 3$, yields $v_c = 0.340$, and for $d = 2$, $v_c = 0.680$, which is, of course, in good agreement with computer data.

The dependences Re $\sigma(\omega)$ and Im $\sigma(\omega)$ calculated by means of formulae (6.6.9) and (6.5.23) (Movaghar et al., 1980) are shown in Fig. 6.7. These curves agree qualitatively with all results described above and, in particular, with the computer data obtained by McInnes et al. (1980).

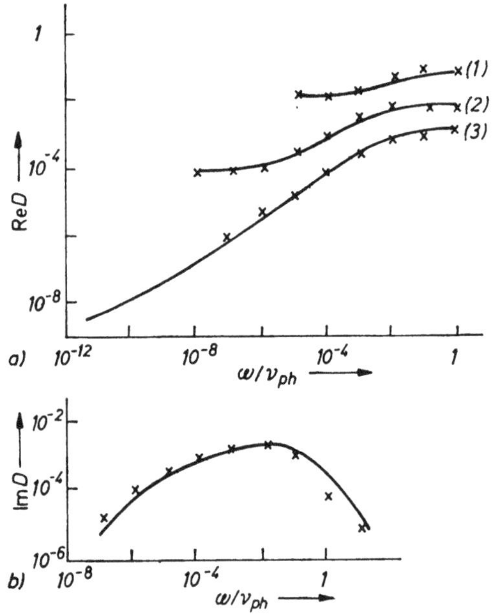

Fig. 6.7. Frequency dependence of the real part (a)) and the imaginary part (b)) of the conductivity for R-percolation, evaluated by means of formulae (6.6.9), (6.6.8) and (6.6.7). The dimensionless quantity D is related to the conductivity σ by $\sigma = (e^2/\alpha kT)\, f(1 - f)\, D$. In a), curve (1) corresponds to $\alpha\mathcal{N}^{-1/3} = 5$, (2) to 9, and (3) to 16. In b), $\alpha\mathcal{N}^{-1/3} = 9$. The hopping probability is given by $W(R) = \nu_{ph}(\alpha R)^{3/2} \exp(-2\alpha R)$. The crosses denote computer data obtained by solving the equations for a lattice with resistors and capacitors

6.6.3. *Results for R-ε-percolation*

In the presence of a finite spread of the energy levels, the quantity (6.6.3) to be averaged does not only depend on the difference between the random parameters associated with sites m and m', but also on ε_m and $\varepsilon_{m'}$ separately. This requires introducing a quantity $W(\varepsilon_m)$ which depends on the energy of the initial site. Within the framework of the self-consistent field approximation, the quantity $W(E)$ is subjected to the following integral equation:

$$W(E) = 4\pi\mathcal{N} \int_0^\infty \mathrm{d}R R^2 \int_0^\infty \mathrm{d}E' N(E')\, \Gamma(E, E', R)$$
$$\times \left\{ \mathrm{i}\omega f(E') \left(1 - f(E')\right) + W(E') \right\}$$
$$\times \left\{ \mathrm{i}\omega f(E') \left(1 - f(E')\right) + \Gamma(E, E', R) + W(E) \right\}^{-1}, \tag{6.6.10}$$

where

$$\Gamma(E, E', R) = \nu_{\mathrm{ph}} \exp\{-2\alpha R - (|E - \mu| + |E' - \mu| + |E - E'|)/2kT\}. \tag{6.6.11}$$

The conductivity is related to $W(E)$ as

$$\sigma = \frac{e^2 \mathcal{N}}{6kT} \langle R^2 \rangle \int_{-\infty}^{\infty} \mathrm{d}E N(E)\, W(E). \tag{6.6.12}$$

Here $N(E)$ is the density of states per site.

A numerical investigation of the integral equation (6.6.10) was performed (for $\omega = 0$) by Movaghar and Schirmacher (1981). Here we examine this equation only in a manner similar to that used by Mott in studying variable-range hopping. Let $W(E)$ be non-zero only in a narrow energy range (of order kT) near the Fermi level μ. Then (6.6.12) becomes

$$\sigma \approx \frac{e^2 \mathcal{N}}{6} \langle R^2 \rangle\, N(\mu)\, W. \tag{6.6.13}$$

Assuming that the major contribution arises from hops near the Fermi level, so that $E' > \mu, E < \mu$. Then, we have $f(E') \left(1 - f(E')\right) \to 1$ and $\Gamma(E, E', R) \to \Gamma\left((E' - E), R\right)$, where $\Gamma(E, R) = \nu_{\mathrm{ph}} \exp(-2\alpha R - |E|/kT)$. Furthermore, we take $N(E')$ out of the integral at the point $E' = \mu$. In this way, we obtain for W the following equation:

$$\widetilde{W} = \frac{4T}{T_0} (\mathrm{i}\tilde{\omega} + \widetilde{W}) \int_0^\infty \mathrm{d}x \frac{x^3 \exp(-x)}{\mathrm{i}\tilde{\omega} + \widetilde{W} + \exp(-x)}, \tag{6.6.14}$$

where $T_0 = 24\alpha^3/\pi N(\mu) k$ is the parameter involved in the Mott $T^{-1/4}$ law (5.1.19), $\widetilde{W} = W/\nu_{\mathrm{ph}}$, and $\tilde{\omega} = \omega/\nu_{\mathrm{ph}}$. On deriving (6.6.14), the quantity $2\alpha R + E/kT$ has been considered as the length of a four-dimensional vector having the components $2\alpha \boldsymbol{R}$ and E/kT. Note that an investigation of variable-range hopping as a problem of nearest-neighbour hopping in the four-dimensional space leads to a parameter T_0 which somewhat differs from that obtained by means of the percolation approach (cf. Section 5.1., see also Pike and Seager, 1974).

Though the above assumptions concerning $W(E)$ are rather crude and not confirmed by a numerical solution of (6.6.10), the dependence $\sigma(\omega, T)$ evaluated by means of (6.6.13) and (6.6.14) is in fairly good agreement with numerical data, at least for $\omega = 0$ (MOVAGHAR and SCHIRMACHER, 1981).

Note that (6.6.14) is almost a trivial generalization of (6.6.9) for the case of changing over from the three-dimensional R-space to the four-dimensional R-ε-space. The analysis of (6.6.14) is analogous to that of (6.6.9) (cf. Section 6.5.).

It is easy to show that in the dc limit, equation (6.6.14) gives Mott's law $\sigma(0) \sim \exp\left(-(T_0/T)^{1/4}\right)$, $(T_0/T \gg 1)$, but in the high-frequency limit, formula (6.6.13) (cf. also the expression for $\langle R^2 \rangle$ (6.6.8)) leads to the results of the two-site model on account of an ε-spread. The frequency dependence of the conductivity calculated by means of (6.6.14) and (6.6.13) (MOVAGHAR et al., 1980) is shown in Fig. 6.8. As

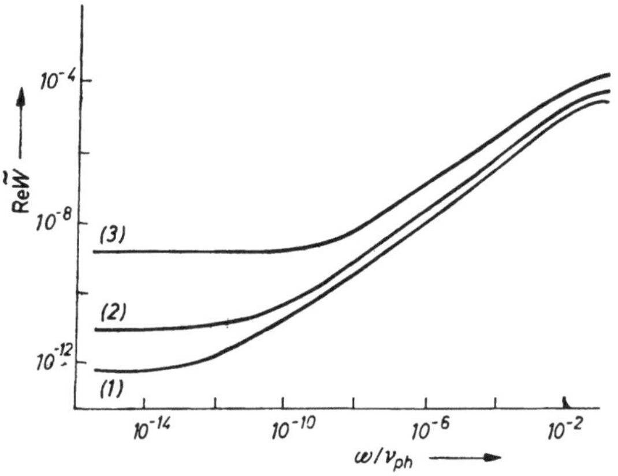

Fig. 6.8. Frequency dependence of the effective hopping probability $\tilde{W} = W/\nu_{\text{ph}}$, evaluated for R-ε-hopping by means of formula (6.6.14); curve (1) $T = 80\,\text{K}$, (2) $100\,\text{K}$, (3) $300\,\text{K}$; $T_0 = 5 \times 10^7\,\text{K}$. Adopting the approximation Re $\tilde{W} \sim \omega^s$, then in the region of intermediate frequencies, $s \approx 0.7$ holds. For not too high frequencies, the frequency dependence of σ is governed by $\tilde{W}(\omega)$, since for such frequencies, $\langle R^2 \rangle$ is virtually constant

can be seen from this figure, after replacing the dimensionless parameter $\alpha\mathcal{N}^{-1/3}$ by $(T_0/T)^{1/4}$, the dependences $\sigma(\omega)$ for R- and R-ε-percolation do not differ. Analogously as in the case of R-percolation, for R-ε-percolation, $\sigma(\omega)$ can be analyzed in terms of hopping across an infinite cluster, multiple hopping, hopping across medium-sized clusters, and the two-site approximation (clusters are now formed by bonds which are defined in the four-dimensional R-ε-space). Let us draw attention to a pecularity of R-ε-hopping, which was qualitatively predicted by BÖTTGER et al. (1979b) on the basis of the cluster approximation. As can be seen from Fig. 6.8, in the range of moderate frequencies, Re $\sigma(\omega)$ behaves approximately as ω^s with $s < 1$. For R-ε-percolation, s decreases with increasing temperature, while for R-percolation it is independent of temperature. The results obtained will be discussed in more detail in the next section.

6.7. Discussion of theoretical results for the ac conductivity and comparison with experimental findings

6.7.1. *Four characteristic frequency regimes*

The basic theoretical result concerning the ac conductivity in disordered systems is the assertion that in such systems the conductivity increases with increasing frequency. This increase is above all a consequence of the disorder and not of the hopping mechanism. To see this, note that with increasing frequency during half a period of the external electrical field, charge carriers can move through clusters of decreasing size, so that with increasing frequency well conducting regions of finite size become more and more effective. Just due to this fact in a multitude of disordered materials an increase of $\sigma(\omega)$ has been observed (see, e.g., the reviews by Böttger and Bryksin, 1976b, 1982, Long, 1982, and references therein), wherby in many of these materials the transport has possibly not hopping character.

Experimental ac conductivity data can often be described by a dependence of the type

$$\mathrm{Re}\,\big(\sigma(\omega) - \sigma(0)\big) \sim \omega^{s'}, \quad \mathrm{Im}\,\sigma(\omega) \sim \omega^{s''}, \tag{6.7.1}$$

where the exponents s' and s'' are in general somewhat smaller than unity. It is not surprising that in the course of nearly two decades experimental findings obtained on various materials in a broad frequency range have been interpreted on the basis of the two-site model and that good agreement between theory and experiment could be stated. However, dependences of the type (6.7.1) do not necessarily indicate hopping conduction. Such indication requires a more detailed study of the dependence $\sigma(\omega)$. With a view to the interpretation of experimental ac data, in what follows we discuss basic theoretical results concerning $\sigma(\omega)$.

Since, so far, the ac hopping conductivity has been theoretically studied in more detail for R-percolation than for R-ε-percolation, our main emphasis will be on materials where the former situation is realized, i.e. ac conductivity in the variable-range hopping regime will be discussed to a less extent.

As pointed out above, in studying ac hopping conductivity, four frequency regimes may be distinguished:

1. *The regime of extremely low frequencies*, where hopping is realized across the cluster with infinite size. In this regime, $\sigma(\omega)$ may be represented as a regular expansion in powers of ω. Note that such a regime does not exist in one dimension, and that the critical dimension is $d = 2$ (in two-dimensional systems the coefficients of the expansion depend on ω as $\ln \omega$, cf. Section 6.4.). In this regime, the change of $\mathrm{Re}\,\sigma(\omega)$ is small, namely of order $\sigma(0)$. The upper limiting frequency is of order $\nu_{\text{ph}}(\alpha\mathcal{N}^{-1/3})^{-1} \times \exp(-\eta\alpha\mathcal{N}^{-1/3})$. It rapidly decreases with increasing $\alpha\mathcal{N}^{-1/3}$. On approaching the upper limiting frequency, the dependence of σ on $\mathcal{N}$ diminishes. In the dc limit, this dependence has exponential character (cf. (6.3.28)), while in the following frequency regime 2, it is power-like. A characteristic feature indicating the end of the regime 1 is the decrease of the exponent s', which is equal to 2 in regime 1, while it is less than unity in the regime 2. The exponent s'', which is equal to unity in the

regime 1, also begins to diminish at the end of this regime but to a less extent as s' (cf. Fig. 6.1) (for $\sigma(\omega)$ in the regime 1, see also Section 11.4.1.).

2. Multiple-hopping regime. In this regime, the characteristic cluster size is large (but finite) and decreases logarithmically with increasing frequency (cf. (6.3.35)). The frequency dependence of the conductivity is given in (6.4.34), whereby $\text{Re}\,\sigma(\omega) \gg \sigma(0)$. A characteristic feature of the multiple-hopping regime is the fact that after an initial decrease, the exponents s' and s'' increase with increasing frequency, i.e. $\ln \text{Re}\,\sigma$ and $\ln \text{Im}\,\sigma$ plotted versus $\ln \omega$ gives a concave curve (cf. Fig. 6.1). According to (6.4.34), we have

$$s' = 1 - 2/\ln\left(\frac{\pi}{2}\tilde{\Omega}\right), \qquad s'' = 1 - 1/\ln\left(\frac{\pi}{2}\tilde{\Omega}\right), \tag{6.7.2}$$

i.e. s' and s'' increase monotonously and approach unity. Since in (6.4.34) the critical hopping probability occurs only in the argument of the logarithm, σ depends on $\mathcal{N}$ only power-like.

The upper limiting frequency of the regime 2 is determined by the restriction (6.3.20). However, owing to the uncertainty of the value of $\varkappa'$, this restriction has only guiding character, i.e. there is no sharp line between the regimes 2 and 3. Near the upper limit of the regime 2, the parameters s' and s'' reach their limiting values, which are of order $1 - a(\alpha\mathcal{N}^{-1/3})^{-1}$, where a is a number, whose determination is an intricate problem because of the uncertainty of the value of $\varkappa'$.

Since the shape of the dependence $\sigma(\omega)$ is rather smooth, it is not easy to distinguish experimentally between the different frequency regimes. Much more distinct information than from $\sigma(\omega)$ can be obtained from an experimental inspection of the cotangens of the angle of the dielectric losses, $\cot \delta = \text{Im}\,\sigma/\big(\text{Re}\,\sigma - \sigma(0)\big)$ (note that in the common definition of $\cot \delta$ there appears $\text{Re}\,\sigma$ instead of $\text{Re}\,\sigma - \sigma(0)$ however, for $\text{Re}\,\sigma \gg \sigma(0)$, this difference is irrelevant).

In the multiple-hopping regime we have $\sigma \sim \mathrm{i}\omega \left[\ln\left(\frac{\pi}{2}\,\mathrm{i}\tilde{\Omega}\right)\right]^{-p}$ and accordingly

$$\cot \delta = \frac{2}{\pi p} \ln\left(\frac{\pi}{2}\tilde{\Omega}\right). \tag{6.7.3}$$

The effective-medium approximation yields $p = 1$, while the cluster approximation $p = t\left(\frac{1}{r} - 1\right)$ (cf. (6.3.33)). Equation (6.7.3) holds for $\tilde{\Omega} \gg 1$, which is fulfilled in the multiple-hopping regime. Accordingly, in this regime, $\cot \delta$ is a linear function of $\ln \omega$ and always larger than unity.

3. Regime of hopping across medium-sized clusters. This is an intermediate regime placed between the multiple-hopping and the two-site regimes. The corresponding dependence $\sigma(\omega)$ can be obtained by joining the results of the neighbouring regimes because in the regime 3, $\sigma(\omega)$ does not exhibit any dramatic change (cf. Fig. 6.7). A characteristic feature of the regime 3 is the change from a convex to a concave dependence of $\ln \sigma$ on $\ln \omega$. The latter dependence is realized in the two-site regime (see below). In the regime 3, we have $\sigma \sim (\mathrm{i}\omega)^s$, where $s \approx 1 - a(\alpha\mathcal{N}^{-1/3})$; s is independent of frequency and slightly less than unity. Accordingly, here the quantity $\cot\delta = \cot\left(\frac{\pi}{2}s\right)$ does not depend on frequency. Owing to this fact, this regime can be readily found experimentally by studying $\cot\delta$ in dependence on $\ln\omega$. After a linear increase in the regime 2, this dependence passes into a plateau in the regime 3.

4. *Regime of two-site hopping.* This is the high-frequency regime. It is characterized by the fact that with increasing frequency the exponents s' and s'' decrease according to a logarithmic law. At very high frequencies, of order ν_{ph}, the dependence $\mathrm{Re}\,\sigma(\omega)$ passes into the high-frequency plateau (cf. Fig. 6.2). In other words, in this frequency regime, $\ln\sigma$ in dependence on $\ln\omega$ has convex character. The lower limiting frequency of the regime 4 is determined by the restriction (6.2.33). Analogous to the case of the upper limiting frequency of the regime 2, the definition of the lower limiting frequency of the regime 4 has only guiding character. An increase of the parameter $\alpha\mathcal{N}^{-1/3}$ rapidly shifts the latter limiting frequency to the low-frequency region. In the low-frequency range of the regime 4, $\cot\delta$ decreases with increasing frequency as

$$\cot\delta = \frac{2}{5\pi}\ln\left(\frac{2\nu_{ph}}{\omega}\right). \tag{6.7.4}$$

At higher frequencies this decrease speeds up and at some frequency of order ν_{ph} the dependence $\cot\delta$ becomes power-like, $\cot\delta \sim 1/\omega$. Within the regime 4, at frequencies below ν_{ph}, $\cot\delta$ is smaller than unity.

It is worth noticing that the two-site regime and the multiple-hopping regime differ in the characteristic dimensionless frequency parameter. If in the regime 2 the characteristic parameter is given by $\omega/W_c = (\omega/\nu_{ph})\exp(\eta\alpha\mathcal{N}^{-1/3})$, then in the regime 4 the corresponding parameter is ω/ν_{ph}. Both these parameters may differ by many orders of magnitude. Furthermore, note that starting with the regime 2, the concentration dependence of σ is power-like.

Let us now discuss the temperature dependence of the ac conductivity. If ν_{ph} is independent of temperature (or does only weakly depend on T), then, for any frequency, it holds that $\sigma \sim \exp(-\varepsilon_3/kT)$. This dependence results from thermal excitation of electrons from the Fermi level into the region near the maximum of the density of states. We emphasize that in this case the exponent s is independent of temperature. Consider now the situation where ν_{ph} depends on temperature. Such situation occurs for over-barrier hopping. For instance, in the small-polaron model we have $\nu_{ph} \sim \exp(-E_a/kT)$, since the two-site hopping probability is proportional to $\exp(-E_a/kT)$ (cf. (3.2.7)). In this case, it holds that $\sigma(0) \sim \exp[-(\varepsilon_3 + E_a)/kT]$. Furthermore, according to (6.4.34), in the regime 2 up to frequencies of order ν_{ph}, we have $\sigma \sim \exp(-\varepsilon_3/kT)$. Near the high-frequency plateau, the activation energy again becomes equal to $E_a + \varepsilon_3$ (cf. (6.4.39)). In the case that the ac conductivity strongly depends on temperature, the exponent s decreases with temperature in the regime 2, but increases with T in the regime 4.

Hitherto, the case of R-ε-hoppping has been studied to a less extent. According to Section 6.6., the change from R- to R-ε-hopping requires replacing the parameter

$\alpha\mathcal{N}^{-1/3}$ by $(T_0/G)^{1/4}$ (this procedure is not valid for the two-site hopping regime). Hence, in the transition region between the regimes 1 and 2, with increasing frequency, the characteristic exponential $T^{-1/4}$ dependence of σ vanishes, and at yet higher frequencies, σ depends on T only power-like. In particular, for hops near the Fermi level, the two-site model predicts non-activated behaviour (cf. (6.2.31) and (6.2.35)). Furthermore, for such hops the exponent s decreases with increasing ω in the regime 2 and becomes independent of frequency in the regime of two-site hops.[1])

6.7.2. *Experimental findings on doped semiconductors and amorphous materials*

Until some few years back, results of ac experiments were almost exclusively interpreted in terms of the two-site model. However, the inadequacy of such a procedure became evident by physical considerations and by the experimental fact that the exponent s either does not depend on frequency or increases with ω, while the two-site model predicts that s decreases with increasing ω. The latter fact was discussed in detail by HILL and JONSCHER (1979) by considering a large number of materials. BÖTTGER et al. (1979b) compared experimental findings of the ac conductivity on n-Si (POLLAK and GEBALLE, 1961) and p-Ge (GOLIN, 1963) with theoretical results obtained for the multiple-hopping regime. From the agreement found by these authors with regard to frequency, temperature, and concentration dependences of the ac conductivity, the conclusion may be drawn that in the major part of the realized experimental situations hopping is observed in these materials multiple.

Let us here discuss the experimental data of GOLIN (1963) on p-Ge, which are shown in Fig. 6.9. First, note that in the frequency range from 10^2 to 10^5 Hz, for an impurity concentration above 7.8×10^{14} cm^{-3}, the temperature dependence of σ may be well described by $\sigma \sim (1/T) \exp(-\varepsilon_3/kT)$, where $\varepsilon_3/k = 3.5$ K. This is in good agreement with formula (6.5.23) and indicates realization of R-hopping. Note further that in the weakly doped samples ($\mathcal{N} = 3.2 \times 10^{14}$ cm^{-3}), the critical hopping probability W_c (6.4.31) has the rather small value $W_c \approx 0.2$ s^{-1} (for $\alpha^{-1} = 8.7$ nm and $\nu_{ph} \approx 10^{12}$ s^{-1}). Hence, in this case, even for the lowest frequencies used in experiment (10^2 Hz), the criterion (6.3.20) is badly satisfied, so that in all likelihood the intermediate regime 3 is realized. On the other hand, for the heavily doped samples ($\mathcal{N} = 7.8 \times 10^{14}$ cm^{-3}), we have $W_c = 3.7 \times 10^2$ s^{-1}, so that the criterion (6.3.20) is satisfied, though without a large reserve. This indicates transport in the multiple-hopping regime near the upper limiting frequency. As can be seen from Fig. 6.9, the experimental frequency dependence $\sigma(\omega)$ may be well fitted by $\sigma \sim \omega^s$, where s increases with decreasing $\mathcal{N}$, which is in accordance with predictions of the theory. In the high-temperature range, the parameter s' (cf. Fig. 6.9b) is independent of temperature. In the low-temperature limit, s' increases with decreasing temperature, which is also in agreement with the theory, because with lowering T, a change to

[1]) Let us draw attention to the paper of FISHCHUK and RUDKO (1980), where for R-ε-hopping in the multiple-hopping regime, by means of the cluster approximation an exponential $T^{-1/4}$ dependence of Mott's type has been obtained, but with a parameter T_0 being $4^4 = 256$ times smaller than in the dc limit.

R-ε-hopping occurs, for which the theory predicts such an increase. In the frequency range from 10^2 to 10^5 Hz, realized in the experiment, the conductivity only weakly depends on $\mathcal{N}$, which also agrees with theoretical results and which indicates that this frequency range is lying above the regime 1, where σ strongly depends on $\mathcal{N}$ (for the samples in question the upper limiting frequency of the regime 1 is below 10 Hz). The absence of a convex curvature of the plot $\ln \sigma$ versus $\ln \omega$ indicates hopping across medium-sized clusters.

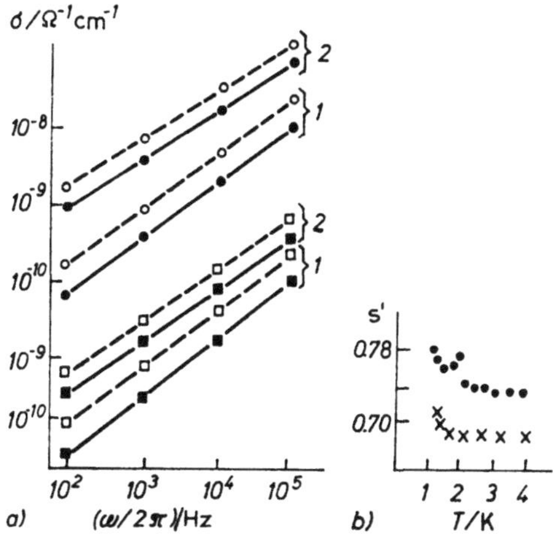

Fig. 6.9. a) Frequency dependence of Re σ (solid lines) and Im σ (dashed lines) in p-Ge with impurity concentrations $\mathcal{N} = 3.2 \times 10^{14}$ cm^{-3} (1) and 7.8×10^{14} cm^{-3} (2) at temperatures $T = 1.2$ K (squares) and 4 K (circles) (after GOLIN, 1963); b) Temperature dependence of the exponent s' (cf. (6.7.1)) for p-Ge with $\mathcal{N} = 3.2 \times 10^{14}$ cm^{-3} (dots) and 7.8×10^{14} cm^{-3} (crosses).

For R-ε-hopping, MOVAGHAR et al. (1980) calculated the dependence $\sigma(\omega)$ for different temperatures on the basis of formula (6.6.14) and compared the results with experimental data obtained on amorphous germanium by LONG and BALKAN (1980). Choosing $T_0 = 3.6 \times 10^8$ K and $\nu_{ph} = 10^{21}$ s^{-1}, then the theoretical curves well agree with the experimental results. However, it should be noted that this value of ν_{ph} much exceeds the widely used value of about 10^{12} to 10^{13} s^{-1}.

BRYKSIN et al. (1981) studied experimentally the ac conductivity in transition metal oxides (Ta_2O_5 and Nb_2O_5) and compared the results with theoretical predictions for the multiple-hopping regime and the regime of medium-sized clusters. They found good agreement between theory and experiment. Note that in these materials the parameter $\alpha\mathcal{N}^{-1/3}$ is very large (larger than 20), since the Bohr radius is very small ($\alpha^{-1} \approx 0.3$ nm). Correspondingly, the upper limiting frequency of the hydrodynamic regime is very small (about 10^{-4} Hz), so that it is very difficult to find this regime. The above authors studied the frequency dependence of $\cot \delta$ (cf. Fig. 6.10). As can be seen from this figure, for low frequencies, $\cot \delta$ depends linearly on $\ln \omega$, which is in agreement with (6.7.2), i.e. multiple hopping is realized in this frequency range. The slope of this dependence is equal to 0.64, which is close to $2/\pi$. Hence, the experimental results indicate that the number p in (6.7.2) is equal to unity. Such a value of p results from the effective-medium approximation (and the diagram technique). At higher frequencies, the plot $\cot \delta$ versus $\ln \omega$ passes into a plateau, which indicates the transitions to the regime of hopping across medium-sized clusters. In the whole frequency range studied, the magnitude of $\cot \delta$ is larger than unity. In the range of very low frequencies (not shown in Fig. 6.10), a tendency to an increase of $\cot \delta$ with decreasing ω was observed, which indicates the transition to the hydrodynamic regime.

Concluding this section, we note that for sufficiently high frequencies, the phononless mechanism might dominate, so that we get $\operatorname{Re} \sigma(\omega) \sim \omega^2 \ln^4 \omega$, which corresponds to $s' \leqq 2$. For large ω, such a dependence is observed sometimes, for instance in certain polymers.

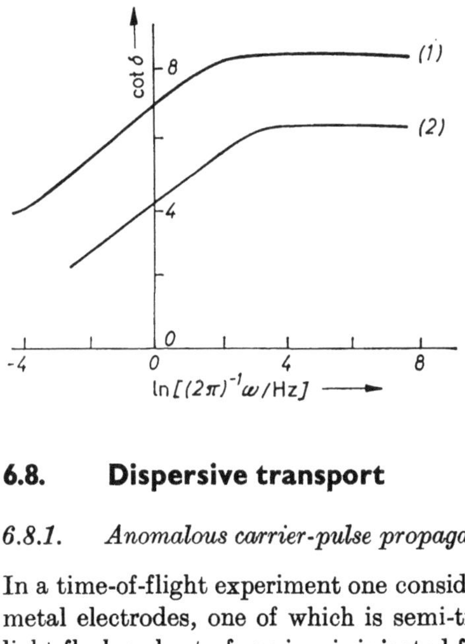

Fig. 6.10. Frequency dependence of $\cot \delta$ in amorphous Ta_2O_5 before (1) and after annealing (2). The annealing reduces the concentration of localization centres

6.8. Dispersive transport

6.8.1. *Anomalous carrier-pulse propagation in amorphous semiconductors*

In a time-of-flight experiment one considers a sample sandwiched between two plane metal electrodes, one of which is semi-transparent. By a short, strongly absorbed, light flash a sheet of carriers is injected from the semi-transparent electrode and one monitors the external current induced by the propagation of the excess carrier pulse through the sample. A widespread characteristic of such experiments is an appreciable dispersion of the transit carrier pulse. Such dispersive (non-Gaussian) behaviour has been observed in a broad class of low-mobility materials such as amorphous inorganic semiconductors (e.g. a-As_2Se_3), polymers and organic semiconductors (for a review, see PFISTER and SCHER, 1978).

Consider first a transit carrier pulse in a crystal where free carriers interact with a single discrete set of shallow traps or achieve conventional thermal diffusion (cf. Fig. 6.11). If the carriers reach dynamic equilibrium with their environement in a time short compared with the inter-electrode transit time, then in such a crystal the transit carrier pulse will be described by Gaussian statistics, i.e. with increasing time t the pulse becomes Gaussian shaped, while its width Δ becomes negligible compared with the mean distance l traversed. This Gaussian behaviour is due to the fact that in the crystal in question the dispersions of the distributions of the event times (trap release times) and the lengths of the individual displacement events of the carriers are small, so that a narrow carrier pulse remains narrow as it traverses the crystal.

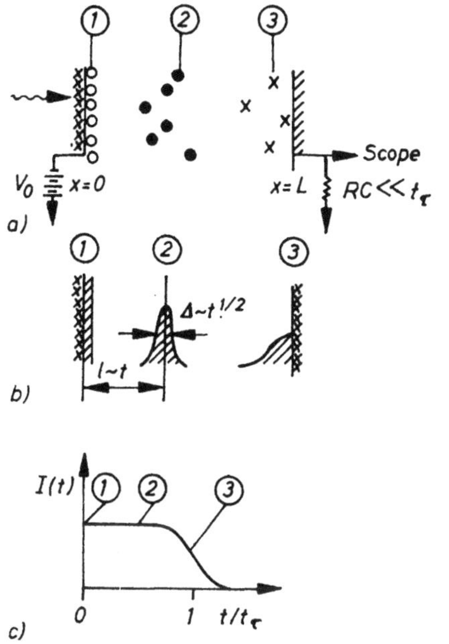

Fig. 6.11. Schematic representation of Gaussian carrier-pulse propagation. a) Position of representative carriers in the sample at $t = 0$ (o), $t < t_\tau$ (•), and $t \sim t_\tau$ (×) (t_τ — transit time); b) Charge distribution in the sample at $t = 0$, $t < t_\tau$, and $t \sim t_\tau$; c) Current $I(t)$ in the external circuit (after Pfister and Scher, 1978) (see text)

In the Gaussian case we have $l \sim t$ (i.e. the mean drift velocity is constant), $\Delta \sim t^{1/2}$ (diffusive broadening), and the current $I(t)$ is approximately constant up to the transit time t_τ, when the leading edge of the pulse reaches the absorbing electrode. After that the current drops to zero. The rounding of the "transit edge" of the current, characterized by $\Delta/l \sim t_\tau^{-1/2}$, reflects the spread of the transit times of the carriers of the pulse.

Let us now turn to the anomalously dispersive non-Gaussian behaviour mentioned above (cf. Fig. 6.12). Under non-Gaussian conditions, the current decreases in time even before the carriers encounter the absorbing electrode. An initial sheet of charge carriers rapidly disperses, instead of travelling as a well defined carrier pulse. The distribution of the carrier packet grows asymetrically. Its leading edge penetrates deep into the bulk, while its maximum moves only slowly out of the region of the initial carrier injection. This is due to the fact that the carriers are immobilized in localized states which have a broad distribution of release times, some of which are longer than the time of observation. Dispersive transport may be viewed as the transient approach to steady state. For observation times much longer than the longest release time from localized states, the carriers would move under the influence of an electric field at a constant drift velocity and spread by diffusion.

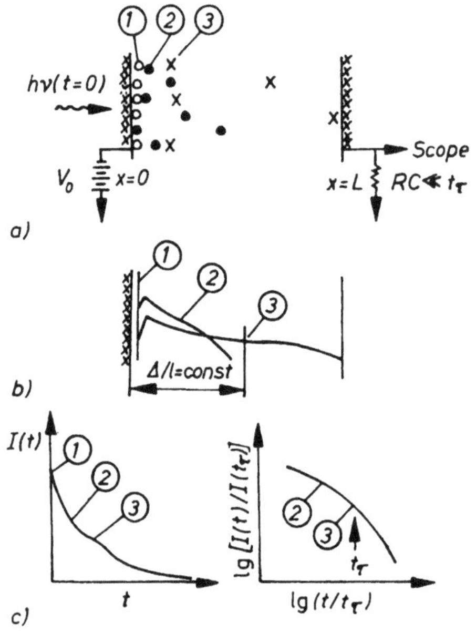

Fig. 6.12. Schematic representation of non-Gaussian carrier-pulse propagation. a) Position of representative carriers in the sample at $t = 0$ (○), $t < t_\tau$ (●), and $t \sim t_\tau$ (×) (t_τ — transit time); b) Charge distribution in the sample at $t = 0$, $t < t_\tau$, and $t \sim t_\tau$; c) Current $I(t)$ in the external circuit (after PFISTER and SCHER, 1978) (see text)

For the non-Gaussian case, the time dependence of the transit current can be characterized by

$$I(t) \sim \begin{cases} t^{-(1-\alpha)} & \text{for } t < t_\tau, \\ t^{-(1+\alpha)} & \text{for } t > t_\tau, \end{cases} \tag{6.8.1}$$

where t_τ is the transit time of the fastest carriers and $0 < \alpha < 1$.[1]) This implies $l \sim t^\alpha$ $\left(I(t) \sim \mathrm{d}l/\mathrm{d}t\right)$ for $t < t_\tau$, i.e. the injected carries appear to slow down continuously. For dispersive carrier propagation, the spread Δ has the same time dependence as the mean position l, i.e. $\Delta/l = \text{const.}$, which is in striking contrast with the Gaussian relationship $\Delta/l \sim t^{-1/2}$. The behaviour (6.8.1), which is embodied in $\Delta/l = \text{const.}$, has been termed the "universality" of the current shape: the transit current $I(t)/I(t_\tau)$ is independent of the transit time when plotted against t/t_τ (cf. Fig. 6.13), unlike the case of Gaussian transport.

---

[1]) Note that in a variety of experimental studies on disordered systems, instead of (6.8.1), one finds that

$$I(t) \sim \begin{cases} t^{-(1-\alpha_1)} & \text{for } t < t_\tau \\ t^{-(1+\alpha_2)} & \text{for } t > t_\tau \end{cases}$$

where $\alpha_1 \neq \alpha_2$ (cf. MARSHALL et al., 1983, and references therein).

To illustrate dispersive behaviour, let us consider multiple-trapping transport. A theoretical study of this transport mechanism (see, e.g., NOOLANDI, 1977, SCHMIDLIN, 1977, SCHIRMACHER, 1981, RUDENKO and ARKHIPOV, 1982) is based on the conception of two fractions of carriers: (i) mobile carriers in (extended) transport states and (ii) immobile carriers in localized states. Between the two fractions of carriers transitions occur. Mobile carriers can become immobile by trapping on localized states. On the other hand, immobile carriers can be released from traps and consequently become mobile. Extended and localized states are separated by some energy ε_0 (mobility edge). The current is carried in the extended states less than about kT above ε_0. The immobile fraction of carriers affects the transport via the parameters of the localized states.

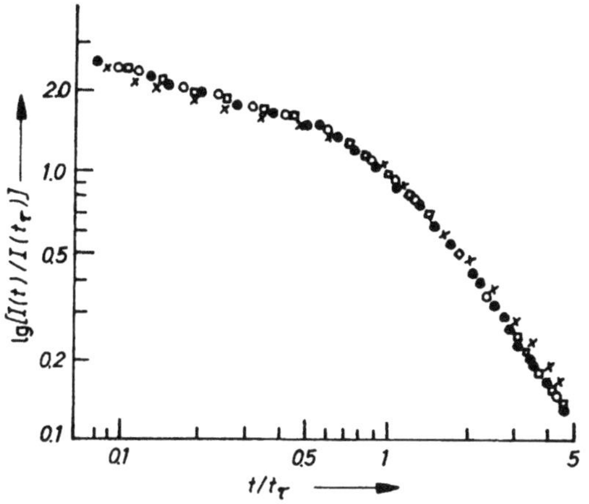

Fig. 6.13. Normalized logarithmic display of the transit current $I(t)$ for amorphous selenium at 143 K. The field E (in V/μm) applied and the transit time t_τ (in ms), respectively, are ×: 6.6, 40; •: 10,13; ○: 13.4, 6; □: 20, 1.7 (after PFISTER, 1976)

Neglecting diffusion and assuming planar geometry, the multiple-trapping transport can be described by the following set of equations:

$$\partial\varrho(x, t)/\partial t + u_c E\, \partial\varrho_c(x, t)/\partial x = 0, \tag{6.8.2}$$

$$\partial\varrho_t(x, t, \varepsilon)/\partial t = \nu_0(\varepsilon)\, N(\varepsilon)\, \varrho_c(x, t) - \nu_0(\varepsilon)\, N_c \exp\left(-(\varepsilon_0 - \varepsilon)/kT\right) \varrho_t(x, t, \varepsilon), \tag{6.8.3}$$

$$\varrho(x, t) = \varrho_c(x, t) + \varrho_t(x, t), \tag{6.8.4}$$

$$\varrho_t(x, t) = \int_{-\infty}^{\varepsilon_0} d\varepsilon \varrho(x, t, \varepsilon). \tag{6.8.5}$$

Here $\varrho(x, t)$ is the total density of carriers at the position x and the time t, $\varrho_c(x, t)$ the density of carriers in the conduction states, $\varrho_t(x, t, \varepsilon)$ the density of carriers trapped on localized states with energy ε, $\varrho_t(x, t)$ the total density of trapped carriers, $N(\varepsilon)$ the density of localized states, $\nu_0(\varepsilon)$ the capture rate, N_c the density of conduction states, u_c the mobility of carriers in conduction states and E is the electrical field applied. Equation (6.8.2) represents the drift equation (continuity equation) for the mobile carriers, and (6.8.3) the equation governing the transitions between conduction and localized states.

The transit current observed in time-of-flight experiments may be expressed as follows

$$I(t) = (eu_cE/L) \int_0^L dx\, \varrho_c(x, t), \tag{6.8.6}$$

where e is the carrier charge and L the distance between the electrodes. Multiplying (6.8.2) by x and integrating the resulting expression from $x = 0$ to $x = L$, then the expression (6.8.6) may be rewritten as

$$I(t) = (e/L)\,(d/dt) \int_0^L dx x\varrho(x, t) + eu_cE\varrho_c(L, t). \tag{6.8.7}$$

The value $\varrho_c(L, t)$ may be found by integrating (6.8.2) from $x = 0$ to $x = L$. Then, (6.8.7) becomes

$$I(t) = (e/L)\,(d/dt) \int_0^L dx x\varrho(x, t) - e(d/dt) \int_0^L dx\varrho(x, t). \tag{6.8.8}$$

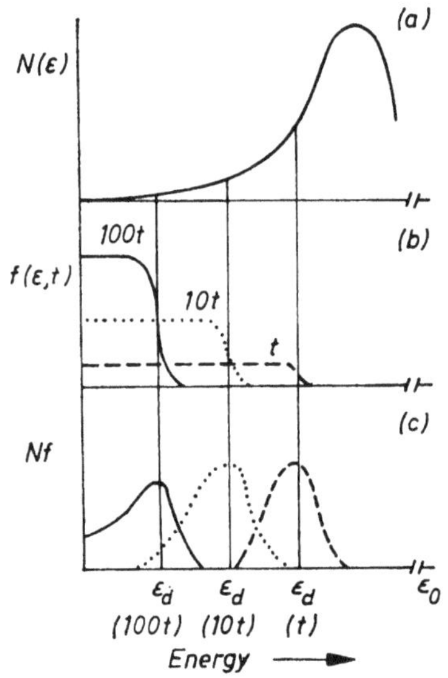

Fig. 6.14. Schematic representation of the density of states $N(\varepsilon)$ of trapping levels and the time evolution of the mean occupation number of a trap with energy ε, $f(\varepsilon, t)$, for a simple model for dispersive multiple trapping transport (see text) (after ORENSTEIN et al., 1982)

Following ORENSTEIN et al. (1982), we now present a simple theory of dispersive multiple-trapping transport. Consider a system with trapping levels distributed within a wide energy range, whereby the density of states of these levels, $N(\varepsilon)$, has either a maximum, as shown in Fig. 6.14(a), or a monotonously decreasing shape below the energy ε_0. The former shape of $N(\varepsilon)$ might be due to the broading of a discrete level, the latter one might be the tail of an energy band in a disordered system.

The carriers (electrons) injected into the system are assumed initially to occupy states above ε_0 (conduction states). They will rapidly be captured by the trapping levels.[1]) If the capture cross section for all these levels is the same, the traps will be occupied uniformly. However, steady state has not yet been attained at the trapping time. The carriers in shallow traps will be released preferentially, while the capture process will remain independent of the energy of a trap. Consequently, in this non-equilibrium transport regime there are two kinds of states: (i) states which are in dynamic equilibrium with the conduction states and (ii) states which are too deep to have come into equilibrium. This fact is illustrated in Fig. 6.14(b), which shows the time evolution of $f(\varepsilon, t)$, the mean occupation number of a trap with energy ε. We see that the energy ε_d separating these two kinds of states depends on the time-delay t after pulsed excitation.

The time dependence of ε_d may be found from (cf. Section 11.4.1.)

$$\nu(\varepsilon_d)\, t = 1\,, \tag{6.8.9}$$

where (cf. (6.8.3))

$$\nu(E) = \nu_0 \exp\left[(\varepsilon - \varepsilon_0)/kT\right] \tag{6.8.10}$$

is the release rate from states of energy ε. Owing to our assumption of a constant capture cross-section, ν_0 is independent of ε (cf. (6.8.3)). Thus, by (6.8.9) and (6.8.10), we find

$$\varepsilon_0 - \varepsilon_d(t) = kT \ln \nu_0 t\,. \tag{6.8.11}$$

Note that a rigorous mathematical treatment of the multiple-trapping problem under consideration (Orenstein et al., 1982) shows that $f(\varepsilon, t)$ is independent of ε for $\varepsilon \ll \varepsilon_d$ and that $f(\varepsilon, t)$ is a Boltzmann distribution in equilibrium with the free carriers for $\varepsilon \gg \varepsilon_d$. Note further, that the trapped carriers form a packet localized in energy about ε_d (Fig. 6.14(c)).

Let us now turn to the time dependence of the transit current $I(t)$. According to (6.8.6), $I(t)$ is proportional to the concentration $\varrho_c(t)$ of excess carriers at $\varepsilon \geqq \varepsilon_0$. By means of (6.8.3), we find that

$$\varrho_c(t) \approx \varrho_0 \frac{N(\varepsilon_0)}{N(\varepsilon_d)}\, \gamma \exp\left[-\left(\varepsilon_0 - \varepsilon_d(t)\right)/kT\right], \tag{6.8.12}$$

where ϱ_0 is the total injected carrier density and γ denotes the fraction of the trapped carriers which are in thermal equilibrium with the carriers at ε_0. The fraction γ is only weakly dependent on time (cf. Fig. 6.14(c)) and temperature.

---

[1]) Let us draw attention to a pecularity of the trapping problem. Intuitively, one could expect that the number of carriers not yet trapped would decay exponentially at long times. However, one finds (Movaghar et al., 1982, Grassberger and Procarccia, 1982) that at long times it decays as $\sim \exp(-\text{const. } t^{d/d+2})$, where d is the dimensionality of the system. This is due to the fact that at long times the surviving carriers are in large trap-free regions (see also Section 11.4.1.).

Inserting (6.8.11) into (6.8.12), we obtain

$$\varrho_c(t) \approx \varrho_0 \frac{N(\varepsilon_0)}{N(\varepsilon_0 - kT \ln \nu_0 t)} \gamma(\nu_0 t)^{-1}. \tag{6.8.13}$$

If we assume that the density of states decreases exponentially away from ε_0, i.e.

$$N(\varepsilon) \sim \exp\left[-(\varepsilon_0 - \varepsilon)/kT_0\right] \tag{6.8.14}$$

then, by (6.8.13), we find that

$$I(t) \sim (\nu_0 t)^{-(1-\alpha)}, \tag{6.8.15}$$

where $\alpha = T/T_0$. This is the dependence (6.8.1) we have looked for.

Thus, if the energies of the trapping levels are spread over many kT, i.e. $T_0 > T$, there will be a large spread of release times, and one will be able to measure the current before quasi-equilibrium is attained.

The multiple-hopping behaviour is not the only possible source of dispersive behaviour. A broad distribution of event times can also arise from hopping, since a random distribution of localization centres gives rise to an enormous spread of hopping times. In the case of hopping, fluctuations in site energies will tend to increase the dispersion in event times and give rise to a temperature dependence of the exponent α in (6.8.1), similarly as in the multiple-trapping case. Another source of dispersive transport may be the so-called trap-controlled hopping, where the carriers are interacting with hopping sites corresponding to distinct sets of energy levels.

6.8.2. *Relationship between* ac *hopping conductivity and dispersive transport*

If we are interested in times shorter than the characteristic time t_τ at which in a time-of-flight experiment the first carriers reach the absorbing boundary, then the transit current $I(t)$ may be readily expressed by the ac conductivity as

$$I(t) = \frac{E}{2\pi \mathrm{i}} \int_{c-\mathrm{i}\infty}^{c+\mathrm{i}\infty} \mathrm{d}s\, \mathrm{e}^{st} \frac{\sigma(s)}{s}, \tag{6.8.16}$$

where, after the replacement $s \to \mathrm{i}\omega$, $\sigma(s)$ represents the ac conductivity, and E is the electrical field.

To obtain (6.8.16), note that in the linear response theory the current $I(t)$ is related to $E(t)$ as

$$I(t) = \int_{-\infty}^{t} \mathrm{d}t' \sigma(t - t')\, E(t'). \tag{6.8.17}$$

Taking the Laplace transform

$$I(s) = \int_0^{\infty} \mathrm{d}t\, \mathrm{e}^{-st} I(t), \tag{6.8.18}$$

and accounting for $E(t) = E\theta(t)$ ($\theta(t)$ — Heaviside step function), which holds for a time-of-flight experiment, then (6.8.17) becomes

$$I(s) = \frac{\sigma(s)}{s} E. \tag{6.8.19}$$

Finally, taking the inverse Laplace transform of (6.8.19), we obtain (6.8.16).

According to (6.8.16), for $t < t_{\tau}$, $I(t)$ is directly related to $\sigma(\omega)$. In particular, for hopping transport there exist four time regimes (cf. Section 6.7.1.): hopping across infinite, large, medium-sized, and two-site clusters. The characteristic times separating the different regimes from each other may be obtained from the corresponding characteristic frequencies presented above by replacing $\omega \to 1/t$. $I(t)$ was investigated in detail in the multiple-hopping regime on the basis of the cluster approach (ZVYAGIN, 1979), and in a large time interval on the basis of the effective-medium method (BRYKSIN and YASHIN, 1981). According to the latter paper, in the multiple-hopping regime we have $I(t) \sim t^{-1} \ln^{-2} (tW_c)$ (W_c — critical hopping probability), which is close to the dependence $t^{-(1-\alpha)}$, which results from the approximation $\sigma \sim \omega^s$, $0 < s < 1$, where $\alpha = 1 - s$ (BUTCHER and CLARK, 1980, MOVAGHAR et al., 1980, 1983).

With respect to the latter fact, note that for ac conductivity of the form (cf. Section 6.7.)

$$\sigma(\omega) = \sigma_0 + \sigma_1 \left(\frac{i\omega}{W_c}\right)^s, \qquad 0 < s < 1, \tag{6.8.20}$$

the transit current (6.8.16) becomes

$$I(t) = \sigma_0 E \left(1 + \frac{\sigma_1}{\sigma_0} \frac{1}{\Gamma(\alpha)} (W_c t)^{-(1-\alpha)}\right), \tag{6.8.21}$$

where $\alpha = 1 - s$, and $\Gamma(\alpha)$ denotes the Γ-function. Accordingly, for times less than $W_c^{-1}(\sigma_1/\Gamma(\alpha)\,\sigma_0)^{1/(1-\alpha)}$, we find a $t^{-(1-\alpha)}$ behaviour of $I(t)$. For later times, the first term of (6.8.21), which is determined by the dc conductivity, dominates and the current becomes constant. According to POLLAK (1977) and BRYKSIN and YASHIN (1981), at sufficiently large times, $I(t)$ passes into this plateau as $t^{-3/2} \exp(-tW_c)$.

It should be noticed that such a smooth transition to a plateau must be absent in one-dimensional disordered systems, which lack a region of hopping across an infinite cluster. In such hopping systems the dependence of the type $t^{-(1-\alpha)}$ must range to any large time (but, of course, $t < t_{\tau}$).

Today, there is a consensus that in multiple-trapping transport a broad distribution of trap energies can give rise to dispersive behaviour. However, there is still some controversy revolved around whether or not hopping, in particular isoenergetic hopping (R-hopping), would produce dispersive behaviour, too. Using the mathematical formalism of continuous-time random walk (CTRW) on a lattice (see Section 6.8.3) SCHER and MONTROLL (1975) predicted anomalously dispersive transport for R-hopping. On the other hand, using a percolation approach, POLLAK (1977) pointed out that substantial transit-pulse dispersion should not occur for hopping, except at times much shorter than those appropriate to an experimental study. According to

Pollak, dispersive behaviour of the hopping current is limited to a time scale corresponding to hops to a few sites only. This is due to the fact that, unlike trapping, in hopping the difficulty to enter a site is strongly correlated to the difficulty to leave it, so that "slow states" can never capture enough carriers to obscure the transit of particles via "fast states" (cf. SCHMIDLIN, 1980).

Several Monte Carlo studies have attempted to answer the question whether hopping can give rise to dispersive behaviour which can be experimentally observed. For hopping between isoenergetic sites in a random array, MARSHALL (1978, 1981, 1983) did not obtain a transit current $I(t)$ that exhibits specific dispersive character. This may be due to the choice of the parameters ($\alpha \mathcal{N}^{-1/3}$, size of the specimen, time scale, etc.) involved, so that the calculations already relate to the regime of hopping across the infinite cluster (MOVAGHAR et al., 1980). On the other hand, SILVER et al. (1979) and ADLER and SILVER (1982) found that hopping in a spatially disordered system can lead to anomalous dispersion. Furthermore, for a Gaussian energy distribution of hopping sites located on a cubic lattice, SCHÖNHERR et al. (1981) found a reasonable amount of dispersion, which was primarily due to the energy relaxation of the hopping carriers. According to SILVER et al. (1982), an exponential energy distribution of hopping sites should even give rise to dispersive transport which is not experimentally distinguishable from multiple trapping.

As noted above, in the case of multiple trapping, the probability to enter a trap is in general much larger than the probability to leave it, which may essentially broaden the regime of times where dispersive transport can be observed (POLLAK, 1977). Similarly, asymmetrical transition probabilities also occur for hopping in strong electrical fields, which may likewise lead to a broadening of the latter regime (ZVYAGIN, 1979, BRYKSIN and YASHIN, 1981).

As mentioned above, anomalously dispersive behaviour of the transit current (cf. (6.8.1)) is intimately related to the non-Gaussian character of the diffusive spread of the initial charge carrier packet. Diffusion is suitably described by means of the conditional-probability function $P_{m'm}$, which is governed by equation (6.4.2). In the case of disordered systems, the basic problem is the averaging of this equation. As pointed out in Section 6.5., for R-hopping, it is convenient to introduce, instead of $P_{m'm}$, the conditional probability function $P(\boldsymbol{r}', \boldsymbol{r})$ (6.5.1), which depends on the continuous variables $\boldsymbol{r}'$ and $\boldsymbol{r}$ and which is governed by equation (6.5.2).

The quantity $P(\boldsymbol{r}) \equiv \langle P(\boldsymbol{r}' + \boldsymbol{r}, \boldsymbol{r}') \rangle$ (where $\langle \dots \rangle$ indicates the configuration average) represents the concentration of charge carriers at point $\boldsymbol{r}$, if these where at the initial time at point $\boldsymbol{r} = 0$. The finding of the equation for $P(\boldsymbol{r})$ from (6.5.2) is the major problem of the theory of diffusion processes and electrical conduction in disordered systems. So far, this problem is not completely solved, though for the special case of a lattice model it could be successfully treated by the projection operator technique of Zwanzig (KLAFTER and SILBEY, 1980b, MUKAMEL, 1983) (see below), self-consistent effective-medium approximations (WEBMANN, 1981, WEBMAN and KLAFTER, 1982, ODAGAKI and LAX, 1981b, HAUS et al., 1982, 1983), a real-space renormalization group method (MACHTA, 1981, 1983), and, without assuming a lattice model, by diagram techniques (GOCHANOUR et al., 1979; Section 6.5. of the present book; see also Section 11.4.1.).

Intuitively, it seems reasonable to expect that $P(\boldsymbol{r})$ should satisfy a diffusion equation whose Fourier transform with respect to $\boldsymbol{r}$ has the form (cf. (6.5.29))

$$sP(\boldsymbol{q}, s) = -q^2 D(s)\, P(\boldsymbol{q}, s) + 1, \tag{6.8.22}$$

where s is the Laplace parameter and D the diffusion coefficient which is related to σ by the Einstein relation $\sigma = e^2 D\, \mathrm{d}n/\mathrm{d}\mu$ (n — electron density, μ — chemical potential). The existence of such an equation for the problem quoted was postulated by BUTCHER (1978) and BUTCHER and CLARK (1980), and a derivation of it, without using a regular lattice, was put forward by BUTCHER and CLARK (1981) and in Section 6.5.

In the case of disordered systems, a characteristic feature of (6.8.22) consists in the strong frequency dependence of $D(\omega)$ down to low frequencies. This means that in time representation, (6.8.22) becomes an integro-differential equation describing random walks with memory. In the CTRW model (see Section 6.8.3.) this memory arises from random waiting times between consecutive hops. If $D(s)|_{s=0} \to D_0 \neq 0$ (which applies to hopping in three-dimensional systems) for sufficiently large times, according to (6.8.22), the random-walk process takes on Markovian character, and the spread of the charge-carrier packet becomes Gaussian-like. However, for not too large times, the process under consideration has non-Markovian character (BUTCHER and CLARK, 1980).

6.8.3. *Continuous-time random-walk theory*

The continuous-time random-walk (CTRW) conception was first suggested by MONTROLL and WEISS (1965). To discuss a random walk with random jump times on a regular lattice, Montroll and Weiss introduced a distribution function $\psi(t)$ which describes the probability for a jump event to happen at time t after the preceding such event. The CTRW approach was elaborated later by SCHER and LAX (1973), who realized that this approach is also appropriate for describing transport in disordered systems. To handle the latter problem, Scher and Lax mapped the irregular system onto a regular lattice, which greatly simplifies the mathematical analysis of the problem, but which implies the need to get $\psi(t)$ by a suitable averaging procedure. In this way the ac hopping conductivity can be explained. The dc hopping conductivity becomes inversely proportional to the mean waiting time.

The formalism of CTRW was first applied to the transit current in a time-of-flight experiment by SCHER and MONTROLL (1975), who were able to show that an algebraic time dependence of $\psi(t)$ of the form (cf. BLUMEN et al., 1983)

$$\psi(t) \sim t^{-(1+\alpha)}, \qquad 0 < \alpha < 1, \tag{6.8.23}$$

leads to the dispersive behaviour (6.8.1) of the transit current. Note that such a slowly decaying distribution function $\psi(t)$ sharply contrasts with the exponential time dependence

$$\psi(t) \sim \exp\,(-t/\tau), \tag{6.8.24}$$

which describes random walk processes characterized by a single event time τ, and which is sufficient (but not necessary) for obtaining non-dispersive (Gaussian) transport. In the case (6.8.23), the basic transport equation is non-Markovian, while in the case (6.8.24) it is Markovian.

The fact that for a disordered system, the CTRW approach in general leads to a non-Markovian transport equation becomes clear if one realizes that the basic idea of this approach is to observe a carrier move along some path, whereby the fluctuations of the event times encountered on the path are "folded back" to the probability function $\psi(t)$, which, of course, depends on the time history of the path.

Scher and Montroll originally constructed their theory to describe hopping transport. Later they proposed that the theory also holds for trap-controlled transport. In the case of hopping, $\psi(t)$ describes the distribution of hopping times, in the case of trapping, the distribution of trap-release times (cf. PFISTER and SCHER, 1978, and references therein.).

The CTRW model is an attractive approach to the transport problem in disordered systems. It is mathematically tractable, since it contains all information on the disorder in a single function $\psi(t)$, which requires calculation only of a single parameter α, provided that (6.8.23) is valid. Furthermore, it very well accounts for experimental transport data obtained in a broad class of materials.

In the literature, there has been much discussion on the points to what degree the CTRW model is exact and what transport mechanisms can be adequately described by this model. As for the former point, KLAFTER and SILBEY (1980b) proved the validity of the CTRW approach by performing the configuration average of the basic transport equation (see also HALPERN, 1981). The latter point is still some matter of controversy. According to ALEXANDER (1981), the CTRW method is not reliable in the limit of long times or low frequencies for random-hopping (R-hopping) systems, while it may be reasonable for random-trapping ones (see also KIEVELSON, 1980).

Let us now present the mathematical formalism of CTRW on a discrete lattice. The basic quantity in a CTRW is the conditional-probability function $P_{\bar{m}m}(t)$ representing the probability for a walker (carrier) to occupy the site m at time t if he started from site $\bar{m}$ at $t = 0$. We assume that $P_{\bar{m}m}(t)$ is determined by the following rate equation (master equation) (cf. (2.6.1))

$$\frac{\mathrm{d}P_{\bar{m}m}(t)}{\mathrm{d}t} = \sum_{m''} [P_{\bar{m}m''}W_{m''m} - P_{\bar{m}m}W_{mm''}], \tag{6.8.25}$$

which describes a wide class of dynamical processes in random systems. In particular, for nearest-neighbour transfer rates

$$W_{mm'} = W_{m'm} \tag{6.8.26}$$

and a suitable distribution of these rates, (6.8.25) may be taken for a random hopping model. Furthermore, for

$$W_{mm'}/W_{m'm} = \exp\left[-(\varepsilon_m - \varepsilon_{m'})/kT\right], \tag{6.8.27}$$

with random variables (trap depths) ε_m, (6.8.25) represents a model for random trapping, where transitions occur by thermal excitation of a carrier from a trap m to a conduction state and its subsequent capture in another trap m', so that the transition rates mainly depend on the energy of the initial trap m (see, e.g., ALEXANDER et al., 1981).

In what follows, for brevity, we will suppress the label $\overline{m}$ of the initial site, i.e. $P(m, t) \equiv P_{\overline{m}m}(t)$. To generate step-by-step the probability to arrive at a given site, we follow SCHER and LAX (1973) by defining $R_n(m, t)\,\mathrm{d}t$ as the probability for a walker to arrive at site m within the time interval $(t, t + \mathrm{d}t)$ in n steps, given that he was at site $\overline{m}$ at time $t = 0$ when the observation was started. It holds that

$$P(m, t) = \int_0^t \mathrm{d}t' R(m, t')\, \Phi_m(t - t'), \tag{6.8.28}$$

where

$$R(m, t) = \sum_{n=0}^{\infty} R_n(m, t) \tag{6.8.29}$$

is the probability per unit time to reach m in time t, independent of the number o steps to get to m, and $\Phi_m(t)$ is the probability that the walker remains fixed on site m in the time interval $(0, t)$.

The probability of reaching the site m in $n + 1$ steps, $R_{n+1}(m, t)$, is related to $R_n(m', t)$, the probability of reaching site m' in n steps, by

$$R_{n+1}(m, t) = \sum_{m'} \int_0^t \mathrm{d}t' \psi(m, m', t - t')\, R_n(m', t'), \tag{6.8.30}$$

where $\psi(m, m', t)\,\mathrm{d}t$ is the probability that the transition from m' to m occurs within the time interval $(t, t + \mathrm{d}t)$[1]. Accounting for the initial condition

$$R_0(m, t) = \delta_{m,\overline{m}}\delta(t - 0^+), \tag{6.8.31}$$

and summing (6.2.30) over n, one obtains

$$R(m, t) - \sum_{m'} \int_0^t \mathrm{d}t' \psi(m, m', t - t')\, R(m', t') = \delta_{m,\overline{m}}\delta(t - 0^+), \tag{6.8.32}$$

where $\boldsymbol{m}$ denotes the position vector of the mth site

$$\boldsymbol{m} = m_1\boldsymbol{a}_1 + m_2\boldsymbol{a}_2 + m_3\boldsymbol{a}_3 \tag{6.8.33}$$

with the components $m_i (i = 1, 2, 3)$ equal to an integer and the unit primitive translation vectors $\boldsymbol{a}_i$ of the lattice.

The probability $\Phi_m(t)$ for a walker to stay on site m in the disordered system

---

[1]) TUNALEY (1974) casts doubt on the CTRW model of Scher and Lax by suggesting that the waiting-time distribution for the first observed jump is generally different from that for all later jumps. However, later on KUMAR and HEINRICHS (1980) pointed out that (for the averaged system) the waiting time distribution for the first jump coincides with that between consecutive jumps.

under consideration obeys the equation

$$\frac{\mathrm{d}\Phi_m}{\mathrm{d}t} = -\sum_{m'} W_{mm'}\Phi_m(t), \tag{6.8.34}$$

from which we obtain

$$\Phi_m(t) = \mathrm{e}^{-\sum_{m'} W_{mm'}t}. \tag{6.8.35}$$

The total probability $\psi_m(t)\,\mathrm{d}t$ that a walker will leave site m in the time interval $(t, t + \mathrm{d}t)$ is given by

$$\psi_m(t) = \sum_{m'} \psi(m', m, t) \tag{6.8.36}$$

and is related to $\Phi_m(t)$ as

$$\Phi_m(t) = 1 - \int_0^t \mathrm{d}t' \psi_m(t') \tag{6.8.37}$$

or

$$\psi_m(t) = -\frac{\mathrm{d}\Phi_m(t)}{\mathrm{d}t}. \tag{6.8.38}$$

Thus, by (6.8.38), (6.8.34) and (6.8.36), we get

$$\psi(m, m', t) = W_{m'm}\, \mathrm{e}^{-\sum_{m''} W_{m'm''}t}. \tag{6.8.39}$$

In the CTRW method the quantities $\psi(m, m', t)$ and $\psi_m(t)$ are replaced by their configuration averages: $\psi(m, m', t) \to \langle\psi(m, m', t)\rangle \equiv \psi(m - m', t)$ and $\psi_m(t) \to \langle\psi_m(t)\rangle \equiv \psi(t)$, so that

$$\psi(t) = -\frac{\mathrm{d}}{\mathrm{d}t}\left\langle \mathrm{e}^{-\sum_{m'} W_{mm'}t} \right\rangle = \sum_m \psi(m, t). \tag{6.8.40}$$

Here $\langle\ldots\rangle$ indicates the average over the possible configurations of the (compositionally) disordered system under consideration. Note that SCHER and LAX (1973) considered the more complicated problem of mapping the configuration averaged solution for a n-step random walk on a positionally disordered system onto a n-step CTRW on a discrete lattice. In what follows we use the average quantities $\psi(m - m', t)$ and $\psi(t)$.

Thus, equation (6.8.32) becomes

$$R(m, t) - \sum_{m'} \int_0^t \mathrm{d}t' \psi(m - m', t - t')\, R(m', t) = \delta_{m,\overline{m}}\, \delta(t - 0^+). \tag{6.8.41}$$

Taking now the Laplace transform of (6.8.41), then we obtain

$$R(m, s) - \sum_{m'} \psi(m - m', s)\, R(m', s) = \delta_{m,\overline{m}}, \tag{6.8.42}$$

where

$$\psi(m, s) = \int_0^\infty \mathrm{d}t\, \mathrm{e}^{-st}\psi(m, t). \tag{6.8.43}$$

Equation (6.8.42) is solved with the use of the Fourier transform (this is the reason for mapping the random system onto a regular lattice)

$$U(\boldsymbol{q}, s) = \sum_{\boldsymbol{m}} R(\boldsymbol{m}, s)\, \mathrm{e}^{-\mathrm{i}\boldsymbol{q}\boldsymbol{m}}, \tag{6.8.44}$$

where $q_i = 2\pi m_i/a_i N^{1/3}$, m_i is an integer, $a_i = |\boldsymbol{a}_i|$, N labels the number of lattice points in the system, and periodic boundary conditions are assumed. We get

$$R(\boldsymbol{m}, s) = (1/N) \sum_{\boldsymbol{q}} \frac{\exp(\mathrm{i}\boldsymbol{q}\boldsymbol{m})}{1 - \Lambda(\boldsymbol{q}, s)}, \tag{6.8.45}$$

where

$$\Lambda(\boldsymbol{q}, s) = \sum_{\boldsymbol{m}} \psi(\boldsymbol{m}, s) \exp(-\mathrm{i}\boldsymbol{q}\boldsymbol{m}), \tag{6.8.46}$$

which is called the generalized-structure function of the CTRW.

Taking now the Laplace transform of (6.8.28), then by (6.8.37) we obtain

$$P(\boldsymbol{m}, s) = R(\boldsymbol{m}, s)\, [1 - \psi(s)]/s, \tag{6.8.47}$$

where we have replaced $\psi_{\boldsymbol{m}}$ by its configuration average ψ, and where $R(\boldsymbol{m}, s)$ is given by (6.8.45). By (6.8.47), we find from (6.8.42) the basic equation of the CTRW method

$$P(\boldsymbol{m}, s) - \sum_{\boldsymbol{m}'} \psi(\boldsymbol{m} - \boldsymbol{m}', s)\, P(\boldsymbol{m}', s) = \{[1 - \psi(s)]/s\}\, \delta_{\boldsymbol{m},\overline{\boldsymbol{m}}}. \tag{6.8.48}$$

If we assume

$$\psi(\boldsymbol{m} - \boldsymbol{m}', t) = \left\langle \frac{W_{\boldsymbol{m}'\boldsymbol{m}}}{\sum\limits_{\boldsymbol{m}''} W_{\boldsymbol{m}'\boldsymbol{m}''}} \psi_{\boldsymbol{m}'}(t) \right\rangle$$

$$\approx \left\langle \frac{W_{\boldsymbol{m}'\boldsymbol{m}}}{\sum\limits_{\boldsymbol{m}''} W_{\boldsymbol{m}'\boldsymbol{m}''}} \right\rangle \langle \psi_{\boldsymbol{m}'}(t) \rangle \equiv p(\boldsymbol{m} - \boldsymbol{m}')\, \psi(t), \tag{6.8.49}$$

then from (6.8.48) we get

$$\frac{\mathrm{d}P(\boldsymbol{m}, t)}{\mathrm{d}t} = \int_0^t \mathrm{d}t' \varphi(t - t') \sum_{\boldsymbol{m}'} [p(\boldsymbol{m} - \boldsymbol{m}')\, P(\boldsymbol{m}', t') - p(\boldsymbol{m}' - \boldsymbol{m})\, P(\boldsymbol{m}, t')], \tag{6.8.50}$$

where

$$\varphi(s) = s\psi(s)/[1 - \psi(s)]. \tag{6.8.51}$$

Equation (6.8.48) and (6.8.50) have a non-Markovian character. In the case of an ordered system with the nearest-neighbour transition rate W, we get a Markovian behaviour, since $\psi(t) = zW e^{-zWt}$ and thus $\varphi(s) = zW$, $\varphi(t) = 2\delta(t)\, zW$ (z — coordination number). However, a non-exponential $\psi(t)$, i.e. a spread of the transition rates, yields a non-Markovian behaviour.

Let us now turn to the problem of calculating the ac conductivity $\sigma(\omega)$. We start from equation (2.6.14), which relates $\sigma(\omega)$ to the conditional-probability function.

Note that in a disordered system, at large times (small s), a walker loses his memory of the initial state, i.e. in (2.6.14) the dependence of the mean-square displacement on m' may be neglected. Then, for a cubic system and Boltzmann statistics ($f_m \ll 1$), equation (2.6.14) may be written as

$$\sigma(\omega) = \frac{e^2 n s^2}{6kT} \sum_{\boldsymbol{m}} \boldsymbol{m}^2 P(m, s)|_{s=\mathrm{i}\omega}, \tag{6.8.52}$$

where $n = \sum_m f_m/\Omega$ is the concentration of charge carriers.

By (6.8.47), (6.8.52) becomes

$$\sigma(\omega) = \frac{e^2 n s^2}{6kT} \frac{1 - \psi(s)}{s} \sum_{\boldsymbol{m}} \boldsymbol{m}^2 R(m, s)|_{s=\mathrm{i}\omega}. \tag{6.8.53}$$

We perform the sum in (6.8.53) by accounting for

$$\sum_{\boldsymbol{m}} \boldsymbol{m}^2 R(m, s) = -\frac{\partial^2}{\partial \boldsymbol{q}^2} U(\boldsymbol{q}, s)\Big|_{\boldsymbol{q}=0}, \tag{6.8.54}$$

where U is defined in (6.8.44).

Inserting (6.8.54) into (6.8.44) and accounting for

$$\frac{1}{N} \sum_{\boldsymbol{m}} \mathrm{e}^{\mathrm{i}\boldsymbol{m}(\boldsymbol{q}'-\boldsymbol{q})} = \delta_{\boldsymbol{q},\boldsymbol{q}'}, \tag{6.8.55}$$

then we obtain

$$\begin{aligned} -\frac{\partial^2}{\partial \boldsymbol{q}^2} U(\boldsymbol{q}, s)|_{\boldsymbol{q}=0} &= -\frac{\partial^2}{\partial \boldsymbol{q}^2} [1 - \Lambda(\boldsymbol{q}, s)]^{-1}|_{\boldsymbol{q}=0} \\ &= -\frac{\partial^2}{\partial \boldsymbol{q}^2} \Lambda(\boldsymbol{q}, s)|_{\boldsymbol{q}=0} [1 - \Lambda(0, s)]^{-2} \\ &\quad - \left(\frac{\partial}{\partial \boldsymbol{q}} \Lambda(\boldsymbol{q}, s)\right)^2\Big|_{\boldsymbol{q}=0} [1 - \Lambda(0, s)]^{-3}. \end{aligned} \tag{6.8.56}$$

From (6.8.40) and (6.8.46) we get

$$\Lambda(0, s) = \psi(s), \tag{6.8.57}$$

$$\frac{\partial}{\partial \boldsymbol{q}} \Lambda(\boldsymbol{q}, s)|_{\boldsymbol{q}=0} = \sum_{\boldsymbol{m}} \boldsymbol{m} \psi(m, s), \tag{6.8.58}$$

$$-\frac{\partial^2}{\partial \boldsymbol{q}^2} \Lambda(\boldsymbol{q}, s)|_{\boldsymbol{q}=0} = \sum_{\boldsymbol{m}} \boldsymbol{m}^2 \psi(m, s). \tag{6.8.59}$$

The expression (6.8.58) vanishes, since the system is assumed to lack any intrinsic asymmetry. Thus, inserting (6.8.54), (6.8.56), (6.8.57) and (6.8.59) into (6.8.53), the conductivity becomes

$$\sigma(\omega) = \frac{e^2 n}{6kT} \frac{s\psi(s)}{1 - \psi(s)} \Delta^2(s)|_{s=\mathrm{i}\omega}, \tag{6.8.60}$$

with the frequency dependent mean-square displacement

$$\Delta^2(s) = \sum_m m^2 \psi(m, s)/\psi(s). \tag{6.8.61}$$

In particular, if (6.8.49) holds, than $\Delta^2 = \sum_m m^2 p(m)$, i.e. in this case Δ^2 is independent of ω so that the frequency dependence of σ is governed solely by $\psi(\mathrm{i}\omega)$. SCHER and LAX (1973) argue that, in general, the dominant ω-dependence is in $\psi(\mathrm{i}\omega)$, while Δ^2 only weakly depends on ω.

From (6.8.60), the expression for the dc conductivity can readily be obtained. To this end, let us assume that the carrier makes a hop at some time, i.e. that $\psi(t)$ is normalized to unity.

Thus, $\psi(s) \to 0$ as $s \to 0$, and we find that

$$\sigma(0) = \frac{e^2 n}{kT} \Delta^2(0)/6\bar{t}, \tag{6.8.62}$$

where

$$\bar{t} = \int_0^\infty \mathrm{d}t \, t\psi(t) \tag{6.8.63}$$

is the mean waiting time. Note that in (6.8.62) the diffusion coefficient $\Delta^2(0)/6\bar{t}$ has the familiar form.

On the basis of the CTRW method SCHER and LAX (1973) studied the hopping conductivity in doped semiconductors, in particular, the transition of $\sigma(\omega)$ from an ω^s dependence, where $s = 0.6 - 0.8$, to essentially dc behaviour, depending on temperature and concentration. They obtained reasonable agreement with experimental findings. The concentration dependence of $\sigma(0)$ resulting from (6.8.62) turns out to be the same as found by Miller and Abrahams (1960), i.e. $\sigma(0) \sim \exp(\alpha\mathcal{N}^{-1/3})^{3/2}$, instead of $\sigma(0) \sim \exp(\alpha\mathcal{N}^{-1/3})$ as predicted by percolation theory, i.e. the CTRW approach underestimates the dc conductivity (cf. KIEVELSON, 1980) (α — reciprocal Bohr radius, $\mathcal{N}$-concentration of sites).

For times smaller than the transit time t_τ, the transit current $I(t)$ in a time-of-flight experiment can be obtained by inserting (6.8.60) into (6.8.16), i.e.

$$I(t) = \frac{E}{2\pi\mathrm{i}} \frac{e^2 n}{6kT} \int_{c-\mathrm{i}\infty}^{c+\mathrm{i}\infty} \mathrm{d}s \, \mathrm{e}^{st} \frac{\psi(s)}{1 - \psi(s)} \Delta^2(s). \tag{6.8.64}$$

For times $t > t_\tau$, the effect of the absorbing electrode must yet be accounted for. As mentioned above, a detailed study of the transit current by means of the CTRW method was performed by MONTROLL and SCHER (1975) (cf. PFISTER and SCHER, 1978), who pointed out that $\psi \sim t^{-(1+\alpha)}$ gives for the current $I(t)$ the time dependence (6.8.1), i.e. $I(t) \sim t^{-(1-\alpha)}$ for $t < t_\tau$ and $I(t) \sim t^{-(1+\alpha)}$ for $t > t_\tau$. Note that the CTRW method has also been used to study recombination processes of carriers hopping in disordered solids in the presence of molecular recombination centers (cf., e.g., BLUMEN et al., 1983). Furthermore, atomic migration on disordered surfaces (cf., e.g., KLAFTER and SILBEY, 1980b) and exciton trapping and sensitized luminescence in molecular crystals (cf., e.g., KRENKRE and PARRIS, 1983, LORING et al., 1983, and references therein) are mathematically closely related to the transport problems under consideration (see also Section 11.4.1.).

Following KLAFTER and SILBEY (1980b) (cf. KLAFTER and SILBEY, 1983), we now give a firm theoretical foundation of the CTRW method. To this end, consider a lattice whose sites are fractionally occupied by atoms. We assume that the transitions between occupied sites are governed by the rate equation (6.8.25) and that the transition probabilities $W_{mm'}$ depend only on the spatial separation between the corresponding sites.

Let us rewrite (6.8.25) as follows

$$\frac{\mathrm{d}}{\mathrm{d}t} P(m, t) = \sum_{m'}' \widetilde{W}_{mm'} P(m', t) = \sum_{\text{all } m'} \widetilde{W}_{mm'} \eta_m \eta_{m'} P(m', t), \tag{6.8.65}$$

where

$$\widetilde{W}_{mm'} = (1 - \delta_{m,\overline{m}'}) W_{m'm} - \delta_{m,\overline{m}'} \sum_{m''(\neq m)} W_{mm''}, \tag{6.8.66}$$

$P(m, t) = P_{\overline{m}m}(t)$ ($\overline{m}$ denotes the initial site), the prime at the summation sign indicates that the summation is only over occupied sites, and the occupation numbers η_m are defined as

$$\eta_m = \begin{cases} 1 & \text{if site } m \text{ is occupied,} \\ 0 & \text{otherwise.} \end{cases} \tag{6.8.67}$$

The formal solution of (6.8.65) may be written as

$$P(m, t) = [\exp(\boldsymbol{V}t)]_{m\overline{m}}, \tag{6.8.68}$$

where

$$V_{mm'} = \widetilde{W}_{mm'} \eta_m \eta_{m'}. \tag{6.8.69}$$

Averaging (6.8.68) over all possible configurations, we find that

$$\langle P(m, t) \rangle = \langle \exp(\boldsymbol{V}t) \rangle_{m\overline{m}}, \tag{6.8.70}$$

and, taking the Laplace transform,

$$\langle P(m, s) \rangle = \langle (s\boldsymbol{1} - \boldsymbol{V})^{-1} \rangle_{m\overline{m}}. \tag{6.8.71}$$

To perform the configuration average, Klafter and Silbey used the Zwanzig projection-operator method. Defining a configuration-averaging operator D by

$$DA \equiv \langle A \rangle, \tag{6.8.72}$$

then one gets

$$\frac{\mathrm{d}}{\mathrm{d}t} [D \exp(\boldsymbol{V}t)] = D\boldsymbol{V} \exp(\boldsymbol{V}t) = D\boldsymbol{V}D \exp(\boldsymbol{V}t) + D\boldsymbol{V}(1 - D) \exp(\boldsymbol{V}t), \tag{6.8.73}$$

$$\frac{\mathrm{d}}{\mathrm{d}t} [(1 - D) \exp(\boldsymbol{V}t)] = (1 - D) \boldsymbol{V}D \exp(\boldsymbol{V}t) + (1 - D) \boldsymbol{V}(1 - D) \exp(\boldsymbol{V}t), \tag{6.8.74}$$

and thus

$$(1 - D)\exp(Vt) = \int_0^t dt' \exp[(1 - D) V(t - t')] (1 - D) VD \exp(Vt'). \quad (6.8.75)$$

The latter equation can readily be verified by differentiating it with respect to time and by using (6.8.74) and (6.8.75).

Inserting (6.8.75) into (6.8.73), we obtain

$$\frac{d}{dt} [D \exp(Vt)] = DVD \exp(Vt) + \int_0^t dt' DV(1 - D)$$

$$\times \exp[(1 - D) V(t - t')] (1 - D) VD \exp(Vt'), \quad (6.8.76)$$

and, by Laplace transform,

$$\langle (s\mathbf{1} - V)^{-1} \rangle = [s\mathbf{1} - \langle V \rangle - \langle \delta V [s\mathbf{1} - (1 - D) V]^{-1} \delta V \rangle]^{-1}$$

$$\equiv [s\mathbf{1} - M(s)]^{-1}, \quad (6.8.77)$$

where $\delta V = V - \langle V \rangle$ and $M(s)$ is the so-called self-energy matrix.

On multiplying (6.8.77) from the left by $(s\mathbf{1} - M(s))$ and using (6.8.71), we find the following generalized master equation

$$s\langle P(m, s) \rangle = \delta_{m,\bar{m}} + \sum_{m'} M(m, m', s) \langle P(m', s) \rangle. \quad (6.8.78)$$

By (2.6.9), i.e. $\sum_m P(m, s) = 1/s$, we get from (6.8.78)

$$M(m, m, s) = -\sum_{m'(\neq m)} M(m', m, s), \quad (6.8.79)$$

and thus

$$s\langle P(m, s) \rangle = \delta_{m,\bar{m}} + \sum_{m'(\neq m)} \{M(m, m', s) \langle P(m', s) \rangle$$

$$-M(m', m, s) \langle P(m, s) \rangle\} \quad (6.8.80)$$

and, by inverse Laplace transform,

$$\frac{d}{dt} \langle P(m, t) \rangle = \sum_{m'(\neq m)} \int_0^t dt' \{M(m, m', t-t') \langle P(m', t') \rangle$$

$$-M(m', m, t - t') \langle P(m, t') \rangle\}. \quad (8.6.81)$$

Thus, the averaged probability $\langle P(m, t) \rangle$ is governed by a non-Markovian equation.

To proceed in establishing the relation between (6.8.80) and the basic CTRW equation (6.8.48), note that after averaging the system at hand is translationally invariant and thus

$$M(m, m', s) = M(m - m', s). \quad (6.8.82)$$

Further, we define

$$M(m - m', s) = \tilde{\psi}(m - m', s) \{s/[1 - \tilde{\psi}(s)]\}, \tag{6.8.83}$$

where

$$\tilde{\psi}(s) = \sum_{m'(\neq m)} \tilde{\psi}(m' - m, s) = \{[1 - \tilde{\psi}(s)]/s\} \sum_{m'(\neq m)} M(m' - m, s) \tag{6.8.84}$$

or

$$\tilde{\psi}(s) = \sum_{m(\neq m')} M(m - m', s) \left[s + \sum_{m(\neq m')} M(m - m', s)\right]^{-1}. \tag{6.8.85}$$

Thus, inserting (6.8.83) into (6.8.80), we obtain

$$\left(s + \tilde{\psi}(s) \frac{s}{1 - \tilde{\psi}(s)}\right) \langle P(m, s)\rangle$$
$$= \delta_{m,\overline{m}} + \sum_{m'(\neq m)} \{s/[1 - \tilde{\psi}(s)]\} \tilde{\psi}(m - m', s) \langle P(m', s)\rangle, \tag{6.8.86}$$

which may be rearranged as

$$\langle P(m, s)\rangle - \sum_{m'(\neq m)} \tilde{\psi}(m - m', s) \langle P(m', s)\rangle$$
$$= \{[1 - \tilde{\psi}(s)]/s\} \delta_{m,\overline{m}}. \tag{6.8.87}$$

The latter equation may be identified as the basic CTRW equation (6.8.48), which establishes the correctness of the CTRW approach. However, an explicit expression for $\psi(m' - m, t)$, the probability of the particle jumping from site m to site m' after waiting time t, requires calculation of $M(m' - m, s)$ from (6.8.77).

7. Non-Ohmic hopping conduction

7.1. Basic formulae and results for hopping in strong electrical fields

7.1.1. *The expression for the* dc *hopping current*

Unlike the band mobility, the hopping mobility in general strongly depends on the strength of the external electrical field $\boldsymbol{E}$. An investigation of this dependence may therefore supply relevant additional information on the transport mechanism. Accordingly, the study of the field dependence of the hopping mobility is a very interesting problem.

For an electrical field of arbitrary strength, the density of the dc hopping current is given by (cf. (2.3.3), (2.4.23) and (2.4.19))

$$\boldsymbol{j} = \frac{1}{2\Omega} \sum_{mm'} (\boldsymbol{R}_m - \boldsymbol{R}_{m'}) \, i(m', m). \tag{7.1.1}$$

Here $i(m', m)$ is the current running from site m' to site m, for which holds that (cf. (2.4.23))

$$\sum_{m'} i(m', m) = 0, \tag{7.1.2}$$

and which is given by

$$i(m', m) = e\widetilde{W}_{m'm} \left[\varrho_{m'}(1 - \varrho_m) \exp\left(\frac{1}{2} \beta V_{m'm}\right) - \varrho_m(1 - \varrho_{m'}) \exp\left(-\frac{1}{2} \beta V_{m'm}\right)\right]. \tag{7.1.3}$$

Here $V_{m'm} = V_{m'} - V_m$, $V_m = \varepsilon_m + eu_m$; ε_m is the energy of the mth site, $u_m = \boldsymbol{E}\boldsymbol{R}_m$ the external field, and $\widetilde{W}_{m'm} = \widetilde{W}_{mm'}$ the symmetrized hopping probability.

Introducing the change of the chemical potential, $\delta\mu_m$, by

$$\varrho_m = [e^{\beta(\varepsilon_m - \delta\mu_m - \mu)} + 1]^{-1}, \tag{7.1.4}$$

then the expression (7.1.3) may be rewritten as

$$i(m', m) = e\widetilde{W}_{m'm} \times \frac{\sinh \frac{\beta}{2} (-\delta\mu_m + \delta\mu_{m'} + eu_{m'} - eu_m)}{\cosh \frac{\beta}{2} (\varepsilon_m + \varepsilon_{m'} - \delta\mu_m - \delta\mu_{m'} - 2\mu) + \cosh \frac{\beta}{2} (\varepsilon_m - \varepsilon_{m'} - \delta\mu_m + \delta\mu_{m'})}. \tag{7.1.5}$$

The symmetrized hopping probability $\widetilde{W}_{m'm}$ is given by (cf. (2.4.19))

$$\widetilde{W}_{m'm} = \int_{-\infty}^{\infty} dt' \cos\left[\frac{1}{\hbar} V_{m'm} t'\right] \varphi_{m'm}(t'), \tag{7.1.6}$$

where

$$\varphi_{m'm}(t) = \frac{1}{\hbar^2} |J_{m'm}|^2 e^{-2S_T(m',m)} \times \left\{\exp\left[\sum_{\boldsymbol{q}} \frac{|\gamma(\boldsymbol{q})|^2 [1 - \cos \boldsymbol{q}(\boldsymbol{R}_{m'} - \boldsymbol{R}_m)]}{N \sinh(\hbar\omega_{\boldsymbol{q}}\beta/2)} \cos \omega_{\boldsymbol{q}} t\right] - 1\right\}. \tag{7.1.7}$$

Here $J_{m'm}$ is the resonance integral between the atoms m and m', $\gamma(\boldsymbol{q})$ the electron-phonon coupling constant, and $\omega_{\boldsymbol{q}}$ the frequency of the phonon with wave vector $\boldsymbol{q}$ (the vibrational branch index has been dropped in (7.1.7), likewise the dependence of γ on the site index m). $S_T(m', m)$ is defined in (A 1.18).

The most important property of $\widetilde{W}_{m'm}$ is $\widetilde{W}_{m'm} \sim |J_{m'm}|^2 \sim \exp(-2\alpha |\boldsymbol{R}_{m'm}|)$, where $\boldsymbol{R}_{m'm} = \boldsymbol{R}_{m'} - \boldsymbol{R}_m$ and α is the reciprocal Bohr radius. In the case of strong electron-phonon interaction (over-barrier hopping), i.e. for $|\gamma(\boldsymbol{q})|^2 \gg 1$, $\widetilde{W}_{m'm}$ only weakly depends on $\boldsymbol{E}$ if $e\boldsymbol{E}\boldsymbol{R}$ is less than the height of the barrier (which is of the order of E_a (3.2.8)). Hence, for strong interaction with phonons and small spread of the energy levels, we have

$$\widetilde{W}_{m'm} = W_0 \exp(-2\alpha |\boldsymbol{R}_{m'm}|), \tag{7.1.8}$$

where W_0 only weakly depends on $\boldsymbol{E}$, $\boldsymbol{R}_m$ and ε_m.

For weak interaction with phonons, i.e. for $|\gamma(\boldsymbol{q})|^2 \ll 1$, in (7.1.7), we may expand the exponential in powers of the coupling constant. In the lowest nonvanishing order, from (7.1.6) and (7.1.7), we find that

$$\widetilde{W}_{m'm} = \frac{1}{\sinh(|V_{m'm}|\,\beta/2)} \frac{|J_{m'm}|^2}{\hbar^2} \frac{\pi}{N} \sum_{\boldsymbol{q}} |\gamma(\boldsymbol{q})|^2 \left[1 - \cos \boldsymbol{q}(\boldsymbol{R}_m - \boldsymbol{R}_{m'})\right]$$
$$\times\, \delta\left(\omega_{\boldsymbol{q}} - \frac{|V_{m'm}|}{\hbar}\right). \tag{7.1.9}$$

Hence, for weak interaction with phonons and small spread of the energy levels, we have

$$\widetilde{W}_{m'm} = W_0'[\sinh(|V_{m'm}|\,\beta/2]^{-1} \exp(-2\alpha |\boldsymbol{R}_{m'm}|), \tag{7.1.10}$$

where W_0' only weakly depends on $\boldsymbol{E}$, $\boldsymbol{R}_m$ and ε_m.

7.1.2. Some problems and results of the dc conduction

In the case of an ordered system (small-polaron model), owing to the existing translational symmetry we have $\delta\mu_m = \varepsilon_m = 0$. Accordingly, the current may be immediately calculated by means of (7.1.7). For a disordered system, however, a calculation of the hopping current is much more complicated, since at first the set of equations (7.1.2), (7.1.5) must be solved in order to find $\delta\mu_m$.

As in the case of weak fields (cf. Section 2.5.), equations (7.1.5) and (7.1.2) may be thought of as Kirchhoff's first and second law, respectively. The expression for $\boldsymbol{j}$ (7.1.1) represents the epuation for the external circuit of the corresponding network. For weak external fields, it is sufficient to expand sinh (in (7.1.5)) in powers of the argument up to the linear term and to put $\delta\mu_m = u_m = 0$ in W. This yields the random-resistor network of Miller and Abrahams, where $u_m + \delta\mu_m/e = U_m$ (cf. (2.5.13)) is the electrical potential at the mth site.

For fields of arbitrary strength, the expression (7.1.5) may in general not be interpreted as the current running through a nonlinear resistor, because it does not depend only on the difference between the potentials U_m and $U_{m'}$, but it depends separately on these quantities. In the general case, the expression (7.1.5) may be thought of as the current-voltage characteristic of a triode (or a transistor) with a grounded grid (or base) with the potentials U_m and $U_{m'}$ on the anode and the cathode (or the emitter and the collector), respectively. So far, an investigation of a network consisting of such elements has not been reported in the literature. In several papers (Austin and Sayer, 1974, Böttger and Bryksin, 1975a, Shklovskii, 1976) the problem was tried to be reduced to a network of nonlinear resistors. However, as pointed out by Böttger and Bryksin (1980), such an approach in general leads to incorrect results (cf. Section 7.2.3.).

Concrete results for non-Ohmic hopping conduction in disordered systems have been obtained by various methods, including also semi-qualitative ones. A method of the latter category is that used by MOTT (1970, 1971) and SHKLOVSKII (1972b) to study the effect of an electrical field on hopping conduction at very low temperatures: At $T \to 0$, in the presence of $\boldsymbol{E}$, almost all levels below (above) the Fermi level are occupied (unoccupied) and an electron may hop from a state below the Fermi level to a state above it by emitting a phonon (cf. Fig. 7.1). The spacing between the final state and the Fermi level is less than $e\boldsymbol{ER}$, where $|\boldsymbol{R}|$ is the hopping distance. For very strong fields ($eE/2\alpha kT \gg 1$), hopping occurs between nearest-neighbour sites, which yields a field-independent current (cf. (7.1.1), (7.1.3), (7.1.10)). However,

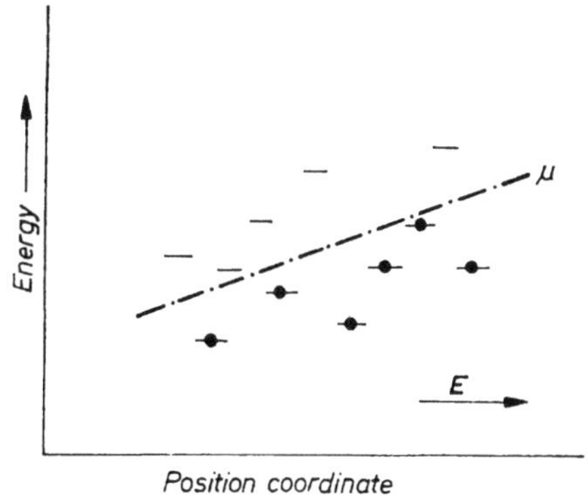

Fig. 7.1. Illustration of hopping by phonon emission at $T \to 0$; the sites below (above) the Fermi level μ are occupied (unoccupied); $\boldsymbol{E}$ — electrical field

for not too strong fields ($eE/2\alpha kT \gtrsim 1$), hopping occurs between more remote sites. Determining the effective hopping distance R by a procedure analogous to that used above for obtaining the Mott $T^{-1/4}$-law (5.1.19), $(4\pi/3)\, R^3 eER\, N(\mu) \approx 1$ ($N(\mu)$ — density of states at the Fermi level), one finds that (SHKLOVSKII, 1972b)

$$j(E) \sim \exp\left(-A/E^{1/4}\right), \tag{7.1.11}$$

where the constant A is temperature independent.

The dependence (7.1.11) was also found by the formalism of averaging in the four-dimensional R-ε-space (APSLEY and HUGHES, 1975) and by using percolation theory (POLLAK and RIESS, 1976, SHKLOVSKII, 1976, VAN DER MEER et al., 1982). A dependence of this type has also been discussed for hopping by photon emission (BEREZIN, 1980, 1981), and by transfer of the excess energy to another localized carmer (BEREZIN, 1983, 1984). Experimental results showing a $E^{-1/4}$ dependence (7.1.11) are shown in Fig. 7.2. Now let us turn to hopping conduction in moderate fields ($eE/2\alpha kT < 1$) (cf. Fig. 7.3.). In the first papers dealing with this problem (MOTT, 1971, HILL, 1971, APSLEY and HUGHES, 1974) the effect to the field was taken into account by replacing $|\varepsilon_{mm'}|$ by $|\varepsilon_{mm'}| - e\boldsymbol{ER}_{mm'}$ in the expression for the hopping probability of the type (5.1.16), which is equivalent to a renormalization of α. It was found that

$$j(E) \sim \exp\left(BE^n\right), \tag{7.1.12}$$

where according to MOTT (1971) and HILL (1971) $n = 1$ and $B \sim T^{-5/4}$ (HILL, 1971), but according to ASSLEY and HUGHES (1974) $n = 2$ and $B \sim T^{-9/4}$. The latter authors used the formalism of averaging in the four-dimensional space. An increase of the hopping current with increasing field of the type (7.1.12) was also found in

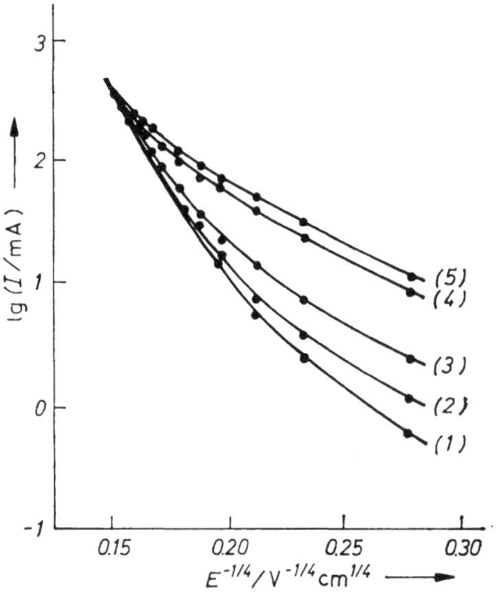

Fig. 7.2. Dependence of the current I on the electrical field $\boldsymbol{E}$ for strongly compensated germanium at different temperatures: (1) 1.54, (2) 4.2, (3) 10, (4) 15, (5) 25 K (after IONOV and SHLIMAK, 1972). As can be seen, in the region where the current is independent of temperature, the experimental data can be described by (7.1.11)

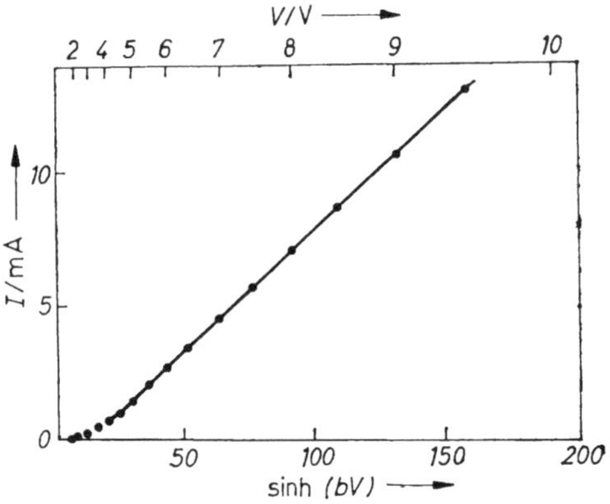

Fig. 7.3. Plot of the current I versus sinh (bV) (V — voltage, $b = 0.364\ V^{-1}$) for a thin amorphous SiC film at room temperature (after NAIR and MITRA, 1977). At low fields, the current tends to be Ohmic, while at relatively high fields, where $eE < 2\alpha kT$ (α — reciprocal Bohr radius), $I \sim \exp(bV)$ holds

more recent papers by making use of ideas of percolation theory: POLLAK and RIESS (1976) predicted $n = 1$, $B \sim T^{-1}$, SHKLOVSKII (1976) $n = 1/(1 + \nu)$, $B \sim T^{-1/(1+\nu)}$ (where ν the critical exponent of the correlation length L) and VAN DER MEER et al. (1982) $n = 2$, $B \sim T^{-9/4}$.

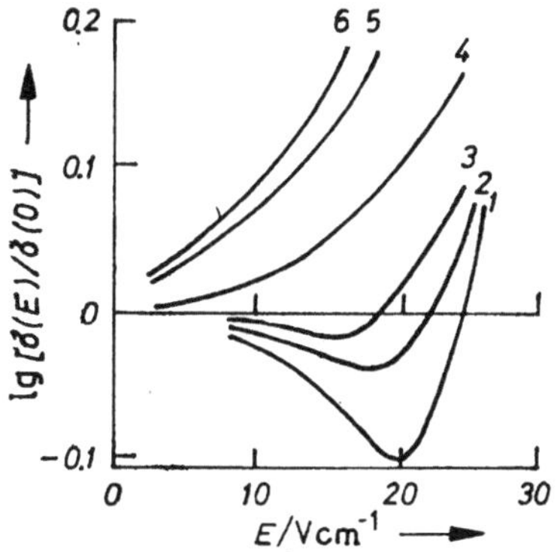

Fig. 7.4. Field dependence of the conductivity in n-Ge with donor concentration $6 \times 10^{15}\,cm^{-3}$ for several temperatures: $T = 4.2$ (1), 2.5 (2), 2.25 (3), 2.0 (4), 1.63 (5), 1.44 K (6) (after ZABRODSKII and SHLIMAK, 1977)

It is interesting to note that there are experimental results obtained on doped semiconductors (REDFIELD, 1975, ZABRODSKII and SHLIMAK, 1977) and amorphous transition metal oxides (BRYKSIN et al., 1980) which exhibit a deviation from Ohmic behaviour even at relatively weak fields (cf. Fig. 7.4). According to these results, at sufficiently low temperatures the conductivity decreases with increasing field. Only at higher fields, the current begins to increase according to a formula of the type (7.1.12). As illustrated inFig. 7.4., at very low temperatures, where in the Ohmic regime the dc conductivity obeys the Mott $T^{-1/4}$ law, the increase of the current with the field has been observed down to lowest fields.

For R-hopping and strong electron-phonon coupling, a decrease of the differential conductivity with increasing field just above the Ohmic region was first predicted by BÖTTGER and BRYKSIN (1979) on the basis of an effective-medium approximation (see, Section 7.3.1.). By means of this approximation it was pointed out that such a decrease of the conductivity occurs up to some critical field E_0, above which the current begins to increase according to a law of the type (7.1.12). According to BÖTTGER and BRYKSIN (1980) this behaviour can be understood as follows: The region of fields where $E < E_0$ is characterized by a continuous transition from "isotropic" to "directed" percolation. At fields $E > E_0$, the current flow is governed by directed percolation. The transition from isotropic to directed percolation originates from the "returns" in the percolation paths, which act as bottlenecks for the current in the case of strong fields (see Section 7.2.).

BÖTTGER and WEGENER (1984a, b) succeeded in solving numerically equations (7.1.1) to (7.1.3) for models with about 1000 sites. For R-hopping results of these computations are shown in Fig. 7.5 (for strong electron-phonon coupling) and Fig. 7.6 (for weak electron-phonon coupling). For strong coupling with phonons, we see that above the Ohmic region the conductivity decreases, before at higher fields it begins to increase. The fact that in the case of weak coupling with phonons, at sufficiently high fields, the current becomes constant, can be readily seen from (7.1.1), (7.1.3) and (7.1.10). Computer studies for R-ε-hopping and strong electron-phonon interaction (BÖTTGER et al., 1985a, b) show that at moderate temperatures the differential conductivity σ_E exhibits a (relative) minimum separated from the Ohmic regime by a relative maximum. On lowering the temperature the hight of the maximum increases, while the depth of the minimum decreases and completely vanishes at sufficiently low temperatures (see also Section 11.4.2.).

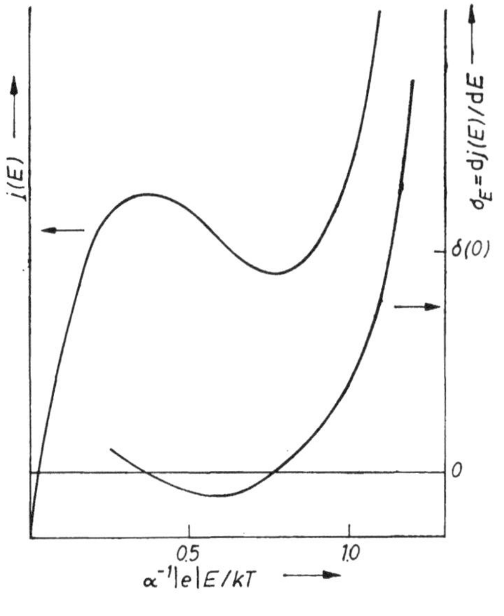

Fig. 7.5. Field dependence of the differential conductivity σ_E for strong electron-phonon coupling; computed for a model with 1000 sites, without spread of energy levels, $\alpha\mathcal{N}^{-1/3} = 15$, and an average site-occupation probability 0.5 (after BÖTTGER and WEGENER, 1984b)

Note that for strong coupling with phonons, and fields just above the Ohmic regime, the percolation treatment predicts a slowing down of the increase of the conductivity rather than a decrease (cf. Section 7.2.). This shortcoming of the percolation treatment may be due to the fact that an essential prerequisite of this approach, a weak field dependence of the number of charge carriers on the percolation paths, is not fulfilled. Results of the numerical examination of the non-Ohmic hopping conduction suggest that in the non-Ohmic regime the percolation treatment has, at least in its present form, above all guiding character.[1]) It gives a clear idea of the physical situation, but its quantitative results must be considered with some precaution.

Note that a decrease of the conductivity with increasing field just above the Ohmic regime has been also predicted for semiconductors with macroscopic inhomo-

[1]) It turns out (BÖTTGER and WEGENER, 1984a, b) that at higher fields (in the regime of the exponential increase of the current) the "current cluster" (defined by the $i(m', m)$ (7.1.3)) much better coincides with the percolation cluster (cf. Section 7.2.) than in the regime where the conductivity decreases with increasing field. Furthermore, in the latter regime the concentration of charge carriers in the current cluster depends much stronger on the strength of the external field than in the former regime.

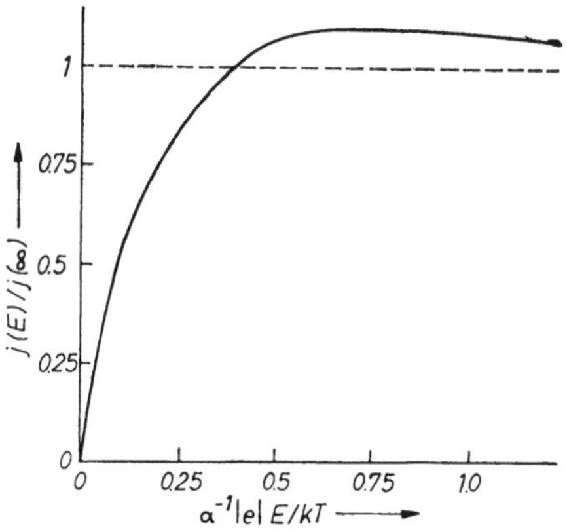

Fig. 7.6. Current-voltage characteristic $j(E)$ for weak electron-phonon interaction; computed for a model with 1000 sites, without spread of energy levels, $\alpha\mathcal{N}^{-1/3} = 15$, and an average site-occupation probability 0.5 (after Böttger and Wegener, 1984b)

geneities (Fishchuk, 1982b), which can be described by a network of random macroscopic barriers. This suggests that such behaviour is characteristic of percolation systems in which there exist Ohmic and diffusion contributions to the current.

In actual disordered systems, the field may not only affect the mobility, but ionization of "donors" by the external field may be also important. As well known, the ionization energy decreases with increasing field (Poole-Frenkel effect), which leads to a field-dependent charge-carrier concentration. The current becomes proportional to $\exp\{e\beta E^{1/2}/kT\}$ (β — Poole-Frenkel constant). A dependence of this type is often observed in experiments. However, this topic is outside the scope of the present book. The reader who is interested in more details on the Poole-Frenkel effect is referred, e.g., to Mott and Davis (1979) and references therein.

It should be noticed that sufficiently strong electrical fields can affect the localization range of electronic states (delocalization) (cf., e.g., Prigodin, 1980, Soukoulis et al., 1983, Lee et al., 1983). To the authors knowledge, so far, this effect has not been taken into account on studying non-Ohmic hopping conduction. Possibly, it also requires account of nondigonal elements (with respect to the site indices) of the density matrix.

7.1.3. *The* ac *conductivity problem*

To study the hopping conductivity in strong alternating electrical fields, Böttger and Bryksin (1975b, 1976e) started from (2.4.10) and accounted for nondiagonal elements of the density matrix $\varrho^m_{m'}$, too. For a field of frequency ω, an estimation yields $\varrho^m_{m'} \sim \tilde{J}_{m'm}(\varrho_m - \varrho_{m'})/(\varepsilon_{m'} - \varepsilon_m \pm \hbar\omega)$ (cf. (2.4.11)). Accordingly, for most of the pairs it holds that $\varrho^m_{m'} \ll \varrho_m, \varrho_{m'}$, provided that we have $\Gamma \gg \tilde{J}$ (Γ — width of the spread of the energy levels). However, there are "resonance pairs" characterized by $|\varepsilon_{m'} - \varepsilon_m| \approx \hbar\omega$. These pairs may not be neglected. Although the number of such pairs is small, their contribution to the conductivity in an alternating field may be considerable. This contribution corresponds to the "phononless contribution" studied for weak fields in Section 6.2.3. The "phonon contribution" to the conductivity arises from non-resonance transitions, for which the nondiagonal elements of the density matrix may be neglected. The phononless contribution turns out to exhibit a strong tendency to a nonlinear behaviour. This tendency arises from intensive transitions in resonance pairs, which try to balance out the occupation probabilities at both sites of a pair. Phonon transitions counteract such a levelling and try to restore an occupation according to the thermal equilibrium. Böttger and Bryksin (1975b) investigated the situation where the phononless contribution dominates at low fields. This contribution was shown to lead to saturation at some field $E = E_c$ due to the heating effect described above. For $E > E_c$, the further increase of the current is due to the phonon contribution. This field dependence of the current is illustrated in Fig. 7.7. E_c was estimated to be of the order of 10^2 to 10^4 V/cm in the infrared-frequency regime $\omega \approx 10^{11}$ to $10^{12}\ s^{-1}$.

Relaxation currents in strong electrical fields will be studied in Section 7.3.2.

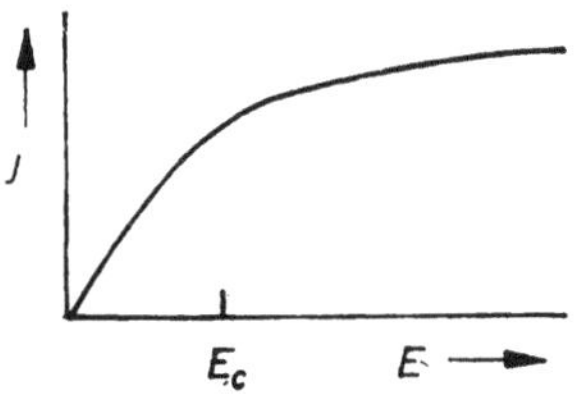

Fig. 7.7. Dependence of the real part of the current, j, on the amplitude, E, of an alternating electrical field; for $E < E_c$, the phononless contribution to the current dominates. The increase of j for $E > E_c$ is due to the phonon contribution (after Böttger and Bryksin, 1976b)

7.2. Percolation approach to non-Ohmic hopping conduction and directed percolation

7.2.1. *Expression for the current in arbitrary electrical fields for hopping along a percolation path*

In the hopping regime the current density for an arbitrary field is given by equations (7.1.1) to (7.1.3). In accordance with the basic ideas of percolation theory, we assume (Böttger and Bryksin, 1980) that in the system under consideration the current runs along certain optimal one-dimensional paths from electrode to electrode. In the regime where Ohm's law holds, this assumption means that the current takes paths with minimal resistance. In the general case, the current paths are determined by the requirement of maximum current flux. Let us assume that such paths exist and that the current along each path is conserved and equal to i. Then we have $j \sim i$. Consider a percolation path and number the sites along it in an increasing manner in the direction of the current flow, say, from the left to the right electrode (the field $\boldsymbol{E}$ is directed along the x-axis). Note that it does not at all hold that $(\boldsymbol{R}_{m+1})_v > (\boldsymbol{R}_m)_x$; however, it does hold that $(\boldsymbol{R}_{m+n})_x > (\boldsymbol{R}_m)_x$ for $n \to \infty$. Restricting ourselves in the following to the Boltzmann statistics ($\varrho_m \ll 1$), then from (7.1.3) we obtain

$$i = -e\widetilde{W}_{m+1,m} \left\{\varrho_{m+1} \exp\left(\frac{1}{2}\beta V_{m+1,m}\right) - \varrho_m \exp\left(-\frac{1}{2}\beta V_{m+1,m}\right)\right\}. \quad (7.2.1)$$

On solving this equation with respect to ϱ_{m+1}, we have

$$\varrho_{m+1} = -i\Gamma_{m+1,m} + \varrho_m \exp\left(-\beta V_{m+1,m}\right), \tag{7.2.2}$$

where

$$\Gamma_{m+1,m} = (e\widetilde{W}_{m+1,m}) \exp\left(-\frac{1}{2}\beta V_{m+1,m}\right). \tag{7.2.3}$$

On iterating (7.2.2), we can express ϱ_m by i:

$$\varrho_m = -i \sum_{n=0}^{\infty} \Gamma_{m-n,m-n-1} \exp\left(-\beta V_{m,m-n}\right). \tag{7.2.4}$$

Now we sum both the sides of (7.2.3) over m and obtain

$$i = -n_e \left\{\frac{1}{N} \sum_m \sum_{n=0}^{\infty} \Gamma_{m,m-1} \exp\left(-\beta V_{m+n,m}\right)\right\}^{-1}. \tag{7.2.5}$$

Here

$$n_e = \frac{1}{N} \sum_m \varrho_m \tag{7.2.6}$$

is the concentration of electrons in the percolation path and N is the total number of sites along the path ($N \to \infty$).

In the following we restrict ourselves to the case of R-percolation, where the percolation paths are determined only by the spacings between sites and not by their energies. Assuming the spread of the energy levels to be small compared with kT, expression (7.2.5) can be somewhat reduced:

$$i = -en_e \left\{\frac{1}{N} \sum_m \sum_{n=0}^{\infty} (\widetilde{W}_{m,m-1})^{-1} \exp\left(\frac{1}{2}\beta e u_{m-1,m}\right) \times \exp\left(-\beta e u_{m+n,m}\right)\right\}^{-1}. \tag{7.2.7}$$

In the case of a one-dimensional system with nearest-neighbour hops only, on the basis of (7.2.7) the current can be calculated rigorously (cf. Section 7.4.). An investigation of the current in two- or three-dimensional systems is much more complicated than in one-dimensional systems. One problem encountered in the former systems arises from the fact that n_e may depend on the field. To see this, consider a percolation path which has "dead ends". For zero field, the electrons are distributed regularly on the sites in this path and in the dead ends (if all energies are equal to each other). On applying a field, a portion of the electrons transfer from the percolation path into dead ends, "stick fast" there, and thus do not contribute to the current. This leads to a reduction of the density of electrons in the percolation path, but to an enhancement of the density in dead ends, where the electrons are distributed according to the barometric formula. In this way the electron density in some dead ends is lowered but in other enhanced. In the case of the Boltzmann statistics (the number of electrons or holes is much less than the number of sites) this effect may be very important, since dead ends may act as traps (cf. Fig. 7.8). It should be noted that such dead ends "enriched" with electrons may lead to large gradients of the electron concentration in the system, because of which a penetration of electrons through dead ends can occur. If the contribution of this latter mechanism to the current becomes of the order of that from percolation paths, the problem ceases to be a percolational one. This situation was studied by NGUEN VAN LIEN and SHKLOVSKII (1981), according to whom the transport mechanism of penetration through dead ends becomes most effective for small concentrations of electrons or holes. In this case, the current runs approximately along a straight line from electrode to electrode and does not circumvent "hard hops". Owing to this fact, the expression for the current looks like the result of MILLER and ABRAHAMS (1960) $\ln j \sim (\alpha\mathcal{N}^{-1/3})^{3/2}$ (for weak electron-phonon coupling and disregarding spread of the energy levels), while for very strong fields the percolation approach predicts $\ln j \sim \alpha\mathcal{N}^{-1/3}$.

7.2.2. *Selection of the optimal percolation path and the expression for the current*

In the following we assume the field to be weak enough so that one-dimensional percolation paths exist and n_e does not strongly depend on $\boldsymbol{E}$. However, in principle, n_e decreases with increasing $\boldsymbol{E}$. Accordingly, the result must be multiplied by a function which decreases with increasing $\boldsymbol{E}$.

With such an assumption if follows from (7.2.7) that the current has to be obtained from the condition that the expression in the denominator of (7.2.7) is a minimum. Disregarding pre-exponential factors, the current may be written as

$$i \sim \left[\widetilde{W}_{m,m-1} \exp\left(-\frac{1}{2}\,\beta e u_{m-1,m}\right) \exp\left(\beta e u_{m+n,m}\right)\right]_{\max}, \tag{7.2.8}$$

where the right-hand side is an element of the optimal percolation path.

An important feature of (7.2.7) is the summation over m (along the path) of terms $\exp\left[\beta\boldsymbol{E}(\boldsymbol{R}_{m+n} - \boldsymbol{R}_m)\right]$. Consider a path with the property $(\boldsymbol{R}_{m+1} - \boldsymbol{R}_m)_x > 0$ for any m. For such a path the series with respect to n consists of monotonically decreasing terms with the largest term at $n = 0$, which is equal to unity. Thus, in (7.2.8) this series may be neglected. However, on constructing the first percolation path, the direction of the current with respect to the external field is in general not accounted for. A real path contains "returns" (cf. Fig. 7.8), which in the Ohmic regime do not play any role. However, on leaving this regime the situation changes strongly. Let us construct in the non-Ohmic regime the percolation path in the usual manner i.e. each site is put into a sphere the radius of which is increased up to a value R_c where an infinite cluster of connected spheres occurs. The corresponding current path is twisted and it contains returns. In the thermodynamic limit, for the length of the maximum return λ in this path, we have $\lambda(R_c) \to \infty$. Consequently, among the terms of the series with respect to n, there is an infinitely large one. That is, for any weak field $\boldsymbol{E}$, this first percolation path gives a vanishing contribution to the current. Therefore, it is necessary to construct spheres with radius $\sim R_c + \alpha^{-1}$ instead of R_c (cf. Section 5.3.1.). Consequently, instead of a single first path, a backbone cluster is obtained with a spacing between points of intersections of the order of the correlation length $L \sim R_c(\alpha R_c)^\nu$ ($\nu = 0.9$ for $d = 3$) (cf. Section 5.3.4.). In such a cluster, the current path can be chosen in a manner that a return in the path can occur only between two consecutive points of intersections in the cluster. Thus, the maximum return is now finite and proportional to L. We see that the non-Ohmic region begins at fields $|e| E\lambda(R_c + \alpha^{-1})/kT \approx 1$.

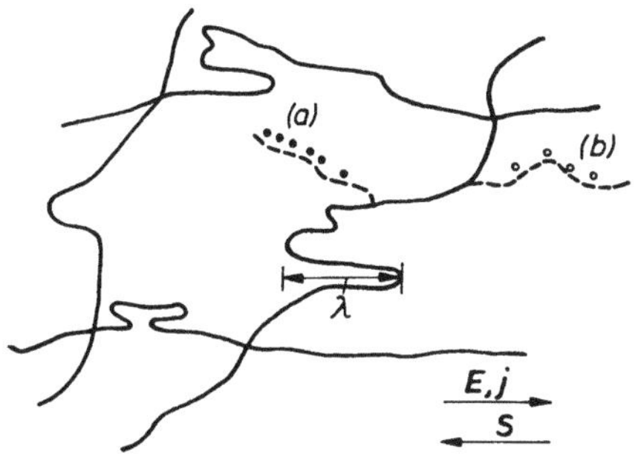

Fig. 7.8. Schematic representation of the percolation cluster with returns and two "dead ends". In the dead end (a) the electron concentration is higher than the average concentration, whereas in the dead end (b) it is lower. The arrows show the direction of the electrical field $\boldsymbol{E}$, the current $\boldsymbol{j}$ and the electron flux $\boldsymbol{S}$, respectively (after BÖTTGER and BRYKSIN, 1980)

Now let us consider fields at which this parameter becomes of the order of unity or greater. At such fields a reconstruction of current paths begins in order to diminish the effect of the returns. This can be realized by a further increase of the radius of the spheres around the sites. With increasing radius the backbone cluster becomes more and more compact, and consequently the number of current paths increases. Therefore, these paths become less twisted. On the other hand, the minimal value $W \sim \exp(-2\alpha R)$ decreases. However, this decrease is compensated by the decrease of the maximal return within a current path due to the larger compactness of the cluster. Let us now construct a backbone cluster using a radius $R > R_c$ for the spheres. We enlarge R as long as the inequality $\exp(|e| E\lambda/kT) \geqq \exp(2\alpha R)$ is fulfilled. At some $R = R_0$, this inequality becomes an equation

$$2\alpha R_0 = \frac{|e| E\lambda(R_0)}{kT}. \tag{7.2.9}$$

The characteristic radius R_0, defined in (7.2.9), is the radius which corresponds to the optimal path for the current at given $\boldsymbol{E}$, where the contribution to the maximum return is equal to the contribution of the hardest hop.

A construction of spheres is justified if the product $\widetilde{W}_{m,m-1} \exp\left(-\frac{1}{2} \beta e u_{m-1,m}\right)$ does not depend on $\boldsymbol{E}$. In the case of weak coupling with phonons and negligible spread of the energy levels, this product is independent of $\boldsymbol{E}$ for hops in the direction against $\boldsymbol{E}$, but proportional to $\exp(-\beta |e u_{m-1,m}|)$ for hops in the opposite direction (which results from (7.1.10) by using $\sinh(|V_{m'm}|\,\beta/2) \approx \exp\left(\frac{\beta}{2}\,|V_{m'm}|\right)$. In the case of strong interaction with phonons and negligible spread of the energy levels (over-barrier hops), the characteristic figure is a rotary ellipsoid with the site in a focus (cf. (7.1.8), (7.2.8))

$$2\alpha R - \frac{|e|\,ER}{2kT}\cos\theta = \varrho, \tag{7.2.10}$$

where θ is the angle between $\boldsymbol{R}_m - \boldsymbol{R}_{m-1}$ and $\boldsymbol{E}$. In this case, (7.2.9) takes the form

$$\varrho_0 = \frac{|e|\,E\lambda(\varrho_0)}{kT}, \tag{7.2.11}$$

where ϱ_0 is the critical value of ϱ.

According to (7.2.8), the current is given by the relation

$$j \sim \exp(-\varrho_0). \tag{7.2.12}$$

An evaluation of ϱ_0 requires knowledge of the dependence $\lambda(\varrho)$. Obviously, λ depends on the parameters ϱ_c and ϱ_m, where ϱ_c is the value of ϱ at which the first infinite cluster of connected ellipsoids occurs, i.e. $\lambda(\varrho_c) \to \infty$, and ϱ_m is the value of ϱ at which the first percolation path without returns occurs, i.e. $\lambda(\varrho_m) = 0$. Assuming for λ a power-like dependence on $\varrho - \varrho_c$ and $\varrho_m - \varrho$, we write (Böttger and Bryksin, 1980)

$$\lambda(\varrho) = 2\delta R_c \left(\frac{\varrho - \varrho_c}{\varrho_c}\right)^{-\nu'} \left(\frac{\varrho_m - \varrho}{\varrho_c}\right)^{\mu}, \tag{7.2.13}$$

where δ, μ, and ν' are positive numbers.

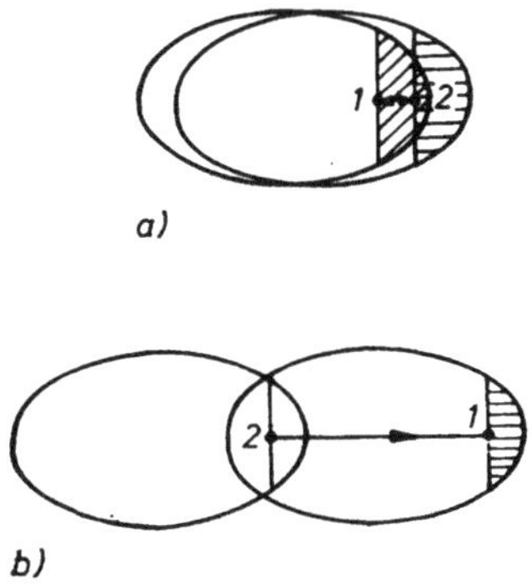

Fig. 7.9. Characteristic figures (ellipsoids) for constructing the percolation paths. In a) an electron may hop from site 1 to site 2 and vice versa, but in b) only from site 1 to 2. The arrows show the allowed direction of the current flow (after Böttger and Bryksin, 1980)

Let us now discuss in some more detail the meaning of the quantity ϱ_m. On constructing the ellipsoides of (7.2.10) the angle θ is admitted to change from 0 to π. Adopting the concept of "inclusive" figures (cf. Section 5.2.1.), a site 2 is connected with a given site 1, when site 2 lies within the figure constructed around site 1 (cf. Fig. 7.9). For $\boldsymbol{E} = 0$, the ellipsoid becomes a sphere, and if site 2 is connected by a bond with site 1, then site 1 is also connected with site 2. For $\boldsymbol{E} \neq 0$, a situation can occur where site 2 is connected with 1, but site 1 is not connected with site 2 (Fig. 7.9(b)), i.e. the current can run only from site 2 to site 1 but not vice versa. (Note that electrons hop from site 1 to site 2). Thus, the construction of the ellipsoids for $\boldsymbol{E} \neq 0$ requires marking the bonds in the critical cluster by arrows. A portion of bonds obtains arrows in both directions (Fig. 7.9(a)) and another portion only arrows directed from left to right. The current can run only in the direction of an arrow. Thus, a portion of bonds acts like diodes. However, there are also a number of bonds which permit a current flow from the right to the left, i.e. against the field. This number increases with decreasing field. In order to prevent such returns, we must retain only bonds with arrows directed from the left to the right. This can be realized by allowing the angle θ at the ellipsoids to change from 0 to $\pi/2$ only (by neglecting the shaded part of the ellipsoids in Fig. 7.9), and by assuming a site 2 to be connected with a given site 1 if site 2 is placed in the ellipsoid which is cut off and drawn around site 1. Such a construction requires neglect of dead ends and also of sites whose related bonds have arrows all either both directed to or away from them, because bonds of this kind do not let the current pass.

Obviously, on performing this procedure at some value $\varrho = \varrho_m$ a first percolation path without any return occurs. Such percolation is directed. ϱ_m is called the threshold of directed percolation. Of course, it holds that $\varrho_m > \varrho_c$. Directed percolation differs from isotropic one by the fact that any elementary hop from site to site occurs only from right to left (opposite to the direction of $\boldsymbol{E}$).

Let us for a moment dwell on the problem of directed percolation. This of type percolation was first suggested by BROADBENT and HAMMERSLEY (1957), who proposed a percolation problem in which neighbouring lattice sites are connected at random by directed bonds (diodes). Directed percolation has been the subject of much current interest, which is stimulated by the fact that directed percolation is known to lie in a universality class different from isotropic percolation (OBUKHOV, 1980) and by its similarity to Reggeon field theory (CARDY and SUGAR, 1980). An important feature of directed percolation is the existence of two different length scales: the longitudinal and transverse correlation length, $L_{||}$ and $L_{\perp}$, which diverge near the percolation threshold with different exponents (cf. (5.2.10)) $\nu_{||}$ and $\nu_{\perp}$ ($\nu_{||} > \nu_{\perp}$), respectively (KINZEL and YEOMANS, 1981; cf. BELL and ESSAM, 1983, and references therein). $L_{||}$ can be defined as the root-mean-square distance from the origin of a particle at site $\boldsymbol{R}$ which is connected to the origin (cf. Section 5.2.1.). In contrast to isotropic percolation, $L_{||}$ and $\nu_{||}$ (as well as the percolation threshold) depend on the angle the vector $\boldsymbol{R}$ makes with the anisotropy axis (the direction of $\boldsymbol{E}$).

The transverse correlation length $L_{\perp}$ can be defined as the average distance $|\boldsymbol{R}_1 - \boldsymbol{R}_2|$ between two sites perpendicular to $\boldsymbol{R}$, which belong to the same cluster, i.e. which are both connected with the origin along the directions $\boldsymbol{R}_1$ and $\boldsymbol{R}_2$ (KLEIN and KINZEL, 1981). Fig. 7.10 shows a typical cluster in directed percolation near (below) the percolation threshold. The lengths $L_{||}$ and $L_{\perp}$, respectively, give the length and the width of the cigar-shaped percolation cone. Above the threshold, percolation is possible in a cone shaped region of finite opening angle about the anisotropy axis (field direction). For more details on the problem of directed percolation the reader is referred e.g. to REDNER (1982); see also Section 11.3.1.

Let us now proceed to study non-Ohmic hopping conduction within the framework of the percolation treatment under consideration. To determine the quantities ϱ_c and ϱ_m involved in (7.2.13), BÖTTGER and BRYKSIN (1980) used the concept of the

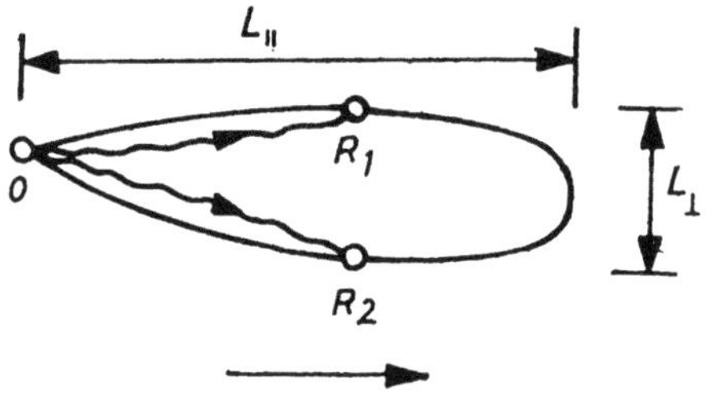

Fig. 7.10. Schematic representation of a typical cluster in directed percolation near (below) the percolation threshold (see text). The circle denotes the origin, the arrow the anisotropy axis

dimensional invariants. However, as pointed out by BÖTTGER et al. (1982) by numerical computations, the concept of dimensional invariants is not suitable for the percolation problem in question. Therefore, the latter authors determined numerically the field dependences of ϱ_m and ϱ_c. The results obtained may be well fitted by the formulae

$$\varrho_c(E) = 2\alpha R_c \left(1 - \frac{|e|\,E}{4\alpha kT}\right)^\tau \tag{7.2.14}$$

and

$$\varrho_m(E) = 2\alpha R_m \left(1 - \frac{|e|\,E}{4\alpha kT}\right)^\tau, \tag{7.2.15}$$

where $\tau = 0.65 \approx 2/3$ and $R_m = (\varrho_m/2\alpha)_{E\to 0}$. Furthermore, it was found that

$$\gamma = \frac{R_m}{R_c} - 1 \approx 0.07, \tag{7.2.16}$$

in accordance with the result of NGUEN and SHKLOVSKII (1981).

Inserting now (7.2.13) into (7.2.11), and accounting for

$$|e|\,E/\alpha kT \ll 1, \tag{7.2.17}$$

we get

$$\varrho_0 = \varrho_c \left\{1 + \left(\frac{2\,|e|\,E\delta R_c}{\varrho_c kT}\right)^{1/\nu'} \left(\frac{\varrho_m - \varrho_c}{\varrho_c}\right)^{\mu/\nu'}\right\}. \tag{7.2.18}$$

Substituting ϱ_c (7.2.14) and ϱ_m (7.2.15) into (7.2.18), and taking into account (7.2.16), we obtain

$$\varrho_0 = 2\alpha R_c \left\{1 - \frac{|e|\,E}{6\alpha kT} + \left(\frac{|e|\,E\delta}{\alpha kT}\right)^{1/\nu'} \gamma^{\mu/\nu'}\right\}. \tag{7.2.19}$$

Thus, by means of (7.2.12) and (7.2.19), we find that the current is given by

$$j_1 \sim \exp(-2\alpha R_c) \exp\left\{-2\alpha R_c \left[\left(\delta \frac{|e| E}{\alpha kT}\right)^{1/\nu'} \gamma^{\mu/\nu'} - \frac{|e| E}{6\alpha kT}\right]\right\}. \tag{7.2.20}$$

On the other hand, inserting (7.2.15) into $j \sim \exp(-\varrho_m)$, we get

$$j_2 \sim \exp[-2\alpha R_c(1 + \gamma)] \exp\left[\frac{|e| ER_c}{3kT} (1 + \gamma)\right], \tag{7.2.21}$$

where we have used (7.2.16) and (7.2.17).

The expression (7.2.21) may be thought of as describing the current at sufficiently high electrical fields, where the current is governed by directed percolation, while the expression (7.2.20) should apply to low electrical fields. The dependence of the return λ on the parameter ϱ was numerically studied on a three-dimensional model with 1000 sites by Böttger and Wegener (1984c). The results obtained for $\lambda(\varrho)$ by examining about 350 configurations were used to determine the parameters δ, ν' and μ involved the expression (7.2.13). On neglecting in each case the value of $\lambda(\varrho)$ for ϱ near ϱ_c (because of the finiteness of the model), by a fitting procedure it was found that $\delta = 0.26$, $\nu' = 1.1$, and $\mu = 1.0$ (with errors of about 50%). Using these values of the parameters δ, ν', μ, and γ from (7.2.16), then in the braces of the exponent of (7.2.20), the second term dominates. Accordingly, for strong interaction with phonons, even at low electrical fields (above the Ohmic regime), the differential conductivity increases with the field, contrary to the predictions of the effective-medium approximation (cf. Section 7.3.1) and numerical computations (Böttger and Wegener, 1984a, b). It should be noted that, owing to the inaccuracy of the numerical values of δ, ν', and μ, we cannot completely rule out the possibility that within the framework of the percolation approach a decrease of the conductivity may occur at low fields.

As mentioned in Section 7.1.2., numerical computations (Böttger and Wegener, 1984b) indicate that at low fields the number n_e (cf. (7.2.6)) of the charge carriers in the percolation path rather strongly depends on the strength of the external field. Accordingly, the expression for the current (7.2.20) must yet be multiplied by a factor $n_e(E)$, which improves the qualitative agreement with results of the effective-medium approximation and the numerical computations.

So far, here we have studied non-Ohmic hopping conduction for the case of R-percolation. Let us now turn to R-ε-percolation. Hitherto, for the latter case, a theoretical study of non-Ohmic conduction in the low-field region does not exist. In the high-field region, where directed percolation governs the current flow, R-ε-hopping may be investigated in the four dimensional R-ε-space. To do this, we use, as usual for R-ε-hopping, the following expression for the probability:

$$\widetilde{W} \sim \exp\left(-2\alpha R + \frac{\varepsilon}{kT} + \frac{|e| ER}{2kT} \cos\theta\right), \tag{7.2.22}$$

equate the exponent of (7.2.22) to some ϱ, and determine the volume of this figure in the four-dimensional space by integrating over θ from 0 to $\pi/2$ (and not from 0 to π, i.e. we account for the condition of directed percolation). Assuming the density of states to be independent of ε and equating the result of the above integration to $v_c\mathcal{N}^{-1}$ (v_c — critical-volume fraction), we obtain the following result for the current (POLLAK and RIESS, 1976, BÖTTGER and BRYKSIN, 1980)

$$j \sim \exp\left\{-2\alpha r_c + \frac{3}{16}\frac{Er_c}{kT}\right\}, \tag{7.3.23}$$

where $2\alpha r_c = (T_0/T)^{1/4}$ with T_0 given in (5.1.19). Note that integration over θ from 0 to π yields $j \sim \exp(AE^2)$, which is the result found by APSLEY and HUGHES (1974) (cf. Section 7.1.2.).

Note that in the case of the Fermi statistics (R-ε-percolation) the expression for the current depends non-linearly on the distribution function (cf. (7.1.3)) and therefore no such simple solution as (7.2.5) occurs. Hence, at present we cannot make an unambiguous statement on non-Ohmic hopping behaviour at very low temperatures from the point of view of the percolation approach presented above, although in the case of the Fermi statistics for sufficiently high fields the mechanism of directed percolation may also be expected to determine the current.

Let us yet draw attention to the percolation approach to non-Ohmic hopping conduction of VAN DER MEER et al. (1982), which pretends also to cover the case of R-ε-percolation. This approach is based on the fact that the pair currents $i(m', m)$ (cf. (7.1.3)) always flow from sites with higher electrochemical potential $U_m + \mu$ (cf. (2.5.13)) to sites with lower ones, so that these currents form a network with well-defined directed bonds. Accordingly, the charge transport is realized through self-avoiding current paths along which the electrochemical potential monotonously decreases. To solve the corresponding problem, characteristic percolation figures are defined by the envelope of all sites which are connected with a given site by pair currents which exceed a given limiting pair current. Some drawback of this method is in that the characteristic figures depend explicitly on the change of the chemical potential, $\delta\mu_m$, which is an unknown quantity. Using the concept of dimensional invariants, VAN DER MEER et al. (1982) obtained the following dependences of the hopping current on the external field (cf. Section 7.1.2):

$$j(E) \sim \exp\left[-a_c(\alpha^{-3}kTN)^{-1/4} + \frac{e^2E^2}{b_c\alpha^{5/4}(kT)^{9/4}N^{1/4}}\right]$$

$$\text{for } \frac{eE}{2\alpha kT} \ll 1, \tag{7.2.24}$$

$$j(E) \sim \exp\left[-c_c\alpha(NeE)^{-1/4}\right] \quad \text{for } \frac{eE}{2\alpha kT} \gtrapprox 1, \tag{7.2.25}$$

$$j(E) \sim \exp\left[-d_c(\alpha^{-3}\Delta\varepsilon N)^{-1/3}\right] \text{ for } \frac{eE}{2\alpha kT} \gg 1. \tag{7.2.26}$$

Here N is the density of localized states, which is assumed to be constant near the Fermi level, $\Delta\varepsilon$ is the width of the band of localized states, and the parameters resulting from the percolation analysis are estimated to be of the order a_c, c_c, $d_c \approx 1$ and $b_c \approx 10$.

We conclude this section by drawing attention to the fact that the field dependence of the dimensional "invariants" B_c and v_c, which was studied by BÖTTGER et al. (1982) and BÖTTGER and WEGENER (1984c), may also be expected to modify the results of the percolation treatment of hopping transport in strong fields given in (7.2.24) to (7.2.26).

7.3. Effective-medium theory for the case of strong fields

7.3.1. Stationary currents

Let us, for simplicity, consider hopping on a cubic lattice in which the hopping probability is the random quantity. Then the expression for the current is given by the relation (7.1.1), where $\boldsymbol{R}_m - \boldsymbol{R}_{m'}$ must be replaced by the vector $\boldsymbol{g}$ connecting nearest neighbour sites (accounting for nearest-neighbour hops only). In the absence of a spread of the energy levels, according to (7.1.3), the current running from site $\boldsymbol{m} + \boldsymbol{g}$ to site $\boldsymbol{m}$ is given by (for $\varrho_{\boldsymbol{m}} \ll 1$)

$$i_{\boldsymbol{m}+\boldsymbol{g},\boldsymbol{m}} = W_{\boldsymbol{m}+\boldsymbol{g},\boldsymbol{m}} \left(\varrho_{\boldsymbol{m}+\boldsymbol{g}}\, e^{-\frac{1}{2}\beta|e|\boldsymbol{E}\boldsymbol{g}} - \varrho_{\boldsymbol{m}}\, e^{\frac{1}{2}\beta|e|\boldsymbol{E}\boldsymbol{g}}\right). \tag{7.3.1}$$

The occupation probability $\varrho_{\boldsymbol{m}}$ is governed by (7.1.2).

According to the effective-medium concept (cf. Sections 5.3.5. and 6.4.) all hopping probabilities $\widetilde{W}_{\boldsymbol{m}+\boldsymbol{g},\boldsymbol{g}}$ are replaced by a uniform one denoted by $\mathscr{W}$. Then all $\varrho_{\boldsymbol{m}}$ are equal to each other: $\varrho_{\boldsymbol{m}} = f_0$, where f_0 is the equilibrium distribution function, which for the Boltzmann statistics reads $f_0 = e^{-\beta|\mu|}$ (μ — chemical potential). The current density becomes

$$j_x = \frac{2e}{a} \mathscr{W}\, e^{-\beta|\mu|} \sinh \lambda, \tag{7.3.2}$$

where $\lambda = \dfrac{1}{2}\, eEa\beta$, $a = |\boldsymbol{g}|$ (a — lattice constant). Here we have assumed the electrical field to be along the x-axis, which lies in (100) direction of the lattice.

Following BÖTTGER and BRYKSIN (1979), the effective probability is determined by the condition that the average value of the local change of $\varrho_{\boldsymbol{m}}$ vanishes when any $\mathscr{W}$ is replaced by the random $\widetilde{W}_{\boldsymbol{m}+\boldsymbol{g},\boldsymbol{m}}$. This approach is completely equivalent to the condition of varishing average t-matrix in the coherent-potential approximation (CPA) (cf. Section 6.4.).

Performing such a replacement, the equation for $\varrho_{\boldsymbol{m}}$ (7.1.2) takes the form

$$\sum_{\boldsymbol{g}} \left(\varrho_{\boldsymbol{m}+\boldsymbol{g}}\, e^{-\frac{1}{2}\beta|e|\boldsymbol{E}\boldsymbol{g}} - \varrho_{\boldsymbol{m}}\, e^{\frac{1}{2}\beta|e|\boldsymbol{E}\boldsymbol{g}}\right) \cdot$$

$$+ \left(\frac{\widetilde{W}_{\boldsymbol{m}_0+\boldsymbol{g},\boldsymbol{m}_0}}{\mathscr{W}} - 1\right) (\delta_{\boldsymbol{m},\boldsymbol{m}_0} - \delta_{\boldsymbol{m},\boldsymbol{m}_0+\boldsymbol{g}_0})\, (\varrho_{\boldsymbol{m}_0+\boldsymbol{g}_0}\, e^{-\lambda} - \varrho_{\boldsymbol{m}_0}\, e^{\lambda}) = 0. \tag{7.3.3}$$

Here $\boldsymbol{m}_0 + \boldsymbol{g}_0$ and $\boldsymbol{m}_0$ denote the sites between which $\mathscr{W}$ is replaced by the actual value of the hopping probability ($\boldsymbol{g}_0$ is directed along $\boldsymbol{E}$; if $\boldsymbol{g}_0$ is perpendicular to $\boldsymbol{E}$, then $\varrho_{\boldsymbol{m}}$ does not change). Making use of the $\boldsymbol{k}$-representation

$$\varrho(\boldsymbol{k}) = \sum_{\boldsymbol{m}} \varrho_{\boldsymbol{m}} \, \mathrm{e}^{\mathrm{i}\boldsymbol{k}\boldsymbol{m}}, \qquad \varrho_{\boldsymbol{m}} = \frac{1}{N} \sum_{\boldsymbol{k} \neq 0} \varrho(\boldsymbol{k}) \, \mathrm{e}^{-\mathrm{i}\boldsymbol{k}\boldsymbol{m}} + \frac{1}{N} \, \mathrm{e}^{-\beta|\mu|}, \tag{7.3.4}$$

(N — number of sites) we obtain from (7.3.3) the following expression for $\varrho_{\boldsymbol{m}}$:

$$\varrho_{\boldsymbol{m}} = f_0 + \left(\frac{\tilde{W}_{\boldsymbol{m}_0+\boldsymbol{g}_0,\boldsymbol{m}_0}}{\mathscr{W}} - 1\right) \left(\varrho_{\boldsymbol{m}_0+\boldsymbol{g}_0} \, \mathrm{e}^{-\lambda} - \varrho_{\boldsymbol{m}_0} \, \mathrm{e}^{\lambda}\right) \frac{1}{N} \sum_{\boldsymbol{k} \neq 0} \times \frac{\mathrm{e}^{\mathrm{i}\boldsymbol{k}(\boldsymbol{m}_0+\boldsymbol{m})} (\mathrm{e}^{\mathrm{i}k_x a} - 1)}{\sum_{\boldsymbol{g}} \left(\mathrm{e}^{-\mathrm{i}\boldsymbol{k}\boldsymbol{g}} \, \mathrm{e}^{-\frac{1}{2}\beta|e|\boldsymbol{E}\boldsymbol{g}} - \mathrm{e}^{\frac{1}{2}\beta|e|\boldsymbol{E}\boldsymbol{g}}\right)}. \tag{7.3.5}$$

Hence, the self-consistency equation reads

$$\left\langle \left(\frac{\tilde{W}_{\boldsymbol{m}_0+\boldsymbol{g}_0,\boldsymbol{m}_0}}{\mathscr{W}} - 1\right) \left(\varrho_{\boldsymbol{m}_0+\boldsymbol{g}_0} \, \mathrm{e}^{-\lambda} - \varrho_{\boldsymbol{m}_0} \, \mathrm{e}^{\lambda}\right) \right\rangle = 0. \tag{7.3.6}$$

The quantity $\varrho_{\boldsymbol{m}_0+\boldsymbol{g}_0} \mathrm{e}^{-\lambda} - \varrho_{\boldsymbol{m}_0} \mathrm{e}^{\lambda}$ may be obtained from (7.3.5) by putting there $\boldsymbol{m} = \boldsymbol{m}_0$, $\boldsymbol{m} = \boldsymbol{m}_0 + \boldsymbol{g}_0$. Inserting the resulting expression into (7.3.6), we find for $\mathscr{W}$ the equation

$$\left\langle \frac{\tilde{W} - \mathscr{W}}{W + \chi \mathscr{W}} \right\rangle = 0, \tag{7.3.7}$$

where

$$\chi = \left\{ \frac{1}{N} \sum_{\boldsymbol{k} \neq 0} \frac{\operatorname{cohs} \lambda - \cos(k_x a - \mathrm{i}\lambda)}{2 + \cosh \lambda - \cos(k_x a - \mathrm{i}\lambda) - \cos k_y a - \cos k_z a} \right\}^{-1} - 1. \tag{7.3.8}$$

Replacing now the summation over $\boldsymbol{k}$ by an integration and neglecting the restriction $\boldsymbol{k} \neq 0$ ($\boldsymbol{k} = 0$ is not a singular point), then we have

$$\chi = \frac{3}{1 + 2\varphi(1 - \cosh \lambda)} - 1, \qquad \varphi(z) = z \int_0^\infty \mathrm{d}t \, \mathrm{e}^{-(3+z)t} I_0^3(t), \tag{7.3.9}$$

where I_0 is the modified Bessel function.

From (7.3.8) we obtain for $E \to 0$, $\chi \to d - 1$, where d is the dimension of the system, and for $E \to \infty$, $\chi \to 0$. Thus, in the low-field regime (7.3.7) agrees with the well-known result for the Ohmic regime (Kirkpatrick, 1973). The fact, that χ vanishes at high fields, shows that for any system at high fields the effective dimension of the system approaches unity. Evidently, at high fields the current cannot choose the optimal path by by-passing hard hops, but it is forced to run from electrode to electrode along a straight line. This supports the hypothesis that with increasing field the percolation paths become less twisted.

For a model where the random quantity $\tilde{W}$ may assume two values, $\mathscr{W}_0$ with probability p and $\mathscr{W}_0\delta$, with probability $(1 - p)$, a calculation of the current by (7.3.2) and (7.3.7) yields the curves presented in Fig. 7.11. We see that at sufficiently small δ (large difference between the magnitude of both the hopping probabilities) and $p = p_c = 1/3$, a N-type characteristic occurs for the field dependence of the current $j(E)$. With increasing δ, the region of negative differential conductivity,

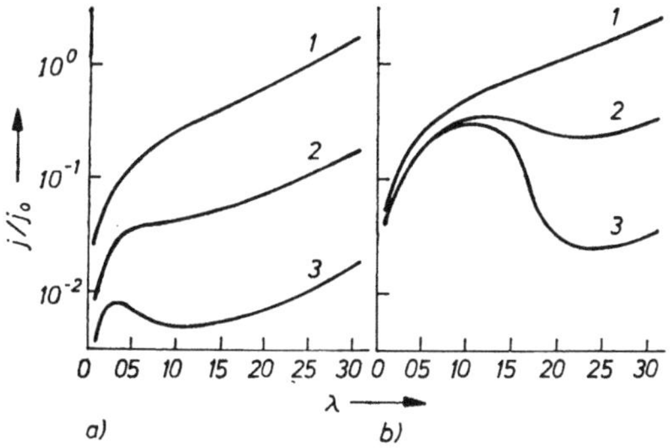

Fig. 7.11. Field dependence of the current. $j_0 = (2|e|/a^2)\,\mathscr{W}_0 \exp(-\beta|\mu|)$ $\lambda = (|e|\,Ea/2kT)$; a) $p = 0.35$, b) $p = 0.6$; (1) $\delta = 10^{-1}$; (2) 10^{-2}, (3) 10^{-3} (after Böttger and Bryksin, 1979)

$\mathrm{d}j/\mathrm{d}E$, vanishes and only a decrease of $\mathrm{d}j/\mathrm{d}E$ with increasing $\boldsymbol{E}$ remains. At still larger δ, $j(E)$ as well as $\mathrm{d}j/\mathrm{d}E$ increases monotonously with the field. With increasing $p(p > p_c)$ the region of negative or decreasing $\mathrm{d}j/\mathrm{d}E$ shifts to higher fields. This is due to the fact that at sufficiently large p the infinite cluster of well conducting elements becomes very compact, which renders possible a good current flow. However, as mentioned before, at high fields the effective dimension of any system approaches unity, i.e. the current is forced also to run through poorly conducting elements. At sufficiently high fields the current is characterized by a slowing down of the increase (or even decrease) of the current with the field.

The increase of the current at high fields is connected with the assumption that $\tilde{W}$ does not depend on the field, which, as mentioned above, is characteristic of over-barrier hops (strong coupling with phonons). In the case that the coupling with phonons is weak, all the above results remain valid, but in (7.3.2) the factor $\sinh \lambda$ must be removed. Owing to this fact, at high fields the current becomes independent of the field, which agrees with the result described in Section 7.1.2.

Böttger and Bryksin (1979) studied the field dependence of the hopping current also by using a model where the sites are spatially distributed at random and $\tilde{W}$ depends on the spacings between the sites. For such a model, at not too high fields, the current depends on E as

$$j = \sigma_0 E \frac{\sinh(|e|\,ER_c/2kT}{|e|\,ER_c/2kT} \exp\left\{-\frac{c\alpha R_c}{9}\left(\frac{|e|\,ER_c}{4kT}\right)^2\right\}. \tag{7.3.10}$$

Here

$$\sigma_0 = c\,\frac{e^2}{R_c{}^2\alpha kT}\,W_c\,e^{-|\mu|/kT} \tag{7.3.11}$$

is the conductivity in a weak field, $W_c = W_0 \exp(-\eta 2\alpha R_c)$ is the hopping probability in the absence of the field (6.4.31), and c is a numerical factor. Such a dependence was also obtained by BRYKSIN and YASHIN (1981) on the basis a more consequent application of the effective-medium approximation to disordered systems with R-hopping (cf. Section 7.3.2.).

7.3.2. *Relaxation currents*

As pointed out in Section 6., at non-zero frequency of the external electrical field, the hopping conductivity is governed by hops across finite clusters whose size increases with decreasing frequency. It is clear that for finite clusters, the field-induced reconstruction of the percolation paths, described in Section 7.2., affects the current to a minor degree than for the infinite cluster. Hence, it follows that the field dependence of the current diminishs with increasing frequency. In the non-Ohmic regime it is more reasonable to study the relaxation process after suddenly turning on the external field, rather than the frequency dependence of the current, since in the case of strong fields, the form of the current does not reflect that of the field applied.

Following BRYKSIN and YASHIN (1981), here we investigate relaxation currents in the presence of an electrical field of arbitrary strenght by using an effective-medium approximation which is analogous to that described in Section 7.3.1. for the dc limit (i.e. for large times). However, unlike Section 7.3.1., here we consider a system with randomly distributed sites. To construct a lattice with nearest-neighbour bonds, we use the procedure illustrated in Section 6.4. The axis [100] of the cubic lattice thus obtained is directed along the field $\boldsymbol{E}$. In the presence of an external field, the probability for hops between nearest neighbours is written as

$$W_{m'm}(E) = W_{m'm}(0)\exp(\gamma\lambda)\,. \tag{7.3.12}$$

Here $W(0)$ is the hopping probability in the absence of a field, $\lambda = eER_c/2kT$, R_c is the characteristic distance for hops in the absence of a field. Furthermore, it holds that $\gamma = 1$ for hops along the field (i.e. for the neighbour placed in the segment of the solid angle directed along the field), $\gamma = -1$ for hops opposite to the field, and $\gamma = 0$ for hops perpendicular to the field. The approximation (7.3.12) is reasonable for not too strong fields, where the field-induced change of the hopping length, δR_c, may be neglected $(eE\delta R_c/kT \ll 1)$.

In this way the problem of evaluating relaxation current becomes similar to the dc problem considered in Section 7.3.1., expect that the term $\mathrm{d}\varrho_m/\mathrm{d}t$ must be included in the rate equation of the type (7.1.2). Using the Laplace representation instead of the time representation, then analogous to (7.3.2), we obtain the following expression for the current in an effective medium with the hopping probability $\mathscr{W}(s)$:

$$j(E, s) = \frac{1}{s}\,2neR_c\mathscr{W}(s)\sinh\lambda\,, \tag{7.3.13}$$

where $n = \mathcal{N} \exp(-|\mu|/kT)$ is the concentration of electrons, and s is the Laplace-transform variable conjugate to t. In the spirit of the above assumptions, the lattice constant of the effective lattice is given by R_c.

For $\mathscr{W}(s)$, the self-consistency equation has the form (7.3.7) or (6.4.24), and the coefficient χ depends on s as well as E:

$$\chi = \left\{\frac{1}{(2\pi)^3} \int \mathrm{d}k \frac{(\cos h\lambda - \cos k_x)}{\frac{s}{2\mathscr{W}} + \cos h\lambda + 2 - \cos k_x - \cos k_y - \cos k_z}\right\}^{-1} - 1. \tag{7.3.14}$$

Analogously as in (6.4.24), the configuration average must be performed with regard to all possible realizations of W. For R-hopping, we have $W = \nu_{\text{ph}} \exp(-2\alpha R)$ and the averaging procedure implies the integration over the intersite spacing R. Since a neighbouring site is placed inside a segment of the solid angle, the Poisson distribution has the form (6.4.22).

The self-consistency equation (6.4.24) was discussed in detail in Section 6.4. Hence, we may immediately proceed to its form (6.4.27), which is valid for not too small times, where the major contribution to the current arises from hopping across clusters of large size. Contrary to Section 6.4., for relaxation currents, in an expression for χ of the type (6.4.28), a field dependent term appears. Hence, for hopping across large clusters, we have

$$\chi = 2 - 6\xi \left(\frac{eER_c}{2kT}\right)^2 + 3\xi \frac{s}{\mathscr{W}}, \qquad \xi \approx 0.253. \tag{7.3.15}$$

On account of (7.3.15), the self-consistency equation (6.4.27) becomes

$$\tilde{\mathscr{W}} \ln \tilde{\mathscr{W}} = \tilde{s}. \tag{7.3.16}$$

Here we have introduced the dimensionless quantities $\tilde{\mathscr{W}} = \mathscr{W}/W_c(E)$, $\tilde{s} = 0.42\alpha R_c s/W_c(E)$; $0.42 = 2\xi/3 \ln 1.5$. Furthermore, $R_c = \frac{1}{2} \eta \mathcal{N}^{-1/3}$ (cf. (6.4.31)), and $W_c(E)$ is the probability of a critical hop on account of the field:

$$W_c(E) = W_c \exp\left(-0.42\alpha R_c \left(\frac{eER_c}{2kT}\right)^2\right), \tag{7.3.17}$$

where W_c is the hopping probability in the absence of a field (6.4.31). In the dc limit $s \to 0$, we have $\mathscr{W} = W_c(E)$, and for the current we recover (up to a numerical factor) the result (7.3.10). The time dependence of the relaxation current is found by taking the inverse Laplace transform of (7.3.2), which after some straightforward transformations leads to the following expression for the current

$$j(E, t) = j(E)\, f\big(tW_c(E)/0.21\alpha R_c\big), \tag{7.3.18}$$

where

$$f(\tau) = \frac{1}{2\pi i} \int\limits_{-i\infty}^{i\infty} \frac{\mathrm{d}\tilde{s}}{\tilde{s}}\, e^{\tilde{s}\tau} \tilde{\mathscr{W}}(\tilde{s}). \tag{7.3.19}$$

Here $j(E)$ is the stationary current

$$j(E) = c\,\frac{en}{kT}\,R_c^2 W_c(E)\,E\,, \tag{7.3.20}$$

where n is the concentration of charge carriers and c is a numerical factor.

According to (7.3.18), in the presence of a field of arbitrary strength, the time dependence of the relaxation current agrees with the corresponding expression in the Ohmic regime after replacing in this expression t by $tW_c(E)/W_c = t \exp\left(-0.42\,\alpha R_c \times (eER_c/2kT)^2\right)$. On account of (7.3.16), from (7.3.19), we get the following asymptotic expressions

$$f(\tau) = \begin{cases} 1 + \left(\dfrac{e}{2\pi}\right)^{1/2} \tau^{-3/2} \exp\,(-\tau/e) & \text{for } \tau \gg 1\,, \\[2ex] \dfrac{\sqrt{2\pi}}{e}\,(\tau \ln^2 \tau)^{-1} & \text{for } -\ln \tau \gg \sqrt{\pi}\,. \end{cases} \tag{7.3.21}$$

Hence, for very large times $\tau \gg 1$, where the current runs through an infinite cluster, the quantity $j(t)$ decreases exponentially with time and its magnitude differs only slightly from that of the stationary current. In this case, the current-voltage characteristic is strongly nonlinear (cf. (7.3.20)), and the differential conductivity decreases with increasing field E. In the multiple-hopping regime, we have $j \sim t^{-1} \ln^{-2} t$, which is equivalent to the frequency dependence (6.4.34). According to (7.3.18), (7.3.20), and (7.3.21), the major feature of this régime is in that the current depends on E algebraically, essentially as $j \sim E$. The experimental data obtained on amorphous transition metal oxides by Bryksin et al. (1982) are in accordance with the theoretical results reported here.

7.4. Non-Ohmic hopping conduction in a one-dimensional model

Following Böttger and Bryksin (1980), let us here study non-Ohmic hopping conduction in a one-dimensional model with nearest-neighbour hops. In the one-dimensional case, the quantity n_e in (7.2.5) represents the mean number of electrons per site and the summation over m in the denominator of (7.2.5) may be replaced by the configurational average. Using for Γ the expression (7.2.3), the relation (7.2.5) becomes

$$\begin{aligned} \frac{|e|\,n_e}{i} = \Big\langle \widetilde{W}^{-1}_{m,m-1} \exp\left\{-\frac{\beta}{2}\,(\varepsilon_m - \varepsilon_{m-1} + eu_m - eu_{m-1})\right\} \\ \times \left[1 + \exp\left\{-\beta(\varepsilon_{m+1} - \varepsilon_m + eu_{m+1} - eu_m)\right\}\right. \\ \left. + \exp\left\{-\beta(\varepsilon_{m+2} - \varepsilon_m + eu_{m+2} - eu_m)\right\} + \cdots\right]\Big\rangle\,, \end{aligned} \tag{7.4.1}$$

where $\langle\ldots\rangle$ indicates the configuration average.

For hops between nearest neighbours and non-correlated random variables $\boldsymbol{R}_m$ and ε_m, the averaging over $\boldsymbol{R}_m$ and ε_m is independently performed. Assuming, as usual, that $\tilde{W}$ does not depend on ε_m, we obtain

$$\frac{|e|\,n_e}{i} = \left\langle \tilde{W}^{-1}_{m,m-1} \exp\left[-\frac{1}{2}\beta(u_m - u_{m-1})\right] \left\langle \exp\left(\frac{1}{2}\beta\varepsilon\right)\right\rangle_\varepsilon \right.$$
$$\times \left[\left\langle \exp\left(-\frac{1}{2}\beta\varepsilon\right)\right\rangle_\varepsilon + \langle \exp(-\beta\varepsilon)\rangle_\varepsilon \left\langle \exp\left(\frac{1}{2}\beta\varepsilon\right)\right\rangle_\varepsilon \right.$$
$$\left.\left. \times \left(\exp\left[-\beta e(u_{m+1} - u_m)\right] + \exp\left[-\beta e(u_{m+2} - u_m)\right] + \cdots\right)\right]\right\rangle_R, \tag{7.4.2}$$

where $\langle\ldots\rangle_\varepsilon$ and $\langle\ldots\rangle_R$ indicate respectively the averaging over the energy and the position of the sites.

On writing

$$u_{m+n} - u_m = (u_{m+n} - u_{m+n-1}) + \cdots + (u_{m+1} - u_m), \tag{7.4.3}$$

and accounting for the fact that $\tilde{W}_{m,m+1}$ depends only on $\boldsymbol{R}_m - \boldsymbol{R}_{m-1}$, we can carry out the averaging over the spacing R between nearest neighbours. Thus the infinite series in (7.4.2) becomes a geometrical one (note that here we have $R > 0$ and $e = -|e|$, so that this series converges). After simple transformations, we finally obtain

$$i = |e|\,n_e(1 - \langle \exp(-\beta\,|e|\,ER)\rangle_R) \left\langle \tilde{W}^{-1}(R) \exp\left(-\frac{1}{2}\beta\,|e|\,ER\right)\right\rangle_R^{-1}$$
$$\times \left\langle \exp\left(\frac{1}{2}\beta\varepsilon\right)\right\rangle_\varepsilon^{-1} \left\{\left\langle \exp\left(-\frac{1}{2}\beta\varepsilon\right)\right\rangle_\varepsilon (1 - \langle \exp(-\beta\,|e|\,ER)\rangle_R)\right.$$
$$\left. + \langle \exp(-\beta e)\rangle_\varepsilon \left\langle \exp\left(\frac{1}{2}\beta\varepsilon\right)\right\rangle_\varepsilon \langle \exp(-\beta\,|e|\,ER)\rangle_R\right\}^{-1}. \tag{7.4.4}$$

In the limit $E \to 0$, the expression (7.4.4) agrees with that derived by Böttger and Bryksin (1975a):

$$i = \frac{e^2\beta l E \exp(\beta\mu)}{\langle \tilde{W}^{-1}(R)\rangle_R \left\langle \exp\left(\frac{1}{2}\beta\varepsilon\right)\right\rangle_\varepsilon^2}, \tag{7.4.5}$$

where $l = \langle R\rangle$ is the average spacing between nearest neighbours, and we have used for n_e the expression $n_e = \exp(\beta\mu)\,\langle \exp(-\beta\varepsilon)\rangle_\varepsilon$ (μ — chemical potential).

However, in the region of nonlinear field dependence, equation (7.4.4) differs from the expression obtained by Böttger and Bryksin (1975a) for a chain of non-linear resistors. Accordingly, the use of non-linear resistors to describe hopping in strong fields (Austin and Sayer, 1974, Böttger and Bryksin, 1975a, Shklovskii, 1976) leads in general to incorrect results. In particular, for a random distribution of the sites in a chain, according to (7.4.4), the current vanishes, because it holds that

$$\left\langle \tilde{W}^{-1} \exp\left(-\frac{1}{2}\beta\,|e|\,ER\right)\right\rangle \to \infty \text{ for } 2\alpha > \frac{1}{l} + \frac{|e|\,E}{2kT}. \tag{7.4.6}$$

On the other hand, for such a distribution, the non-linear resistor model yields a non-vanishing current, due to the "breakdown" of hard hops.

8. The Hall effect

As already mentioned in Section 3.5., in the hopping regime, the Hall mobility u_{H} may strongly differ from the drift mobility u, while for band transport, u_{H} and u are of the same order of magnitude. The examination of the hopping Hall effect in disordered systems is a very interesting problem, because it gives a deeper insight into the hopping mechanism, but it is also a very intricate problem.

The hopping Hall mobility may be calculated by means of the rate equation (3.5.4) and the expression for the current (2.3.3). The most important feature of the rate equation (3.5.4) is the presence of three-site hopping probabilities, which must be included in order to get a non-vanishing Hall effect. Since the Hall current is proportional to the electrical field $\boldsymbol{E}$ and the magnetic field $\boldsymbol{H}$, we must linearize (3.5.4) with respect to $\boldsymbol{E}$ and $\boldsymbol{H}$, similarly as in Section 2.5., where we have linearized the rate equation with respect to $\boldsymbol{E}$ in order to study the hopping transport in the Ohmic regime.

8.1. The random-resistor network and the Hall effect

In what follows, we assume that the three-site hopping probabilities $W_{m_1m_2m}$ are proportional to $\boldsymbol{H}$, i.e. we neglect the zero order-in-$\boldsymbol{H}$ contribution to $W_{m_1m_2m}$, because this contribution is small compared with the two-site probabilities W_{m_1m}.

The calculation of the Hall effect requires knowledge of the linear-in-$\boldsymbol{EH}$ contribution ϱ_m^{EH} to the density matrix ϱ_m. The equation for ϱ_m^{EH} contains the linear-in-$\boldsymbol{E}$ (ϱ_m^{E}), linear-in-$\boldsymbol{H}$ (ϱ_m^{H}) and field independent ($\varrho_m^{(0)}$) contributions to ϱ_m. Therefore, first we must find the latter quantities.

The quantity $\varrho_m^{(0)}$ satisfies equation (2.5.1) and is given by the Fermi function (cf. Section 2.5.):

$$\varrho_m^{(0)} \equiv f_m = \{e^{\beta(\varepsilon_m-\mu)} + 1\}^{-1}. \tag{8.1.1}$$

According to Section 2.5., the equation for ϱ_m^{E} (2.5.16) (in Section 2.5. we used the notation $\varrho_m^{(1)}$ instead of ϱ_m^{E}) is equivalent to Kirchhoff's equations for a random-resistor network with capacitors, which is illustrated in Fig. 2.3.

Let us now determine ϱ_m^{H}. Analogous to (2.5.3), we introduce $\delta\mu_m^{H}$ the change of the chemical potential due to the magnetic field:

$$\varrho_m^{H} = \beta\delta\mu_m^{H} f_m(1 - f_m). \tag{8.1.2}$$

Here, unlike (2.5.3), the factor exp $(i\omega t)$ is lacking, since the magnetic field is assumed to be independent of time. Then, from (3.5.4), we obtain for $\delta\mu_m^H$ the following equation:

$$0 = \frac{1}{e} \sum_{m'} (\delta\mu_{m'}^H - \delta\mu_m^H)/Z_{m'm} + \sum_{m'} (i_{m'm}^{\mathrm{d(e)}} + i_{m'm}^{\mathrm{d(h)}}), \tag{8.1.3}$$

where $Z_{m'm}$ is the resistance of the resistor between sites m and m' in the network of MILLER and ABRAHAMS (cf. (2.5.15) and (2.5.11)) and the quantities i^{d} are defined as

$$\begin{aligned} i_{m_1m}^{\mathrm{d(e)}} &= e \sum_{m_2} f_{m_1}(1 - f_{m_2})(1 - f_m)\, W_{m_1m_2m}^{(\mathrm{e})H}, \\ i_{m_1m}^{\mathrm{d(h)}} &= e \sum_{m_2} (1 - f_{m_1})\, f_{m_2} f_m W_{m_1m_2m}^{(\mathrm{h})H}, \end{aligned} \tag{8.1.4}$$

and have the properties

$$\begin{aligned} i_{m_1m}^{\mathrm{d(e),(h)}} &= -i_{mm_1}^{\mathrm{d(e),(h)}}, \\ \sum_{m_1} i_{m_1m}^{\mathrm{d(e),(h)}} &= 0. \end{aligned} \tag{8.1.5}$$

The quantities $i_{m_1m}^{\mathrm{d(e),(h)}}$ have the dimension of a current and may be interpreted as the electron (e) and hole (h) contributions to the diamagnetic current between sites m_1 and m, which are induced in the system by the magnetic field. According to (8.1.5), the sum of the diamagnetic currents running into a site vanishes. Hence, (8.1.3) may be interpreted as Kirchhoff's laws for a random-resistor network, where the potential at site m is given by $\delta\mu_m^H/e$. Since in the absence of $\boldsymbol{E}$ the potential at the boundaries is equal to zero, we have $\delta\mu_m^H = 0$ and $\varrho_m^H = 0$.

On account of $\varrho_m^H = 0$, from (3.5.4), for ϱ_m^{EH}, we obtain

$$\begin{aligned} i\omega\varrho_m^{EH} &= \sum_{m_1} \{[\varrho_{m_1}^{EH}(1 - f_m) - \varrho_m^{EH} f_{m_1}]\, W_{m_1m}^0 - \varrho_m^{EH}(1 - f_{m_1}) - \varrho_{m_1}^{EH} f_m]\, W_{mm_1}^0\} \\ &\quad + \sum_{m_1m_2} \{f_{m_1}(1 - f_{m_2})(1 - f_m)\, W_{m_1m_2m}^{(\mathrm{e})EH} + [\varrho_{m_1}^E(1 - f_{m_2})(1 - f_m) \\ &\quad - f_{m_1}\varrho_{m_2}^E(1 - f_m) - f_{m_1}(1 - f_{m_2})\,\varrho_m^E]\, W_{m_1m_2m}^{(\mathrm{e})H} \\ &\quad -(1 - f_{m_1})\, f_{m_2} f_m W_{m_1m_2m}^{(\mathrm{h})EH} + [-\varrho_{m_1}^E f_{m_2} f_m \\ &\quad + (1 - f_{m_1})\,\varrho_{m_2}^E f_m + (1 - f_{m_1})\, f_{m_2}\varrho_m^E]\, W_{m_1m_2m}^{(\mathrm{h})H}\}. \end{aligned} \tag{8.1.6}$$

Now we point out that W^{EH}, the linear-in-$\boldsymbol{EH}$ contribution to the three-site hopping probabilities, may be excluded from (8.1.6). To do this, we make use of the symmetry relation (3.5.5) linearized with respect to $\boldsymbol{EH}$:

$$\begin{aligned} W_{m_1m_2m}^{EH} &= \exp\{\mp\beta(\varepsilon_m - \varepsilon_{m_1})\}\,\{W_{mm_1m_2}^{EH} \pm \beta e\boldsymbol{E}(\boldsymbol{R}_m - \boldsymbol{R}_{m_1})\, W_{mm_1m_2}^H\} \\ &= \exp\{\mp\beta(\varepsilon_m - \varepsilon_{m_1})\}\, W_{mm_1m_2}^{EH} \pm \beta e\boldsymbol{E}(\boldsymbol{R}_m - \boldsymbol{R}_{m_1})\, W_{m_1m_2m}^H. \end{aligned} \tag{8.1.7}$$

Here the upper sign corresponds to electrons and the lower one to holes. Inserting (8.1.7) into (8.1.6), we find that the term W^{EH} associated with the electron contribution is proportional to $\sum_{m_1m_2} (1 - f_{m_1})(1 - f_{m_2})\, W_{mm_1m_2}^{(\mathrm{e})EH}$. Replacing the indices of summation as $m_1 \rightleftarrows m_2$, and making use of the property $W^{EH}_{mm_1m_2} = -W^{EH}_{mm_2m_1}$, we see that this contribution is equal to zero. Analogously, we find that the corresponding hole contribution vanishes.

Introducing $\delta\mu^{EH}_m$, the change of the chemical potential in linear approximation with respect to $\boldsymbol{E}$ and $\boldsymbol{H}$, as (cf. (2.5.3))

$$\varrho^{EH}_m = \beta\delta\mu^{EH}_m f_m(1 - f_m), \tag{8.1.8}$$

and using the relation between ϱ^E_m and $\delta\mu^E_m$ (2.5.3), then, on account of (8.1.7), we see that the right-hand side of (8.1.6) depends on $\delta\mu^E_m$ and $e\boldsymbol{E}\boldsymbol{R}_m$ only through the difference $\delta\mu^E_m - e\boldsymbol{E}\boldsymbol{R}_m = eU_m$, where U_m is the potential in the random-resistor network with capacitors describing the (longitudinal) conductivity in the absence of a magnetic field (cf. (2.5.13)).

Thus, (8.1.6) may be written as

$$i\omega C_m \delta\mu^{EH}_m/e = \frac{1}{e} \sum_{m_1} \frac{\delta\mu^{EH}_{m_1} - \delta\mu^{EH}_m}{Z_{m_1m}} + i_m. \tag{8.1.9}$$

Here C_m and Z_{m_1m}, respectively, are the capacities and resistances of the elements in the resistor network illustrated in Fig. 2.3 (cf. (2.5.15)), and the quantity i_m has the dimension of a current and is given by

$$i_m = e\beta \sum_{m_1} (U_{m_1} - U_m)\,(i^{d(e)}_{m_1m} + i^{d(h)}_{m_1m}) = e^2\beta \sum_{m_1m_2} (U_{m_1} - U_m)\, W^H_{m_1m_2m}, \tag{8.1.10}$$

where

$$W^H_{m_1m_2m} = f_{m_1}(1 - f_{m_2})\,(1 - f_m)\, W^{(e)H}_{m_1m_2m} + (1 - f_{m_1})\, f_{m_2} f_m W^{(h)H}_{m_1m_2m}. \tag{8.1.11}$$

According to (2.3.3), in the linear approximation with respect to $\boldsymbol{E}$ and $\boldsymbol{H}$, the current (Hall current) is given by

$$\boldsymbol{j}^{EH} = \frac{i\omega}{\Omega e} \sum_m \boldsymbol{R}_m C_m \delta\mu^{EH}_m. \tag{8.1.12}$$

The set of equations (8.1.9) are equivalent to Kirchhoff's laws for the random-resistor network with capacitors and "extrisic" currents represented in Fig. 8.1.

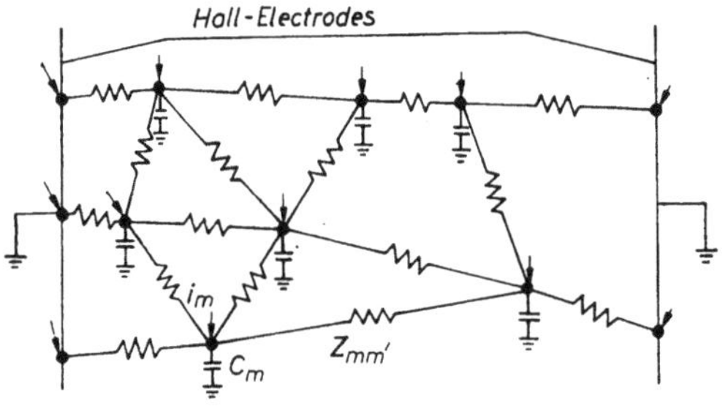

Fig. 8.1. Equivalent scheme of a random-resistor network with capacitors and extrinsic currents, which describes the Hall effect in the hopping regime (cf. equation (8.1.9))

This equivalent scheme corresponds to the situation where at the Hall electrodes all $\delta\mu_m^{EH}$ are equal to each other and vanish (measurement of the Hall current under short-circuit conditions). For $\omega \neq 0$, according to (8.1.12), the density of the Hall current j^{EH} is given by the sum of the currents running through the capacitors and through the external circuit, multiplied by the position vector of the sites, and divided by the volume of the system. The sum rule for the extrinsic currents

$$\sum_m i_m = 0, \tag{8.1.13}$$

which results from (8.1.10) and (8.1.5), guarantees that $\boldsymbol{j}^{EH}$ (8.1.12) does not depend on the choice of the origin.

In particular, for $\omega = 0$ (dc Hall effect), the left hand side of (8.1.9) vanishes, i.e. in the equivalent scheme the capacitors may be ignored. In this case, the density of the Hall current $\boldsymbol{j}^{EH}$ is given by the current in the external circuit, divided by the area of the Hall electrode (cf. the analogous situation in Section 2.5.).

Let us now discuss the origin of the extrinsic currents i_m. Until turning on the electrical field, in the system diamagnetic currents are running and the sum over all diamagnetic currents running into and away from any site is equal to zero. After turning on the electrical field, in the system new currents occur due to the interaction of the diamagnetic currents with the electrical field. The new currents are directed perpendicularly to $\boldsymbol{H}$ as well as to the external electrical field $\boldsymbol{E}$, and they are proportional to the product of the diamagnetic currents and the internal electrical field. The sum of such currents running into and away from any site does in general not vanish (i.e. $i_m \neq 0$). This leads to a change of the distribution function, i.e. to a non-vanishing $\delta\mu_m^{EH}$. The inhomogeneous change of the distribution function causes diffusive currents, which are described by the first term on the right-hand side of (8.1.9). For $\omega = 0$, at each site the sum of the diffusive currents and the currents induced by the external field vanishes (cf. (8.1.9)).

Let us now draw attention to the occurrence of the the p-n anomaly of the sign of the Hall effect (this phenomenon was described in Section 3.5.). According to (8.1.9) and (8.1.10), the Hall current is proportional to the three-site probability $W^H_{m_1m_2m}$ defined in (8.1.11). In the limiting case of electron transport ($f_m \ll 1$), we have $W^H_{m_1m_2m} \approx f_{m_1} W^{(e)H}_{m_1m_2m}$, while for hole transport ($1 - f_m \ll 1$) $W^H_{m_1m_2m} \approx (1 - f_{m_1}) \times W^{(h)H}_{m_1m_2m}$. As pointed out in Appendix 2., the change from $W^{(e)H}$ to $W^{(h)H}$ is realized by the replacement $\boldsymbol{H} \to -\boldsymbol{H}$, $J_{m'm} \to -J_{mm'}$, and $\varepsilon_m \to -\varepsilon_m$, from which the relative sign $-(-1)^n$ results, where n denotes the number of factors J involved in $W^{(e),(h)H}$. Accordingly, in the case of three-site hops, where $n = 3$, the electron and hole Hall coefficients have the same sign. This was first pointed out by Holstein (1973). Note that according to Emin (1977), the sign of the hopping Hall effect depends not only on the local geometry, but also on the nature and the relative orientations of the local orbitals between which the charge carrier moves (above, we have only considered the case of s-states).

For the Hall current, in two cases explicit expressions can be obtained from (8.1.9) and (8.1.12): for ordered systems (small polarons) and for disordered systems in the high-frequency limit (the frequency must be sufficiently low so that the transport has Markovian character). Both the cases are characterized by the absence of diffusive currents — in the ordered case due to the translation symmetry, and in the disordered case due to the presence of displacement currents. This situation is completely analogous to that encountered in calculating the ac hopping conductivity near the high-frequency plateau (cf. Section 6.). In both cases the first term on the right-hand side of (8.1.9) may be ignored and we have $\delta\mu_m^{EH} = ei_m/i\omega C_m$. Inserting this $\delta\mu_m^{EH}$ into the expression for the current (8.1.12), and using the expression for i_m (8.1.10), where U_m must be replaced by $-e\boldsymbol{E}\boldsymbol{R}_m$ (since here we have $\delta\mu_m^E = 0$), then for the Hall current, we find formula (3.5.10), which after symmetrization takes the form (3.5.11). By means of this formula, in Section 3.5. the Hall effect of small polarons in crystals was studied, and by Böttger and Bryksin (1977b) the Hall effect in disordered systems in the region of the high-frequency plateau for strong as well as weak interaction with phonons.

Note that in formula (3.5.11), there is no exponentially small factor of the type $\exp\{-\eta\alpha\mathcal{N}^{-1/3}\}$, which is characteristic of the dc conductivity. The absence of this factors is due to the fact that, according to (3.5.11), $\boldsymbol{j}^{EH}$ is determined by the configuration average of $W^H_{m_1m_2m}$, which on its part is proportional to the product of three resonance integrals $J_{m'm}$ associated with the triad of sites involved. The main contribution to $\boldsymbol{j}^{EH}$ comes from closely placed sites, with a spacing between them of order α^{-1}. This situation is analogous to that encountered for the (longitudinal) ac conductivity in the region of the high-frequency plateau, where the conductivity is governed by pairs of sites with a spacing of order α^{-1} (cf. (6.4.39)). As will be shown below, the dc hopping Hall effect is proportional to an exponentially small factor of the type $\exp\{-\eta\alpha\mathcal{N}^{-1/3}\}$.

8.2. The dc Hall effect

A calculation of the dc Hall effect, by means of (8.1.9) and (8.1.12), requires that we find at first the potentials U_m at the sites of the resistor network of Miller and Abrahams in the absence of a magnetic field $\boldsymbol{H}$. Then, by means of the quantities U_m and the three-site hopping probabilities, given in Appendix 2., the extrinsic currents i_m (8.1.10) can be calculated. Finally, solving the equivalent scheme illustrated in Fig. 8.1, the current density $\boldsymbol{j}^{EH}$ can be obtained. The realization of such a programme is a very intricate problem and requires certain approximations. In the present section we average the random resistor network with extrinsic currents by making use of ideas of the percolation theory (Böttger and Bryksin, 1977c). Such a procedure is reasonable if the spread of the random parameters of the system is sufficiently large. For R-percolation, for instance, this means that the dimensionless parameter $\alpha\mathcal{N}^{-1/3}$ must be sufficiently large.

The averaging of the equivalent scheme represented in Fig. 8.1 runs as follows:

Analogously as in the case of the dc hopping conductivity (cf. Section 5.), the resistors $Z_{m'm}$ are removed first of all from the network and then re-incorporated one after the other, the smallest first. At a certain value $Z_{m'm} = Z_c$ (Z_c — critical resistance) Ohmic and Hall currents simultaneously occur as the critical cluster grows isotropically in all directions. A further replacement of resistors $Z_{m'm} > Z_c$ is assumed not essentially to affect the Hall current in the system, so that we may confine ourselves to a consideration only of the critical cluster (analogous to the situation of calculating the Ohmic current). Strictly speaking, also resistors being somewhat larger than Z_c have to be re-incorporated, so that the critical distance is $\sim R_c + \alpha^{-1}$ but not R_c, i.e. a critical network occurs, and not a first percolation path (cf. Section 5.3.). Such a procedure leads to another numerical factor in σ, but the percolation approach is generally correct only to within such a factor, i.e. the determination of it exceeds the accuracy of the method.

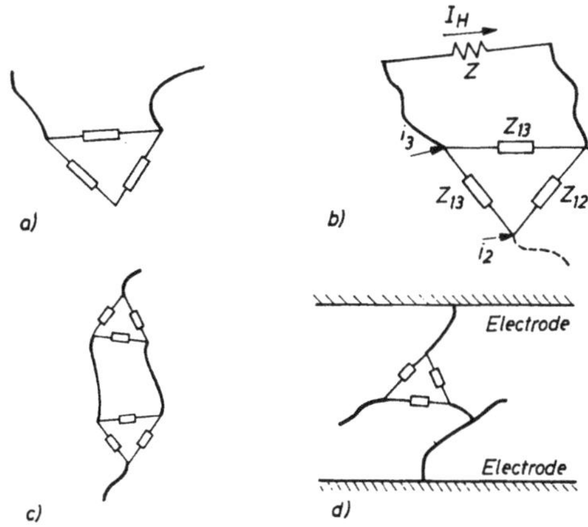

Fig. 8.2. Different branching situations of one-dimensional paths (see text). Bold lines indicate percolation paths, and dashed lines-dead ends. In b), the resistor Z is introduced for evaluating the contribution of this configuration to the Hall current I_H

Extrinsic currents i_m occur in the critical cluster only at those sites where three-site hops are possible. Four such situations might occur: a) three sites "enclodes" in a one-dimensional percolation path (Fig. 8.2a)); b) a triad of sites on a percolation path at the branching-off of a "dead" end (branch of a percolation path not reaching the electrodes) (Fig. 8.2b)); c) a triad at a bifurcation of a percolation path which subsequently reunits (Fig. 8.2c)); d) a triad at a bifurcation of a percolation path, where each new path reaches an electrode or joins with another percolation path (Fig. 2d)). The branching situations of the type a), b), and c) do not contribute to the Hall current. In what follows, this will be shown for the type b) (note that the situation a) is a special case of the situation b)). We assume, for simplicity, that the triads of sites are isolated from each other.

Consider now the contribution to the Hall current I_H from the triad shown in Fig. 8.2b), where Z is some load resistor. Solving Kirchhoff's equation for this triad on account of $i_1 + i_2 + i_3 = 0$ (cf. (8.1.13)), we obtain

$$I_H = \frac{Z_{13}[i_3(Z_{12} + Z_{23}) + i_2 Z_{12}]}{Z_{13}(Z_{12} + Z_{23}) + Z(Z_{12} + Z_{23} + Z_{13})}. \tag{8.2.1}$$

The currents i_2 and i_3 are defined in (8.1.10). Accounting for the symmetry properties for W^H (cf. (3.5.5), (3.5.7),

$$W^H_{m_1m_2m} = -W^H_{m_1mm_2} \text{ and } W^{H'}_{m_1m_2m} = -W^H_{m_2m_1m}, \text{ we get}$$

$$i_2 = e^2\beta(U_3 - U_1)\, W^H_{123}, \qquad i_3 = e^2\beta(U_1 - U_2)\, W^H_{123}. \tag{8.2.2}$$

Now U_i must be calculated. To do this, we consider the same replacement scheme, but where now an Ohmic current I is running through the one-dimensional paths and the extrinsic currents i_m are absent. Then, the U_i's are the potentials at the sites of the triad due to the current I. Solving the corresponding Kirchhoff's equations we obtain

$$U_1 - U_3 = \frac{IZ_{13}(Z_{12} + Z_{23})}{Z_{13} + Z_{12} + Z_{23}}, \quad U_1 - U_2 = \frac{IZ_{23}Z_{13}}{Z_{13} + Z_{12} + Z_{23}}. \tag{8.2.3}$$

Inserting (8.2.2) and (8.2.3) into (8.2.1), we obtain $I_H = 0$. Analogously, likewise for the situation shown in Fig. 8.2c), it may be proved that $I_H = 0$.

The fact that the situations a) to c) do not contribute to the Hall current, may be intuitively understood as follows: The magnetic field deflects the current, but in the resistor network such a deflection can occur only on branchings of one-dimensional percolation paths. The Hall current is proportional to the redistribution of the currents in the branching points. It is easy to see that in the situations a) to c) no effective redistribution of the current occurs. Consequently, a Hall current occurs only due to branchings of the type represented in Fig. 8.2d). In other words, the Hall effect may be investigated by considering only the critical backbone cluster characterized by neglection of dead ends and twisting of the percolation paths. This cluster is schematically shown in Fig. 8.3.

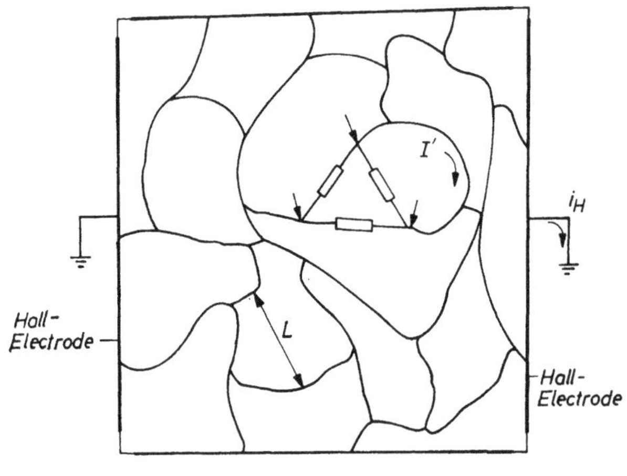

Fig. 8.3. Schematic representation of the backbone cluster. This figure shows a triad of sites with extrinsic currents at the branching point of percolation paths (L — correlation length, I' — current along percolation paths, i_H — Hall current)

The concentration of such branchings (triads) is small and the distance between them is of the order of the correlation length L (cf. Section 5.), which is much larger than the average distance between the sites. Accordingly, we may consider the contribution of a single triad to the Hall current, and after that sum up the contributions of all triads in order to get the total Hall current. This procedure is equivalent to an averaging over all possible triad configurations. Note that two resistances of the triad (Z_{12} and Z_{13}) are less than or equal to the critical resistance Z_c, while the third resistance may in principle be larger than Z_c.

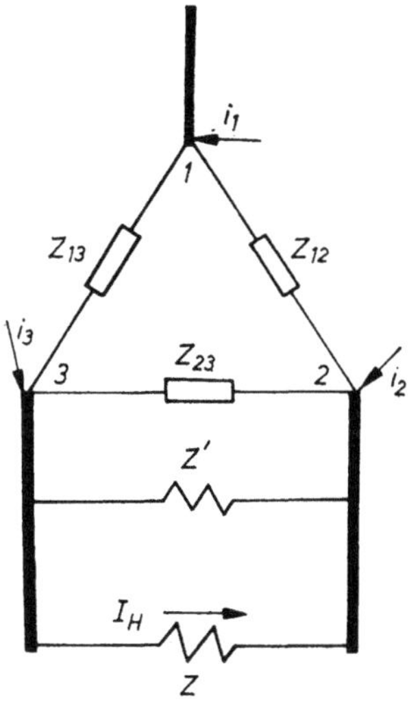

Fig. 8.4. Equivalent scheme for the evaluation of the contribution of the Hall current I_H of a triad of sites (see text). Bold lines indicate one-dimensional percolation paths

Hence, let us consider a triad placed on a branching point in the backbone cluster. The equivalent scheme for such a triad is represented in Fig. 8.4. This scheme contains two equivalent resistors Z' and Z. Z' is the effective resistor for the currents running through the external (with regard to the selected triad) circuit, but which are not running through the Hall electrodes (for instance, the current I' in Fig. 8.3). The circuit with the resistor Z corresponds to the current running through the Hall electrode. Accordingly, in Fig. 8.4, I_H is the contribution to the Hall current from the selected triad. The resistance Z' is of the order of the sum of the resistances along a section of a percolation path of length L, and it holds that $Z' \gtrapprox Z_c$. The resistance Z increases with increasing dimension of the system (it is a macroscopical quantity) and therefore $Z \gg Z_c$ holds. Moreover, Z does not depend on the choice of the triad, because the Hall current runs through the whole system. Hence, Z is a self-averaging quantity. In what follows, we assume that $Z' \gg Z_c$. The possible error arising from this assumption cannot be serious.

Owing to the latter assumption, the triad of resistors with external currents may be replaced by an equivalent voltage generator. After determining the difference of the voltages between sites 2 and 3, the current of interest, I_H, can readily be found. To do this, at first the currents i_1, i_2, and i_3 are assumed to be known, whereby

$i_1 + i_2 + i_3 = 0$. Then, the external circuit is detached and the current i_{23}, running between sites 2 and 3 due to the extrinsic currents, is calculated. The difference of the potentials $v_3 - v_2$ is equal to $i_{23}Z_{23}$ and is given by

$$v_3 - v_2 = Z_{23} \frac{i_2 Z_{12} - i_3 Z_{13}}{Z_{12} + Z_{13} + Z_{23}}. \tag{8.2.4}$$

According to (8.2.2), an evaluation of the currents i_2 and i_3 requires knowledge of $U_2 - U_1$ and $U_3 - U_1$. We assume that

$$U_2 - U_1 \approx Z_{12} I, \qquad U_3 - U_1 \approx Z_{13} I \tag{8.2.5}$$

holds, where I is the Ohmic current in a one-dimensional percolation path. I is assumed to be approximately the same in all sections of the backbone cluster. I is of the order

$$I \approx \frac{L^2}{S} i = L^2 j = L^2 \sigma E, \tag{8.2.6}$$

where S is the area of one of the Ohmic electrodes, i the Ohmic current, j the Ohmic current density, σ the dc conductivity of the system, L the correlation length, and E the strength of the external electrical field. In deriving (8.2.6), we have assumed that the number of one-dimensional paths running into an electrode is of order S/L^2.

Now I_{H} may be determined. We have

$$I_{\mathrm{H}} = \frac{v_3 - v_2}{Z} \approx e^2 \beta L^2 \sigma \frac{W_{123}^H}{Z} \frac{Z_{12} Z_{13} Z_{23}}{Z_{12} + Z_{13} + Z_{23}} E. \tag{8.2.7}$$

The resistance Z is proportional to $1/\sigma$, and it depends on the size of the system and on L. Evidently, Z does not depend on the extent of any plane parallel to the Hall electrode, if the area of the Hall electrode is sufficiently large. Consequently, $Z = (l/\sigma) \times f(L)$ holds, i.e. from a consideration of the dimension it follows that

$$Z = l/\sigma L^2. \tag{8.2.8}$$

By multiplying I_{H} by the number of triads in the volume Ω, which is equal to Ω/L^3, we obtain i_{H}, the total Hall current. The density of the Hall current $j^{EH} \equiv j_{xy}$ is related to i_{H} by $j_{xy} = i_{\mathrm{H}}/S_{\mathrm{H}}$, where S_{H} is the area of a Hall electrode. Thus, for the transverse (Hall-)conductivity $\sigma_{xy} = j_{xy}/E$, we obtain

$$\sigma_{xy} = a_{\mathrm{H}} e^2 \frac{\sigma^2 L}{kT} \left\langle W_{123}^H \frac{Z_{12} Z_{13} Z_{23}}{Z_{12} + Z_{13} + Z_{23}} \right\rangle_{\mathrm{c}}. \tag{8.2.9}$$

Here a_{H} is a numerical factor which cannot be determined within the framework of the theory presented here. The symbol $\langle \ldots \rangle_{\mathrm{c}}$ indicates the average over all possible configurations. The average must be performed on account of the fact that two of the resistances of the triad are less than the critical resistance Z_{c}. The latter requirement implies that in the triangle, whose vertices are given by the sites of the triad, two legs are shorter than or equal to the critical hopping distance R_c. This condition holds for systems with R-percolation as well as with R-ε-percolation.

The probability W_{123}^H involved in (8.2.9) is defined in (8.1.11). By means of the three-site hopping probabilities $W_{123}^{(e),(h)H}$ determined in Appendix 2, W_{123}^H can be obtained for strong as well as weak electron-phonon coupling.

For strong electron-phonon coupling, from (A 2.7) we get

$$W_{123}^H = \frac{\pi}{16\sqrt{3}} \frac{\cosh\left\{\left[\frac{1}{3}(\varepsilon_1 + \varepsilon_2 + \varepsilon_3) - \mu\right]\Big/ 2kT\right\}}{\cosh\{(\varepsilon_1 - \mu)/2kT\} \cosh\{(\varepsilon_2 - \mu)/2kT\} \cosh\{(\varepsilon_3 - \mu)/2kT\}}$$
$$\times \frac{J_{12}J_{23}J_{31}}{\hbar E_a kT} \frac{eH}{\hbar c} [(\boldsymbol{R}_3 - \boldsymbol{R}_1) \times (\boldsymbol{R}_2 - \boldsymbol{R}_1)]$$
$$\times \exp\left\{-\left[\frac{4}{3} E_a + \frac{1}{24E_a}(\varepsilon_{12}^2 + \varepsilon_{13}^2 + \varepsilon_{23}^2)\right]\Big/ kT\right\}, \tag{8.2.10}$$

where $\varepsilon_{mm'} = \varepsilon_m - \varepsilon_{m'}$. For weak electron-phonon coupling from (A 2.8) we obtain

$$W_{123}^H = \frac{eH}{32c} [(\boldsymbol{R}_3 - \boldsymbol{R}_1)\times(\boldsymbol{R}_2 - \boldsymbol{R}_1)] \frac{J_{12}J_{23}J_{31}}{J_0^4} \nu_{\text{ph}}^2$$
$$\times \left\{\left[\cosh\left((\varepsilon_1 - \mu)/2kT\right) \cosh\left((\varepsilon_2 - \mu)/2kT\right) \cosh(\varepsilon_{13}/2kT)\right.\right.$$
$$\left.\left.\times \cosh(\varepsilon_{23}/2kT)\right]^{-1} + 3 \rightleftarrows 2 + 3 \rightleftarrows 1\right\}. \tag{8.2.11}$$

Here, we have introduced the notation

$$\widetilde{W}_{m'm} = \frac{|J_{m'm}|^2}{J_0^2} \nu_{\text{ph}}. \tag{8.2.12}$$

According to (A 2.9), ν_{ph} only weakly depends on the random quantities ε_m. As usual, this dependence will be neglected. The frequency ν_{ph}, which characterizes the probability of two-site hops, was introduced in Section 6. (cf. (6.3.15)).

Let us now consider in more detail the case of R-percolation, where the spread of the energy levels may be neglected and where $|\varepsilon_0 - \mu|/kT > 1$ holds (ε_0 denotes the position of the maximum of the density of states). Then, according to (8.2.10) and (8.2.11), for weak as well as strong interaction with phonons, W_{123}^H may be written as

$$W_{123}^H = W_0^H \exp\{-\alpha[|\boldsymbol{R}_1 - \boldsymbol{R}_2| + |\boldsymbol{R}_1 - \boldsymbol{R}_3| + |\boldsymbol{R}_2 - \boldsymbol{R}_3|]\}. \tag{8.2.13}$$

The quantitiy W_0^H only weakly (power-like) depends on the random quantities — the lengths of the legs of the triangle formed by the triad of sites. As will be shown below, the lengths of the three legs are of the order of $\mathcal{N}^{-1/3}$, the mean distance between the sites. Hence, in what follows, we replacc in W_0^H all distances by $\mathcal{N}^{-1/3}$. Then, according to (8.2.10), for strong electron-phonon coupling, we find to within a numerical factor

$$W_0^H = \frac{eH}{\hbar c} \frac{J_0^3}{\hbar E_a kT} \mathcal{N}^{-2/3} \exp\left\{-\left(\frac{4}{3} E_a + \varepsilon_3\right)\Big/ kT\right\}, \tag{8.2.13}$$

and, from (8.2.11), for weak coupling

$$W_0{}^H = \frac{eH}{c} \frac{v_{\text{ph}}^2}{J_0} \mathcal{N}^{-2/3} \exp\left(-\varepsilon_3/kT\right). \tag{8.2.14}$$

Inserting now W^H_{123} (8.2.13) and $Z_{m'm}$ (6.3.2) into σ_{xy} (8.2.9), we obtain

$$\sigma_{xy} \approx \frac{e^2}{kT} \sigma^2 L W_0{}^H Z_0{}^2 e^{2\varepsilon_3/kT} R_c{}^{-6} \int\limits_0^{R_c} \mathrm{d}R_1 R_1{}^2 \int\limits_0^{R_c} \mathrm{d}R_2 R_2{}^2$$

$$\times \Pi\left(\frac{R_1}{R_c}\right) \Pi\left(\frac{R_2}{R_c}\right) \int\limits_{-\pi}^{\pi} \mathrm{d}\varphi \sin\varphi$$

$$\times \frac{\exp\left\{\alpha\left(R_1 + R_2 + \sqrt{R_1{}^2 + R_2{}^2 - 2R_1R_2\cos\varphi}\right)\right\}}{\exp(2\alpha R_1) + \exp(2\alpha R_2) + \exp\left\{2\alpha\sqrt{R_1{}^2 + R_2{}^2 - 2R_1R_2\cos\varphi}\right\}}. \tag{8.2.15}$$

Here $\Pi(R/R_c)$ denotes the probability of finding no more than two sites of the backbone cluster within a sphere of radius R drawn about a site placed on a branching point of a percolation path. It should be noted that Π depends only on the dimensionless parameter R/R_c (and not αR!), i.e. Π is determined solely by the topology of the backbone cluster. The integration is over two distances and one angle, because the triad forms a triangle (R_1 is the distance between sites 1 and 2, R_2 the distance between sites 1 and 3, and φ the angle between the corresponding position vectors). The factor $R_c{}^{-6}$ has been introduced for normalization. As usual, (8.2.15) holds to within a numerical factor.

In the limiting case of sufficiently large values $\alpha\mathcal{N}^{-1/3}$, the main contribution to the integral (8.2.15) comes from the region around the point $R_1 = R_2 = R_c$, $\varphi = \pi/6$, i.e. from triads forming equilateral triangles with length R_c of a leg. Expanding in (8.2.15) the exponent in a series around this point, we find to within a numerical factor (in particular, replacing $\Pi(1)$ by unity):

$$\sigma_{xy} \approx \frac{e^2}{kT} (\alpha R_c)^{-3} \sigma^2 L W_0{}^H Z_0{}^2 \exp\left(\frac{2\varepsilon_3}{kT} + \alpha R_c\right). \tag{8.2.16}$$

Now the Hall mobility can be readily found by using the formula $u_H = (c/H)\,\sigma_{xy}/\sigma$ (3.5.3). With the aid of the relation $\sigma \approx (Z_cL)^{-1} = (Z_0L)^{-1} \exp\{-2\alpha R_c - \varepsilon_3/kT\}$ (cf. Section 5.), we find that

$$u_H \approx \frac{e^2}{kT} \frac{c}{H} W_0{}^H Z_0 \exp\left\{\frac{\varepsilon_3}{kT} - \alpha R_c\right\} (\alpha R_c)^{-3}. \tag{8.2.17}$$

For strong electron-phonon coupling, we have $Z_0 \approx \left(\hbar kT \sqrt{E_a kT}/e^2 J_0{}^2\right) \exp(E_a/kT)$ (cf. (5.1.8)), where E_a is the activation energy (3.2.8). By means of this expression,

W_0^H from (8.2.13), and (8.2.17), we obtain for strong electron-phonon coupling,

$$u_{\mathrm{H}} = b_{\mathrm{H}} \frac{e\mathcal{N}^{-2/3}}{\hbar} \frac{J_0}{\sqrt{E_{\mathrm{a}}kT}} (\alpha\mathcal{N}^{-1/3})^{-3} \exp\left\{-\frac{\eta}{2}\alpha\mathcal{N}^{-1/3} - \frac{E_{\mathrm{a}}}{3kT}\right\}, \qquad (8.2.18)$$

where b_{H} is a numerical factor whose magnitude remains open, and R_{c} may be expressed as $R_{\mathrm{c}} = \eta\mathcal{N}^{-1/3}/2$, with $\eta = 1.7$.

For weak coupling with phonons, we have $Z_0 \approx kT/e^2\nu_{\mathrm{ph}}$ and W_0^H is given in (8.2.14). Thus, by means of (8.2.17), for weak electron-phonon coupling, we obtain

$$u_{\mathrm{H}} = b_{\mathrm{H}}' \frac{e\mathcal{N}^{-2/3}}{\hbar} \frac{\hbar\nu_{\mathrm{ph}}}{J_0} (\alpha\mathcal{N}^{-1/3})^{-3} \exp\left\{-\frac{\eta}{2}\alpha\mathcal{N}^{-1/3}\right\}, \qquad (8.2.19)$$

where b_{H}' is a numerical factor.

On account of (3.5.3), for strong coupling with phonons, the Hall coefficient becomes

$$R = \frac{c_{\mathrm{H}}}{enc} \frac{kT}{J_0} (\alpha\mathcal{N}^{-1/3})^{-3+\nu} \exp\left\{\frac{\eta}{2}\alpha\mathcal{N}^{-1/3} + 2E_{\mathrm{a}}/3kT\right\}, \qquad (8.2.20)$$

and for weak coupling,

$$R = \frac{c_{\mathrm{H}}'}{enc} \frac{kT}{J_0} (\alpha\mathcal{N}^{-1/3})^{-3+\nu} \exp\left(\frac{\eta}{2}\alpha\mathcal{N}^{-1/3}\right), \qquad (8.2.21)$$

where c_{H} and c_{H}' are numerical factors. Recall that in the critical network the correlation length is given by $L = \mathcal{N}^{-1/3}(\alpha\mathcal{N}^{-1/3})^{\nu}$, where $\nu = 0.9$ (cf. Section 5.3.). Furthermore, in (8.2.20) and (8.2.21), n is the effective concentration of charge carriers, $n = \mathcal{N} \exp(-\varepsilon_3/kT)$.

Hence, for strong as well as weak interaction with phonons, the Hall mobility exponentially depends on the concentration of sites $\mathcal{N}$, as $\sim \exp(-\eta\alpha\mathcal{N}^{-1/3}/2)$. Note that the numerical factor in this exponent is two times smaller than that in the exponent of the (longitudinal) conductivity σ. With regard to the temperature dependence of u_{H} in the case of R-percolation, note that for strong electron-phonon coupling, we have $u_{\mathrm{H}} \sim \exp\left[-\frac{1}{3}(E_{\mathrm{a}}/kT)\right]$ i.e.(the activation energy is three times smaller than that of the drift mobility), while for weak coupling, u_{H} may only weakly depend on temperature (power-like) through ν_{ph}.

According to (8.2.20) and (8.2.21), for strong as well as weak interaction with phonons, the Hall coefficient R strongly differs from $1/enc$. Hence, the ratio u_{H}/u may differ from unity by several orders of magnitude, and it strongly (exponentially) depends on $\mathcal{N}$ and, for strong electron-phonon coupling, it also depends exponentially on temperature.

A somewhat different method for averaging the expression (8.2.9) was proposed by Friedman and Pollak (1978), who used an explicit expression for the function $\Pi(R/R_{\mathrm{c}})$, which in our opinion, however, is not well substantiated. The corresponding expression decreases exponentially with increasing R, $\Pi \sim \exp\left(-\frac{4\pi}{3}\mathcal{N}R^3\right)$, so that the integrand in (8.2.15) has a maximum with respect to R_1 and R_2. For not too large $\alpha\mathcal{N}^{-1/3}$ ($\alpha\mathcal{N}^{-1/3} < 27$) this maximum is placed within the limits of integration, which allows the upper limits of integration to be replaced by ∞. In this manner, for weak electron-phonon coupling one gets

$$u_{\mathrm{H}} = 1.69 \times 10^{-2} \frac{e\mathcal{N}^{-2/3}}{\hbar} \frac{\hbar\nu_{\mathrm{ph}}}{J_0} (\alpha\mathcal{N}^{-1/3})^{1.75} \times \exp\left[-\eta\alpha\mathcal{N}^{-1/3} + 0.11 (\alpha\mathcal{N}^{-1/3})^{3/2}\right]. \tag{8.2.22}$$

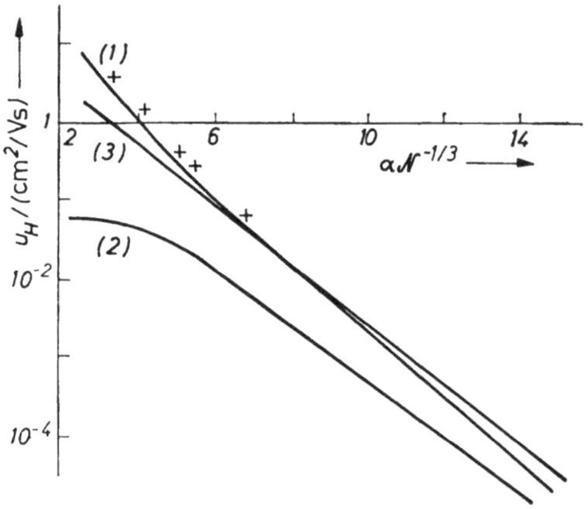

Fig. 8.5. Hall mobility u_{H} in dependence on the parameter $\alpha\mathcal{N}^{-1/3}$: (1) is calculated by (8.2.19) with $b'_{\mathrm{H}} = 28$, (2) by (8.2.22), and (3) by (8.2.29); $\frac{e}{\hbar\alpha^2}\frac{\hbar\nu_{\mathrm{ph}}}{J_0} = 5.66\ \mathrm{cm^2/Vs}$, $\eta = 1.73$; crosses represent computer results of BUTCHER and MC INNES (1981)

Results of evaluations of u_{H} on the basis of (8.2.19) and (8.2.22) are illustrated in Fig. 8.5. We learn from this figure that for not too small values $\alpha\mathcal{N}^{-1/3}$ both dependences practically run parallel, i.e. the results differ only by a numerical factor. It should be noticed that a percolation approach may not pretend to give the correct numerical factor. Hence, using (8.2.19), the number b_{H} was set equal to unity. The authors of the result (8.2.22) give some value b_{H}, which, however, is hardly correct, since already in the starting formula (8.2.9) a numerical factor was dropped. Note that the percolation approach does not apply to the case of small values $\alpha\mathcal{N}^{-1/3}$.

An interesting expression for the Hall current $\boldsymbol{j}^{EH}$ was derived by BUTCHER and KUMAR (1980), who succeeded in expressing $\boldsymbol{j}^{EH}$ only through the potentials U_m in the resistor network with capacitors (cf. Fig. 2.3.), after excluding the quantities $\delta\mu_m^{EH}$, which occur in (8.1.9) and (8.1.12). The corresponding procedure may be performed for R-hopping, where the capacitors $C_m = C$ do not depend on m.

In what follows, we outline this method. To do this, let us introduce the Green function $G_{m'm}$ of equation (8.1.9):

$$i\omega C G_{m'm} = \delta_{m'm} + \sum_{m_1} (G_{m'm_1} - G_{m'm}) Z_{m_1 m}^{-1}. \tag{8.2.23}$$

Then, it holds that $\delta\mu_m^{EH} = e \sum_{m'} i_{m'} G_{m'm}$, and the expression for the current (8.1.12) takes the form

$$\boldsymbol{j}^{EH} = \frac{i\omega C}{\Omega} \sum_{m'm} i_{m'} G_{m'm} \boldsymbol{R}_m . \tag{8.2.24}$$

Note that $G_{m'm}$ is also the Green function of equation (2.5.16), so that $\frac{U_m}{i\omega C} = -\sum_{m'} \times (\boldsymbol{E}\boldsymbol{R}_{m'})\, G_{m'm}$ holds. Because the Green function is symmetric, i.e. $G_{m'm} = G_{mm'}$, is also holds that $\frac{U_m}{i\omega C} = -\sum_{m'} (\boldsymbol{E}\boldsymbol{R}_{m'})\, G_{mm'}$. The latter equation allows $G_{m'm}$ to be excluded from (8.2.24). To do this, we introduce the vector $\boldsymbol{U}_m$, whose projections on the x-, y-, and z-axes (U_m^x, U_m^y, and U_m^z) give the electrochemical potentials for the case that the external field $\boldsymbol{E}$ is directed along the x-, y-, and z-axes, respectively. Then, we have

$$j_y = -\frac{1}{\Omega E} \sum_m i_m U_m^y . \tag{8.2.25}$$

Inserting the expression for the extrinsic currents i_m (8.1.10) into (8.2.25) and making use of the symmetry relations holding for $W^H_{m_1m_2m}$, finally we find that

$$\boldsymbol{j}^{EH} = \frac{e^2}{6\Omega kTE} \sum_{m_1m_2m} W^H_{m_1m_2m} [(\boldsymbol{U}_{m_1} - \boldsymbol{U}_{m_2}) \times (\boldsymbol{U}_{m_2} - \boldsymbol{U}_m)] . \tag{8.2.26}$$

Note that, because the quantity C does not appear in (8.2.26), this relation also holds for R-ε-percolation, but only in the dc limit.

Evidently, formula (8.2.26) is very suitable for computer calculations, because it does not require the solution of the equation for $\delta\mu_m^{EH}$ (8.1.9). It is worth noticing that (8.2.26) exactly passes into the expression for the Hall mobility on the high-frequency plateau (3.5.11) if we replace $\boldsymbol{U}_m \to -\boldsymbol{E}\boldsymbol{R}_m$, which may be done in the high-frequency limit.

Butcher and McInnes (1981) evaluated the Hall mobility on the basis of (8.2.26) analytically by using some ansatz for $\langle(\boldsymbol{U}_{m_1} - \boldsymbol{U}_{m_2}) \times (\boldsymbol{U}_{m_2} - \boldsymbol{U}_m)\rangle'$, the average of $(\boldsymbol{U}_{m_1} - \boldsymbol{U}_{m_2}) \times (\boldsymbol{U}_{m_2} - \boldsymbol{U}_m)$ over the positions of all sites except the triad m_1, m_2, and m. This ansatz reads (cf. Section 5.3.5.)

$$\begin{aligned} \langle(\boldsymbol{U}_{m_1} - \boldsymbol{U}_{m_2}) \times (\boldsymbol{U}_{m_2} - \boldsymbol{U}_m)\rangle' &= E^2 [(\boldsymbol{R}_{m_1} - \boldsymbol{R}_{m_2}) \times (\boldsymbol{R}_{m_2} - \boldsymbol{R}_m)] \\ &\times f(\boldsymbol{R}_{m_1} - \boldsymbol{R}_{m_2})\, f(\boldsymbol{R}_m - \boldsymbol{R}_{m_2})\, f(\boldsymbol{R}_{m_1} - \boldsymbol{R}_m) , \end{aligned} \tag{8.2.27}$$

where

$$f(\boldsymbol{R}) = \begin{cases} 1 & \text{if } |\boldsymbol{R}| > R_c \\ \exp [2\alpha(|\boldsymbol{R}| - R_c)] & \text{if } |\boldsymbol{R}| < R_c , \end{cases} \tag{8.2.28}$$

i.e. outside a sphere with radius R_c the elctrochemical potential is equal to the external potential and inside this sphere it decreases exponentially. After introducing this ansatz, an evluation of $\boldsymbol{j}^{EH}$ on the basis of (8.2.26) reduces to a simple averaging over a triad of sites m_1, m_2, and m, which, for weak electron-phonon coupling, leads to the following result

$$u_{\mathrm{H}} = \frac{24.3}{8} \frac{e\mathcal{N}^{-2/3}}{\hbar} (\alpha\mathcal{N}^{-1/3})^{-2} \frac{\hbar\nu_{\mathrm{ph}}}{J_0} \exp\left(-\frac{1}{2}\eta\alpha\mathcal{N}^{-1/3}\right), \tag{8.2.29}$$

which is very close (except a power-like dependence on $(\alpha\mathcal{N}^{-1/3})^{-1}$) to (8.2.19).

The dc Hall mobility was also studied by MOVAGHAR et al. (1981) (also for the case of R-hopping) by using the Green-function method proposed by MOVAGHAR et al. (1980) for studying ac hopping conductivity (cf. Section 6.). However, as the hopping Hall effect is a three-site phenomenon, one must go beyond the two-site self-consistent theory, appropriate for the ac conductivity, and formulate a three-site self-consistent approximation. The theory proposed by MOVAGHAR et al. (1981) is based on an exact solution of a disordered triangular tree-lattice model and the generalization of it to treat a random network. The resulting expression for the Hall mobility is in agreement with (8.2.19), but this method also allows the numerical factor b_{H} to be determined, which was found to be equal to 28.

Fig. 8.5 shows the concentration dependence of the Hall mobility evaluated on the basis of formulae (8.2.19), (8.2.22) and (8.2.29). We see that the dependence (8.2.19) well agrees with the computer data of BUTCHER and MC INNES (1981) if we use $b_{\mathrm{H}} = 28$.

So far, we have considered only the case of low densities ($\alpha\mathcal{N}^{-1/3} \gg 1$), which may be investigated by using methods of the percolation theory. BUTCHER and MC INNES (1981) and MOVAGHAR et al. (1981) have also studied the Hall mobility u_{H} in the limit of high densities of sites, where $\alpha\mathcal{N}^{-1/3}$ is not a large quantity. In the latter case we may expect the disorder to affect the Hall mobility only weakly, so that the dc Hall mobility does not strongly differ from the value on the high-frequency plateau. To obtain the latter value, we must average formula (3.5.11) over triads. As mentioned above, the main contribution to such an average comes from triads with a spacing between the sites of order α^{-1}. Using the expression for the (longitudinal) conductivity on the high-frequency plateau (cf. (6.4.38), we obtain the following expression for u_{H} on the high-frequency plateau:

$$u_{\mathrm{H}} = c \frac{e\mathcal{N}^{-2/3}}{\hbar} \frac{\nu_{\mathrm{ph}}}{J_0} (\alpha\mathcal{N}^{-1/3})^{-5}, \tag{8.2.30}$$

where c is some numerical factor. The expressions obtained for the dc Hall mobility by BUTCHER and MC INNES (1981) and MOVAGHAR et al. (1981) agree with (8.2.30) (strictly speaking, the expressions obtained by these authors differ from each other and from (8.2.30) by a numerical factor).

Let us now outline the evaluation of the Hall mobility in the case of R-ε-percolation. Formula (8.2.9) also applies to this case. However, the average now involves distances as well as the energy levels ε_m. For hops near the Fermi level, the three-site hopping probabilities may be also written in the form (8.2.13), but $W_0{}^H$ is now a quantity depending on the random variables ε_m. For instance, for weak electron-phonon coupling, according to (8.2.11), $W_0{}^H$ must be replaced by $W_{123}^{H(0)}$, defined as

$$W_{123}^{H(0)} = W_0{}^H\{e^{\beta|\varepsilon_1|}\Delta_{13}\Delta_{12} + e^{\beta|\varepsilon_2|}\Delta_{23}\Delta_{12} + e^{\beta|\varepsilon_3|}\Delta_{13}\Delta_{23}\}, \tag{8.2.31}$$

where W_0^H is given by (8.2.14), and

$$\Delta_{ik} = \exp\left\{-\frac{\beta}{2}\left(|\varepsilon_i - \mu| + |\varepsilon_k - \mu| + |\varepsilon_i - \varepsilon_k|\right)\right\}. \tag{8.2.32}$$

Furthermore, Z_0^{-1} must be replaced by $Z_{ik} = Z_0^{-1}\Delta_{ik}$ (cf. (5.1.7). To get the Hall mobility, we have to insert $W_{123}^{H(0)}$ and Z_{ik} into (8.2.9) and to perform the configuration average over the triads on account of the following restriction:

$$\frac{R_{ij}}{R_{\max}} + \frac{1}{2E_{\max}}\{|\varepsilon_i - \mu| + |\varepsilon_j - \mu| + |\varepsilon_i - \varepsilon_j|\} < 1, \tag{8.2.33}$$

where R_{ij} denotes the distance between sites i and j. Here $E_{\max} = kT(T_0/T)^{1/4}$, $R_{\max} = (1/2\alpha)\,(T_0/T)^{-1/4}$, and the temperature T_0 is given in (5.1.19). The restriction (8.2.33) results from the requirement that the resistances in the triad must be smaller than the critical resistance Z_c. If one excludes the possibility that the density of states vanishes at the Fermi level μ, then, on performing the configuration average, for hops near μ, a constant density of states may be assumed. On averaging expression (8.2.9), GRUNEWALD et al. (1981) evaluated numerically the occurring multiple integral. The resulting dependence $u_H(T)$ can be well fitted by $u_H \sim \exp\left(-\frac{3}{8}(T_0/T)^{1/4}\right)$, where T_0 is defined in (5.1.19). According to GRUNEWALD et al. (1981), the latter expression may be analytically obtained by assuming that the main contribution to u_H results from triads of sites forming an equilateral triangle with the length of a leg equal to $\frac{3}{8}\,\alpha^{-1}(T_0/T)^{1/4}$ and with the energy at each site equal to $\frac{1}{4}\,kT(T_0/T)^{1/4}$. Note for the dc Hall mobility at low temperatures and low site densities, an approximate $T^{-1/4}$ behaviour was also found by FRIEDMAN and POLLAK (1981) (by a percolation approach) and by MOVAGHAR et al. (1983) (by the Green-function method).

Let us now consider the experimental situation. In some chalcogenid glasses (As—Te—I, As—Te—Ge, As—Te) the dc Hall effect was measured by SEAGER et al. (1973), who found an activated temperature dependence of u_H characterized by an activation energy which is three times smaller than that for the drift mobility. Such a result is in accordance with (8.2.18). It reflects the formation of small polarons (cf. (3.5.17)), but not any specific of disorder. In the paper in question the phenomenon of the p-n-anomaly is reported: the sign of the Hall effect suggests electron conduction, while the thermpower indicates hole transport. In certain other chalcogenid glasses, the experimental Hall mobility depends weakly on temperature, while the drift mobility exhibits an activated behaviour (see, e.g., MOTT and DAVIS, 1979), which, in principle, agrees with the above result for weak electron-phonon coupling. ZABRODSKII (1980) measured the Hall effect in heavily doped Ge at low temperatures. However, in this material the doping level was so high that charge carrier localization is probably not realized in the impurity band. More recently, BISKUPSKI (1982) measured the Hall effect on doped InP in dependence of $\alpha\mathcal{N}^{-1/3}$ for $\alpha\mathcal{N}^{-1/3} \lessapprox 4$, i.e. in the high-density limit. The latter experimental data well agree with the theory of Movaghar et al. (1981) (cf. Movaghar et al., 1983a), which, as mentioned above, also covers the high-density region. Generally, the measurement of the dc hopping Hall mobility is a very intricate problem, because its magnitude is very small. We have learned above that in disordered systems, σ_{xy} (analogous to σ) increases with increasing frequency and reaches at sufficiently high frequencies a plateau (cf. (6.2.30)). Hence, it should be easier to study experimentally the ac Hall effect than the dc Hall effect. For weak electron-phonon coupling, the ac hopping-Hall effect was studied theoretically by Holstein (1961) within the framework of a three-site model. It was found that $u_H \sim \omega^{1/2}T^{1/2}$, whereby u_H much exceeds the drift mobility. However, the ac Hall effect measurements of Amitay and Pollak (1966) did not confirm this result[1]) (see also Section 11.4.2.).

Concluding this section, let us draw attention to evaluations of the Hall mobility within the framework of the random-phase model mentioned in Section 4.1. For this model, the Hall effect was investigated by Friedman (1971) by making use of methods from the small-polaron theory. If the Hall mobility is governed by three-site processes, then $u_H/u \approx kT/3J$ (u — drift mobility, J — resonance integral) holds, and this ratio is in general less than unity. In this case, the sign of the Hall effect always exhibits electron character, i.e. the p-n-anomaly occurs. The four-site Hall effect in the random-phase model was studied by Kaneyoshi (1976). It may be important in system with "cellular" disorder.

Let us yet draw attention to the papers of Fishchuk and Rudko (1978, 1979, 1980a, b) devoted to a study of the high-frequency Hall effect and the Faraday effect within the framework of the three-site model on account of phononless intersite transitions, and to an investigation of the Hall effect in disordered systems in strong electrical fields by the effective-medium approximation by Fishchuk (1982a).

9. Some other hopping phenomena

9.1. Thermoelectric power

Despite numerous experimental studies of the thermoelectric power in the hopping regime, a first-principles theory of this phenomenon is still lacking. A calculation of the Seebeck coefficient S on the basis of the generalized Kubo formula (Kubo et al., 1957), i.e. in terms of the correlator of the electrical current and the heat flux, encounters the difficulty that, in the presence of electron-phonon interaction, the

---

[1]) Amitay and Pollak (1966) found that their experimental data were two orders of magnitude smaller than those predicted by the three-site theory of Holstein (1961). According to Movaghar et al. (1983b), who applied their self-consistent theory (cf. Movaghar et al., 1983a) to the ac hopping Hall effect, this discrepancy between experiment and theory is spurious, because the comparison performed by Amitay and Pollak (1966) involves a two-site expression for Re $\sigma_{xx}(\omega)$ which does not apply to the frequency range studied.

operator of the heat flux cannot be defined uniquely. Accordingly, at present there are various attempts to study the thermopower in the hopping regime by making use of intuitive considerations. However, the results of these studies are often controversial, so that further work is required.

The investigation of the thermoelectric power in disordered systems has been much stimulated by the work of CUTLER and MOTT (1969), who proposed the following approach. A function $\sigma(\varepsilon)$ is introduced by

$$\mathrm{d}\boldsymbol{j} = \sigma(\varepsilon)\, \boldsymbol{E}\, \mathrm{d}\varepsilon\,, \tag{9.1.1}$$

where $\mathrm{d}\boldsymbol{j}$ denotes the differential current, $\boldsymbol{E}$ the electrical field, and ε the electronic energy. Thus, $\sigma(\varepsilon)$ is related to the conductivity σ by

$$\sigma = \int \mathrm{d}\varepsilon \sigma(\varepsilon)\,, \tag{9.1.2}$$

and the differential energy flux is given by

$$(\varepsilon - \mu)\, \frac{\mathrm{d}\boldsymbol{j}}{e} = \frac{1}{e}\, \sigma(\varepsilon)\, (\varepsilon - \mu)\, \boldsymbol{E}\, \mathrm{d}\varepsilon\,, \tag{9.1.3}$$

where μ is the chemical potential ($e = |e|$). Integrating (9.1.3) over ε, we obtain on the left-hand side the total heat flux, which is equal to $j\Pi$, where Π is the Peltier coefficient being connected with S by the relation $S = \Pi/T$. This yields the following expression for the Seebeck coefficient:

$$S = -\frac{k}{e} \int \mathrm{d}\varepsilon\, \frac{\varepsilon - \mu}{kT}\, \frac{\sigma(\varepsilon)}{\sigma}\,. \tag{9.1.4}$$

In deriving (9.1.4), two questions remain open. Firstly, the physical meaning of $\sigma(\varepsilon)$ and the way of calculating this quantity are not clear. Secondly, the relation (9.1.4) defines the heat flux only for the case that an electron does not undergo inelastic collisions during its motion.

According to (9.1.2), $\sigma(\varepsilon)$ represents the contribution to the conductivity of electrons whose energy is between ε and $\varepsilon + \mathrm{d}\varepsilon$. However, such a quantity is well defined only if an electron does not leave this energy interval during its motion, i.e. if the scattering is elastic. In this case, (9.1.4) represents the Pisarenko formula, which results from the Boltzmann equation. However, in hopping processes the collisions with phonons have inelastic character and the energy of the final state differs from the energy of the initial one. This problem was first studied by BRENIG et al. (1973a) and BUTCHER (1974) on the basis of the rate equations for hopping conduction (2.4.23) by introducing the conception of the heat flux in the sense of (9.1.3). The expression for the heat flux was obtained from the energy conservation equation. However, the corresponding expression contains unknown contributions from the power dissipation density, so that the approach of these authors is not satisfactory. More appealing is the treatment of OVERHOF (1975), who assumes that in a hop between sites m and m' the energy $\frac{1}{2}\,(\varepsilon_m + \varepsilon_{m'} - 2\mu)$ is transported, so that the heat flux $q_{mm'}$

through a resistor $Z_{mm'}$ becomes (GRÜNEWALD and THOMAS, 1979)

$$q_{mm'} = j_{mm'}(\varepsilon_m + \varepsilon_{m'} - 2\mu)/2e, \tag{9.1.5}$$

where $j_{mm'}$ is the current through $Z_{mm'}$.

Let us now turn to the second problem mentioned above. According to (9.1.3), an electron carries the energy $(\varepsilon - \mu)$. However, in the presence of interaction with phonons, the motion of an electron may be accompanied by a heat transport due to phonons. In this case, it may be expected that in (9.1.3), and consequently in (9.1.4), $(\varepsilon - \mu)$ must be replaced by $(\varepsilon - \mu + D(\varepsilon, T))$ (FRITZSCHE, 1971). The study of the thermoelectric power by BRENING et al. (1973a) and BUTCHER (1974) did not lead to such a term D. This may be due to the fact, mentioned above, that the expression for the heat flux given by these authors is not defined uniquely.

EMIN (1975b) studied the thermoelectric power in a two-site hopping model by assuming that each site m and m' may be associated with an individual temperature T_m and $T_{m'}$, respectively (an analogous method was used by EFROS (1967) to evaluate the thermoelectric power of small polarons). In this model a term D occurs and a concrete expression can be derived for it. It turns out that the term D remains finite for any weak interaction with phonons. It vanishes, however, in the limit of elastic scattering. A prescription for a calculation of the local temperatures T_m and $T_{m'}$ was not given by EMIN (1975b). Thus, it is not clear in what way the two-site problem may be related to the corresponding problem of the total system. This seems to require use of the heat flux conception.

In the what follows, we analyse the expression (9.1.4). To do this, it is convenient to rewrite this expression as

$$S = -\frac{1}{e}\frac{\langle\varepsilon\rangle - \mu}{T}, \tag{9.1.6}$$

where

$$\langle\varepsilon\rangle = \int \mathrm{d}\varepsilon \, \frac{\sigma(\varepsilon)}{\sigma} \, \varepsilon. \tag{9.1.7}$$

At high temperatures, kT exceeds the spread of the energy levels (for a disordered system) or the width of the polaron band (for small polarons), i.e. $kT \gg \langle\varepsilon\rangle$, and we find that

$$S = \frac{1}{e}\frac{\mu}{T} = \frac{k}{e} \ln\{n/(\mathcal{N} - n)\}. \tag{9.1.8}$$

Here n denotes the concentration of charge carriers, and $\mathcal{N}$ the concentration of sites (for instance, for hops across donors in compensated semiconductors, $n = \mathcal{N}_\mathrm{D} - \mathcal{N}_\mathrm{A}$, $\mathcal{N} = \mathcal{N}_\mathrm{D}$). The relation (9.1.8) is sometimes called the Morin formula (MORIN, 1958) (it was derived for small polarons by SCHOTTE, 1966, EFROS, 1967). It is interesting to note that for hopping between deep Cr-donor levels in InSb, a temperature independent thermopower was observed (KOSAREV et al., 1976), in accordance with (9.1.8).

According to (9.1.8), the sign of the thermoelectric power uniquely determines the type of conduction. If $n < \mathcal{N}/2$, then $S < 0$, and the conduction is due to electrons, while for $n > \mathcal{N}/2$, $S > 0$, it is due to holes. Thus, for $n = \mathcal{N}/2$, the sign of the thermoelectric power changes (whereas the Hall effect in the hopping regime may preserve its sign, cf. Section 8.1.).

At intermediate temperatures, where the conductivity is described by (5.1.9), the Seebeck coefficient becomes (KOSAREV, 1974)

$$S = \eta \frac{1}{e} \frac{\varepsilon_3}{T}, \tag{9.1.9}$$

where ε_3 is the activation energy of the conductivity and η is a number of order unity, whereby for weak compensation, $\eta > 0$ holds, but for strong compensation, $\eta < 0$. Thus, also in this case, S changes its sign by passing-over from electron to hole conduction.

In this context, let us draw attention to the work of GRÜNEWALD and THOMAS (1979) and WHALL (1981), in which Kirkpatrick's effective-medium approximation (EMA) (KIRKPATRICK, 1973; cf. Sections 5.2.4, 5.3.5, 6.4.) was used to study the thermoelectric power in the hopping regime on the basis of the formula (9.1.5) for the heat current between sites m and m'. According to GRÜNEWALD and THOMAS (1979), the EMA is especially accurate in the high-temperature region, but less so at lower temperatures and probably incorrect at low temperatures, where conduction is by variable-range hopping. WHALL (1981) used the EMA to calculate the Seebeck coefficient for nearest-neighbour hopping at intermediate temperatures and found that $S = A/T + BT$, in qualitative agreement with experimental data obtained on cerium sulphide ($Ce_{3-x}S_4$), fluorine-substituted magnetite ($Fe_3O_{4-x}F_x$) and "pure" magnetite[1]) (Fe_3O_4), which are assumed (cf. GRAENER et al., 1979) to be examples of Fermi glasses (in which hopping occurs between localized states near the Fermi level; cf. Section 4.1.).

According to an early proposal of CUTLER and MOTT (1969), at sufficiently low temperatures, $\sigma(\varepsilon)$ should be sharply peaked near $\varepsilon = \mu$. Then, (9.1.4) takes the form of the standard metallic formula

$$S = -\frac{\pi^2}{3} \frac{k^2 T}{e} \left[\frac{\mathrm{d} \ln (uN)}{\mathrm{d}\varepsilon}\right]_{\varepsilon=\mu}. \tag{9.1.10}$$

Here $N(\varepsilon)$ is the density of states per unit energy and $u(\varepsilon)$ the energy dependent mobility defined as $\sigma(\varepsilon) = eN(\varepsilon)\, u(\varepsilon)\, f(\varepsilon)\, [1 - f(\varepsilon)]$ ($f(\varepsilon)$-Fermi function). The expression (9.1.10) immediately follows from (9.1.4) by noting that $f[1 - f] = -kT(\partial f/\partial \varepsilon)$ and by expanding $(\varepsilon - \mu)\, N(\varepsilon)\, u(\varepsilon)$ in a power series about $\varepsilon = \mu$. It should be noted that the validity of this expansion is restricted to cases for which $\left[\left(\mathrm{d}(uN)/\mathrm{d}\varepsilon\right)/uN\right]_{\varepsilon=\mu} \ll 1/kT$ holds, which in general requires low temperatures (cf. FRIEDMAN, 1977). The formula (9.1.10) has been used to explain the thermoelectric power in the hopping regime at low and intermediate temperatures (see, e.g., WHALL et al., 1977). However, its validity has been called in question (cf. KOSAREV, 1974, MOTT and DAVIS, 1979).

---

[1]) In "pure" magnetite, localization is believed to be due to electron-electron interaction (cf. WHALL et al., 1977, MOTT, 1979, WHALL, 1981).

In the work of ZVYAGIN (1973), KOSAREV (1974) and OVERHOF (1975) the thermoelectric power in the variable-range hopping regime was investigated by interpreting the quantity $\langle \varepsilon \rangle$ in (9.1.6) as the average energy along the one-dimensional current paths predicted by percolation theory. According to ZVYAGIN (1973) and KOSAREV (1974), $\langle \varepsilon \rangle$ is defined as

$$\langle \varepsilon \rangle = \int \mathrm{d}\varepsilon N(\varepsilon)\, \varepsilon / \int \mathrm{d}\varepsilon N(\varepsilon) . \tag{9.1.12}$$

Here $N(\varepsilon)$ is the density of states, and the integration is over an energy range near the Fermi level with a width of the order $\Delta\varepsilon \approx kT_0(T/T_0)^{3/4}$, within which variable-range hopping occurs (cf. (5.1.17), (5.1.18)).

Note that in the case that $\sigma(\varepsilon) = \text{const.} \times N(\varepsilon)$ within $\Delta\varepsilon$ and $\sigma(\varepsilon) = 0$ otherwise, the two definitions of $\langle \varepsilon \rangle$, given above in (9.1.7) and (9.1.12), coincide. In this case, we find that

$$S \sim -\frac{1}{e} \frac{(\Delta\varepsilon)^2}{T} \frac{\mathrm{d} \ln N(\mu)}{\mathrm{d}\mu}, \tag{9.1.13}$$

i.e. $S \sim T^{1/2}$ holds. An analogous result was obtained by OVERHOF (1975). Such a $T^{1/2}$ dependence of the Seebeck coefficient is observed, for example, on fluorine-substituted magnetic in the variable-range hopping regime (GRAENER et al., 1979). According to EMIN (1975b), on account of the heat flux due to interaction with phonons, one gets a finite Seebeck coefficient also for $\mathrm{d}N(\mu)/\mathrm{d}\mu = 0$.

For more details on the hopping thermoelectric power the reader is referred, e.g., to BONCH-BRUEVICH et al. (1981) (see also WYSOKINSKI and BRENIG, 1985).

9.2. Magnetoresistance

As pointed out in Section 8., in a hopping system a magnetic field $\boldsymbol{H}$ gives rise to a Hall effect, owing to the fact that the current paths are deflected by such a field. On the other hand, a magnetic field causes the Bohr radius to shrink, which produces a positive magnetoresistance in the hopping regime (SLADEK, 1958). The shrinkage of the Bohr radius may also cause an increase in the activation energy ε_3, owing to a reduction of electron correlation (cf. Section 5.4.3.). Furthermore, in doped semiconductors (e.g. n-InSb) the shrinkage of the Bohr radius can also give rise to a metal-non-metal transition at a fairly well defined magnetic field, which depends on the impurity concentration (ISHIDA and OTSUKA, 1977).

The main purpose of the present section is to outline the application of the percolation theory to the magnetoresistance in the hopping regime. Thereby we follow SHKLOVSKII and EFROS (1979), who described the percolation approach to the magnetoresistance in detail. Furthermore, we will discuss the anomalous (negative) magnetoresistance observed in some hopping systems (amorphous germanium and silicon), which is supposed to be due to spin-flip transitions.

A calculation of the hopping transition rate requires knowledge of the localized wavefunctions, i.e. requires examination of the Schrödinger equation in the presence of a magnetic field. A magnetic field (directed along the z-axis) gives rise to an additional "magnetic" potential $\hbar^2\varrho^2/8m\lambda^4$ (oscillator potential), where λ is the magnetic length

$$\lambda = \left(\frac{c\hbar}{eH}\right)^{1/2}, \tag{9.2.1}$$

m the electron mass, and $\varrho = \sqrt{x^2 + y^2}$ the distance from the z-axis. For weak magnetic fields, i.e. $\alpha\lambda \gg 1$ (α — reciprocal Bohr radius in a zero magnetic field) the effect of the magnetic potential on the wavefunction is small. However, for strong fields, i.e. $\alpha\lambda \ll 1$, in the direction perpendicular to the z-axis the elctron is much stronger localized by the magnetic potential than by the Coulomb potential.

In the presence of a magnetic field, the Schrödinger equation cannot be solved analytically. Approximate solutions have been proposed by several authors (e.g. YAFET et al., 1956, ELLIOTT and LOUDON, 1960, HASEGAWA and HOWARD, 1961; cf. SHKLOVSKII and EFROS, 1979, CHROBOCZEK, 1980). For $\alpha\lambda \gg 1$ (weak fields) and $\alpha^{-1} \ll \varrho \ll \lambda^2\alpha$, the asymptotic behaviour of the wavefunction (which is suitable for calculating the magnetoresistance; cf. SHKLOVSKII and EFROS, 1979) is given by

$$\chi(\boldsymbol{r}) \sim \exp\left(-\alpha r - \frac{\varrho^2 r}{24\lambda^4\alpha}\right), \tag{9.2.2}$$

while for $\alpha\lambda \gg 1$ and $\varrho \gg \lambda^2\alpha$, or for $\alpha\lambda \approx 1$ (moderate fields) and $\alpha\lambda \ll 1$ (strong fields), by

$$\chi(\boldsymbol{r}) \sim \exp\left(-\frac{\varrho^2}{4\lambda^2} - \alpha_{\mathrm{H}}\,|z|\right), \tag{9.2.3}$$

where

$$\alpha_{\mathrm{H}} = \frac{\sqrt{2mE_{\mathrm{H}}}}{\hbar}. \tag{9.2.4}$$

Here E_{H} is the ionization energy of the state. For sufficiently strong fields, it holds that $E_{\mathrm{H}} \sim H^{1/3}$, while for weak fields, $E_{\mathrm{H}} = \hbar^2\alpha^2/2m$, i.e. $\alpha_{\mathrm{H}} = \alpha$.

The exponential dependence of the conductance $Z_{mm'}^{-1} \sim \exp(-\zeta_{mm'})$ between sites m and m' is governed by the overlap integral of the wavefunctions belonging to sites m and m'. Using (9.2.2) and (9.2.3), respectively, and accounting for the additional phase factor arising from the magnetic field (cf. Section 2.1.), one obtains (MIKOSHIBA and GONDA, 1962, MIKOSHIBA, 1962), for $\alpha\lambda \gg 1$ and $(X_{mm'}^2 + Y_{mm'}^2)^{1/2} \ll \lambda^2\alpha$,

$$\zeta_{mm'} = 2\alpha R_{mm'} + \frac{(X_{mm'}^2 + Y_{mm'}^2)\,R_{mm'}}{12\alpha\lambda^4}, \tag{9.2.5}$$

and for $\alpha\lambda \gg 1$ and $(X_{mm'}^2 + Y_{mm'}^2)^{1/2} \gg \alpha\lambda^2$, or for $\alpha\lambda \ll 1$ and arbitrary $R_{mm'}$,

$$\zeta_{mm'} = \frac{X_{mm'}^2 + Y_{mm'}^2}{2\lambda^2} + 2\alpha_{\mathrm{H}}\,|Z_{mm'}|. \tag{9.2.6}$$

Here $R_{mm'} = |\boldsymbol{R}_{mm'}| \equiv |\boldsymbol{R}_m - \boldsymbol{R}_{m'}|$, where $\boldsymbol{R}_m = (X_m, Y_m, Z_m)$ denotes the position vector of the mth site (impurity).

The hopping magnetoresistance in the low and high magnetic field case was theoretically studied first by MIKOSHIBA (1962), who provided an important contribution to the understanding of this phenomenon. The percolation approach to the magnetoresistance was elaborated by SHKLOVSKII (1971, 1973). More recently, MOVAGHAR and SAUER (1980) studied the magnetoresistance by means of an effective-medium approximation, which makes use of the exact solution of a Bethe lattice (cf. Section 6.4.). The latter authors practically found exact agreement with the percolation results.

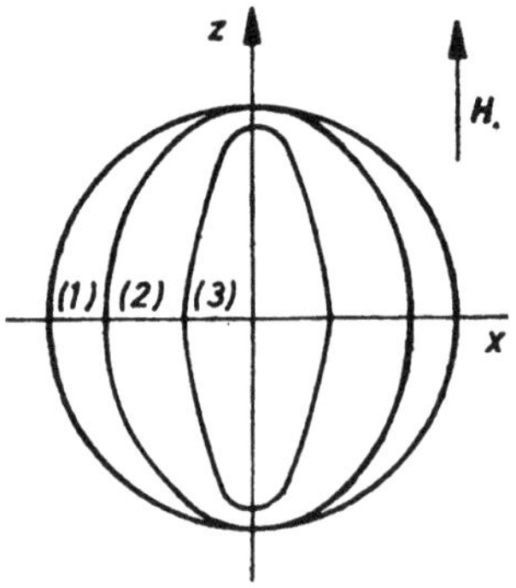

Fig. 9.1. The shape of the percolation figures for hopping magnetoresistance: (1) for $\boldsymbol{H} = 0$, (2) for a weak magnetic field, and (3) for a strong magnetic field (after SHKLOVSKII and EFROS, 1979)

The shape of the (inclusive) characteristic percolation figures (cf. Section 5.2.1), given by $\zeta_{mm'} = \zeta$, is shown in Fig. 9.1. For weak magnetic fields, the characteristic figure is approximately an ellipsoid, while for strong magnetic fields, it represents two cut-off rotational paraboloids composed on their bases. According to SHKLOVSKII (1971), the critical value ζ_c of the parameter ζ (for which an infinite cluster of connected sites occurs) is given by

$$\mathcal{N} V_c = B_c, \tag{9.2.7}$$

where

$$V_c = (2/3)\, \pi \zeta_c{}^3 \int_0^\pi d\theta \sin\theta\, R^3(\theta). \tag{9.2.8}$$

Here $\mathcal{N}$ is the concentration of sites, θ the angle between the direction of the magnetic field and the radius vector, $R(\theta)$ the angle-dependent radius of the characteristic figure, V_c the volume of this figure, and B_c the average number of bonds per site, which has the same value as for spheres, i.e. $B_c = 2.7$ (cf. Section 5.2.1.).

Consider first the weak magnetic field case. Squaring both sides of (9.2.5), setting $\zeta_{mm'} = \zeta$, and replacing approximately R^2 by $(\zeta/2\alpha)^2$, then one gets the following equation for an ellipsoid

$$\frac{Z^2}{A^2} + \frac{X^2 + Y^2}{C^2} = 1, \tag{9.2.9}$$

where

$$A = \zeta/2\alpha \text{ and } C = (\zeta/2\alpha)\,(1 - \zeta^2/96\lambda^4\alpha^4)\,. \tag{9.2.10}$$

The volume of this ellipsoid is given by $V = \dfrac{4}{3}\,\pi AC^2$.

Inserting the expression for the volume into (9.2.7), and solving this equation by an expansion with respect to $\zeta_c{}^2/96\lambda^4\alpha^4$, we find that

$$\zeta = \eta\alpha\mathcal{N}^{-1/3} + \frac{B_c}{24\pi\alpha\mathcal{N}\lambda^4}, \tag{9.2.11}$$

where $\eta = 2(3B_c/4\pi)^{1/3} = 1.73$. The second term on the right-hand side of (9.2.11) is small compared with the first one. Nevertheless, it may be larger than unity and thus contribute to the exponential dependence of the magnetoresistance. Accordingly, for weak magnetic fields, the magnetoconductivity is given by an expression like (5.1.9), where ϱ_3 depends on H as

$$\varrho_3(H) = \varrho_3(0)\exp\left\{t\,\frac{e^2H^2}{\alpha\mathcal{N}c^2\hbar^2}\right\} \tag{9.2.12}$$

with

$$t \equiv B_c/24\pi = 0.036\,. \tag{9.2.13}$$

Note that the effective-medium theory of MOVAGHAR and SAUER (1980) yields (9.2.12) with $t = 0.034$.

The first term on the right-hand side of (9.2.11) agrees with the value of ζ_c obtained without a magnetic field. Hence, in the case of a weak magnetic field, the magnetoresistance is governed by hops across distances of the order of $\mathcal{N}^{-1/3}$. Consequently, the result (9.2.12) is valid for

$$\alpha\lambda \gg 1, \quad \alpha^2\lambda^2 \gg \alpha\mathcal{N}^{-1/3} \gg 1\,. \tag{9.2.14}$$

Using $\alpha^2\lambda^2 \gg \alpha\mathcal{N}^{-1/3}$, one can define a critical magnetic field

$$H_c = \frac{\mathcal{N}^{1/3}c\hbar}{e\alpha^{-1}}, \tag{9.2.15}$$

which separates the low magnetic field limit from the high field one.

Let us now turn to the high-field case, where the characteristic figure is given by (9.2.6) with $\zeta_{mm'} = \zeta$. Inserting the expression for the volume of this figure, $V = \pi\lambda^2 \times \alpha_H^{-1}\zeta^2$, into (9.2.7), we find that

$$\zeta_c = q/(\mathcal{N}\alpha_H^{-1}\lambda^2)^{1/2}, \tag{9.2.16}$$

where

$$q \equiv (B_c/\pi)^{1/2} = 0.92\,. \tag{9.2.17}$$

Using (9.2.1) and (9.2.16) we obtain the dependence

$$\varrho_3(H) = \varrho_0\exp\{q[\lambda^2\alpha_H^{-1}\mathcal{N}]^{-1/2}\}, \tag{9.2.18}$$

which is valid for $H \gg H_c$, where H_c is given in (9.2.15). Note that the effective-medium theory of MOVAGHAR and SAUER (1980) yields the value 0.93 for the quantity q in (9.2.18). Taking into account that E_H (cf. (9.2.4)) becomes independent of H at sufficiently low fields, and behaves as $E_H \sim H^{1/3}$ at high fields, we may expect that at $H \gg H_c$ the quantity $\varrho_3(H)$ is approximately given by

$$\varrho_3(H) = \varrho_0 \exp \{\text{const } H^m\}, \tag{9.2.19}$$

where the exponent m somewhat exceeds 0.5.

It is interesting to note that the exponential dependence predicted by formulae (9.2.12) and (9.2.18) is valid for the longitudinal as well as the transverse magnetoresistance. Solely the pre-exponential factor differs in both the cases. This is a consequence of the exponential dependence of the hopping probability, owing to which in strong fields (i.e. in the case of strongly anisitropic characteristic figures; cf. Fig. 9.2) hops with a small angle θ between $\boldsymbol{R}_{mm'}$ and $\boldsymbol{H}$, but with $|\boldsymbol{R}_{mm'}|$ larger than $\mathcal{N}^{-1/3}$, are more effective than those with a large angle but with a hopping distance of the order of $\mathcal{N}^{-1/3}$. This fact was mathematically proved by SHKLOVSKII (1977).

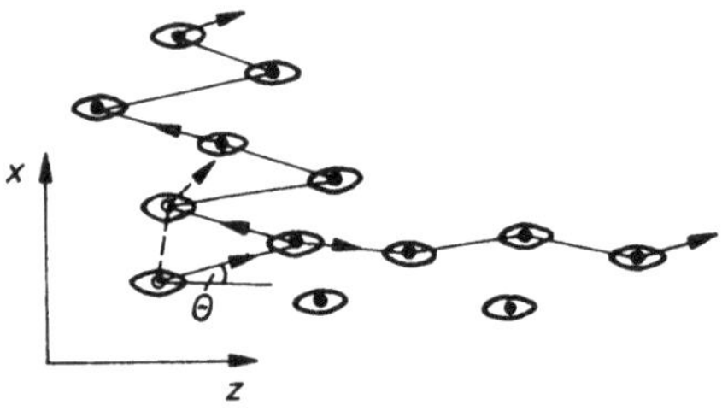

Fig. 9.2. Illustration of the fact that the longitudinal and transverse hopping magnetoresistance exhibits the same exponential behaviour. The magnetic field is directed along the z-axis. The ellipsoids represent the percolation figures. Because of the exponential dependence of the hopping probability, for an electron moving in x-direction the path along the dashed line is much less effective than that along the solid line (after SHKLOVSKII and EFROS, 1979)

The observation of SLADEK (1958), that in n-InSb with an increase of the magnetic field up to H = 28 kOe the transverse and longitudinal magnetoresistances increase nearly by a factor 10^5, but their ratio only by a factor 2.5, experimentally confirmes the isotropy of the exponential dependence of the magnetoresistance.

Let us consider further experimental findings. For $H < H_c$, the factor t (cf. (9.2.12)) assumes the values: n-GaAs (0.03–0.06) (EMEL'JANENKO et al., 1973, KAHLERT et al., 1976), n-Ge(0.048) (CHROBOCZEK and SLADEK, 1966), p-Ge(0.011–0.013) (CHROBOCZEK et al., 1974), and n-InP(0.04) (EMEL'JANENKO et al., 1975). For high fields, KAHLERT et al. (1976) found for n-GaAs, $q = 0.68$ (cf. (9.2.18)). Accordingly, in general, the theoretical values of t and q are in rather good agreement with experimentally determined data (cf. CHROBOCZEK, 1980). But further theoretical effort is necessary to explain the small t values found for p-Ge.

So far, we have here considered only the case of R-percolation. It is worth noticing that percolation theory has also been used to study the magnetoresistance in the variable-range hopping regime. Here, according to SHKLOVSKII (1973) (cf. SHKLOVSKII and EFROS, 1979, TOKUMOTO et al., 1982), for weak magnetic fields, the specific magnetoresistance $\varrho(H)$ is given by

$$\ln \frac{\varrho(H)}{\varrho(0)} = t_1 \frac{e^2H^2}{\alpha^4c^2\hbar^2} \left(\frac{T_0}{T}\right)^{3/4}, \tag{9.2.20}$$

where T_0 is defined in (5.1.19) and t_1 is a numerical factor, $t_1 = 5/2016$. For high fields, it was found that

$$\varrho(H) = \varrho_0 \exp \left\{\left[\frac{T_0(H)}{T}\right]^{1/3}\right\}, \tag{9.2.21}$$

where

$$T_0(H) = \frac{2.1\, eH\alpha_{\mathrm{H}}}{N(\mu)\, c\hbar k} \tag{9.2.22}$$

($N(\mu)$-density of states at the Fermi level). The low- and high-field regimes are separated by the magnetic field

$$H_c \approx \frac{\hbar c}{e} \left[N(\mu)\, kT\right]^{1/4} \alpha^{5/4}, \tag{9.2.23}$$

which can be obtained by equating the exponents of (9.2.20) and (9.2.21) and setting $\alpha_{\mathrm{H}} \approx \alpha$.

Using the wavefunctions of YAFET et al. (1956), which have a Gaussian decay, one obtains a $T^{2/5}$ dependence instead of the $T^{1/3}$ dependence (9.2.21) (SUPRATO and BUTCHER, 1975, TOKUMOTO et al., 1982). However, the use of these functions for studying hopping transport has been criticized by SHKLOVSKII and EFROS (1979). The magnetoresistance at very low temperatures has been measured in n-InSb (GERSHENSON et al., 1974, MANSFIELD and KUSZTELAN, 1978, TOKUMOTO et al., 1982, cf. MANSFIELD and TOKUMOTO, 1983).

Let us now turn to the anomalous magnetoresistance which appears in amorphous germanium and silicon in the temperature range where the conductivity is dominated by hopping processes. This phenomenon was first observed by MELL and STUKE (1970). Later MELL (1974) found both positive and negative contributions to the magnetoresistance in amorphous germanium and silicon and showed that the former depends strongly on sample preparation. The anomalous magnetoresistance in amorphous germanium and silicon has been confirmed by several authors (e.g. CLARK et al., 1974, YOSHIDA et al., 1980, MEHRA et al., 1981, KUIVALAINEN et al., 1982). Its origin, however, remained unexplained for several years after the first observation of this effect.

Note that negative magnetoresistance is a rather common phenomenon in degenerate crystalline semiconductors. In these materials it is explained on the basis of spin-disorder scattering, i.e. exchange scattering of the extended state electrons due to localized spins in shallow donor states (Toyazawa, 1962, Khosla and Fischer, 1970) and by a mechanism in which multiple scattering of electrons by impurities is suppressed by the magnetic field (Kawabata, 1980). However, these models as well as the two-band model of Hedgcock and Raudorf (1970) (cf. Mehra et al., 1981), which assumes the existence of a band with high carrier mobility above a band with low mobility separated by a sharp mobility edge between them, cannot succesfully applied to the anomalous magnetoresistance in amorphous semiconductors (cf. e.g. Kuivalainen et al., 1982).

A model which explains in detail the magnetic field, temperature, and spin-density dependence of the magnetoresistance measured in amorphous semiconductors is the Movaghar-Schweitzer-Osaka hopping model, which was proposed by Movaghar and Schweitzer (1978) and further developed by Osaka (1979). In this model the localized gap states (dangling bond-type states) near the Fermi level μ are spread in energy with singly occupied states below μ, doubly occupied (spin singlet) states above μ and empty states distributed at random. The doubly occupied states are separated from the singly occupied ones by a Hubbard-type repulsion energy U.

In contradiction to the common hopping model, in the Movaghar-Schweitzer-Osaka model a fraction N_s/N of sites has a spin, which is considered frozen or slowly changing in comparison to the hopping times (N_s — number if singly occupied states, N — total number of sites). An electron which hops to a singly occupied state must have the correct spin orientation otherwise the hop is forbidden. One has to distinguish between normal hops which involve no spin flip and anomalous hops which involve spin flips. It turns out that the anomalous hopping process is on the average faster than the second nearest-neighbour normal process. Consequently, an electron on the average takes a path through the specimen which also involves anomalous processes. Thus, the anomalous hopping magnetoresistance results from the magnetic field dependence of spin-flip transitions and is closely related to the hopping induced spin-lattice relaxation (cf. Mell et al., 1980). The latter phenomenon was theoretically studied by Movaghar and Schweitzer (1977) and Movaghar et al. (1978).

The expression for the hopping magnetoresistance obtained in the Movaghar-Schweitzer-Osaka model consists of a positive and a negative term. The sign of the magnetoresistance depends on the magnitude of the quantity $\mu_B \bar{H}_I \nu_s/\hbar$, where H_I is the average internal magnetic field, which is a superposition of exchange, dipolar, hyperfine, and spin-orbit contributions, μ_B is Bohr's magneton, and $\nu_s = \nu_0 \times \exp\left(-(T_0^h/T)^{1/4}\right)$ is the hopping frequency (for variable-range hopping) without any spin flip, where T_0^h is of the same order of magnitude as Mott's parameter T_0 (cf. (5.1.19)). Hence, a change of the sign of the magnetoresistance (cf. Fig. 9.3) due to changing the conditions of preparation of the sample (e.g. doping) can result either from a change of the density of states at the Fermi level (note that $T_0 \sim 1/N(\mu)$), or from a change of the internal magnetic field H_I (cf. Mell et al., 1980, Kuivalainen et al., 1982).

Note that in random hopping systems a negative contribution to the magnetoresistance may also be due to the energy dependence of the localization length along with intra-site electron-electron interaction (Fukuyama and Yosida, 1979, Kamimura et al., 1983).

Concluding this section let us draw attention to the phenomenon of ESR-induced hopping transport, which is due to absorption of energy from an excited spin by the combined action of the exchange interaction and spin-orbit interaction (Kamimura and Mott, 1976, cf. Morigaki, 1980). This phenomenon is closely related to the hopping magnetoresistance.

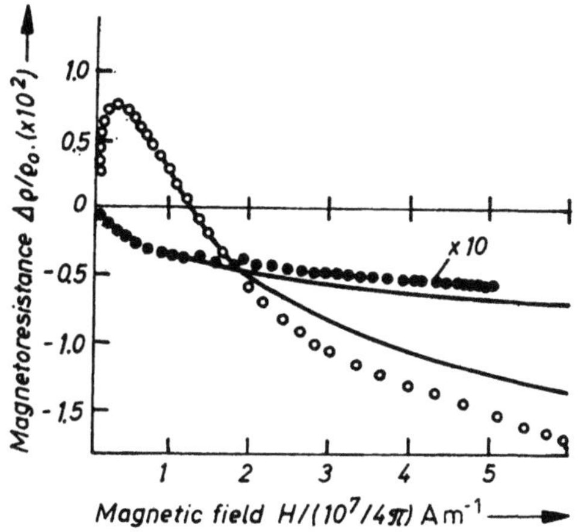

Fig. 9.3. Magnetoresistance in rf-sputtered hydrogenated amorphous silicon in dependence of the magnetic field for two different conditions of preparation: (•) partial hydrogen pressure 0%, deposition rate 6,5 nm/min, spin density 1.5×10^{20} cm^{-3}; (o) partial hydrogen pressure 2%, deposition rate 5,5 nm/min, spin density 3.1×10^{19} cm^{-3}. The solid curves are calculated by using the Movaghar-Schweitzer-Osaka model (see text) (after Kuivalainen et al., 1982)

9.3. Optical phenomena

The study of optical properties of solids may provide important information on the character of the charge transport, the strength of the electron-phonon coupling, the density of the electron states, etc.

9.3.1. *Absorption of light in disordered materials with strong electron-phonon coupling*

As pointed out in Section 3.3., a characteristic feature of small-polaron transport in crystals is the occurrence of a Gaussian peak in the absorption spectrum of light, whose maximum is at $\hbar\omega = 4E_a$, where E_a is the activation energy of the small-polaron drift mobility (cf. (3.3.9)). Hence, from an experimental investigation of the frequency dependence of the absorption coefficient one may obtain a strong argument in favour of small-polaron formation. The presence of disorder may strongly deform the small-polaron absorption peak quoted. This effect will be studied in the following.

To evaluate the real part of the conductivity, $\sigma'(\omega)$, we start from the expression for the current (2.5.7) to (2.5.9). Here we may neglect the diffusive contribution to the current (2.5.9), because this contribution vanishes in the high-frequency regime. Accordingly, restricting ourselves to the Ohmic contribution to the current, from (2.5.8) and (2.5.6) we obtain the following closed expression for $\sigma'(\omega)$:

$$\sigma'(\omega) = \frac{e^2\mathcal{N}}{\hbar^2\omega} \sinh\left(\frac{\hbar\omega}{2kT}\right) \sum_{mm'} |J_{mm'}|^2 (\boldsymbol{R}_m - \boldsymbol{R}_{m'})_x^2 f_m(1 - f_{m'})$$

$$\times \exp\{-(\varepsilon_m' - \varepsilon_m)/2kT\} K(\varepsilon_m' - \varepsilon_m), \tag{9.3.1}$$

where

$$K(\varepsilon) = \int_{-\infty}^{\infty} \mathrm{d}t \exp\left\{-2S_\mathrm{T} + i\left(\omega - \frac{\varepsilon}{\hbar}\right)t + \sum_{\boldsymbol{q}} \frac{|\gamma(\boldsymbol{q})|^2 \cosh(\omega_{\boldsymbol{q}}t)}{\sinh(\hbar\omega_{\boldsymbol{q}}/2kT)}\right\}. \tag{9.3.2}$$

Here $\mathcal{N}$ is the concentration of sites, S_T the factor of the polaron narrowing of the electron band defined in (A 1.18), and the other notations are given in Section 2. In what follows, we neglect the dependence on the random parameter $\boldsymbol{R}_m - \boldsymbol{R}_{m'}$ (the difference between the positions of the sites) of the polaron shift, the activation energy etc., which allows the factor $[1 - \cos \boldsymbol{q}(\boldsymbol{R}_m - \boldsymbol{R}_{m'})]$ below the summation sign over $\boldsymbol{q}$ to be replaced by unity.

Replacing in (9.3.1) the sum over the sites by a configuration average (over $\boldsymbol{R}_m - \boldsymbol{R}_{m'}$ and ε_m), and assuming that there is not correlation between the random quantities, then we obtain

$$\sigma'(\omega) = \frac{e^2\mathcal{N}^{1/3}}{\hbar^2\omega} \bar{J}^2 \sinh\frac{\hbar\omega}{2kT}$$

$$\times \int \mathrm{d}\varepsilon\, \mathrm{d}\varepsilon' \frac{P(\varepsilon)\, P(\varepsilon')\, K(\varepsilon - \varepsilon')}{4\cosh\left(\frac{\varepsilon - \mu}{2kT}\right) \cosh\left(\frac{\varepsilon' - \mu}{2kT}\right)}, \tag{9.3.3}$$

where $\bar{J}^2$ denotes $|J_{mm'}|^2$ averaged over $\boldsymbol{R}_m - \boldsymbol{R}_{m'}$,

$$\bar{J}^2 = \mathcal{N}^{2/3}\left\langle \sum_{m'} (\boldsymbol{R}_m - \boldsymbol{R}_{m'})_x^2 |J_{mm'}|^2 \right\rangle, \tag{9.3.4}$$

and $P(\varepsilon)$ is the distribution function of the site energies, for which in the following a Gaussian shape will be assumed:

$$P(\varepsilon) = \frac{1}{\Gamma\sqrt{2\pi}} \exp\{-\varepsilon^2/2\Gamma^2\}. \tag{9.3.5}$$

Here Γ labels the width of the spread of the energy levels.

For large values of the electron-phonon coupling costant γ, the time integral in K (9.5.2) may be evaluated by means of the saddle-point method, analogously as in Section 3.3. in studying the absorption of light by small polarons in crystals. Then we get

$$K(\varepsilon) = \frac{\hbar \sqrt{\pi}}{\sqrt{4E_a kT}} \exp\left\{-\frac{E_a}{kT} - \frac{(\hbar\omega - \varepsilon)^2}{16E_a kT}\right\}, \tag{9.3.6}$$

where the activation energy E_a is defined in (3.2.8).

In (9.3.3), the integration over ε and ε' may be rigorously performed in the case that the absorption is governed by hops near the maximum of the density of states, and that the Fermi level is sufficiently far from this maximum, i.e. if $|\mu| > kT$ and $kT > \Gamma$. This situation corresponds to the case of the Boltzmann statistics for the charge carriers. It allows in (9.5.3) $\cosh [(\varepsilon - \mu)/2kT]$ to be replaced by $\frac{1}{2} \times \exp |(\varepsilon - \mu)/2kT|$, so that $\sigma'(\omega)$ becomes (Böttger and Bryksin, 1974)

$$\sigma'(\omega) \sim \exp\left\{-\frac{(\hbar\omega - 4E_a - \Gamma^2/kT)^2}{16E_a kT + 4\Gamma^2}\right\}, \tag{9.3.7}$$

where we have written down only the strongest (exponential) dependence on ω. According to (9.3.7), in disordered systems for hops near the maximum of the density of states, the spread of the energy levels leads to a broadening of the absorption peak, but the Gaussian shape of the peak is preserved (compare (9.3.7) with (3.3.9)). The position of the peak is displaced to the short-wave region by Γ^2/kT.

Somewhat more complicated is the situation for $\Gamma \gg kT$ and hops near the Fermi level. For $\Gamma \gg kT$ and $2kT \ll \hbar\omega$, from (9.5.3), we obtain

$$\sigma'(\omega) \sim \int\limits_{-\infty}^{\infty} \mathrm{d}x \int\limits_{\mu - x/2}^{\mu + x/2} \mathrm{d}x' \exp\left\{-\frac{E_a}{kT} - \frac{\hbar\omega + |x|}{2kT} - \frac{x'^2 + 4x^2}{4\Gamma^2} - \frac{(\hbar\omega - x)^2}{16E_a kT}\right\}, \tag{9.3.8}$$

where $x = \varepsilon - \varepsilon'$ and $x' = (\varepsilon' + \varepsilon)/2$.

If we are interested only in the exponential dependence of σ' on ω, then (9.3.8) may be analyzed as follows: We look for the position of the maximum of the expressions being quadratic in x and x' in the exponent in (9.5.8). If this maximum is placed within the limits of intergration, then we take the integrand at the point of the maximum, otherwise x and x' must be replaced by the corresponding values at the boundaries of the region of integration. Accordingly, the exponential dependence of σ' on ω is given by (Bryksin, 1982)

$$\sigma'(\omega) = \begin{cases} \exp\left\{-\dfrac{\mu^2}{\Gamma^2} - \dfrac{(\hbar\tilde{\omega})^2}{16E_a kT}\right\}, \text{ for } \hbar\tilde{\omega} < -\dfrac{8E_a |\mu| kT}{\Gamma^2} & \text{(9.3.9a)} \\[2ex] \exp\left\{-\dfrac{\mu^2}{\Gamma^2} - \dfrac{(\hbar\tilde{\omega} - |\mu|)^2}{16E_a kT + 2\Gamma^2}\right\}, \text{ for } -\dfrac{8E_a |\mu| kT}{\Gamma^2} < \hbar\tilde{\omega} < \dfrac{8E_a |\mu| kT}{\Gamma^2} + 2|\mu| & \text{(9.3.9b)} \\[2ex] \exp\left\{-\dfrac{(\hbar\tilde{\omega})^2}{16E_a kT + 4\Gamma^2}\right\}, \text{ for } \hbar\tilde{\omega} > \dfrac{8E_a |\mu| kT}{\Gamma^2} + 2|\mu|, & \text{(9.3.9c)} \end{cases}$$

where $\hbar\omega = \hbar\tilde{\omega} - 4E_a$.

The dependence $\sigma'(\omega)$, evaluated on the basis of (9.3.9), is illustrated in Fig. 9.4. We see that this dependence exhibits a peak at $\hbar\omega = 4E_a + |\mu|$ (in the frequency range (9.3.9b)) with a longer wing at the high frequency side than at the low frequency one: for small μ, the descent of the low frequency wing (described by (9.3.9a)) is of order $4\sqrt{E_a kT}$ (as in crystals), while that of the high frequency wing (cf. (9.3.9c)) is of order $\sqrt{16E_a kT + 4\Gamma^2}$. The peak is Gaussian-like only in the vicinity of the maximum on a distance of order $8E_a |\mu| kT/\Gamma^2$ in both directions from the maximum. Accordingly, the asymmetry of the peak is the larger the smaller $|\mu|$ (for a half-filled band, $\mu = 0$) and the larger the spread of the energy levels, Γ, compared with $4\sqrt{E_a kT}$.

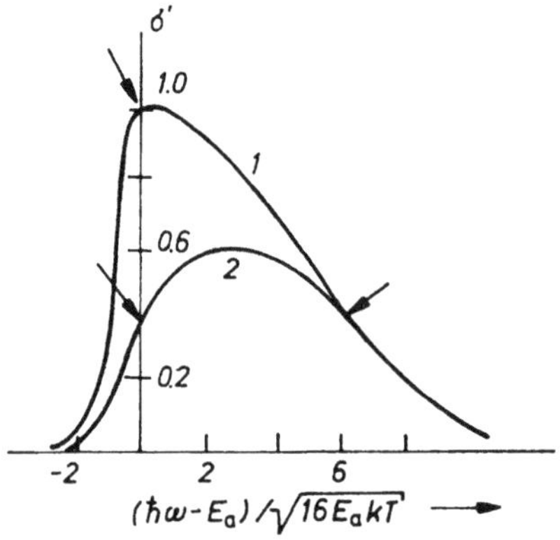

Fig. 9.4. Frequency dependence of the real part of the conductivity σ' evaluated by means of formula (9.3.9) and plotted in arbitrary units; $\Gamma/\sqrt{16E_a kT} = 3$; $|\mu|/\sqrt{16E_a kT} = 0$ (curve 1), $= 3$ (curve 2). The arrows indicate the boundaries of the frequency regimes. For $\mu = 0$, the frequency regime (9.3.9b) is lacking

The result (9.3.9) may be understood within the framework of the two-site model (cf. Fig. 3.5) as follows: In the presence of a spread of the energy levels, the shape of the adiabatic levels remains unchanged, but for neighbouring sites, the minima of the adiabatic levels differ by $\varepsilon' - \varepsilon$. Therefore, in the frequency region $\hbar\omega < 4E_a$, intersite optical transitions are most probable if the energy of the final state ε is smaller than that of the initial state ε'. Since such transitions are forbidden by the Pauli principle, in this frequency region the major contribution comes from hops between sites whose energies are equal to each other and equal to μ. Accordingly, in this frequency regime the spread of the levels does not affect the shape of the absorption curve. Here this shape agrees with that in crystals (frequency regime (9.3.9a)).

On the other hand, in the frequency regime $\hbar\omega > 4E_a$, for hops between neighbouring sites with equal energy, the transition probability decreases with increasing ω. Here the major contribution to the absorption comes from hops between pairs of sites with energies near the Fermi level, whereby the energy of the final state exceeds that of the initial one by an amount $\hbar\omega - 4E_a$. Therefore, at high frequencies, the decrease of σ' with increasing frequency is only due to the decrease of the density of states, and not due the reduction of the transition probability owing to an increase of the difference between level spacing and photon energy. This leads to an extension of the high-frequency wing of σ', compared with the low frequency one.

For the same reason, for $\mu \neq 0$, the peak of σ' is shifted to higher frequencies. In this case, up to $\hbar\omega = 4E_a$, the absorption is governed by hops between pairs of sites with energies at the Fermi level. At higher frequencies, the absorption is mainly due to hops from sites with energies at the Fermi level to sites associated with higher energies. Since the density of states increases with the distance from the Fermi level, the absorption continues to increase even for $\hbar\omega > 4E_a$. This increase continues up to frequencies at which the level of the optimal final state reaches the maximum of the density of states, i.e. up to $\hbar\omega = 4E_a + |\mu|$. At still higher frequencies, σ' decreases due to the decrease of the density of states.

If is worth noticing that a strongly asymmetric absorption peak has been observed on strongly non-stochiometric tungsten dioxide WO_2 (Gavrilyuk et al., 1982)

9.3.2. *Resonant-Raman scattering in materials with localized charge carriers*

For processes of secondary radiation, the differential scattering cross-section may be evaluated on the basis of the following formula (Enderlein and Peuker, 1977, Enderlein et al., 1979)

$$\frac{\mathrm{d}^2\sigma}{\mathrm{d}\omega_s\,\mathrm{d}\Omega} = \left(\frac{e^2}{mc^2}\right)^2 \frac{\omega_s}{\omega_i} \lim_{T\to\infty} \frac{1}{2T} \operatorname{Re} \int\limits_0^\infty \mathrm{d}\omega \int\limits_{-T}^{T} \mathrm{d}\tau\, e^{i(\omega_s-\omega)\tau} \langle A^+(\omega)\, A(\omega_s)\rangle, \tag{9.3.10}$$

where

$$A(\omega) = \frac{i}{2\pi m\hbar} \int\limits_{-\infty}^{\infty} \mathrm{d}t \int\limits_{-\infty}^{t} \mathrm{d}t'\, e^{i(\omega t-\omega_i t')} [P^i(t', -\boldsymbol{q}_i), P^s(t, \boldsymbol{q}_s)]. \tag{9.3.11}$$

Here $A(\omega)$ is the operator of the scattering amplitude; $\omega_i(\omega_s)$, $\boldsymbol{q}^i(\boldsymbol{q}^s)$, $\boldsymbol{e}^i(\boldsymbol{e}^s)$ is the frequency, wave vector, and polarization vector of the incident (scattered) light, respectively. Furthermore, $P^{i,s} = \boldsymbol{e}^{i,s}\boldsymbol{P}$, where $\boldsymbol{P}$ is the momentum operator, and the bracket $\langle\ldots\rangle$ indicates the thermal and configurational average. For instance, for intraband absorption, the commutator of the dipole moment (2.1.18) and the total Hamiltonian (2.1.14) yields (cf. Böttger et al., 1979a)

$$\boldsymbol{P}(\boldsymbol{q}) = \frac{im}{\hbar} \sum_{mm'} J_{mm'}(\boldsymbol{R}_m - \boldsymbol{R}_{m'}) \exp\left(-i\boldsymbol{q}\, \frac{\boldsymbol{R}_m + \boldsymbol{R}_{m'}}{2}\right) a_{m'}^+ a_m. \tag{9.3.12}$$

To study (9.3.10) for systems with localized charge carriers, it is convenient to use the polaron canonical transformation described in Section 2.2. Then we obtain

$$\tilde{\boldsymbol{P}}(\boldsymbol{q}) = \frac{im}{\hbar} \sum_{mm'} J_{mm'}\hat{\Phi}_{mm'}(\boldsymbol{R}_m - \boldsymbol{R}_{m'}) \exp\left(-i\boldsymbol{q}\, \frac{\boldsymbol{R}_m + \boldsymbol{R}_{m'}}{2}\right) a_{m'}^+ a_m. \tag{9.3.13}$$

Analogous to the evaluation of the hopping current described in Section 2., the differential scattering cross-section may be calculated on the basis of (9.3.10) by means of a perturbation expansion with respect to powers of the operator $\mathcal{H}'$ (2.2.8) (BÖTTGER et al., 1979a). For optical properties, it holds that $\hbar\omega \gg \mathcal{W}$, where $\mathcal{W}$ is a characteristic hopping probability. Accordingly, in this case, it is in general sufficient to restrict oneself to the lowest non-vanishing order of the perturbation theory. Hence, in the evolution operators in (9.5.10), the operator $\mathcal{H}'$ may in general be ignored, so that the scattering cross-section becomes proportional to fourth power of the resonance integral J. If, howevei, the lowest order terms exhibit a resonance denominator, then an infinite series of terms (diagrams) must be summed up, which in general corresponds to an account of a finite lifetime of the electronic states (broadening of the electron levels).

In the lowest order of the perturbation expansion, the differential cross-section is proportional to $J_{m_1m_2}J_{m_2m_3}J_{m_3m_4}J_{m_4m_1}$, i.e. up to four sites may be involved in the elementary scattering process. It is reasonable to separate two-site processes, for which $m_1 = m_3$ and $m_2 = m_4$ hold. A characteristic of the two-site processes is the fact that the initial and final electron states coincide, so that the scattering spectrum becomes proportional to $\delta(\omega_l - \omega_s - n\omega_{op})$, where ω_{op} is an optical phonon frequency, and $n = 0, \pm 1, \pm 2 \dots$ is the number of phonons participating in the scattering process. This is the so-called "discrete" contribution to the scattering spectrum. It consists of discrete lines, even in disordered systems on neglection of phonon dispersion.

For electron-phonon coupling of arbitrary strength and zero temperature, the differential scattering cross-section for two-site processes is given by (BÖTTGER et al., 1978)

$$\frac{d^2\sigma}{d\omega_s\, d\Omega} = \frac{3\pi}{4} (2\alpha)^{-7}\, \Omega \mathcal{N}^2 J_0^4 \left(\frac{e^2}{c^2\hbar^3}\right) \frac{\omega_s}{\omega_l} [1 + 2(\boldsymbol{e}^s\boldsymbol{e}^l)]\, K, \tag{9.3.14}$$

$$K = \int\limits_{-\infty}^{\mu} d\varepsilon \int\limits_{\mu}^{\infty} d\varepsilon' P(\varepsilon)\, P(\varepsilon') \int\limits_{-\infty}^{\infty} \frac{dt_1}{2\pi} \int\limits_{-\infty}^{0} dt\, dt' \exp \Big\{ -i(\omega_l - \omega_s)\, t_1$$

$$+ i\omega_l(t - t') + W(t + t') + \frac{i}{\hbar} (\varepsilon - \varepsilon')\, (t - t') - \sum_{\boldsymbol{q}} \frac{|\gamma(\boldsymbol{q})|^2}{N} [2 - e^{-i\omega_{\boldsymbol{q}}t}$$

$$- e^{-i\omega_{\boldsymbol{q}}t'} - e^{i\omega_{\boldsymbol{q}}t_1} (1 + e^{i\omega_{\boldsymbol{q}}(t'-t)} - e^{i\omega_{\boldsymbol{q}}t'} - e^{\omega_{\boldsymbol{q}}t})] \Big\}. \tag{9.3.15}$$

Here Ω is the volume of the system, $J_{mm'} = J_0 \exp(-\alpha\, |\boldsymbol{R}_m - \boldsymbol{R}_{m'}|)$, and W denotes the reciprocal lifetime of an electron state.

For strong coupling with phonons ($\gamma^2 \gg 1$), the time integral in (9.3.15) may be evaluated by means of the saddle-point method. There are two regions of t, t', and t_1-values where the exponent becomes maximum: (i) about $t = t' = 0$, t_1 arbitrary; (ii) about $t_1 = 0$, $t' = t$ arbitrary.

Expanding the exponent in (9.3.15) in powers of t and t' about $t = t' = 0$ (region (i)) up to the second order, we obtain

$$K = \int_{-\infty}^{\mu} \mathrm{d}\varepsilon \int_{\mu}^{\infty} \mathrm{d}\varepsilon' P(\varepsilon)\, P(\varepsilon') \int_{-\infty}^{\infty} \frac{\mathrm{d}t_1}{2\pi} \int_{-\infty}^{0} \mathrm{d}t\, \mathrm{d}t' \exp\Big\{-\mathrm{i}(\omega_\mathrm{i} - \omega_\mathrm{s})\, t_1$$

$$+ \mathrm{i}\omega_\mathrm{i}(t - t') + W(t + t') + \frac{\mathrm{i}}{\hbar} (\varepsilon - \varepsilon')\, (t - t') + \frac{\mathrm{i}}{\hbar}\, 2E_\mathrm{p}(t' - t)$$

$$- \tilde{\Omega}^2(t^2 + t'^2) + tt' \sum_{\boldsymbol{q}} \frac{|\gamma(\boldsymbol{q})|^2}{N}\, \omega_{\boldsymbol{q}}^2 \mathrm{e}^{\mathrm{i}\omega_{\boldsymbol{q}} t_1}\Big\}, \tag{9.3.16}$$

where

$$\tilde{\Omega}^2 = \frac{1}{2} \sum_{\boldsymbol{q}} \frac{|\gamma(\boldsymbol{q})|^2}{N}\, \omega_{\boldsymbol{q}}^2 \approx \frac{\omega_\mathrm{op} E_\mathrm{p}}{\hbar}. \tag{9.3.17}$$

Here E_p is the polaron shift (2.2.7) and $\tilde{\Omega}$ characterizes the width of the Gaussian peak of the intraband optical absorption in the small-polaron model.

Expanding in (9.3.16) the exponential with respect to the last term, we get

$$K = \frac{1}{\tilde{\Omega}^2} \sum_{n=0}^{\infty} \delta(\omega_\mathrm{i} - \omega_\mathrm{s} - n\omega_\mathrm{op})\, K_n, \tag{9.3.18}$$

where

$$K_n = \frac{1}{n!} \Bigg\{ \left[\frac{\mathrm{d}}{\mathrm{d}\alpha} - \frac{\mathrm{d}}{\mathrm{d}\alpha'}\right]^n \mathrm{Re} \int_{-\infty}^{\mu} \mathrm{d}\varepsilon \int_{\mu}^{\infty} \mathrm{d}\varepsilon' P(\varepsilon)\, P(\varepsilon') \int_{0}^{\infty} \mathrm{d}t\, \frac{\sqrt{\pi}}{2\sqrt{\alpha'}}\, \mathrm{e}^{-\alpha t^2}$$

$$\times \left[1 - \Phi\!\left(t\sqrt{\alpha'}\right)\right] \exp\left[-\frac{i\sqrt{2}\, t}{\tilde{\Omega}} \left(\omega_\mathrm{i} - \frac{2E_\mathrm{p}}{\hbar} - \frac{\varepsilon' - \varepsilon}{\hbar}\right)\right]\Bigg\}\Bigg|_{\alpha=\alpha'=1} \tag{9.3.19}$$

Here Φ is the error function. In deriving (9.3.19), the dispersion of phonons has been neglected. Furthermore, we have put $W = 0$, i.e. a finite lifetime is not accounted for.

The expression (9.3.19) describes multiphonon Raman transitions: $n = 0$ describes the Rayleigh scattering, $n = 1, 2, \ldots$, the one-, two- ... phonon Raman scattering. Since $T = 0$ has been assumed, only Stokes lines appear.

From an inspection of the integrals occuring in (9.3.19), for not too large values n, we get

$$K_n \sim \left[\frac{\tilde{\Omega}^2}{\tilde{\Omega}^2 + \Gamma^2 + \left(\omega_\mathrm{i} - \dfrac{2E_\mathrm{p} + \Gamma}{\hbar}\right)^2}\right]^n, \tag{9.3.20}$$

i.e. for $\Gamma \ll \tilde{\Omega}$, the amplitude of the peak depends weakly on n, and the peak is centered about $\hbar\omega_\mathrm{i} \approx 2E_\mathrm{p} + \Gamma$. However, far from resonance, K_n decreases strongly with increasing n. An increase of Γ leads to a rapid reduction of the amplitude of the peak (especially for large n) and to a broading of the peak. The latter fact is characteristic of disordered systems. The scattering amplitude becomes maximum when the chemical potential μ coincides with the maximum of the density of states. Note that the weak dependence of the amplitude of the scattering peak on n (for strong coupling with phonons) was first predicted by TRIFONOV and PEUKER (1965).

Let us now turn to the inspection of the second saddle point in (9.3.15) ($t_1 = 0$, $t' = t$ arbitrary). The corresponding contribution to the spectrum describes luminescence, in contrast to (9.3.18), which describes resonant-Raman scattering. It turns out that the contribution from the second saddle point is above all important in the case that the density of states $P(\varepsilon)$ exhibits two maxima, i.e. if the interaction with light leads to electron transitions between two narrow bands. In what follows for simplicity, we assume, that $P(\varepsilon) = \frac{1}{2}\{\delta(\varepsilon) + \delta(\varepsilon - \hbar\omega_g)\}$, where $\hbar\omega_g$ is the width of the gap between the bands. Expanding the exponent in (9.3.15) in powers of t_1 and $(t - t')$ up to the second order, we obtain

$$K = \frac{1}{2}\int_0^\infty \mathrm{d}t\big(\tilde{\Omega}^4 - \tilde{\Omega}^4(t)\big)^{-1/2} \exp\left\{-\frac{\left[\omega_s - \omega_l + \frac{4}{\hbar}\big(E_p - E_p(t)\big)\right]^2}{8[\tilde{\Omega}^2 - \tilde{\Omega}^2(t)]} - \frac{\left[\omega_l + \omega_s - 2\omega_g + \frac{4}{\hbar}E_p(t)\right]^2}{8[\tilde{\Omega}^2 + \tilde{\Omega}^2(t)]}\right\}, \tag{9.3.21}$$

where

$$E_p(t) = \sum_q \frac{\hbar\omega_q}{2N}|\gamma(q)|^2 \cos\omega_q t,$$
$$\tilde{\Omega}^2(t) = \sum_q \frac{\omega_q^2}{2N}|\gamma(q)|^2 \cos\omega_q t. \tag{9.3.22}$$

The integrand in (9.3.21) has two characteristic times: τ — the time during which $E_p(t)$ and $\tilde{\Omega}^2(t)$ go to zero, and W^{-1} — the lifetime of the polaron state. In contrast to the situation in resonant-Raman scattering, in the present situation the transition from the lower to the upper electron state may be considered as a real transition, provided that $W^{-1} \gg \tau$, i.e. if the time needed for polaron formation is much shorter than the lifetime of the upper electron state. In this case, the integral (9.3.21) can readily be evaluated by setting $E_p(t) = 0$ and $\tilde{\Omega}^2(t) = 0$. Then, we get

$$K = \frac{1}{4W\tilde{\Omega}^2}\exp\left[-\frac{(\omega_s - \omega_g + 2E_p/\hbar)^2}{4\tilde{\Omega}^2}\right]\exp\left[-\frac{(\omega_l - \omega_g - 2E_p/\hbar)^2}{4\Omega^2}\right]. \tag{9.3.23}$$

The result (9.3.23) may be interpreted as luminescence with a time between absorption and emission of the order of W^{-1}. The second term on the right-hand side of (9.3.23) describes the absorption of the incident light and the first term the luminescence, i.e. the maximum of the luminescence is at the frequency $\omega_s = \omega_g - 2E_p/\hbar$ and this process is most effective when the frequency of the incident light is given by $\omega_i = \omega_g + 2E_p/\hbar$. Between the maxima of absorption and luminescence, there occurs a large Stockes shift given by $4E_p/\hbar$. Such a Stokes shift is well known in the literature (cf., e.g., STREET, 1977). As pointed out by BÖTTGER et al. (1978), a Stokes shift may be also expected to occur for luminescence transitions between a narrow polaron band and broad band with weak electron-phonon coupling. In this case the polaron shift is given by $2E_p/\hbar$.

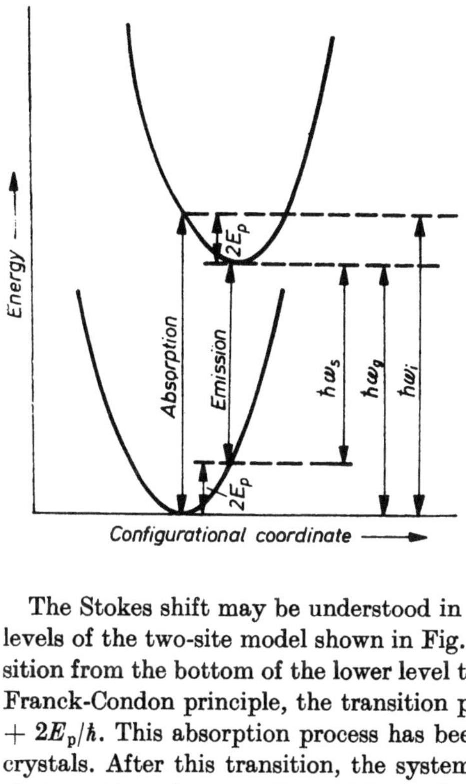

Fig. 9.5. Schematic representation of optical transitions (absorption and luminescence) between the two-site adiabatic levels in systems with strong electron-phonon coupling, illustrating the occurrence of a Stokes shift of luminescence

The Stokes shift may be understood in terms of transitions between the adiabatic-levels of the two-site model shown in Fig. 9.5. Absorption of a photon causes a transition from the bottom of the lower level to the upper one, whereby, according to the Franck-Condon principle, the transition probability becomes maximum for $\omega_i = \omega_g + 2E_p/\hbar$. This absorption process has been discussed in Section 3.3. for polarons in crystals. After this transition, the system thermalizes, i.e. at the initial site of the electron, the polaron cavity vanishes and at final site, such a cavity develops, whereby the electron falls down to the bottom of the upper level. If the lifetime on the upper level, W^{-1}, exceeds the time of thermalization, a radiative transition from the bottom of the upper level to the lower level at frequency $\omega_s = \omega_g - 2E_p/\hbar$ occurs, i.e. the corresponding frequency is, by $4E_p/\hbar$, smaller than the frequency of the light intially absorbed.

Let us now concisely discuss three- and four-site transitions. It turns out that for strong coupling with phonons, the corresponding transition probabilities are proportional to $\exp(-\delta S_T)$, where δ is a numerical factor of order unity, and S_T is defined in (A1.18). Accordingly, such transition are above all important in systems with weak electron-phonon coupling (BÖTTGER et al., 1979b). The major feature of three-site (and also four-site) hops is the fact that the energies of the initial and the final states do not coincide. For a n-phonon process, the corresponding scattering spectrum is proportional to $\delta(\omega_i - \omega_s - n\omega_{op} + \varepsilon_i/\hbar - \varepsilon_f/\hbar)$, where ε_i denote the energy of the initial and the final electron states, respectively. Owing to the continous spread of the site energies, the spectrum arising from three-and four site processes is strongly broadened, i.e. these processes produce a continous spectrum.

The reader who is interested in more details on the scattering cross-section in disordered systems with localized charge carriers is referred to BÖTTGER et al. (1978, 1979a, b). Furthermore, the theory of interband absorption and Raman scattering in disordered systems having a smooth random potential is described in detail in the monograph by BONCH-BRUEVICH et al. (1981).

Chapter IV. Classical Hopping Transport

10. Diffusive atomic motion

10.1. Some basic features of diffusive atomic motion

10.1.1. Materials exhibiting diffusive atomic motion

In recent years, there has been considerable interest in the migration of atoms in solids such as in superionic conductors (see, e.g., BOYCE and HUBERMAN, 1979, DIETRICH et al., 1980, DIETERICH, 1981), metal hydrides (see, e.g., ALEFELD and VÖLKL, 1978), non-stoichiometric crystals, and on solid surfaces (see, e.g., KLAFTER and SILBEY, 1980a).

Let us first consider some features of superionic conductors. These conductors are solid state systems that are characterized by ionic conductivity σ which is typical of molten salts, i.e. $\sigma \approx 1(\Omega\ \text{cm})^{-1}$. For comparison, the ionic conductivity of the "ordinary" ionic conductor NaCl is about 10^{-15} $(\Omega\ \text{cm})^{-1}$ at room temperature. There is a variety of superionic materials. The common structural feature of these materials is the existence of conduction paths connecting fractionally occupied sites.

Fig. 10.1 shows the temperature dependence of some typical superionic conductors. Note that there are different types of transitions into the superionic state. For example, in AgI this transition is accompanied by a well-defined first-order phase transition, in PbF_2 by a second-order phase transition, while in Na-β-Al_2O_3 there is an exponential growth in the conductivity without much cooperative behaviour.

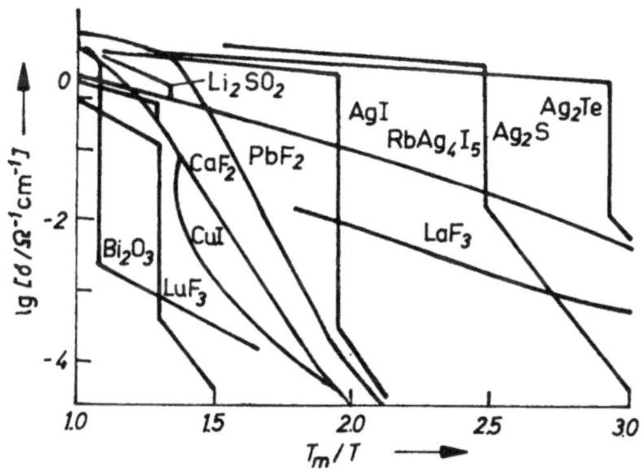

Fig. 10.1. Logarithm of the conductivity of some typical superionic conductors plotted against T_m/T, where T_m is the melting temperature of the material (after BOYCE and HUBERMAN, 1979)

Let us consider in some more detail AgI, which is a paradigm of a superionic material (see, e.g. MELLANDER, 1982, and references therein). In this material at the transition temperature $T_c = 147\,°C$ the conductivity jumps by four orders of magnitude. The entropy changes at T_c and at the melting point T_m are both about half the change of entropy at the melting point of an ordinary ionic crystal, which suggests that in AgI at T_c one sublattice is melting.

Experimentally, the diffusion coefficient D of the high temperature phase of AgI is found to be $D = 10^{-5}\,cm^2\,s^{-1}$, which is the value one gets for water at room temperature. Using the Einstein relation

$$\sigma = n(Ze)^2\,D/kT\,, \tag{10.1.1}$$

(Ze — charge of the carriers) one finds for the superionic phase of AgI a charge-carrier density $n \approx 10^{22}\,cm^{-3}$, which is essentially the entire content of one sublattice. This again suggests that at T_c one sublattice is melting. In other words, in the high temperature phase of AgI, the ions of one sublattice seem to diffuse liquid-like through the cage provided by the ions of the other sublattice.

However, this picture is not quite correct. By modern methods (neutron scattering, EXAFS, molecular-dynamics technique) the mobile ions (Ag^+) in AgI at $T > T_c$ were shown to be essentially located on equivalent crystallographic sites, whereby each Ag^+-ion disposes on the average of six such sites. The elementary diffusion process is a jump of a single Ag^+-ion from one site to another. Between such jumping events the ion oscillates in the potential minima provided by the lattice. Hence, in the high temperature phase of AgI a mobile ion moves like a diffusing oscillator rather than a particle in a liquid.

The motion of a single ion may be characterized by two times: τ_0 — the time spent in the minimum of a potential well and τ_1 — the time spent in free flight from one minimum to another. These times are given by

$$\tau_0 = a^2/2dD(T) \tag{10.1.2}$$

and

$$\tau_1 = a(M/kT)^{1/2}\,, \tag{10.1.3}$$

where a is the distance between neighbouring sites, d the dimensionality, and M the ion mass. For $\tau_1/\tau_0 = 0$, the motion is solid-like, for $\tau_1/\tau_0 = 1$, gas-like, and for intermediate values of τ_1/τ_0, liquid-like. What kind of motion is realized in a given material, may depend on the properties of the material and the temperature range considered.

Diffusive atomic motion is often interpreted by a hopping model, i.e. in terms of random (over-barrier) hopping from site to site. Such a model is based on the assumption that the time of flight τ_1 is short compared with the mean residence time τ_0 and may be neglected. A description of oscillations in a potential well and the motion during the escape over a barrier is excluded by this assumption.

A hopping model will be reasonable if the temperature is small compared to the hight of the potential barriers. In particular, for AgI such a model appears to be not adequate, because in this material (as can be estimated from EXAFS) at 470 K τ_1

is about a third of τ_0. However, there are other superionic materials (for example, Na-β-Al_2O_3, cf. KIMBALL and ADAMS, 1978) which can be described by a hopping model.

Note that at present the origin of the superionic state is not completely understood. Effects of interaction between the mobile ions are apparently very important for the existence of such a state (see, for example, LAM and BUNDE, 1978, SCHOMMERS, 1980, DIETERICH et al., 1980).

Now let us turn to the problem of atomic diffusion in metal hydrides. From the physical point of view, the great interest in these materials is due to the fact that they allow quantum diffusion of light (interstitial) atoms in crystals to be studied, which is a challenging problem of solid state physics. Note that deviations from the predictions of the classical theory of atomic diffusion (cf. Section 10.1.3.) have been observed for all metal-hydrogen systems, which suggests that (at sufficiently low temperatures) quantum phenomena are important, such as discrete levels in the potential wells and tunnelling through potential barriers (cf. Section 10.1.2.).

With respect to particle diffusion, positive muons (μ^+) may be considered as light isotope of hydrogen (mass ratio 1:9). At high temperatures, where diffusion is by over-barrier hopping (see below), the mass difference is not very important, but it becomes crucial at low temperatures, where tunnelling dominates the diffusion process.

Muon diffusion can be measured by the motional narrowing effect in muon spin rotation experiment: when the muon is stationary, it is depolarized by the surrounding (nuclear) magnetic moments; when it is moving, the rate of depolarization is reduced by motional narrowing. This method is sensitive enough for studying diffusion at low temperatures, where it is not easy to obtain reliable experimental data on diffusion of hydrogen (for more details on muon spin relaxation, see, e.g., KARLSSON, 1982, SEEGER and MONACHESI, 1982, and references therein).

Note that quantum effects may also be important in migration of self-interstitials (e.g. Au self-interstitials) and heavy-impurity interstitials (e.g. Fe interstitials in Al) in some metals at very low temperatures (see, e.g., SCHOBER and STONEHAM, 1982) and of physicosorbed and weakly chemisorbed atoms on solid surfaces. Furthermore, quantum tunneling of atoms dominates the low temperature behaviour of glasses (and also of certain superionics and ferroelectrics) (ANDERSON et al., 1972, PHILLIPS, 1972 see also BÖTTGER, 1983, and references therein) and the behaviour of certain defects in alkali halides (NARAYANAMURTI and Pohl, 1970).

10.1.2. Classical diffusion versus quantum diffusion

The problem of quantum diffusion of light interstitial atoms is closely related to the small-polaron problem: elastic interaction between the interstitial and its surroundings causes lattice relaxation and favours thus self-trapping of the interstitial. This relation allows the theory of quantum diffusion of atoms to be formulated in rather close analogy to the small-polaron theory (FLYNN and STONEHAM, 1970, KAGAN and KLINGER, 1974, FUJII, 1979, TONKS and SILVER, 1982, PETZINGER, 1982; for a review, see KAGAN, 1981; for some criticism, see KEHR et al., 1981).

The major pecularity of the interstitial problem, as compared with the problem of electron transport in a polar medium, is the fact that the individual potential well of an interstitial contains many levels whose width is very small. For light interstitials, the excitation energies are such that at low temperatures, thermal population is almost limited to the ground state. In this situation an interstitial can escape from its potential well only by tunnelling through the potential barrier.

As in the case of small polarons (see Chapter II) the tunnelling processes of light interstitials can be classified into coherent tunnelling (band-like motion) and incoherent tunnelling (under-barrier hopping[1])), according to whether the particle tunnels to the neighbouring well without or with changing the phonon number.

With the increase of temperature, the higher energy levels become populated and the diffusion transition is determined by the optimum of thermal activation and tunnelling penetration. At sufficiently high temperatures, the diffusion on levels close to the barrier top dominates, i.e. the elementary jump process is classical over-barrier hopping with an activation energy given by the barrier height. In this case the hopping rate is (see Section 10.1.3.)

$$W = \frac{\omega_0}{2\pi} \, e^{-U_0/kT} \tag{10.1.4}$$

where ω_0 is a characteristic (optical phonon) frequency. In terms of W the diffusion coefficient reads $D = a^2 W = a^2/2d\tau_0$, where a is the distance between the minima of neighbouring potential wells, τ_0 the residence time and d the dimensionality of the system.

Let us now in some more detail consider the tunnelling process of an interstitial at low temperatures. We assume that the interstitial has formed a small polaron i.e. the interstitial is assumed to be surrounded by a locally relaxed lattice. As described in Chapter II, owing to the translational symmetry of the crystal, at low temperatures the particle, dressed by a phonon cloud,moves coherently through the lattice. The width of the corresponding small-polaron band is roughly given by $z\tilde{J}$, where $\tilde{J} = J \exp(-S_T)$, with z the coordination number, J the resonance integral (tunnelling matrix element), and $\exp(-S_T)$ is due to the renormalization of J by strong interaction with phonons (cf. Section 3.1.). For example, for positive muons in aluminium one finds $S_{T=0K} \approx 2$ (Kehr et al., 1982).

At finite temperatures, phonon scattering processes give rise to a finite mean-free path of the particle and the resulting diffusion coefficient becomes

$$D \sim v^2 \tau, \tag{10.1.5}$$

where v is the velocity of the particle and τ an inverse scattering rate. It has been estimated

$$v \approx \begin{cases} (3kT/m^*)^{1/2}, & z\tilde{J} \gg kT, \\ a\tilde{J}/\hbar, & z\tilde{J} \ll kT, \end{cases} \tag{10.1.6}$$

[1]) Note that in the present chapter the term "over-barrier hopping" is used solely for classical hops over a potential barrier separating neighbouring sites.

(a — distance between neighbouring interstitial sites, m^* — effective particle mass) and $\tau \sim T^{-9}$ (for phonon scattering) (cf. KEHR et al., 1982)[1]).

With increasing temperature, the mean-free path $l \approx v\tau$ becomes as short as the distance a so that the conception of band-like motion is no longer applicable.[2]) The particle is now essentially localized at intersitial sites and the elementary jump process can be induced by absorption (or emission) of phonons. At low temperatures ($kT \ll \hbar\omega_{op}$, ω_{op} — optical phonon frequency) the rate of this (under-barrier) thermally activated hopping process depends on temperature as $W \sim T^7$ (due to two-phonon processes) and at higher temperatures ($kT \lesseqgtr \hbar\omega_{op}$) as (cf. (3.2.10))

$$W = \left[\frac{\pi}{4E_a kT}\right]^{1/2} \frac{J^2}{\hbar} \exp\left[-E_a/kT\right]. \tag{10.1.7}$$

Here the activation energy E_a arises from the need to equalize (by multiphonon processes) the levels of the initial and the final site so that tunnelling can occur.

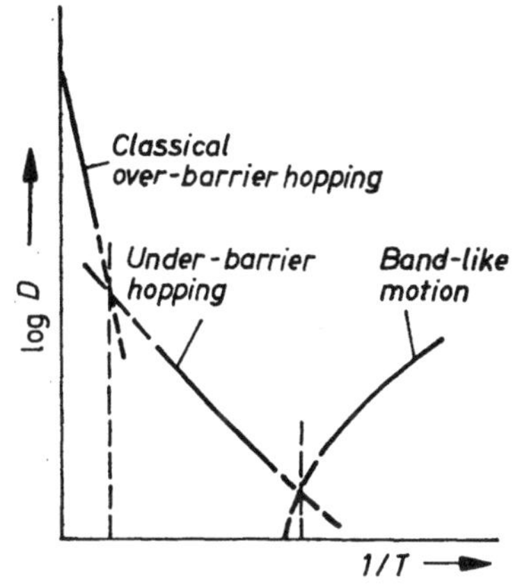

Fig. 10.2. Schematic representation of the temperature dependence of the logarithm of the diffusion constant of light particles in crystals (after FUJII, 1979)

Let us compare formulae (10.1.7) and (10.1.4) for under- and over-barrier hopping, respectively. For under-barrier hopping, the activation energy E_a is of the order of $\hbar\omega_{op}$. The prefactor is proportional to the square of the tunnelling integral J, the small energy parameter of the system, and is usually small as compared with $\hbar\omega_{op}$. For over-barrier hopping, the activation energy usually satisfies $U_0 \gg \hbar\omega_{op}$, and the prefactor is of the order of $\hbar\omega_{op}$. The three elementary jump processes are schematically summarized in Fig. 10.2.

---

1) For the diffusion of positive muons in aluminium, $\tau \sim T^{0.6}$ has been observed (HARTMANN et al., 1980). JÄCKLE and KEHR (1983) suggest that electron-muon scattering should dominate, which yields $\tau \sim T^{-1}$.

2) Note that for interacting atomic particles, carrier localization has been predicted for $T \to 0$, which with increasing temperature is gradually removed due to interaction with phonons, but without restoration of the band motion (KAGAN, 1981, KAGAN and MAKSIMOV, 1983). This phenomen was observed in $^3He-^4He$ solid state solutions by MICHEEV et al. (1982).

For example, muon-spin-rotation experiments (GRAF et al., 1980) indicate for positive muons in iron that coherent tunnelling becomes important below 35 K and that under-barrier hopping (with $E_a = 14$ meV) occurs between 35 and 90 K, while above 150 K over-barrier hopping sets in (with $U_0 = 86$ meV). The tunnelling integral is of the order $J \approx 1$ meV. Furthermore, for hydrogen in FCC (face centred cubic) metals, magnetic relaxation measurements (HOHLER and SCHREYER, 1982) indicate over-barrier hopping for high temperatures, and at low temperatures tunnelling transitions with smaller activation energies and smaller pre-exponential factors, whereby the transition from high- to low-temperature behaviour is a gradually one and begins at room temperature.

For ionic conductors, the masses are larger than about two orders of magnitude than the hydrogen mass. Accordingly, in such a material we have a continuous energy level spectrum in a potential well and thus in general completely classical diffusive atomic motion (over-barrier hopping).

10.1.3. *Hopping rate for over-barrier hopping*

In the simple hopping problem one assumes that a particle (ion) at some site m will hop to a nearby vacant site m' and that this process depends neither on the occupancy of sites other than m and m' nor on the length of time that the particle has already spent at site m. For this hopping problem, let us determine the rate of over-barrier transitions by making use of the so-called absolute-rate theory (ART) (see, e.g., VINYARD, 1957, ICHE and NOZIERES, 1976; cf. CARMELI and NITZAN, 1983, and references therein).

The ART is based on the assumption that quasi-equilibrium conditions prevail during the escape of a particle over the barrier, which implies a clear separation of the time scales of particle and heat bath motion. For hopping motion of ions, this means that thermal equilibrium among the levels in a potential well is very much faster established than a hopping transition to a neighbouring potential well occurs. In other words, at each step of the diffusion process the lattice is assumed to be entirely relaxed.

The term hopping rate (transition rate) refers to the transition of the representative point in the configuration space of a many-particle system between two neighbouring configurations. Within the ART, the hopping rate is defined as the forward current per particle over the saddle point of the surface of the potential energy of interaction. In particular, for the simple hopping problem, the corresponding current is that over the barrier of the height U_0 separating two neighbouring minima of the single-particle potential formed by the crystalline cage.

Let us now study the hopping rate for the latter problem by using the ART. To do this, we consider a one-dimensional system, confine ourselves to the classical regime (valid for $\hbar\omega_{op} \ll kT$) and assume that the thermal equilibrium distribution

$$f(\varepsilon) = e^{-\beta(\varepsilon-\mu)} \tag{10.1.8}$$

($\beta = 1/kT$, μ — chemical potential, ε — energy of the particle) extends to all energies including the states above the top of the potential well.

The net number of trapped particles is

$$n = \sum_\alpha f(\varepsilon_\alpha) = \int_{-U_0}^{0} \mathrm{d}\varepsilon N(\varepsilon)\, f(\varepsilon), \tag{10.1.9}$$

where the summation is over the states in the potential well, $N(\varepsilon) = \sum_\alpha \delta(\varepsilon - \varepsilon_\alpha) = 1/\hbar\omega(\varepsilon)$ is the density of states. The origin of the energy is chosen at the top of the well.

The integral in (10.1.9) is governed by a region of the width $\sim kT$ near the bottom of the well. For deep wells ($U_0 \gg kT$), we may write $N(\varepsilon) \approx \int \mathrm{d}\alpha\delta(\varepsilon - \alpha\hbar\omega_0) = 1/\hbar\omega_0$, where ω_0 is the vibrational frequency $\omega(\varepsilon)$ near $\varepsilon = 0$, the bottom of the well. Thus, we obtain

$$n = \frac{1}{\beta\hbar\omega_0}\, \mathrm{e}^{\beta(\mu + U_0)}. \tag{10.1.10}$$

The rate W at which ions leave the well is determined through

$$S = nW, \tag{10.1.11}$$

where S is the flux of particles escaping from the potential well. The kinetic theory yields

$$S = \frac{1}{L} \sum_\alpha v_\alpha f(\varepsilon_\alpha) = \int_0^\infty \frac{\mathrm{d}p}{h}\, v f(\varepsilon) = \frac{1}{h\beta}\, \mathrm{e}^{\beta\mu}, \tag{10.1.12}$$

where the summation is over the states above the top of the barrier, $v = p/M$ is the velocity of the particle, p its momentum, M its mass, $\varepsilon = p^2/2M$ its energy, and L is the length of the system. Inserting (10.1.10) into (10.1.11) and comparing the resulting expression with (10.1.12), we find the so-called absolute hopping rate (10.1.4).

Note that the rate (10.1.4) is independent of the dynamics of the particle-bath interaction. According to KRAMERS (1940), this result is valid when the friction coefficient, representing the effect of the bath, lies in a certain range. As pointed out by ICHE and NOZIERES (1972), for deep wells ($U_0 \gg kT$), the ART is valid if δ, the average energy that a particle loses in each collision with the crystalline cage, is large compared with kT. On the other hand, if $\delta \ll kT$, the hopping rate becomes

$$W = \frac{\delta}{kT} \frac{\omega_0}{2\pi}\, \mathrm{e}^{-\beta U_0}. \tag{10.1.13}$$

A small value of δ implies that near the top of the well the collisions between the mobile particle and the cage atoms are nearly elastic. Therefore, a noticable departure from the ART result (10.1.4) may be expected when the mass ratio between the mobile particle and the cage atoms is very small. Such a departure has been observed in Li-ion conductors with heavy skeleton structure, such as $Li_2Ti_3O_7$ (HUBERMANN and BOYCE, 1978). The reader who is interested in more details on the validity of the ART and its extensions is referred, e.g., to CARMELI and NITZAN (1983) and references therein (see also Section 11.5.).

If the conditions of the ART are met, a microscopic theory of the hopping rate (see, e.g., PEACOCK-LÓPEZ and SUHL, 1982) is based on the following general formula

$$W = \frac{2\pi}{\hbar} \sum_{f,i} |\langle f| \, T \, |i\rangle|^2 \, \frac{e^{-\beta E_i}}{Z} \, \delta(E_f - E_i), \tag{10.1.14}$$

where $|i\rangle$ ($|f\rangle$) denotes the direct product of the initial (final) particle and phonon states; $E_i(E_f)$ is the corresponding energy, $\langle f| \, T \, |i\rangle$ the T-matrix element (in the first Born approximation $\langle f| \, T \, |i\rangle$ is the matrix element of the particle-phonon interaction Hamiltonian), and Z the particle-plus-bath partition function. For over-barrier hopping, we have $E_i^0 < U_0$ and $E_f^0 > U_0$, where E_i^0 and E_f^0, respectively, denote the initial and final energies of the particle. The δ-function in (10.1.14) indicates that the transition occurs by absorption of one, two, three, etc., phonons with total energy exceeding the barrier energy U_0.

Note that at present there are no reliable theories of the hopping rate which are not based on a separation of time scales. On the other hand, for hopping of atomic particles such a separation is in general not evident, because the vibrational frequencies of the particle and the cage (heat bath) atoms are not significantly different. In particular, for hopping of a heavy impurity atom, REZAYI and SUHL (1982) showed that the ART overestimates the rate by about an order of magnitude.

Let us yet draw attention to the study of DA FANO and JACUCCI (1977) and FLYNN and JACUCCI (1982) of the question how long the memory of prior hopping events lasts. By molecular-dynamics technique these authors showed that in an extremely short time interval (of about 10% of the Debye period) after a hopping event there is a strong enhanced probability for a second hopping event (Fig. 10.3). In this time

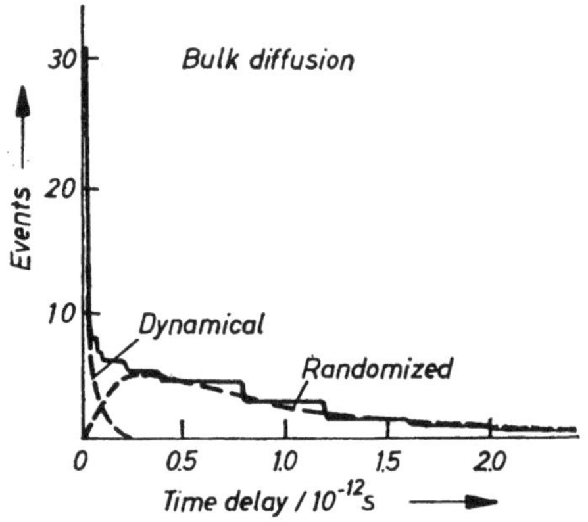

Fig. 10.3. Time delay between successive jumps of a particle (vacancy) (after DA FANO and JACUCCI, 1977 and FLYNN and JACUCCI, 1982). The figure shows for two-particle processes the distribution of second jump events subsequent to the first jump. The broken line separates the double events into the exponential long-time behaviour of the randomized system and the short-time behaviour (peak), where the second event is a direct dynamical consequence of the first one

interval a second event is obviously a dynamical consequence of the first one. It occurs prior to randomization (thermalization) of the system in its new configuration. It turns out that the probability for correlated hops grows rapidly with temperature. At low temperatures the rate theory gives correct predictions.

Note that in the neighbourhood of an impurity atom, the hopping rate can be altered (RICHARDS, 1982).

10.2. Single-particle hopping

Here we study classical hopping in some array of sites where the mobile atomic particles can sit. The probability per unit time that a particle at site m will hop to nearby vacant site m' is given by the hopping rate $W_{mm'}$, which is thermally activated with an activation energy given by the height of the potential barrier between the sites m and m'. In principle, the effective hopping rate $W_{mm'}$ could also depend on the occupancy of sites other than m and m' and on the length of time that the particle already spent on site m. However, both these latter effects are neglected here.

Note that in contrast to hopping of electrons, in the case of atomic diffusive motion, one may also ask for hopping behaviour of individual particles. The latter behaviour can be experimentally studied, for example, by incoherent neutron scattering, tracer ciffusion, and NMR relaxation due to motionally altered dipole interaction. On the other hand, properties of all particles can be obtained, for example, from the bulk diffusion coefficient, the conductivity and NMR relaxation rates due to quadrupolar processes.

10.2.1. A one-dimensional hopping model

To describe hopping motion of a single atomic particle between nearest neighbour sites, we assume that the occupation probability of site m, $\varrho_m(t)$, obeys the rate equation

$$\frac{\mathrm{d}}{\mathrm{d}t}\varrho_m = \sum_{m'}[W_{m'm}\varrho_{m'} - W_{mm'}\varrho_m], \tag{10.2.1}$$

where $W_{mm'}$ is the hopping rate from site m to site m'.

Let us consider hopping in a one-dimensional system with N equally spaced sites with lattice constant a and periodic boundary conditions. In the absence of an external field, we have $W_{m'm} = W_{mm'}$ and (10.2.1) becomes

$$\frac{\mathrm{d}}{\mathrm{d}t}\varrho_m = W[\varrho_{m+1} + \varrho_{m-1} - 2\varrho_m], \tag{10.2.2}$$

where W is the nearest-neighbour hopping rate.

To calculate experimental quantities such as the dynamical structure factor, the diffusion coefficient, etc., we need the conditional-probability function $P_{m'm}(t)$

(cf. Section 2.6.), which is the solution of the rate equation (10.2.1) (in the absence of a field) under the condition that the particle was at site m' at time $t = 0$. As can be readily verified, the corresponding solution of (10.2.2) reads

$$P_{m'm}(t) = \frac{1}{N} \sum_q \exp\left[iqa(m - m')\right] \exp\left[-\lambda(q)\, t\right], \tag{10.2.3}$$

where

$$q = \frac{2\pi}{Na}\, j, \qquad j = 0, \pm 1, \pm 2, \ldots, + \frac{N}{2} \tag{10.2.4}$$

and

$$\lambda(q) = 2W(1 - \cos qa). \tag{10.2.5}$$

Note that the solution (10.2.3) can be found similarly as the solution of the Schrödinger equation or the equation of motion of a harmonically vibrating lattice by means of the Green-function technique.

According to (10.2.3), the hopping process may be considered as a superposition of relaxation modes (alternative to phonons!), which have relaxation rates $0 \leqq \lambda(q) \leqq 4W$, where q labels the modes. For $t \to \infty$, only the mode with $q = 0$ survives. This mode determines the equilibrium distribution $\varrho_m(\infty) = 1/N$.

Let us now turn to experimental quantities. First, consider the incoherent neutron scattering cross section, which is determined by the dynamical structure factor S_{inc}, defined as (see, e.g., Böttger, 1983)

$$\begin{aligned} S_{\text{inc}}(q, \omega) &= \int \frac{dt}{2\pi}\, e^{i\omega t} \left\langle e^{iq(x(t) - x(0))} \right\rangle \\ &\equiv \frac{1}{N} \sum_{mm'} \int \frac{dt}{2\pi}\, e^{i\omega t} P_{m'm}(t)\, e^{iqa(m-m')}, \end{aligned} \tag{10.2.6}$$

where $x(t)$ denotes the position of the particle at time t. By (10.2.3), we find from (10.2.6)

$$S_{\text{inc}}(q, \omega) = \frac{1}{\pi} \frac{\lambda(q)}{\omega^2 + \lambda(q)^2}, \tag{10.2.7}$$

i.e. neutron scattering with momentum transfer q gives a single Lorentzian with half-width $\lambda(q)$. Note that for a single particle there is no coherent scattering.

Now consider the diffusion constant, defined by

$$D = \lim_{t \to \infty} \frac{1}{2t} \left\langle \left(x(t) - x(0)\right)^2 \right\rangle, \tag{10.2.8}$$

where first the thermodynamic limit ($N \to \infty$ at constant density of sites) and after that the limit $t \to \infty$ must be taken. On calculating (10.2.8) by means of (10.2.3), we may consider q as a continous variable, i.e. we may put $\sum_q \to (L/2\pi) \int dq$ and

$(L/2\pi)\,\delta_{q,0} \to \delta(q)$, where L is the length of the system. Thus, we find

$$D = \lim_{t\to\infty} \lim_{N\to\infty} \frac{1}{t} \frac{1}{N^2} \sum_q \sum_{mm'} a^2(m - m')^2 \, e^{iqa(m-m')} \, e^{-\lambda(q)t}$$

$$= - \lim_{t\to\infty} \frac{1}{2t} \int dq \, e^{-\lambda(q)t} \frac{d^2}{dq^2} \delta(q) = \frac{1}{2} \frac{d^2\lambda(q)}{dq^2}\bigg|_{q\to 0}, \qquad (10.2.9)$$

where the last line was obtained by integrating two times by parts. According to (10.2.9), we have $D = Wa^2 = (1/2\tau_0)\, a^2$, where $\tau_0^{-1} = 2W$ is the mean residence time, and in the limit $q \to 0$,

$$\lambda(q) = Dq^2. \qquad (10.2.10)$$

This provides a method to measure D by neutron scattering.

To study the conductivity σ in the system at hand, we add an electrical field $\boldsymbol{E}(t)$, which changes the over-barrier hopping probability as

$$W_{m'm} = \frac{\omega_0}{2\pi} \, e^{-\left(U_0 + \frac{1}{2} Ze\boldsymbol{E}\boldsymbol{R}_{m'm}\right)/kT} \approx W \left[1 - \frac{1}{2} Ze\boldsymbol{E}\boldsymbol{R}_{m'm}/kT\right], \qquad (10.2.11)$$

where Ze is the charge of the particle, $\boldsymbol{R}_{m'm} = \boldsymbol{R}_{m'} - \boldsymbol{R}_m$, $\boldsymbol{R}_m$ is the position of the mth site, and $W = (\omega_0/2\pi) \exp(-U_0/kT)$ is the hopping rate in the absence of a field.

The current $\boldsymbol{j}$ is calculated by inserting (10.2.1) into (2.3.3), where e must be replaced by Ze, and by linearizing ϱ_m and $W_{m'm}$ as $\varrho_m \approx \varrho_m^{(0)} + \varrho_m^{(1)}$ and $W_{m'm} \approx W_{m'm}^{(0)} + W_{m'm}^{(1)}$, where the upperscripts 0 and 1, respectively, denote the zeroth- and first-order terms with respect to the field.

For an ordered system, owing to the translational symmetry of the lattice, $\varrho_m^{(0)}$ is independent of m, $\varrho_m^{(0)} = \varrho_0 = 1/N$, and $\varrho_m^{(1)} = 0$ (analogous to the case of small polarons in a crystal, cf. Section 2.5.). Thus, we obtain

$$\boldsymbol{j} = \frac{eZ\varrho_0}{2\Omega} \sum_{mm'} \boldsymbol{R}_{mm'} [W_{m'm}^{(1)} - W_{mm'}^{(1)}]. \qquad (10.2.12)$$

By (10.2.11), (10.2.12), and $\sigma = j_x/E$ ($\boldsymbol{E}$ is directed along the x-axis), for the periodic one-dimensional system in question, we find that

$$\sigma = n(Ze)^2 \, D/kT, \qquad (10.2.13)$$

where $D = Wa^2$, and n is the concentration of charge carriers. Note that σ is independent of frequency.

The above considerations can readily be generalized to higher-dimensional non-primitive lattices. The corresponding results have been used to explain experimental findings obtained on certain superionic materials and metal hydrides (see, e.g., Dieterich et al., 1980, and references therein).

10.2.2. Classical hopping in disordered systems

The rate equation (10.2.1) has also been used for studying diffusive atomic motion in disordered systems, in particular, in one-dimensional disordered systems (for a review, see Alexander et al., 1981).

The problem of hopping of classical particles in disordered systems is mathematically analogous to the problem of electron hopping in disordered systems, at least in the single-particle case. To see this, note that the equation for the conditional-probability function

$$\frac{\mathrm{d}P_{m'm}(t)}{\mathrm{d}t} = \sum_{m''} [P_{m'm''}(t)\, W^{(0)}_{m''m} - P_{m'm}(t)\, W^{(0)}_{mm''}] \tag{10.2.14}$$

($W^{(0)}_{m'm}$ is the zero-field hopping probability) agrees with (2.6.1) (cf. (2.6.5), (6.4.2)), the equation for the conditional-probability function in the electronic case.

Furthermore, generalizing (10.2.11) to disordered systems, we get

$$W^{(1)}_{m'm} = -\frac{\beta}{2} (Ze)\, \boldsymbol{R}_{m'm}\boldsymbol{E}\, W^{(0)}_{m'm}. \tag{10.2.15}$$

Thus, by (10.2.1), (10.2.15), and $\boldsymbol{E}(t) = \boldsymbol{E}\mathrm{e}^{\mathrm{i}\omega t}$ we obtain for $\varrho^{(0)}_m$ and $\varrho^{(1)}_m$ (cf. Section 10.2.1.), in analogy to equations (2.5.1) and (2.5.3) of the electron problem, the following equations for hopping of classical particles:

$$W^{(0)}_{m'm}\varrho^{(0)}_{m'} = W^{(0)}_{mm'}\varrho^{(0)}_{m}, \tag{10.2.16}$$

$$\mathrm{i}\omega f_m(U_m + \boldsymbol{E}\boldsymbol{R}_m) = \sum_{m'} (U_{m'} - U_m)\, \Gamma_{m'm}, \tag{10.2.17}$$

where

$$U_m = -e\boldsymbol{E}\boldsymbol{R}_m + \delta\mu_m/(Ze) \tag{10.2.18}$$

and

$$\Gamma_{m'm} = f_{m'} W^{(0)}_{m'm}. \tag{10.2.19}$$

Here, $f_m \equiv \varrho^{(0)}_m = \exp\left[-(\varepsilon_m - \mu)/kT\right]$, where ε_m is the energy of the mth site (the depth of the mth potential well), and $\delta\mu_m$ is defined as (cf. (2.5.3))

$$\varrho^{(1)}_m = \beta\delta\mu_m f_m\, \mathrm{e}^{\mathrm{i}\omega t}. \tag{10.2.20}$$

Taking the Laplace transform of (10.2.14) and comparing the resulting equation with (10.2.17), one finds (cf. (2.6.13)) that

$$\delta\mu_m = -(eZ)\, s \sum_{m'} \boldsymbol{E}(\boldsymbol{R}_{m'} - \boldsymbol{R}_m)\, P_{m'm}f_{m'}/f_m, \tag{10.2.21}$$

where s is the Laplace parameter. Thus, by means of (2.3.3), (10.2.20), and (10.2.21), the conductivity can be expressed by the conditional-probability function as (cf. (2.6.14))

$$\sigma_{\gamma\gamma'}(\omega) = s^2 \frac{\beta(Ze)^2}{2\Omega} \left\langle \sum_{mm'} (\boldsymbol{R}_{m'} - \boldsymbol{R}_m)_{\gamma'} (\boldsymbol{R}_{m'} - \boldsymbol{R}_m)_{\gamma} f_{m'}P_{m'm} \right\rangle \Big|_{s=\mathrm{i}\omega}, \tag{10.2.22}$$

where $\langle \ldots \rangle$ indicates the cofiguration average. Equation (20.2.22) is the starting point for most theoretical studies of hopping transport of classical particles.

Owing to the close relation between hopping of classical particles and electrons in disordered systems, formulae and results obtained above for the electron problem (cf. Section 2.6. and 6.4.) analogously apply to hopping of classical particles. In particular, also for classical particles, $\sigma(\omega)$ is at low frequencies a non-decreasing function of ω.

The simplest model for hopping of classical particles in a disordered system is a single particle in a one-dimensional model with equally spaced sites and nearest-neighbour hopping rates $W^{(0)}_{mm'} = W^{(0)}_{m'm}$ which are assumed to be mutually independent random variables distributed according to a probability density $P(W^{(0)})$. This model has extensively been studied for different types of densities $P(W^{(0)})$ by different methods, such as scaling assumptions (ALEXANDER et al., 1981), a real-space renormalization group method (MACHTA, 1982), the replica procedure (STEPHEN and KARIOTIS, 1982) and effective-medium approximations (see, e.g., ALEXANDER et al., 1981, HALEY, 1982; see also Sections 6.4. and 11.5.).

Note that the formulae and results obtained in Section 6.8. for dispersive transport analogously apply to the migration of atomic particles in disordered ionic conductors with a large spread of event times.

10.2.3. Single-particle diffusion versus bulk diffusion

As mentioned above, in the case of atomic diffusion we may ask either for single-particle (individual particle) motion, as obtained by microscopic probes such as NMR, or for bulk properties such as conductivity. Both these properties are in general not related in a simple way.

For atomic diffusion, a diffusion coefficient is usually defined by either one of the following two different ways: (i) we pay attention to the average motion of an individual particle (single-particle diffusion) or (ii) we consider the macroscopic change in concentration (bulk diffusion) (cf. OHTSUKI and OKANO, 1982, MURCH, 1982, and references therein).

The single-particle diffusion coefficient (self-diffusion coefficient) D_s is defined as (for diffusion in x-direction)

$$D_s = \lim_{t\to\infty} \frac{1}{2t} \frac{1}{N_p} \sum_\alpha \langle [x_\alpha(t) - x_\alpha(0)]^2 \rangle, \tag{10.2.23}$$

where N_p is the number of particles, $\langle \ldots \rangle$ indicates the thermodynamic average, $x_\alpha(t)$ is the x-component of the position vector of the αth particle at time t, and we use the convention that Greek indices label particles and Roman one label sites. The right-hand side of (10.2.23) contains the mean-square displacement of a single particle, averaged over all the N_p particles.

On the other hand, the bulk-diffusion coefficient (concentration-diffusion coefficient) D is defined as (for diffusion in x-direction)

$$D = \lim_{t\to\infty} \frac{1}{2t} \frac{1}{N_p} \left\langle \left(\sum_\alpha [x_\alpha(t) - x_\alpha(0)] \right)^2 \right\rangle \equiv \lim_{t\to\infty} \frac{1}{2t} \frac{1}{N_p} \langle X(t)^2 \rangle, \tag{10.2.24}$$

where $X(t)$ denotes the net displacement of all the particles. The right-hand side of (10.2.24) contains the mean-square of the latter quantity. Accordingly, D is determined by the square of the sum of all the individual displacements, while D_s is determined by the sum of the squares of these displacements.

In his classical paper KUBO (1957) pointed out that

$$\langle X(t)^2 \rangle = 2 \int_0^t \mathrm{d}t'(t - t') \langle V(t') \, V(0) \rangle, \tag{10.2.25}$$

where V is the x-component of the total velocity of all particles. This equation can readily be verified by differentiating both the sides of it with respect to time, and by accounting for $\mathrm{d}X/\mathrm{d}t = V$. Thus, by means of (10.2.24) and (10.2.25), the bulk diffusion coefficient can be expressed as

$$D = \frac{1}{N_p} \int_0^\infty \mathrm{d}t \langle V(t) \, V(0) \rangle. \tag{10.2.26}$$

The latter relation and the Kubo formula (KUBO, 1957) for the conductivity,

$$\sigma \equiv \sigma_{xx} = \frac{(Ze)^2}{\Omega kT} \int_0^\infty \mathrm{d}t \langle V(t) \, V(0) \rangle, \tag{10.2.27}$$

may be combined to give the Einstein relation for N_p (interacting) particles

$$\sigma = \frac{N_p (Ze)^2 \, D}{\Omega kT}, \tag{10.2.28}$$

which relates the conductivity and the bulk diffusion coefficient.

On comparing (10.2.23) and (10.2.24), it can readily be seen that only in the absence of correlations (i.e. for non-interacting particles), where $\big\langle (x_\alpha(t) - x_\alpha(0)) (x_{\alpha'}(t) - x_{\alpha'}(0)) \big\rangle = 0$ for $\alpha \neq \alpha'$, we have $D_s = D$.

To describe bulk and single-particle properties by the hopping model, we now introduce two types of stochastic variables. One type of variables called n_m denotes the occupancy of the site m by any particle. Thus, n_m has the value of 1 if site m is occupied (by any particle) and 0 if it is unoccupied. The other type of variables called p_m^α denotes the occupancy of the site m by a specific (distinguishable) particle labeled α. Thus, $p_m^\alpha = 1$ if the specific particle α is at the site m and 0 otherwise.

The net displacement of all particles, $X(t)$, involves only site occupancies n_m. This stems from the relations

$$x_\alpha(t) = \sum_m x_m p_m^\alpha(t), \tag{10.2.29}$$

$$n_m(t) = \sum_\alpha p_m^\alpha(t), \tag{10.2.30}$$

and thus

$$\sum_\alpha x_\alpha(t) = \sum_m x_m n_m(t). \tag{10.2.31}$$

Here x_m denotes the x-component of the position vector of the mth site.

Note that

$$\langle n_m(t)\rangle = \varrho_m(t), \tag{10.2.32}$$

where ϱ_m is the concentration of particles at the site m or, equivalently, the probability that the site m is occupied. Further, the ensemble average of products of two, three ... n's represents two-point, three-point and higher correlation functions. Such correlation functions can be measured by neutron scattering and they can contribute to NMR-relaxation rates due to quadrupolar processes (cf. FEDDERS, 1978). On the other hand, the ensemble average of products of p's describes correlations between specific particles. Such correlation function are measured in tracer experiments and they govern NMR-relaxation times due to motionally altered dipolar interaction (cf. FEDDERS and SANKEY, 1978, SANKEY and FEDDERS, 1980; for a review, see, e.g., KANERT, 1982).

In particular, the two-point distinguishable-particle correlation function defined as

$$D^\alpha(\boldsymbol{R}_m, \boldsymbol{R}_{m'}, t) = \theta(t)\,\langle p_m^\alpha(t)\, p_{m'}^\alpha(0)\rangle \tag{10.2.33}$$

($\Theta(t)$ — Heaviside step function, $\boldsymbol{R}_m$ — position of the mth site) is proportional to the probability that the particle α is at the site m' at time zero and at the site m at time t. It describes the motion of a single particle and can be used to calculate nuclear-spin relaxation rates due to the dipolar interaction between a hopping spin and other spins of fixed particles. Furthermore, the four-point, two-time distinguishable-particle correlation function defined as

$$D^{\alpha\beta}(\boldsymbol{R}_m, \boldsymbol{R}_{\overline{m}}; \boldsymbol{R}_{m'}, \boldsymbol{R}_{\overline{m}'}, t) = \Theta(t)\,\langle p_m^\alpha(t)\, p_{\overline{m}}^\beta(t)\, p_{m'}^\alpha(0)\, p_{\overline{m}'}^\beta(0)\rangle \tag{10.2.34}$$

(α and β refer to two different specific particles) is proportional to the probability that particles α and β are at the sites m' and $\overline{m}'$, respectively, at time zero and at the sites m and $\overline{m}$, respectively, at time t. This correlation function can be used to calculate nuclear-spin relaxation rates due to dipolar interaction between pairs of hopping spins.

A calculation of correlation functions of the n's or p's requires equations governing the time dependence of these stochastic variables. Assuming a simple hopping model, the n's and p's, respectively, obey the following rate equations (master equations) (cf. FEDDERS and SANKEY, 1977, RICHARDS, 1977)

$$\left(\frac{\mathrm{d}n_m}{\mathrm{d}t}\right)_0 = \sum_{m'} [W_{m'm} n_{m'}(1 - n_m) - W_{mm'} n_m(1 - n_{m'})], \tag{10.2.35}$$

$$\left(\frac{\mathrm{d}p_m^\alpha}{\mathrm{d}t}\right)_0 = \sum_{m'} [W_{m'm} p_{m'}^\alpha(1 - n_m) - W_{mm'} p_m^\alpha(1 - n_{m'})], \tag{10.2.36}$$

where $W_{mm'}$ is the hopping rate for a particle at the site m to jump to an unoccupied site m'. Both these equations are based on the usual assumption underlying the simple hopping model, that the hopping process from the site m to the site m' depends neither on time nor on the occupancy of sites other than m and m'. Note that the operation $(\mathrm{d}/\mathrm{d}t)_0$ in (10.2.35) and (10.2.36) is no true derivation with respect to time but rather the change in probability of the quantity operated on in a small time interval Δt, divided by Δt, which requires some care on applying this operation (the chain rule for computing time derivatives does not apply in its usual form) (SANKEY and FEDDERS, 1977).

In the special case where $W_{mm'} = W_{m'm}$, the right-hand side of (10.2.35) is linear, while that of (10.2.36) is quadratic in occupancy variables. This indicates that the motion of a specific (distinguishable) particle does not only require solution of more than just a single set of rate equations, its equations also are more complex (cf. SANKEY and FEDDERS, 1977). Note that for $W_{mm'} = W_{m'm}$ the problem of calculating the two-point correlation function $\langle n_m(t)\, n_{m'}(0)\rangle$ becomes trivial. However, for $W_{mm'} \neq W_{m'm}$, this problem cannot be solved trivially and it requires application of many-body techniques (SANKEY and FEDDERS, 1977).

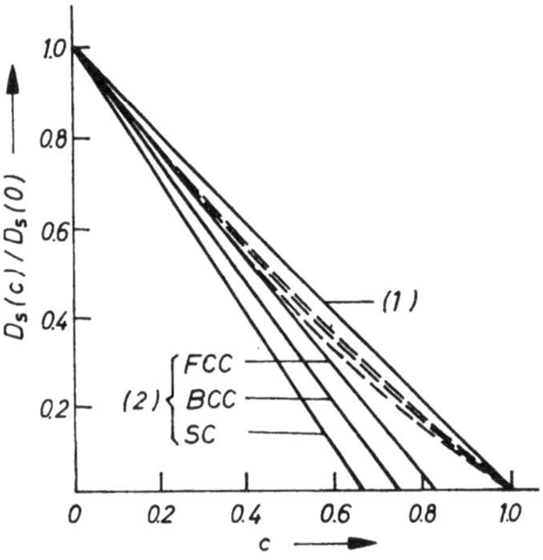

Fig. 10.4. Self-diffusion coefficient D_s as a function of the concentration calculated by the static approximation for different cubic lattices (after KASKI et al., 1982): curve (1) obtained by the VCA, curves (2) obtained by the CPA (FCC: face-centered cubic, BCC: body-centered cubic, SC: simple cubic). The broken curves represent results from a theory (SANKEY and FEDDERS, 1977) that accounts for the dynamic nature of the background

The solution of the rate equation (10.2.36) for a specific particle requires the determination of the time dependence of $n_m(t)$. As a first approximation to this difficult problem KASKI et al. (1982) assumed the n's to be fixed, independently of t. Thus, in this approximation the background particles (the particles except the specific one whose diffusion is studied) occupy fixed lattice sites at random, which allows methods of the theory of disordered systems (see, e.g., BÖTTGER, 1983) such as the VCA (virtual-crystal approximation) and the CPA (coherent-potential approximation) to be applied to the problem under consideration. In this approximation the effect of site blocking in the rate equation becomes evident: jumps to sites that are already occupied are blocked (cf. e.g., MC KEE, 1980, WOLF, 1980, and references therein).

The Fig. 10.4 shows the concentration dependence of the self-diffusion coefficient D_s, calculated for nearest-neighbour hopping by the static approximation and by a theory (SANKEY and FEDDERS, 1977) that takes the dynamic nature of the background explicitely into account (cf. also TAHIR-KHELI and ELLIOTT, 1983). We see that the static (CPA) approximation yields $D_s = 0$ at concentration $c \geqq c_p < 1$, where c_p well agrees with the percolation (threshold) concentration. Thus, in this approximation, beyond c_p there are no continuous unblocked channels in the lattice, and no diffusion is possible. Unlike that, the dynamic theory also allows diffusion near $c = 1$ owing to the hopping-vacancy mechanism.

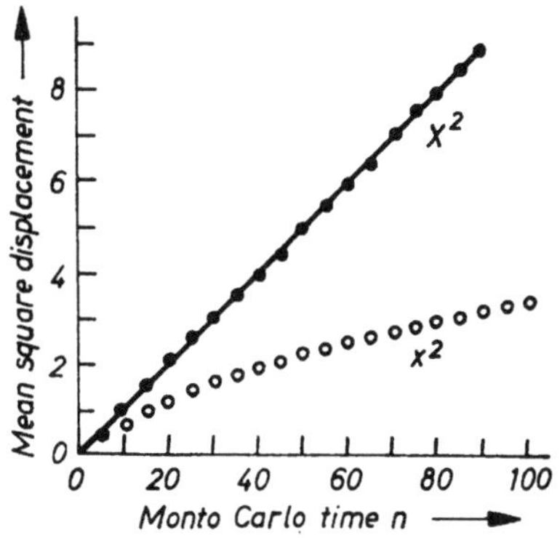

Fig. 10.5. Time dependence of the mean-square displacement of an individual particle, x^2, and the mean-square displacement of all particles, X^2, computed for 4000 particles on 8000 sites, averaged over 1600 different thermal-equilibrium initial conditions (after RICHARDS, 1977). The mean-square displacements are in units of the lattice constant squared; $n = t/\delta t$, where t is the time and δt the duration of a Monte Carlo cycle

As mentioned above, single-particle and bulk properties are in general not related in a simple way. This fact is illustrated in Fig. 10.5, where computer results (RICHARDS, 1977b, cf. ALEXANDER and PINCUS, 1978) for the mean-square displacements $X^2 \equiv N_p^{-1} \left\langle \left(\sum_\alpha [x_\alpha(t) - x_\alpha(0)] \right)^2 \right\rangle$ (cf. (10.2.24)) and $x^2 \equiv N_p^{-1} \sum_\alpha \langle [x_\alpha(t) - x_\alpha(0)]^2 \rangle$ (cf. (10.2.23)) are represented as a function of time of a one-dimensional model with equal sites and nearest-neighbour hopping. One finds that $X^2 \sim t$ but $x^2 \sim t^{1/2}$, rather than $x \sim t$, indicating that no self-diffusion coefficient can be defined for this model. This behaviour is due to the fact that in the one-dimensional hopping system the particles cannot pass each other. The latter fact also leads to anomalous nuclear relaxation times (RICHARDS, 1977, FEDDERS, 1978, LAGENDIJK, 1982).

Note that diffusive atomic motion in one-dimensional systems is of much interest, because among the superionic conductors are materials, such as β-eucrypite (Li Al SiO_4), for which fast ion motion takes place along certain one-dimensional channels.

10.3. Continuous single-particle diffusion

A hopping model is based on the assumption that the duration of a jump between different sites is short compared with the mean residence time and may be neglected. It excludes a description of oscillations around the potential minima and of the motion during the jump process. Such a model will be appropriate for describing diffusive atomic motion if the thermal energy is small compared to the height of the potential barriers which the diffusive particles have to overcome. Otherwise, a continuous diffusion model should be appropriate.

An example of a system to which a continous diffusion model should apply rather than a hopping model is the superionic material AgI. In this material it was estimated from EXAFS and molecular-dynamics studies that at 470 K the time of flight is about a third of the residence time. Furthermore, at sufficiently high frequencies, in this material the ionic conductivity decreases with increasing frequency, which cannot be explained within a hopping model.

In a simple (one-dimensional) phenomenological model for continuous single-particle diffusion in superionic materials, the action of the vibrating lattice on a mobile ion is replaced by frictional and stochastic forces, which leads to the picture of a Brownian motion in a periodic potential provided by the crystalline sublattice (Fulde et al., 1975, Vollmer and Risken, 1980, see also Dieterich et al., 1980, and references therein). The corresponding Langevin equation is given by

$$M\ddot{x} + M\gamma\dot{x} + \frac{\partial V}{\partial x} = f(t), \tag{10.3.1}$$

where M is the mass of the particle, γ the friction constant and $V(x)$ the periodic (sinusoidal) potential. The stochastic force $f(t)$ is assumed to have a white-power spectrum, i.e.

$$\langle f(t)\, f(0)\rangle = 2M\gamma kT\delta(t), \tag{10.3.2}$$

where $\langle\ldots\rangle$ indicates the thermal average.

Note that the solution of (10.3.1) is equivalent to the solution of the Fokker-Planck equation for the distribution function $W(x, v, t)$ ($v = \mathrm{d}x/\mathrm{d}t$)

$$\frac{\partial W}{\partial t} = \left\{-v\frac{\partial}{\partial x} + \frac{1}{M}\frac{\partial}{\partial v}\left[\frac{\partial V}{\partial x} + M\gamma v - f\right] + \gamma\frac{kT}{M}\frac{\partial^2}{\partial v^2}\right\} W. \tag{10.3.3}$$

The Fig. 10.6 shows the frequency dependence of the conductivity calculated by (10.3.1) for the case that the friction is small, i.e. $\gamma < \omega_0$, where ω_0 is the frequency for small oscillations in a potential well. The figure shows a transition from a resonance structure at low temperatures (damped harmonic oscillations) to a Drude from at high temperatures (free diffusion).

A detailed discussion of continous single-particle diffusion is outside the scope of this book. For more details on this subject, the reader is referred, e.g., to the review by Dieterich et al. (1980).

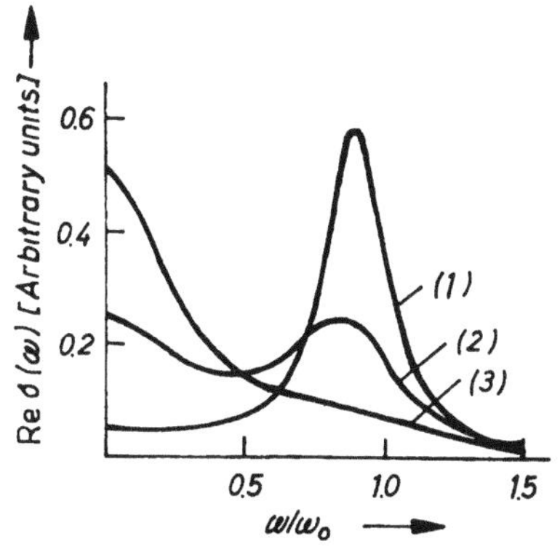

Fig. 10.6. Frequency dependence of the real part of the conductivity $\sigma(\omega)$ calculated by (10.3.1) for a cosine potential of amplitude V_0 for low friction, $\gamma < \omega_0$, at three different temperatures: $kT/V_0 = 0.25$ (1), 0.5 (2), 1.0 (3) (ω_0 — frequency for small oscillations, γ — friction constant) (after DIETERICH et al., 1977)

10.4. Many-particle diffusion

In real ionic crystals, the single-particle hopping picture may be too simplified. Coulomb-forces between the mobile ions can also give rise to hopping events involving the motion of larger numbers of ions and to correlations between subsequent hops.

Examples of correlation effects, quoted in the literature for explaining experimental findings, are the "bounce back" effect and the "caterpillar" mechanism (see, e.g., KIMBALL and ADAMS, 1978). The former effect consists in that an ion which hops from site m to site m' may be expected to be "bounced back" to site m fairly often when interaction is present. This effect tends to reduce the dc conductivity relative to the ac conductivity. Hence, an increase of the conductivity with frequency may be due to disorder as well as interaction. The caterpillar mechanism consists in that (owing to interaction) a hopping ion should tend to push other ions in the same direction in subsequent hops. It may be argued that in some superionic systems the latter mechanism is the leading mechanism giving rise to the high ionic conductivity observed. Since the effect of interaction on diffusive atomic motion is still a problem far from being completely solved, we discuss here only some aspects of it.

One approach to the many-particle problem at hand is based on the lattice-gas model, in which each possible configuration of the system is specified by a set of occupation numbers referring to the different lattice sites assumed to be available for the mobile ions. In this model hops between sites occur through interactions with the vibrating crystalline cage. It is convenient to define a spin variable for each site such that $\sigma_m = +1(-1)$ means that site m is occupied (empty). A configuration of the system is then specified by the set $\sigma = \{\sigma_m\}$.

If $p(\sigma, t)$ denotes the probability of finding a particular configuration σ at time t, we can describe the dynamics of the system by the master equation

$$\frac{\partial p(\sigma, t)}{\partial t} = \sum_{\sigma'} [W(\sigma' \to \sigma)\, p(\sigma', t) - W(\sigma \to \sigma')\, p(\sigma, t)], \tag{10.4.1}$$

which generalizes the simple rate equations discussed in Section 10.2. Here the first term corresponds to the rate at which a configuration σ' changes into the configuration σ and the second term corresponds to the rate at which the configuration σ changes into another configuration σ'. The transition rate $W(\sigma \to \sigma')$ is the probability per unit time that the spin configuration changes from σ to σ'.

The transition rates are determined by the interaction with the crystalline cage and by the interaction between the mobile ions. They must be chosen in such a way that the total spin is conserved, since the total number of ions is fixed. Furthermore, they must satisfy detailed balance so that any initial configuration finally goes over into the equilibrium configuration with the probability

$$p(\sigma) \sim \exp\left(-\mathcal{H}(\sigma)/kT\right), \tag{10.4.2}$$

where $\mathcal{H}(\sigma)$ represents the lattice-gas Hamiltonian (spin-1/2 Ising Hamiltonian).

Note that the above spin formalism may be also generalized to the case that the motion of an individual particle is of interest. To this end, one has to introduce an additional label for the marked particle (tracer ion), which can be done by letting σ_m assume three values. For more details on the spin approach and its application (for instance, to a study conductivity and neutron scattering) the reader is referred to Dieterich et al., 1980, and references therein).

As mentioned in Section 10.3., in certain ionic conductors, such as AgI-type materials, diffusive ionic motion is described by continuous diffusion rather than by hopping. In the former case, the system of interacting mobile ions can be regarded as a liquid embedded in a periodic medium provided by the crystalline cage. The equations of motion then become coupled Langevin equations, instead of (10.3.1).

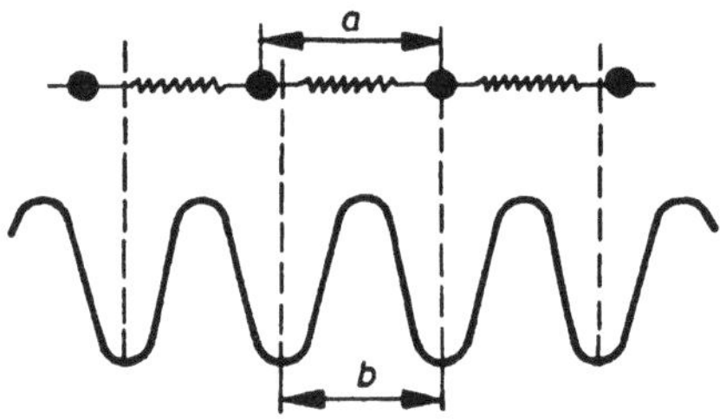

Fig. 10.7. Illustration of the Frenkel-Kontorova model (cf. (10.4.3))

In the strong-coupling limit, where the interaction between the mobile ions is strong compared to the potential of the periodic medium, the problem can be much simplified by expanding the interaction potential up to quadratic terms about the equidistant positions of the ions. In the one-dimensional case, this maps the system on the Frenkel-Kontorova model, which consists of a chain of ions connected by springs situated in a sinusoidal potential (see Fig. 10.7) and which has the potential energy

$$V = \frac{\gamma}{2} \sum_m (x_{m+1} - x_m - a)^2 + V_0 \sum_m \cos \frac{2\pi}{b} x_m. \tag{10.4.3}$$

Here γ is the second derivate of the ion-ion interaction potential and x_m is the position of the mth ion. This model describes two interpenetrating sublattices with chain structure, where one sublattice (with period a) experiences an external periodic potential due to the other sublattice (with period b). This model was originally introduced as dislocation model and has also been used in investigations of charge-density wave systems and disordered solids (cf., e.g., Böttger, 1983, and references therein).

In regard of diffusive atomic motion, the Frenkel-Kontorova model has relevance if a/b is close to unity, i.e. if the concentration of defects (vacancies for $a/b > 1$, interstitials for $a/b < 1$) is small. Because of ion-ion interaction, such a defect is always associated with a distortion of the position of the neighbouring ions (Fig. 10.8), which may appreciably reduce the potential barrier for the motion of the defects.

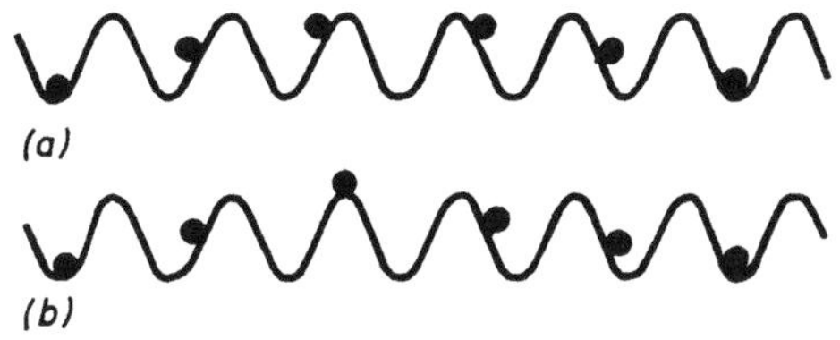

Fig. 10.8. Illustration of the equilibrium positions of the particles in the Frenkel-Kontorova model (10.4.3) for $a/b > 1$. The vacancy represented in (a) moves in (b) to the right.

In the continuous limit the Frenkel-Kontorova model provides soliton (kink) solutions, which suggests the possibility that a defect plus its associated distortion can propagate as a free particle. In real systems, however, one may expect that the soliton becomes pinned by the discretness of the lattice and by anharmonic terms of the ion-ion interaction, which may lead to continous diffusion of the collective coordinate defined by the defect and its associated distortion of the neighbouring atoms, rather than to a soliton-like propagation. Thus, the problem of the dynamics of the collective excitation reduces to that of the dynamics of an effective single particle (collective coordinate) which may be described by a Langevin equation (Pietronero and Strässler, 1978, see also Dieterich et al., 1980).

The results obtained by the Frenkel-Kontorova model may be relevant for charge transport in quasi one-dimensional superionic materials such as hollandite. Further aspects of correlated hopping and liquid-like diffusion in low-dimensional systems were discussed, e.g., by Jacobson et al. (1983).

Chapter V. Recent Developments

The subject of this book is still a very active field of research. After having accomplished the manuscript in summer 1983 some remarkable progress has been achieved in this field. To bring the book up to date, in this chapter, added in proof, we concisely discuss some recent developments.

11. Survey of some recent developments in the field of small polarons, disordered systems and hopping conduction

11.1. Polaron effects

In the past few years the research in the field of small polaron has mainly been concentrated on the phenomenon of polaron pairing (formation of bipolarons) and the polaron effect in organic quasi one-dimensional systems.

As pointed out in Section 2.2., electron-phonon interaction can cause an effective polaron-polaron attraction (cf. formula (2.2.11)). Physically, this attraction is due to the fact that, on replacing on a given site a single electron by two electrons, the coupling constant of electron-phonon interaction γ is doubled and the polaron shift E_p (2.2.7), being proportional to γ^2, is raised by a factor of four, whereas the energy gain due to formation of two isolated polarons amounts to $2E_p$. If this interpolaron attraction is stronger than the Coulomb repulsion, polaron pairing (bipolaron formation) becomes energetically favourable (cf. Section 4.1.).

Chakraverty et al. (1978) have studied bipolaron formation by using a two-site model. They have found four possible pairing states: the Heitler-London singlet and triplet states for pairing of polarons on near-neighbour sites and two singlet states for pairing on a single site (Anderson bipolaron). On account of intercell tunnelling at zero temperature (where only zero-point vibrations are operative), Alexandrov and Ranninger (1981) have shown that a bipolaron band with bandwidth $(\Delta\varepsilon_p)^2/U(\Delta\varepsilon_p \ll U)$ can occur, where $\Delta\varepsilon_p$ is the small-polaron bandwidth (cf. Section 3.1.) and U is the bipolaron-dissociation energy. These authors have also pointed out that a bipolaron system can be described by an anisotropic spin-1/2 Heisenberg Hamiltonian. Furthermore, for sufficiently high bipolaron concentrations, the ground state of the system is a charge-density wave of bipolarons. In the low bipolaron-density limit, the excitation spectrum of a bipolaron system is superfluid-like so that localized bipolarons might be superconducting (for a discussion of superconductivity in the case of strong electron-phonon coupling, see also Mc MILLAN, 1968).

Transport properties of Anderson bipolarons (on-site pairing) has been studied by BRYKSIN et al. (1983) and BRYKSIN and VOLOSHIN (1984). These authors have pointed out that in the optical-absorption spectrum, in addition to the single-polaron peak at frequency $4E_a$ (cf. Section 3.3.), there occur two high-frequency peaks at $4E_a + U$ (arising from the dissociation of the bipolaron state) and $16E_a$ (transition between the adiabatic bipolaron levels on near-neighbour sites), where U is the bipolaron-dissociation energy. The height of the low-frequency single-polaron peak is exponentially small ($\sim \exp(-U/2kT)$) as compared to those of the other peaks, i.e. (by a factor $\sim \exp(-U/2kT)$) the number of isolated polarons is much less than that of bipolarons.

However, the activation energy for near-neighbour hops of Anderson bipolarons is given by $(E_a + U/4)^2/E_a$, i.e. it much exceeds the activation energy of a single-polaron hop, E_a. Consequently, the major contribution to conductivity comes from single-polaron hops, in spite of the small number of these carriers. The conductivity is described by a formula of the type (3.2.10), where n denotes the concentration of single-polaron carriers.

In the past few years, polarons in low-dimensional systems (especially in quasi one-dimensional materials such as polyacethylene) have attracted a great deal of interest. Polarons in these materials are usually large-size polarons so that, strictly speaking, polaron effects in such materials are somewhat outside the scope of this book. However, in view of the great interest they have recently attracted, in what follows we briefly outline this subject.

Recently, theoretical studies of polarons in quasi one-dimensional systems have been performed above all on the basis of two models (cf. CAMPBELL et al., 1982): the continuum version of Holstein's diatomic molecular-crystal model (HOLSTEIN, 1981) and the coupled electron-phonon model by SU et al. (1980). The polarons in the two models are qualitatively similar but differ quantitatively. In particular, in the former model, the polarons are somewhat more extended and less bounded than their counterparts in the latter model.

On the basis of the molecular-crystal model, HOLSTEIN (1981) has shown that along a given chain, charge transport is due to band motion of the large-size (strongly anisotropic) polarons, while between adjacent chains, possibly by hopping. Thereby, the transport along chains is governed by non-trivial scattering processes (cf. SCHÜTTLER and HOLSTEIN, 1983, WILSON, 1983). The formation of bipolarons due to pairing of large-size polarons has been studied by COHEN et al. (1984), who obtained phase diagrams of the polaron-bipolaron system for different effective masses and constants of electron-phonon coupling. The study of these authors is based on the method of evaluating the radius of self-trapped states due to interaction of charge carriers with an elastic medium proposed by EMIN and HOLSTEIN (1976).

Let us now turn to the electron-phonon model by Su et al. (1980). For this model, in contrast to Holstein's model, the ground state contains a dimerized lattice (spontaneous symmetry breaking due to the Peierls effect) and a filled valence band separated by a gap of width 2Δ from an empty conduction band. The degenerate nature of the dimerized ground-state conformation leads to elementary excitation taking the form of kink-like phase boundary solitons and anti-solitons. These excitations provide localized electronic levels at midgap, i.e. at the Fermi level. In addition to these exitations, the model under consideration admits intrinsic polaron states, which might loosely be described as soliton-anti-soliton pairs. These polarons provide electronic states at $\Delta/\sqrt{2}$ apart from the middle of the gap (cf. Brazovskii and Kirova, 1981, Campbell et al., 1982, Brédas et al., 1982 a,b). Moreover, this model also admits bipolaron states, which might be associated with pairing of two soliton-antisoliton pairs (Onodera and Okuna, 1983). These bipolarons provide additional two pairs of electronic states in the gap (cf. Chance et al., 1984, Kaufman et al., 1984). The distribution of electronic states in the Peierls gap may be experimentally studied by means of optical absorption (cf. Fesser et al., 1983).

Note that the soliton-description only works when there are degenerate ground states as in trans-polyacetylene. However, there are many other organic quasi one-dimensional systems which do not have a degenerate ground-state structure, but show very much the same experimental results as polyacetylene. To describe excitations, charge transport, etc. in such systems different models and theoretical concepts have been developed (cf. Cade and Movaghar, 1983, Howard and Conwell, 1984, Chance et al., 1984, and references therein). In particular, hopping conduction in polyacetylene has recently been discussed by Kivelson and Epstein (1984).

Concluding this section, let us draw attention to some further polaron problems. Feinberg and Ranninger (1984) have mapped Holstein's molecular-crystal model for small polarons onto a model of a spin 1/2 coupled to an external plus reaction field. In this picture the spin represents the electron and the field the local lattice deformation. These authors found that this picture allows a deeper insight into the dynamics of the small polaron to be obtained. De Raedt and Lagendijk (1984) have performed a numerical study of electron self-trapping on the basis of Holstein's model. In particular, they studied the influence of the phonon dispersion on the transition between the untrapped and the self-trapped state of an electron. Riseborough (1984) has generalized Holstein's model by inclusion of nonlinear electron-phonon interaction. He has found that an electron subjected to purely quadratic electron-phonon coupling (model for aromatic hydrocarbon crystals) shows high-temperature behaviour that markedly differs from that of the usual small polaron.

Polaron formation in disordered systems near a mobility edge, has recently been studied by Cohen et al., (1984), who have shown that incipient localization facilitates small-polaron formation. Furthermore, there is convincing evidence of the existence of polarons and (Anderson-) bipolarons (associated with intrinsic bond defects) in non-crystalline dielectrics (cf. Chakraverty, 1982, Simikawa and Kondo, 1983). To explain conductivity data of such materials, correlated over-barrier hopping of polarons and/or bipolarons is often invoked. In this mechanism, the height of the barrier is correlated with the separation between the initial and final defect sites via Coulomb interaction (cf. PRAMANIK et al., 1983, and references therein).

Finally, the phase diagram of the ground state of a system with inter-polaron interaction and bipolaron formation has recently been studied by HIRAMOTO and TOYOZAWA (1985), and small-polaron hopping on account of correlation effects due to intersite Coulomb interaction by IHLE (1985).

11.2. Electron states and charge transport in disordered systems

11.2.1. Anderson localization and related problems

The study of Anderson localization in disordered systems received a great impetus from the scaling theory of localization by ABRAHAMS et al. (1979) (cf. Section 4.3.). Especially the problem of weak localization in two-dimensional systems has attracted great attention in the past years, owing to the fact that on such systems one can compare theoretical predictions and experiments to great accuracy. A detailed knowledge of weak localization will be a key to a better understanding of the localization phenomenon in general. Experiments on metallic films (cf. the review by BERGMANN, 1984, and references therein) and silicon-inversion layers (cf., e.g., BISHOP et al., 1984, and references therein) have shown convincing quantitative support for the scaling theory of localization by ABRAHAMS et al. (1979).

Recall that, according to this theory, the conductance g of a system of size l is described by a universal function

$$\beta(g) = \mathrm{d} \ln g / \mathrm{d} \ln l \tag{11.2.1}$$

For large g, the function $\beta(g)$ can be expanded in powers of $1/g$, whereby in the weak-scattering perturbation theory of LANGER and NEAL (1966) the leading correction comes from so-called maximally crossed diagrams.
Therefore,

$$\beta(g) = d - 2 - a/g, \tag{11.2.2}$$

where a is a numerical factor.

For two dimensions ($d = 2$), from (11.2.1) and (11.2.2) we get

$$g(l) = g_0 - a \ln (l/l_0) \tag{11.2.3}$$

$\left(g_0 = g(l_0)\right)$. According to (11.1.3), for weak scattering (large g), the conductance of a square should decrease logarithmically with ist size, i.e. in disordered two-dimensional systems all states should be localized no matter how weak the disorder. Hence, according to the scaling theory of localization, the conductance of a two-dimensional system exhibits with decreasing disorder a smooth and continuous transition from the exponential localization regime (small g) to the weak localization regime (large g).

For the conductivity in three dimensions, $\sigma(l) = g(l)/l$, from (11.2.1) and 11.2.2) we get

$$\sigma(l) = \sigma + g_c/l, \tag{11.2.4}$$

where g_c is the critical conductance and $\sigma = \sigma(\infty)$ is the macroscopic conductivity. According to (11.2.4), for strong enough disorder, scale-dependent diffusion should become significant in three-dimensions (cf. MUTTALIB, 1983, ANDERSON et al., 1983).

On comparing predictions of the scaling theory of localization by ABRAHAMS et al. (1979) with experiment, it must be realized that this theory applies only for length scales less than those set by inealastic scattering events (such as electron-electron and electron-phonon scattering) or magnetic fields (see, e.g., THOULESS, 1984, and references therein). In a metal, at sufficiently low temperatures the inelastic lifetime τ_i (lifetime in an eigenstate of energy) much exceeds the elastic lifetime τ_0 (lifetime in an eigenstate of momentum), so that a conduction electron in a given eigenstate of momentum can be scattered by defects without loosing its phase coherence. Weak localization is a coherence effect. It is produced by quantum interference at defects (cf. BERGMANN, 1984, KHMEL'NITSHII, 1984, and references therein). Inealstic scattering gives a limiting length scale l_i beyond which coherence effects are destroyed. Similarly, beyond the radius of the lowest quantized cyclotron orbit, the magnetic field dephases the wavefunction and destroys localization (GORKOV et al., 1979, HIKAMI et al., 1980).

In the presence of inelastic scattering, the length l_i is the value of l that should be used in scaling theory. In the case of weak localization, for short times, the electron diffuses due to repeated elastic scattering, so that l_i is given by $(D\tau_i)^{1/2}$, where D his the diffusion constant on short lengt scales. The corresponding contribution to the conductance is temperature dependent because the inelastic liefetime depends on temperature (e.g. $\tau_i^{-1} \sim T^p$) (cf., e.g., THOULESS, 1982, and references therein). In the case of strong (exponential) localization, the localization length is less than the inelastic length scale l_i and the conductance is governed by transitions between localized states induced by inelastic scattering events.

In disordered two-dimensional systems, weak localization is suppressed by a magnetic field applied perpendicular to the film. This gives a negative magnetoresistance in a field range where usual orbital effects are negligible, i.e. $\omega_c\tau_0 \ll 1$ (ω_c-cyclotron frequency). Magnetoresistance measurements on two-dimensional systems have yielded determinations of inelastic lifetimes τ_i, and allowed critical comparisons with theory. Furthermore, it has been shown that the localization and correlation effects can be differentiated by magnetoresistance measurements (see, e.g., LIN et al., 1984, MARKIEWICZ and ROLLINS, 1984, POON et al., 1985, BERGMANN, 1984, and references therein). In this manner, many observations of localization have been made.

Recently, the problem of localization in strong magnetic fields (in the quantum limit $\omega_c\tau_0 \gg 1$) has attracted much attention in connection with the observation that in strong magnetic fields in two-dimensional systems (realized with different types of semiconductor devices) the Hall conductance $\sigma_{xy} = I_x/E_y$ shows, with an extremely high precision, a quantized plateau behaviour:

$$\sigma_{xy} = \nu e^2/h \tag{11.2.5}$$

where in the case of the integral quantum Hall effect (VON KLITZING et al., 1980) ν is an integer, while in the case of the fractional quantum Hall effect (TSUI et al., 1982, STORMER et al., 1983) $\nu = n/m$, where n and m are integers with m being odd. A plateau in the Hall conductance is always accompanied by a dip in the longitudinal conductance. This observation suggests that a necessary condition for the quantum Hall effect is the existence of a Fermi gap (energy gap or mobility gap in which the Fermi energy lies).

In the strong magnetic field limit, mixings between different Landau levels can be neglected. To the present knowledge, the existence of the integral quantum Hall effect implies the existence of extended states near the centers of the Landau level and the existence of localizad states (arising from potential fluctuations) in the band tails of the Landau levels (Aoki and Ando, 1981, 1985, Halperin, 1982; cf. Ando, 1984, Levine et al., 1983, Hikami, 1984, Cai and Ting, 1985): The Hall conductance σ_{xy}, as a function of the magnetic field or the carrier density, exhibits plateaus with quantized values, because the occupation of localized states does not change σ_{xy}, while the progressive quantum steps of the Hall conductance necessiate mobile carriers (see also the reviews by von Klitzing and Ebert, 1983, and von Klitzing, 1984). The fractional quantum Hall effect is thought of as one of the most remarkable many-body phenomena in recent years (cf. Laughlin, 1983, Girvin and Jach, 1984, Girvin et al., 1985, Niu et al., 1985, and references therein).

The results obtained from scaling theory of localization have been supported by more systematic approaches that are based on mapping the Anderson-localization problem onto field-theory models, which allows results from field theory and the theory of phase transitions to be applied (Wegner, 1979, Schäfer and Wegner, 1980, Hikami, 1981, Mc Kane and Stone, 1981, Efetov, 1983, Theumann, 1983, Hikami, 1984, cf. Verbaarschot and Zirnbauer, 1985, and references therein).

As mentioned above, predictions of scaling theory of localization have been experimentally well confirmed in two-dimensional systems. Observations of localization effects have also been made in one-dimensional systems, e.g. in narrow silicon-inversion layers (Skocpol et al., 1983) and narrow aluminium wires (Santhanam et al., 1984), and in three-dimensional systems, e.g. in Si:P (Rosenbaum et al., 1983) and InSb (Dynes et al., 1983). There is strong experimental evidence for both localization and interaction effects in disordered systems. For instance, millikelvin measurements of the conductivity on Si:P as a function of donor concentration (Rosenbaum et al., 1983) establish that the transition from metal to insulator is continuous, with no sign of the "minimum-metallic conductivity" (cf. Section 4.1.) has been found, but sharper than predicted by scaling theory of localization, suggesting the important role of electron-electron interaction effects. Note that Mott (1984) now believes that a minimum-metallic conductivity exists only when a transition is induced by a strong magnetic field (for recent discussions of Mott's prediction of a minimum-metallic conductivity see, e.g., Mott and Kaveh, 1983, Kaveh and Mott, 1983, Ghibaudo, 1984, Biskupski et al., 1984, Dubois et al., 1985).

The theory of weak localization in non-interaction two-dimensional electronic systems is now well established. The theory of the metal-insulator transition in three-dimensional systems, however, is yet a very active field of research. It has recently become clear that electron-electron interactions play a much more important role in disordered conductors than in ordered ones. Interactions are especially important near the Anderson transition, which is not surprising because the interactions can by themselves induce a Mott-Hubbard type metal-insular transition. In the last few years, important progress has been made in elaborating scaling theories of interacting disordered electronic systems (Mc MILLAN, 1981, GEFEN and IMRY, 1983, GREST and LEE, 1983, MA and FRADKIN, 1983, ARONOV, 1984, FINKELSTEIN, 1983, 1984, CASTELLANI et al. 1984) but the final word is not yet in.

Concluding this section, let us draw attention to studies of the relation between percolation theory and Anderson localization (see, e.g., SHAPIR et al., 1982, GEFEN et al. 1983, ODAGAKI and CHANG, 1984, DEUTSCHER et al., 1985), the fractal character of eigenstates in disordered systems (AOKI, 1983, SOUKOULIS and ECONOMOU, 1984, ECONOMOU, 1984), the Anderson localization in positionally disordered systems (PURI and ODAGAKI, 1984, LOGAN and WOLYNES, 1985) and to mathematically rigorous investigations of Anderson localization (FRÖHLICH and SPENCER, 1984, MARTINELLI and HOLDEN, 1984).

11.2.2. *Dynamical electron-phonon effects on transport near a mobility edge*

In the last few years effects of dynamical electron-phonon interaction on transport near a mobility edge E_c have attracted growing interest. This problem has been studied in order to explain the negative temperature coefficients of resistivity observed on dirty high resistivity metals (JONSON and GIRVIN, 1979, GIRVIN and JOHNSON, 1980, IMRY, 1980, 1981, BELITZ and SCHIRMACHER, 1983) and to get a deeper insight into the transport mechanisms in amorphous semiconductors (COHEN, 1982, MÜLLER and THOMAS, 1983, 1984).

MOOIJ (1973) has found that metals with resistivity less then $\varrho^* \approx 150$ μΩcm tend to have the usual positive temperature coefficient of resistivity (TCR), defined as $\alpha \equiv \mathrm{d} \ln \varrho(T)/\mathrm{d}T$, while most dirty metals with resistivity greater than ϱ^* are anomalous in that over a large temperature range they exhibit negative TCR's. Numerous models have been proposed to explain this observation (for a recent review see, e.g., NAUGLE, 1984; cf. BELITZ and SCHIRMACHER, 1983, CARINI et al., 1983). These include the Ziman-Faber theory, a Kondo-type model based on electron scattering from two-level states, Mott s—d scattering, and theories based on Anderson localization or incipient localization.

The latter mechanism was first studied by JONSON and GIRVIN (1979) and GIRVIN and JONSON (1980), who pointed out that for large disorder (where the time spent around a given ion by a diffusing electron is of the order of the characteristic time of ionic motion, the adiabatic (or static phonon) approximation breaks down and, in the region of extended states just before the onset of localization, the dynamic electron-phonon interaction causes an increase of the mobility due to random phonon-assisted tunnelling.

IMRY (1980, 1981) applied the scaling theory of localization by ABRAHAMS et al. (1979) to the problem of inelastic processes by introducing a coherence length ξ which diverges as the critical (dimensionless) conductance g_c is approached from either the localized or extended state side. The physical meaning of ξ is that at $T = 0$, $g(l)$ will for $l \geqq \xi$ essentially behave as for a macroscopic sample. Consequently, in the extended state regime, the electrical conductivity (in units $e^2/\hbar$) is given by (cf. (11.2.4))

$$\sigma \approx Ag_c/\xi, \tag{11.2.6}$$

where A is a constant of order unity.

The condition

$$\xi = l_i, \tag{11.2.7}$$

where l_i is the elastic electron mean-free path, defines a temperature T_0, above, which the $T = 0$-theory breaks down because l_i becomes the relevant length scale rather than ξ. For $l_0 < l_i < \xi$, where l_0 is the microscopic length scale below which it makes no sense to define conductivity, (11.2.6) must be replaced by

$$\sigma \approx Bg_c/l_i, \tag{11.2.8}$$

where B is a constant of order unity.

Before onset of localization, ξ can become very large so that $\xi \gg l_0$ applies, and above some temperature T_0, according to (11.2.8), the temperature coefficient of resistivity becomes negative. KAVEH and MOTT (1982) proposed an explanation for Mooij's correlation similar to that of Imry.

The conduction in dirty metals was studied by BELITZ and SCHIRMACHER (1983) by incorporating electron-phonon interaction into the mode-coupling theory of Anderson localization by GÖTZE (1981) and BELITZ and GÖTZE (1982). These authors showed that the conductivity may be written as

$$\sigma = \sigma_1 + \sigma_2, \tag{11.2.9}$$

where σ_1 arises form usual random scattering and the additional term σ_2 from random phonon-assisted tunnelling processes. σ_2 dominates if the electronic mean-free path approaches the De Broglie wavelength. The competition between scattering and tunnelling can cause a negative temperature coefficient of resistivity. This explanation of Mooij's correlation is in the spirit of that by IMRY (1980, 1981).

To study phonon-assisted transport near a mobility edge E_c in disordered semiconductors, MÜLLER and THOMAS (1983, 1984) followed the approach by BELITZ and SCHIRMACHER (1983). However, in contrast to the latter authors, they studied a tight-binding model in site representation. They analysed the transport in term of a continuous function $\sigma(E)$, defined (for non-degenerate statistics) as

$$\sigma = \beta \int dE \sigma(E) \exp\left[-\beta(E - \mu)\right] \tag{11.2.10}$$

($\beta = 1/kT$, μ — chemical potential), i.e. $\sigma(E)$ is the contribution of states with energy E to the conductivity. It turns out that the conductivity is composed of two terms (cf. (11.2.9)): a coherent tunnelling contribution σ_1 and a hopping contribution σ_2. Well above E_c (in the region of extended states) σ is dominated by σ_1, while well below E_c (in the region of localized states) σ is dominated by σ_2. For finite electron-phonon coupling and $T \neq 0$, $\sigma(E)$ as a function of energy turns out to be a smooth function increasing gradually from low hopping-like values to high band-like ones. Upon approaching E_c from below, the conductivity becomes increasingly dominated by the coherent contribution σ_1. The fact that σ_1 does not vanish below E_c is called the phonon-induced delocalization. The hopping contribution σ_2 is nonzero everywhere in the band of electron states. Consequently, for finite electron-phonon coupling and $T \neq 0$, a mobility edge is not defined and near E_c the distinction between hopping and extended-state transport becomes meaningless.

Recently, HOWSON (1984) has ascribed the negative temperature coefficient of resistivity observed on metallic glass alloys (cf. HOWSON and GREIG, 1984) to a combination of effects of incipient localization and electron correlation. For a discussion of further aspects of conduction near a mobility edge, see also KAVEH (1984).

11.3. Problems of percolation theory

11.3.1. Percolation models

Percolation models have become increasingly popular in recent years due to the wide variety of possible applications (ranging from the evolution of galaxies to the spread of disease in an orchard) and the theoretical interest in the relation between the percolation transition and thermodynamic phase transitions (for a recent review see DEUTSCHER et al., 1983). Among the many recent developments in this field there are studies of correlated percolation (see, e.g., TUTHILL and KLEIN, 1983, PERUGGI et al., 1984, BENZONI and CARDY, 1984), continuum and directed percolation (see below), quantum percolation (cf. Section 11.2.1.), and new percolation phenomena such as the spread of epidemic processes (GRASSBERGER, 1983, CARDY, 1983, cf. MAC KAY and JAN, 1984, CARDY and GRASSBERGER, 1985) and invasion percolation (cf. WILLEMSEN, 1984) describing a cluster growth into a sample through selection of paths of least resistance. Here we somewhat dwell only on recent developments in continuum and directed percolation, since these percolation models are of special interest for hopping conduction (cf. Sections 5.3.2., 5.3.3., and 7.2.2.).

A variety of methods have been employed to compute percolation thresholds and critical exponents for continuum percolation in two and three dimensions (cf. Sections 5.2.1. and 5.2.3.). Recent studies of continuum percolation of random-sized objects have been performed by GAWLINSKI and REDNER (1983) and PHANI and DHAR (1984) with the aid of computer simultation, and by KERTESZ and VICSEK (1982) by means of the Monte Carlo renormalization-group technique. These studies gave further evidence that the critical-volume fraction is only approximately a dimensional invariant. KERSTEIN (1983) (cf. ELAM et al., 1984) has proposed to relate the percolation problem of randomly located overlapping spheres to a percolation problem on a random network by identifying this problem with the percolation problem on the edges of the Voronoi tesselation (defined by partition of the space analogous to the construction of Wigner-Seitz cells) of the sphere centres. The edges of the tesselation are either blocked or conducting, depending on whether or not they pass through a sphere. This procedure provides an efficient algorithm for Monte Carlo computation of the percolation threshold and the critical exponents. BALBERG et al. (1984a) have studied the percolation threshold of random stick systems by determining the excluded volume of the sticks (defined as the volume around a strick in which the center of another stick should not be present if interpenetration of the two sticks is to be avoided). The latter quantity is a "dimensional invariant" like the critical-volume fraction (cf. BALBERG et al., 1984b, BALBERG, 1985, BUG et al., 1985). A continuum-percolation model having percolation threshold $p_c = 0$ (containing at $p = p_c = 0$ ($p \neq 0$) arbitrary large conducting loops of vanishing (finite) width) has recently been proposed by TRUGMAN and WEINRIB (1985).

In the last few years, numerous papers have been devoted to directed percolation. In particular, application of the real-space RG (renormalization group) method (cf. Section 5.2.3.) to directed percolation has attracted great interest. It is now well understood that such application requires some care, which is due to the fact that directed percolation is characterized by two different length scales. Recall that the introduction of a preferred direction gives rise to two independent correlation lenghts (cf. Section 7.2.2.): $L_\parallel$, along this direction, and $L_\perp$, perpendicular to it. Close to the percolation threshold, p_c, both the correlation lengths diverge as

$$L_\parallel \sim (p - p_c)^{-\nu_\parallel}, \qquad L_\perp \sim (p - p_c)^{-\nu_\perp}, \tag{11.3.1}$$

with different exponents $\nu_\parallel$ and $\nu_\perp$. Therefore, it holds that

$$L_\parallel^{1/\nu_\parallel} \sim L_\perp^{1/\nu_\perp}. \tag{11.3.2}$$

On partitioning the lattice into blocks, at each renormalization step the relation (11.3.2) must be satisfied. This requires introducing two different rescaling factors $b_\parallel$ and $b_\perp$, so that at each each step we have $L_\parallel \to L'_\parallel = b_\parallel^{-1} L_\parallel$ and $L_\perp \to L'_\perp = b_\perp^{-1} L_\perp$, which by means of (11.3.2) leads to

$$b_\parallel \approx b_\perp^{\nu_\parallel/\nu_\perp}. \tag{11.3.3}$$

This implies that the shape of a block must be changed in a definite way at each renormalization step (the lattice constants $a_\parallel$ and $a_\perp$ change as $a_\parallel \to a'_\parallel = a_\parallel b_\parallel$ and $a_\perp \to a'_\perp = a_\perp b_\perp$). Since $\nu_\parallel/\nu_\perp$ is not known, it must be obtained self-consistently. However, in practice this method is not very feasible because only certain ratios of $b_\parallel/b_\perp$ can be realized with small block sizes.

The problem of changing block sizes was avoided by HERRMANN et al. (1983), by patching square blocks together to form in one direction on infinite "strip" and using finite-size scaling (cf. VANNIMENUS and NADAL, 1984, and references therein), and by ZHANG and YANG (1984), by using an isotropic block and introducing anisotropy by two effective lenghts $S_\parallel$ and $S_\perp$ for both the original and the renormalized lattices, which are taken as basic units for the measurements of $L_\parallel$ and $L_\perp$, respectively.

For directed percolation, the critical data (percolation threshold, critical exponents) have also been studied by other methods such as the Monte Carlo method (see, e.g., WILLIAMS and MACKENZIE, 1984, BENZONI, 1984), series expansions (DE BELL and ESSAM, 1983) and analytical methods.

For a review of the critical data of directed percolation in two dimensions, see BENZONI (1984), and of the critical data of percolation in general, see HARRIS (1983) and SAHIMI (1983).

11.3.2. Fractals and the structure of percolation clusters and their backbones

Recently, various aspects of the geometrical properties of percolation clusters and their backbones (cf. Section 5.2.) on length scales less than the correlation length $L \sim (p - p_c)^{-\nu}$ have received considerable attention. It is now recognized that on length scales $a \ll l \ll L$ (a is the microscopic lattice distance) percolation clusters and their backbones (biconnected part of a percolation cluster) are fractal objects, while on length scales $l \gg L$ they look homogeneous. This has been schown by different methods such as Monte Carlo simulations (KAPITULNIK et al., 1983) and conductivity measurements (PALEVSKI and DEUTSCHER, 1984).

A first operative measure of the fractal geometry is the fractal dimension $\bar{d}$, defined as follows. On mapping a fractal object on an Euclidian space of dimension d, in any volume l^d of length l of this Euclidian space there are $\sim l^d$ sites, but only

$$N(l) \sim l^{\bar{d}} \tag{11.3.5}$$

of these sites pertain to the fractal object.

The fractal dimension $\bar{d}$ of a percolation cluster at the threshold p_c can be expressed in terms of the Euclidian dimension d and the critical exponents β and ν of the percolation transition as

$$\bar{d} = d - \beta/\nu \qquad \text{for} \qquad d \leqq 6 \tag{11.3.6a}$$

and

$$\bar{d} = 4 \qquad \text{for} \qquad d \geqq 6. \tag{11.3.6b}$$

To see this, note that for $l \gg L$, homogeneity of the percolation cluster implies that $N(l) \sim P(p)\, l^d \sim L^{-\beta/\nu} l^d$ ($P(p) \sim (p - p_c)^\beta$ is the fraction of sites in the percolation cluster). Assuming that L is the only relevant length in the problem, we may write $N(l, L)$ in the following scaling form (cf. KAPITULNIK et al., 1983)

$$N(l, L) = L^{-\beta/\nu} l^d f(l/L). \tag{11.3.7}$$

For $l/L \to \infty$, $f(l/L)$ approaches a constant value while for $l/L \ll 1$, $f(l/L) \sim (l/L)^{-\beta/\nu}$, because in the latter limit we expect N to be independent of L. Consequently, for $a \ll l \ll L$, $N(l) \sim l^{d-\beta/\nu}$, in accordance with (11.3.5) and (11.3.6a).

With the aid of the hyperscaling relation (5.2.58), $d\nu = 2\beta + \gamma$, (11.3.6a) may be replaced by

$$\bar{d} = (\beta + \gamma)/\nu, \tag{11.3.8}$$

where γ describes the divergence of the mean size of the finite clusters. Above six dimensions, the critical exponents assume their mean-field values $\beta = 1$, $\gamma = 1$, $\nu = 1/2$ (cf. Section 5.2.4). Thus, for $d \geqq 6$, from (11.3.8) we get (11.3.6b). The breakdown of (11.3.6a) at $d > 6$ has been studied by AHARONY et al. (1984) and ALEXANDER et al. (1984).

Recently, it has been shown (ALEXANDER and ORBACH, 1982, RAMMAL and TOULOUSE, 1983, GEFEN et al., 1983) that linear physical problems on fractals (classical diffusion, vibrational behaviour, etc.) are governed by another dimension, the spectral (fracton) dimension $\bar{\bar{d}} = 2\bar{d}/d_w$, where d_w is the fractal dimension of a random walk on a fractal. The spectral dimension $\bar{\bar{d}}$ is associated with the low-frequency density of states of elastic fractal vibrations $N(\omega) \sim \omega^{\bar{\bar{d}}-1}$ (recall that in Euclidian space, $N(\omega) \sim \omega^{d-1}$). For percolation clusters, it holds that

$$\bar{\bar{d}} = 2(d\nu - \beta)/(t - \beta + 2\nu) \qquad \text{for} \qquad d \leqq 6, \tag{11.3.9}$$

and $\bar{\bar{d}} = 4/3$ for $d \geqq 6$, where t is the critical exponent of conductance $G \sim (p - p_c)^t$ (cf. (5.2.91)).

To see this, consider a walker on a percolation cluster which starts from an arbitrary origin and executes a random walk (DE GENNES', 1976, problem of "the ant in a labyrinth"). For $l \ll L$, where $l^2 \equiv \langle r^2(t) \rangle$ is the mean-square displacement after t time steps, in general one expects anomalous diffusion (GEFEN et al., 1983)

$$l^2 \sim t^{2/(2+\Theta)}, \tag{11.3.10}$$

instead of $l^2 \sim t$. Thus, $t \sim l^{d_w}$, where $d_w = 2 + \Theta$.

The total volume available within the distance l (or likewise the number of sites visited by the walker during time t) is given by

$$V(t) \sim l^{\bar{d}} \sim t^{\bar{d}/d_w}. \tag{11.3.11}$$

Thus, the probability of finding the walker at time t on the origin, $\langle P_0(t) \rangle$ (configuration average of the diagonal element of the conditional-probability function (cf. Section 2.6.)) depends on time t as

$$\langle P_0(t) \rangle \sim [V(t)]^{-1} \sim t^{-\bar{d}/d_w}. \tag{11.3.12}$$

Evaluating the eigenvalue density of states of the diffusion problem $N(\varepsilon)$ by means of

$$N(\varepsilon) = -\frac{1}{\pi} \operatorname{Im} \langle \tilde{P}_0(-\varepsilon + iO^+) \rangle, \tag{11.3.13}$$

where the tilde indicates the Laplace transform of $P_0(t)$, then with the aid of (11.3.12) one gets (ALEXANDER and ORBACH, 1982)

$$N(\varepsilon) \sim \varepsilon^{(\bar{d}-d_w)/d_w}. \tag{11.3.14}$$

On replacing in (11.3.14) ε by ω^2 and multiplying by ω, the frequency spectrum of the related vibrational problem (cf. ALEXANDER et al., 1981) becomes

$$N(\omega) \sim \omega^{\bar{\bar{d}}-1}, \tag{11.3.15}$$

where $\bar{\bar{d}} = 2\bar{d}/d_w$.

Using scaling arguments, (11.3.9) can be obtained as follows (GEFEN et al., 1983): For $l \gg L$, by means of the Einstein relation $\sigma = e^2nD/kT$ (σ is the conductivity, D the diffusion coefficient, and n the concentration of charge carriers), $n \sim P(p) \sim (p - p_c)^\beta$, and $\sigma \sim (p - p_c)^t$, we get $D \sim L^{-(t-\beta)/\nu}$. Analogous to (11.3.7), $D(l)$ may be written in the scaling form

$$D(l) = L^{-(t-\beta)/\nu} f(l/L), \tag{11.3.16}$$

where, for $l/L \to \infty$, $f(l/L)$ approaches a constant value, while for $l/L \ll 1$, $f(l/L) \sim (l/L)^{-(t-\beta)/\nu}$, so that $D(l) \sim l^{-(t-\beta)/\nu}$. On comparing the latter expression with $D \sim l^2/t \sim l^{2-d_w}$, with the aid of (11.3.6a), we recover (11.3.9).

On the basis of (11.3.9.) by means of the known estimates of t, ν, and β (cf. HARRIS,

1983), ALEXANDER and ORBACH (1982) conjectured that $\bar{\bar{d}} = 4/3$ for all dimensions $2 \leqq d \leqq 6$. With the aid of $\bar{\bar{d}} = 4/3$, (11.3.9) becomes

$$t/\nu = \frac{3}{2}\,(d - 4/3 - \beta/3\nu). \tag{11.3.17}$$

In the last few years, the validity and the meaning of the "universality" $\bar{\bar{d}} = 4/3$ has been the subject of numerous studies (see, e.g., HONG et al., 1984, RAMMAL et al., 1984, and references therein). Note that, for instance, for $d = 2$, $t/\nu = 0.973$ has bee found by finite-size scaling calculations (ZABOLITZKY, 1984, LOBB and FRANK, 1984), while by means of (11.3.17) one gets $t/\nu = 0.9479$.

Among the fractal problems of current interest are: the density of states and the nature of eigenstates (in particular, the existence of localized states) on fractal objects (e.g. BANAVAR and CIEPLAK, 1983, DOMANY et al., 1983, RAMMAL, 1983), including vibrational states (e.g. ALEXANDER et al., 1983, ENTIN-WOHLMAN et al., 1984, DERRIDA et al., 1984, RAMMAL, 1984a, AHARONY et al., 1985), random walks (Brownian motion) on fractals (e.g. ANGLES D'AURIAC et al., 1983, STEPHEN, 1984), random walks including suitably chosen correlation between steps (e.g. BEN-AVRAHAM and HAVLIN 1984a, ANGLES D'AURIAC and RAMMAL, 1984, DUXBURY and DE QUEIROZ, 1985), random walks on clusters generated by a previous random walks (e.g. HAVLIN et al., 1984c), random walks on fractal objects with traps (e.g. RAMMAL, 1984b, ZUMOFEN et al., 1984, WEBMAN, 1984, KLAFTER et al., 1984, DEN HOLLANDER, 1984), directed random walks (e.g. VANNIMENUS et al., 1984, CHAKRABARTI and MANNA, 1983, NADDAL and VANNIMENUS, 1985, LIU and ZHANG, 1985, OHTSUKI and KEYES, 1985), and various aspects of the geometrical properties of percolation clusters (e.g. VOSS, 1984, KANTOR, 1984, HAVLIN et al., 1984b, HERRMANN and STANLEY, 1984), such as topological and geometrical properties of random fractals (MARTIN, 1985), a general algorithm for generating new families of fractals (MANDELBROT et al., 1985), conformation of branched random fractals formed in equilibrium processes (FAMILY, 1984), substructures of percolation clusters (HAVLIN et al., 1984b, 1985, BARMA, 1985), and generalizations of the fractal dimension (GRASSBERGER, 1985).

The understanding of the geometrical structure of percolation clusters is indispensable for a study of physical properties such as conductivity, elastic behaviour, etc. As discussed in Section 5.2., the infinite percolation cluster may be divided into two parts the backbone (the biconnected part) and dead ends (dangling bonds). In analogy with the critical exponents β, γ, ν, etc. that describe the statistics of percolation clusters one also define the exponent β_B of the fraction of total sites (bonds) belonging to the backbone, the susceptibility exponent γ_B, the correlation-length exponent ν_B and so on (KIRKPATRICK, 1978). SHLIFER et al. (1979) showed that $\nu_B = \nu$. The backbone is recognized to be a fractal object (cf. PUECH and RAMMAL, 1983, SAHIMI, 1985) with a fractal dimension $\bar{\bar{d}}_B$, which is defined by a relation analogous to (11.3.5), and which is given by $\bar{\bar{d}}_B = d - \beta_B/\nu$ (cf. (11.3.6a)). It has been suggested that the backbone exponents are not independent of other percolation exponents (SAHIMI, 1984; this paper also includes a table of backbone exponents). The Alexander-Orbach conjecture ($\bar{\bar{d}} = 4/3$ for $d \geqq 2$) has been shown to fail numerically for the backbone (STANLEY and CONIGLIO, 1984).

Little is known about the distribution (number) of dead ends. Recently, a scaling theory of dead-end distribution has been proposed by OHTSUKI and KEYES (1984b), which leads to scaling relations for the critical exponents describing the dead-end distribution, and which provides relations between these exponents and the critical exponents of the percolation-cluster and the backbone problems.

The relation (11.3.17) between the dynamic (t) and static (β, ν) percolation exponents diagrees with various previous such relations, in particular the relation (5.2.19)

$$t = \zeta + (d - 2)\,\nu, \tag{11.3.18}$$

derived by simple scaling arguments on the basis of the node-link model (ζ is defined by (5.2.18)). The exponent t as well as the static critical percolation exponents can be evaluated by means of a field theory on the basis of the Potts model without recourse to any geometrical assumptions. For ζ defined by (11.3.17), HARRIS et al. (1984) have obtained $\zeta = 1 + (\varepsilon/42)$, to first order in $\varepsilon = 6 - d$ ($\varepsilon = 0$ gives the mean-field result). On the other hand, by means of (11.3.18), from (11.3.17), one gets $\zeta = \frac{1}{2}(\gamma + \beta)$, where to first order in ε, field theory gives $\gamma = 1 + \varepsilon/7$ and $\beta = 1 - \varepsilon/7$, so that the Alexander-Orbach conjecture breaks down to first order in ε. Note that on the basis of a finite-size calculation of t, SAHIMI et al. (1983) argue that the Alexander-Orbach conjecture is exact. The relationship between dynamic and static critical percolation exponents is yet an active field of research (cf. KERTÉSZ, 1983, SAHIMI, 1984b, FAMILY and CONIGLIO, 1985).

By using field theory, critical exponents of backbones has recently been evaluated by HARRIS (1983) and HARRIS and LUBENSKY (1983), who have also critically examined the node-link model and argued that this model breaks down for d less than some critical value $\hat{d}$, which is possibly larger than 3.

Adopting the mean-field value $\zeta = 1$, then for four and higher dimensions, the node-link model (11.3.18) yields satisfactory results for the critical exponent of conductivity t (cf. HARRIS, 1983), while it fails at lower dimensions. HARRIS and LUBENSKY (1983) conclude that the node-link picture should not be taken too literally.

The node-link-blob model for percolation clusters by CONIGLIO (1981, 1982, 1983), which is a hybrid between the fractal and the node-link models, has recently been examined by HERRMANN and STANLEY (1984), who found that the backbone between any two points of the perculation cluster can be described by a "necklace" of volatile fractals (fractals not being stable under length-scale change) of all sizes.

11.4. Hopping transport in disordered systems

11.4.1. AC conductivity, relaxation currents, and particle diffusion

Recently, for R-hopping in two and three dimensions, PRIGODIN and SAMUKHIN (1983a) and BRYKSIN (1984) have pointed out that in the low-frequency regime 1 (cf. Section 6.7.) the ac conductivity $\sigma(\omega)$ is given by

$$\sigma(\omega) = \sigma(0)\begin{cases}(1 + a_2\tilde{\Omega} + ib_2\tilde{\Omega}\ln\tilde{\Omega} + \cdots) & \text{for } d = 2\\(1 + ia_3\tilde{\Omega} + b_3\tilde{\Omega}^2 + c_3\tilde{\Omega}^{3/2} + \cdots) & \text{for } d = 3,\end{cases} \tag{11.4.1}$$

Here $\tilde{\Omega}$ is the dimensionless frequency (6.4.30) and a, b, and c are real numbers, where c_3 is a small quantity. In contrast to (6.4.33), (11.4.1) contains non-analytic terms, which result from a correct account of the non-analytic behaviour of (6.4.25) at frequency $\omega = 0$. For two- and three-dimensional broken-bond (random-network) models, ODAGAKI (1984) has pointed out, by means of an effective-medium approximation, that $\sigma(\omega)$ exhibits at $\omega = 0$ non-analytic behaviour similar to (11.4.1).

The result (11.4.1) suggests that the scaling concept presented in Section 6.3. must possibly be thougt over anew, especially for $d = 2$, where non-analytic behaviour plays an important role, since this concept implies that at small ω, $\sigma(\omega)$ may be expanded in powers of ω. Evidence that $d = 2$ is the upper marginal dimension for ac hopping conduction has also been obtained by DERRIDA and LUCK (1983) and LUCK (1983), by examining drift velocity, diffusion constant and conductivity in hopping systems with random but biased (non-symmetric) hopping probabilities. By means of a weak disorder expansion (DERRIDA and LUCK, 1983) and a perturbative renormalization-group analysis these authors have pointed out that for $d \leqq 2$, logarithmic corrections to classical diffusion and conduction occur, no matter how weak the disorder, while for $d > 2$, any weak disorder is irrelevant (cf. LUCK, 1984, FISHER, 1984).

The non-analyticity of $\sigma(\omega)$ at $\omega = 0$ also leads to specific features in the long-time behaviour of relaxation currents $j(t)$. PRIGODIN and SAMUKHIN (1983a) have shown that $j(t)$ is given by (7.3.21) only if the electrical field E is sufficiently strong, $E > E_c \sim (kT/eR_c)\,(\alpha R_c)^{-1/2}$, while for $E < E_c$, the long-time asymptotic behaviour reads

$$j(t) - j_0 \sim t^{-3/2} \exp\left(-AE^2 t\right), \tag{11.4.2}$$

where j_0 is the stationary current. Thus, for $E \to 0$, $j(t)$ does not decay exponentially with time t, but as $t^{-3/2}$ (for $d = 3$). Such behaviour may be expected only in the limit of very long times, where $\left(j(t) - j_0\right)/j_0 \ll 1$, because in an expansion of the type (11.4.1) the non-analytic terms are associated with small numerical factors. For not too long times, $j(t)$ may be expected to be given by an expression of the type (7.3.21). Possibly, the power-law decay of relaxation currents in disordered systems may be related to the analogous phenomenon in classical gases (ERNST and DORFMAN, 1972). This has recently been discussed by BRYKSIN (1984).

The problem of diffusion in disordered systems is intimately related with the problem of the time dependence of relaxation currents in such systems. Recently, the diffusion problem has attracted much interest. For istance, by Monte Carlo studies GEISEL and THOMAE (1984) have found that in three dimensions the mean-square displacement $\langle r^2(t)\rangle$ varies as $\langle r^2(t)\rangle \sim t^k$ $(0 < k \leqq 1)$ or like $\langle r^2(t)\rangle \sim t/\ln t$ (anomalous diffusion), which is in accordance with the behaviour of $\sigma(\omega)$ in the frequency regimes 2 and 3 (cf. Section 6.7.1.). PANDEY (1984) has performed Monte Carlo studies for biased diffusion. Bewlow some critical value for the bias, he has observed diffusion for short times and drift for long times, while for large bias, $\langle r^2(t)\rangle$ decreases with increasing bias at constant time, possibly due to field- induced trapping of charge carriers in "dead ends" (cf. Section 7.2.1.). Biased diffusion in disordered hopping systems is a problem of considerable current interest (cf. WHITE and BARMA, 1984, SEIFERT and SUESSENBACH, 1984, OHTSUKI and KEYES, 1984a, LUCK, 1984, BALAKRISHNAN and KHANTHA, 1985, and references therein).

Low-frequency hopping conductivity in disordered semiconductors has recently been studied theoretically by Summerfield and Butcher (1983, 1984) and Dyre (1985). The former authors have pointed out that $\sigma(\omega)$ passes from the frequency regime 1 to the regime 2 at $\tilde{\Omega} \approx 1$ ($\tilde{\Omega}$ is defined by (6.4.30)), which is reflected by a slope in the plot of $\ln\left(\mathrm{Re}\, \sigma(\omega) - \sigma(0)\right)$ versus $\ln \omega$ (cf. Fig. 6.1). Summerfield and Butcher (1983, 1984) have also critically compared theoretical results for $\sigma(\omega)$ with experimental data (cf. Eaves et al., 1984).

The hopping transport in one-dimensional disordered systems is a problem of great current interest. By examining relaxation currents in the broken-bond model, which can be exactly solved, Prigodin and Samukhin (1983b) and Movaghar et al. (1984) have obtained the asymptotic behaviour

$$j(t) \sim \exp\{-v^2\tau - 3[\pi(1 - v)/2]^{2/3} \tau^{1/3}\}, \tag{11.4.3}$$

where $v = eEa/2kT\varepsilon$ is the dimensionless electrical field, ε the fraction of broken bonds ($\varepsilon \ll 1$), a the lattice constant, $\tau = tW\varepsilon^2$ the dimensionless time, and W is the nearest-neighbour hopping rate. In particular, for $E \to 0$, from (11.4.3), one finds that

$$j(t) \sim \sqrt{\tau} \exp[-3(\pi/2)^{2/3} \tau^{1/3}], \tag{11.4.4}$$

where the pre-exponent has been taken from Movaghar et al. (1984) and Anlauf (1984). Note that (11.4.4) may readily be obtained by taking the Laplace transform of (6.3.12) with respect to time and by making use of (6.8.16). For infinitely deep traps, the problem of electric-field-dependent charge-carrier trapping in one dimension is intimately related to the problem just discussed. This trapping problem has recently been studied by Seiferheld et al. (1983), who have found that the long-time behaviour of the current is governed by the decay law

$$j(t) \sim \exp[-(E/E_0)^2 (t/t_0) - (t/t_0)^{1/3}], \tag{11.4.5}$$

which agrees with experimental findings on the quasi one-dimensional material polydiacetylene (E_0 and t_0 are parameters). For infinitely deep traps in three-dimensions, Prigodin (1984a) has found that

$$j(t) \sim \exp[-(E/E_0)^2 (t/t_0) - (t/t_0)^{3/5}]. \tag{11.4.6}$$

Thermal release of charge carriers from traps leads (in a one-dimensional model) to a power-law decay of $j(t)$, instead of the exponential decay (11.4.5) (Prigodin, 1984b).

Recently, Prigodin and Samukhin (1983a) and Movaghar et al. (1984) have pointed out that in one-dimensional systems with random site positions the current $j(t)$ exhibits different time dependence for $t > t_c$ and $t < t_c$, no matter how strong the external field. Here $t_c \sim E^{(1-2\alpha a)}$ is a characteristic time, which increase with decreasing field E (a is the mean nearest-neighbour distance between the sites, and α is the reciprocal Bohr radius of a localized state). According to Prigodin and Samukhin (1983a), the long-time asymptotic behaviour of $j(t)$ is given by

$$j(t) \sim \begin{cases} t^{-[1-1/(2\alpha a-\lambda)]} & \text{for } t > t_c \\ t^{-[1-2/(2\alpha a+1)]} & \text{for } t < t_c \end{cases} \tag{11.4.7}$$

where $\lambda = eEa/2kT$. Consequently, the expression for $\sigma(\omega)$ (6.4.18) only applies for $\omega > \omega_c \sim 1/t_c \sim E^{(2\alpha a-1)}$, while for $\omega < \omega_c$, linear-response theory breaks down.

With regard to charge transport in disordered one-dimensional systems, among the theoretical problems of current interest are the range of validity of the Mott's law for phononless hopping $\sigma(\omega) \sim \omega^2 \ln^2 \omega$ (RIESEBOROUGH, 1983, MORAWECK and BRANDT, 1984, SASO, 1984), the singular behaviour of diffusion constant and drift velocity in hopping models with non-symmetric (biased) hopping rates (e.g. BERNASCONI and SCHNEIDER, 1982, DERRIDA, 1983, KHANTHA and BALAKRISHNAN, 1984), the ac conductivity in systems with biased hopping rates (DERRIDA and ORBACH, 1983, BILLER, 1984), and the non-analytic low-frequency (long-time) behaviour of random one-dimensional models (e.g. VAN BEIJEREN, 1982, DENTENEER and ERNST, 1983, VISSCHER, 1984, NIEUWENHUIZEN and ERNST, 1985). In recent years much interest has attracted the phenomenon of $1/f$ noise in disordered systems, including one-dimensional hopping systems (cf. MARINARI et al., 1983, 1984, KARIOTIS and STEPHEN, 1983, LEHR et al., 1984, FISHER, 1984, RAMMAL et al., 1985).

Let us now turn to the problem of anomalous diffusion in disordered system. This problem may be studied, for instance, on the basis of a non-Markovian diffusion equation (cf. (6.5.29); see also BRYKSIN, 1983b), or with the aid of the scaling hypothesis (BERNASCONI et al., 1980)

$$P(\boldsymbol{r}, s) = P(0, s)\, h\big(2rsP(0, s)\big), \tag{11.4.8}$$

where $P(\boldsymbol{r}, s)$ is the Laplace transform of the diffusion function $P(\boldsymbol{r}, t)$, $P(0, t) = P(\boldsymbol{r} = 0, t)$ is the autocorrelation function, and $h(x)$ is some universal function. Using $h(x) = e^{-x}$, then (11.4.8) becomes equivalent to the non-Markovian diffusion equation (6.5.29). Note that (6.5.29) as well as (11.4.8) are hypothesis' whose range of validity is not yet quite clear at present (cf., e.g., AHARONY and STAUFFER, 1984).

The diagram technique presented in Section 6.5. (cf. BRYKSIN and YASHIN, 1983, BRYKSIN, 1984) does, in general, not provide (6.5.29), except on setting $S(\boldsymbol{q}) = 1$. Recently, BODUNOV and MALYSHEV (1984) have derived an integral equation for the diffusion function which does not involve a quantity like $S(\boldsymbol{q})$, but the kernel of this equation has not a translationally invariant form. Note that a quantity of the type $S(\boldsymbol{q})$ does also not occur by formulating the diagram technique in the position space, as proposed by GOCHANOUR et al. (1979).

In the past few years, evaluation of the quantity $P(0, t)$ (the probability of finding a particle at time t at the origin if it was put at this place at time 0) has received considerable interest. This is due to the fact that by means of this quantity, or the underlying master equation of the type of the hopping-rate equations respectively, one can describe a variety of problems of hopping-type excitation (charge carrier or exciton) diffusion in systems with a band of localized states (e.g. amorphous semiconductors or organic solids), which possibly also contain randomly distributed deep traps and recombination centres. Among these problems are fluorescence phenomena (e.g. KRENKRE et al., 1983, HUBER, 1984), radiationless transport of electronic excitations among randomly distributed chromophores (e.g. KORZENIEWSKI et al., 1983,

LORING et al., 1984), and particle diffusion in systems with random static traps (e.g. ALEXANDER et al., 1981, HAVLIN et al., 1984a, and references therein).

On studying hopping in the disconnected-network (broken-bond) model in d dimensions, TANAKA (1980) has found that

$$P(0, t) \sim \exp\left[-(t/t_0)^\beta\right], \tag{11.4.9}$$

where $\beta = d/(d + 2)$. For $d = 1$, (11.4.9) reads $P(0, t) \sim \exp - (t/t_0)^{1/3}$, which is in accordance with the result by ALEXANDER et al. (1980). The latter formula has rather universal character in one dimension. For the broken-bond model in one dimension, BRYKSIN (1985) has obtained $P(0, t) \sim t^{1/6} \exp\left[-(t/t_0)^{1/3}\right]$. Note that for particle diffusion through a medium with randomly distributed deep traps, $P(0, t)$ decays according to the law (11.4.9) with $\beta = d/(d + 2)$ (GRASSBERGER and PROCACCIA, 1982, KAYSER and HUBBARD, 1983, cf. the footnote after (6.8.8)).

TANAKA (1980) has derived (11.4.9) by considering compact clusters of connected bonds. On account of non-compact clusters, the effective dimension diminishes: In the expression for β given above the Euclidian dimension must be replaced by the spectral dimension $\bar{\bar{d}}$. Using $\bar{\bar{d}} = 4/3$ (for $d \geqq 2$) according to the conjecture of ALEXANDER and ORBACH (1982) (cf. Section 11.3.2), one gets $\beta = 2/5$ (SHAPIR, 1984, PANDEY et al., 1984), in accordance with computer results (PANDEY et al., 1984). SHAPIR (1984) argues that the decay of $P(0, t)$ is governed by clusters having the form of self-avoiding walks, which leads to $\beta = 1/3$ for any dimension.

Another interesting problem of particle diffusion is the "long-time tail" (power-law decay instead of an exponential decay) of the velocity auto-correlation function, which is defined as the second time derivate of the mean-square displacement, and whose time-Fourier transform defines a frequency dependent mobility. This phenomenenon is due to disorder. It is a consequence of non-analycity of the type (11.4.1). It is closely related to the corresponding problem in fluids (see, e.g., VAN BEIJEREN, 1982, HAUS et al., 1983, VISSCHER, 1984, MACHTA et al., 1984, MARRO and MASOLIVER, 1985).

Concluding this section, let us draw attention to some recent work on multiple trapping and related problems (cf. Section 6.8.). Equations (6.8.2) to (6.8.5) describe multiple trapping for the case that recombination may be neglected (e.g. for surface-carrier generation). Trap-controlled recombination following pulsed bulk electron-hole generation has recently been studied theoretically by ARKHIPOV et al. (1983). — The assumptions (the intuitive thermalization concept (6.8.9), a constant cross section for all traps, an exponential density of states, etc.) underlying the treatment of transient photo-current for trap-controlled electronic transport (ORENSTEIN et al., 1982) described in Section 6.8.1 has been critically discussed by MONROE and KASTNER (1983), MARSHALL and MAIN (1983), and MICHIEL et al. (1983). Recently, MONROE (1985) has pointed out that a low temperatures energy relaxation of photo-excited carriers is governed both by hopping between localized band-tail states and multiple trapping. Another problem of current interest is the time-dependence of geminate electron-hole pair recombination and dissociation in hopping systems (e.g. RIES et al., 1984, SUHRKE and ESSER, 1984). — The Green function method of MOVAGHAR et al. (1980, 1983) (cf. Section 6.6.) has recently been generalized such as to describe excitation diffusion in positionally and energetically disordered systems in dependence of time and initial energy (GRÜNEWALD et al., 1984).

The concept of the average waiting-time distribution in the continuous-time random-walk (CTRW) method has been critically examined by HAUS and KEHR (1983). KLAFTER and SILBEY (1983) have discussed the dependence of excitation transport in disordered systems on the initial condition within their master-equation concept described in Section 6.8.3. BLUMEN et al. (1984) have proposed a procedure which is dual to the CTRW-method in that they have included disorder into the spatial component (random walk on a fractal), while the temporal aspect is ordered (exponential waiting-time distribution), contrary to the CTRW-method, where temporal aspect is disordered (power-law waiting-time distribution), while the spectral component is ordered (diffusion on a lattice).

Various aspects of dispersive transport in amorphous semiconductors have recently been discussed in the review by MARSHALL (1983).

Let us yet draw attention to recent papers on diffusion on fractals (O'SHAUGHNESSY and PROCACCIA, 1985, WEGNER and GROSSMANN, 1985), "termite" diffusion in super-normal conductor systems (BUNDE et al., 1985, ADLER et al., 1985), ac conductivity in quasi one-dimensional hopping systems (ODAGAKI et al., 1984) and two-dimensional bond-percolation systems (BILLER, 1985), trap-limited hopping in one-dimensional and quasi one-dimensional systems (PRIGODIN et al., 1985), and on the use of the supersymmetry (CARTON, 1983, KHOLODENKO and FREED, 1984) and replica (KARIOTIS, 1984) methods for studying diffusion on random lattices.

11.4.2. Miscellaneous hopping problems

Let us first consider recent work on the hopping Hall effect in disordered systems. BUTCHER et al. (1983) have performed numerical calculations of the dc Hall effect resulting from three-site hopping processes in a model with random energies and positions of sites. It has been found that the dc Hall mobility μ_H decreases with decreasing temperature T, whereby at sufficiently low temperaturet (in the range of variable-range hopping) μ_H appears to fall off linearly with $T^{-1/4}$, while at sufficiently high T, μ_H arrives at the R-percolation value of μ_H. This behaviour is in agreement with existing analytical predictions (cf. Section 8.2.). So far, there are no reliable experimental hopping-Hall data. In particular, experimental studies of the quantum Hall effect on two-dimensional systems have not provided indication for variable-range hopping in transverse conductivity (cf. WYSOKINSKI and BRENIG, 1983, VON KLITZING, 1984, and references therein).

KLEIN (1985) attempted to measure on p-type germanium the ac hopping Hall effect predicted by HOLSTEIN (1961). However, no Hall effect was detected, resulting for the hopping Hall effect in an upper bound, which is a factor of 37 below the value obtained from Holstein's theory and a factor of 6 below the value given by an extension of this theory to the actual conditions used in experiment (higher magnetic field, lower temperatures).

Recently, VOMVAS and ROILOS (1984) has obtained experimental evidence for the small-polaron hopping-Hall effect in vitreous $80\,V_2O_5 - 20\,P_2O_5$. The fact, that in this material the activation energy of the Hall mobility is not equal to one third of that of the drift mobility, is not incompatible with the small-polaron model, because the one-third-relation between the two activation energies is strictly valid only for hexagonal crystalline materials.

Another problem of current interest is hopping transport in strong electrical fields. Recently, this problem has been studied by solving numerically the rate equation for hopping in the presence of an electrical field of arbitrary strength (BÖTTGER and WEGENER, 1984a, b, BÖTTGER et al., 1985a, b, LEVIN, 1984, LEVIN and SHKLOVSKII, 1984). In this way, for R-hopping and strong electron-phonon coupling, it has been pointed out (BÖTTGER and WEGENER, 1984a, b; cf. Section 7.1.3) that, just above the Ohmic regime the differential conductivity decreases with increasing field up to some critical field above which it begins to increase exponentially with E. This field-dependence of the conductivity is accompanied by a strong reconstruction of the current cluster (BÖTTGER and WEGENER, 1984b). Taking into account random distribution of site energies (BÖTTGER et al., 1985a, b) the conductivity minimum above the Ohmic regime vanishes with decreasing temperature. The disappearance of this minimum turns out to be independent of the form of the density of states at the Fermi level. Among further strong-field problems of current interest are networks of nonlinear resistors (e.g. STRALEY and KENKEL, 1984) and field-induced drift and trapping in percolation networks (e.g. WHITE and BARMA, 1984).

The interplay of disorder and electron-electron interaction is yet a very active field of research (e.g. TAKEMORI and KAMIMURA, 1983, NARA, 1983, KUROBE et al., 1984; cf. Section 11.2.1). Various theoretical studies suggest the existance of a Coulomb gap or pseudogap at the Fermi level in a band of localized states, but a first-principle theory of this phenomenon is still lacking. The experimental situation is somewhat controversy. Contrary to several other authors (cf. Section 5.4.3.), BENZAQUEN and WALSH (1984) drew the conclusion recently that the available experimental data on hopping conduction in doped semiconductors do not give convincing evidence for the existence of a Coulomb gap. The data obtained by the latter authors on compensated n-type GaAs show that the Coulomb gap plays at best a minor role in the hopping conduction at $T > 1$ K. — DAVIES (1984) has recently extended the theory of the Coulomb gap by including short-ranged Hubbard interaction in addition to long-ranged Coulomb interaction. POLLAK (1984) studied non-ergodic behaviour of Anderson insulators without (Fermi glass) and with (Coulomb glass) Coulomb interaction in terms of percolation theory of hopping transport.

A very interesting phenomenon (macroscopic quantum effect) observed in quasi one-dimensional (MOSFET's) systems is the low-temperature random-peak-and-valley structure of the conductance as a function of the chemical potential, which has been interpreted (at not too low temperatures) in terms of variable-range hopping (LEE, 1984, STONE and LEE, 1985, AZBEL, 1984, AZBEL and DI VINCENZO, 1984, and references therein).

Many aspects of hopping conduction in disordered systems are discussed in the monograph by ZVYAGIN (1984).

11.5. Atomic hopping transport

Hopping transport of classical particles is still a very active field of research. Among the problems of current interest are the escape rate from a potential well, tracer diffusion, bulk diffusion, hopping of charged particles, etc.

The Kramers theory for the escape of atomic particles moving in a potential well has been critically examined and generalized by several authors: Boughaleb and Gouyet (1983) have generalized the lattice-gas model (system of classical particles whose center-of-mass coordinates are restricted to a lattice) such as to describe the low-friction Kramers regime, where a particle, hopping to a neighbouring site, does not loose all memory of preceding hops. Carmeli and Nitzan (1983, 1984) have derived an expression for the escape rate which is valid for all values of the friction coefficient. The effect of memory (non-Markovian behaviour) on the escape rate has recently also been studied by Hänggi and Mojtabai (1983), Hänggi (1984), and Jacucci et al. (1984), whereby the latter authors have also proposed a procedure for constructing saddle surfaces for diffusion.

The Kramers theory of escape from a potential well by thermal activation applies for sufficiently high temperatures, while at low temperature tunnelling through the potential barrier becomes effective. Recently, Caldeira and Legett (1981, 1983) have pointed out that the latter process is strongly affected by frictional effects due to interaction with the thermal bath. By means of a functional-integral approach, Grabert and Weiss (1984) have studied the escape rate at temperatures between the high- and low-temperature regimes. They have shown that already at intermediate temperatures quantum effects may appreciably enhance the escape rate as compared to the classical rate. In this context let us also draw attention to the theoretical study of coherent and incoherent tunnelling transport of positive muons by Yamada (1984), the critical analysis of tunnelling of light interstitials in the small-polaron picture by Shie (1984), and the study of Kagan and Maksimov (1983) of localization and phonon-stimulated delocalization of interacting atomic particles which diffuse by coherent tunnelling in a crystal.

Tracer (individual particle) migration is a very complex diffusion process. Due to the exclusion of multiple occupancy of atomic sites, successive tracer hops are strongly correlated, even in the case of stochastic motion (through the available sites) of non-interacting particles. When a particle has made a hop, there is a vacant site left behind that can either be occupied by other particles or a backward hop of the particle itself. Thus a forward hop is always more probable than a backward hop.

As mentioned in Section 10.2.3., tracer diffusion on a linear chain is dramatically influenced by the restriction that particles can never pass each other, which causes the tracer mean-square displacement to increase for long timest as $t^{1/2}$. This result has recently been confirmed by van Beijeren et al. (1983), who have studied the mean-square displacement of a tagged particle in a one-dimensional concentrated lattice gas for all times, both by an approximate theory and by Monte Carlo simulations.

A detailed calculation of frequency- and wavevector-dependent correlation functions for tracer diffusion by nearest-neighbour hops in a regular lattice (whose sites are occupied by atoms which may perform hops) has recently been carried out by Tahir-Kheli and Elliott (1983) by solving the rate equations for a tracer in a non-interacting lattice gas by means of a Green function method. This work has been generalized to lattices consisting of two species of atoms (Tahir-Kheli, 1983) and to the tracer diffusion in the presence of nearest- and further-neighbour transitions (Tahir-Kheli, 1984). Chaturvedi (1983, 1984) has studied tracer diffusion in a non-interacting lattice gas by means of the Zwanzig-Mori formalism.

Non-interacting lattice gases (lattice gases with nearest-neighbour interaction) may be used to describe, for instance, migration of hydrogen in metals, adsorbates on surfaces, and ions in mixed electronic-ionic conductors, where the Coulomb interaction between the ions is sereened by the conduction electrons. On the contrary, diffusion of ions in purely ionic conductors requires use of an interacting lattice-gas model which takes explicitely into account Coulomb forces between the ions. Coulomb repulsion enhances the probability for backward hops. This phenomenon has recently been studied by Bunde et al. (1984).

Another problem of current interest is the bulk diffusion in random hopping systems (Kehr et al., 1983, Kehr, 1983, Evans and Hoffman, 1984), including effects of both repulsive interactions (including ordering) and attractive interactions (including clustering) (Kutner et al., 1983). It has been shown by Monte Carlo studies that mean-field approximations are inadequate for the problem quoted. The randomizing effect of hopping on spatial subconfigurations has been studied by Evans and Hoffman (1984).

Another field where atomic hopping is important is adsorption and desorption on a solid surface (Luque and Cordoba, 1982, Summerside et al., 1982, Sommer and Kreuzer, 1982).

In conclusion, let us draw attention to some recent papers on the dynamics of activated rate processes (De Lorenzi et al., 1984, Mel'nikov, 1985, Pregenzer and Suhl, 1985, Munakata, 1985), a generalized hopping model for frequency-dependent transport in a dynamically disordered medium (Druger et al., 1985), Brownian motion in a periodic potential (Marchesoni and Vij, 1985), and tracer diffusion (Kutner et al., 1984, Chaturvedi and Tosi, 1984).

Appendix 1. Diagram technique for strong electron-phonon coupling

The diagram technique presented here agrees in essence with that used in the small-polaron theory (FIRSOV, 1975), except the following two details: here we also admit disorder (non-vanishing ε_m in (2.2.6)) and, instead of the Laplace representation, we make use of the time representation, in order to admit external electrical fields of arbitrary time dependence.

As pointed out in Section 2.4., the perturbation expansion of the density matrix $\rho_m(t)$ with respect to the Hamiltonian $\mathcal{H}'$ (2.2.6) requires that we find thermodynamic averages of the type

$$\begin{aligned}&\langle \mathcal{H}'(t_1)\, \mathcal{H}'(t_2) \ldots \mathcal{H}'(t_n)\, a_m^+ a_m \mathcal{H}'(t_{n+1}) \ldots \mathcal{H}'(t_l)\rangle \\ &= \sum_{\{m_i, m_i'\}} J_{m_1' m_1}(\boldsymbol{H})\, J_{m_2' m_2}(\boldsymbol{H}) \ldots J_{m_n' m_n}(\boldsymbol{H})\, J_{m_{n+1}' m_{n+1}}(\boldsymbol{H}) \ldots \\ &\quad \times J_{m_l' m_l}(\boldsymbol{H})\, \langle (a_{m_1'}^+ a_{m_1})_{t_1} \ldots (a_m^+ a_m) \ldots (a_{m_l'}^+ a_{m_l})_{t_l} \rangle_{\mathrm{e}} \\ &\quad \times \langle \hat{\Phi}_{m_1' m_1}(t_1) \ldots \hat{\Phi}_{m_l' m_l}(t_l) \rangle_{\mathrm{ph}}. \end{aligned} \tag{A 1.1}$$

Here the indices e and ph, respectively, indicate the averages with regard to the electron and phonon operators. Furthermore, $(a_{m'm}^+)_t = U_0^+(t)\, a_{m'm}^+ U_0(t)$ and $\hat{\Phi}(t) = U_0^+(t)\, \hat{\Phi} U_0(t)$, where $U_0(t)$ is the evolution operator of a free particle (2.4.2). When U_0 acts on electron (phonon) operators, the phonon (electron) contribution to $\mathcal{H}_0^E$ may be neglected.

The set of operators $\mathcal{H}'(t_i)$ placed in (A 1.1) on the left of $a_m^+ a_m$, arises from the power-series expansion of $\exp\{-\beta \mathcal{H}^{(0)}\}\, U^+(t)$, and that on the right of $a_m^+ a_m$, from the expansion of $U(t)$ (cf. (2.3.4)). Recall that on expanding $\exp\{-\beta \mathcal{H}^{(0)}\}$, we must set $\boldsymbol{E} = 0$, and the corresponding points are lying on the vertical branch of the contour C in Fig. 2.1. According to Section 2.4., in the stationary regime, the current is independent of the concrete choice of the starting-Hamiltonian $\mathcal{H}^{(0)}$ in the operator $\exp\{-\beta \mathcal{H}^{(0)}\}$. Hence, in what follows, $\mathcal{H}^{(0)}$ in $\exp\{-\beta \mathcal{H}^{(0)}\}$ may be replaced, for instance, by $\mathcal{H}_{\mathrm{ph}} + \sum_m \varepsilon_m a_m^+ a_m$.

Let us now perform the thermodynamic average over the electron operators. To do this, we use the relation

$$U_0^+(t)\, a_{m'}^+ a_m U_0(t) = a_{m'}^+ a_m \exp\left\{\frac{\mathrm{i}}{\hbar} \int_0^t \mathrm{d}\tau [\varepsilon_{m'} - \varepsilon_m - e\boldsymbol{E}(\tau)\, (\boldsymbol{R}_{m'} - \boldsymbol{R}_m)]\right\}. \tag{A 1.2}$$

To verify (A 1.2), we differentiate $(a_{m'}^+ a_m)_t$ with respect to t. After that, we perform the commutation of $a_{m'}^+ a_m$ with $\mathcal{H}_0^E$. Then, with the aid of (2.1.2), we obtain

$$\frac{\mathrm{d}}{\mathrm{d}t} (a_{m'}^+ a_m)_t = \frac{\mathrm{i}}{\hbar} (a_{m'}^+ a_m)_t\, [\varepsilon_{m'} - \varepsilon_m - e\boldsymbol{E}(t)\, (\boldsymbol{R}_{m'} - \boldsymbol{R}_m)].$$

Integrating this differential equation with the initial condition $(a^+_{m'}a_m)_t = a^+_{m'}a_m$ for $t = 0$, we get (A1.2).

Now the electron average in (A1.1) becomes

$$\langle a^+_{m_1'}a_{m_1}a^+_{m_2'}a_{m_2} \dots a^+_m a_m \dots a^+_{m_l'}a_{m_l}\rangle$$
$$\times \exp\left\{\frac{i}{\hbar}\int\limits_0^{t_1} d\tau[\varepsilon_{m_1'} - \varepsilon_{m_1} - e\boldsymbol{E}(\tau)(\boldsymbol{R}_{m_1'} - \boldsymbol{R}_m)] + \cdots\right.$$
$$\left. + \frac{i}{\hbar}\int\limits_0^{t_l} d\tau[\varepsilon_{m_l'} - \varepsilon_{m_l} - e\boldsymbol{E}(\tau)(\boldsymbol{R}_{m_l'} - \boldsymbol{R}_{m_l})]\right\}. \tag{A1.3}$$

The electron average in (A1.3) is evaluated by means of the Wick theorem, which, in the language of the diagram technique, leads to a pairwise connection of all interaction points and the clamp by electron lines. In the diagram language, each interaction point is associated with a pair of operators $a^+_{m_l'}a_{m_l}$, whereby on the contour C, the operator a_{m_l} is placed earlier than $a^+_{m_l'}$. The operator $a_m{}^+$ (indicated by ⊕) of the clamp is the starting point for a line having an arrow directed away from the clamp. The end of this line is connected with the operator a_{m_l} (indicated by ○) belonging to some interaction point, and from the associated operator $a^+_{m_l'}$ (indicated by ⊕) originates another electron line, which runs into another interaction point and so on. Finally, the chain of these electron lines terminates in the operator a_m (indicated by ○) of the clamp. This chain does not necessarily connect all interaction points. A diagram may also contain several closed rings of electron lines, whereby all rings, but a single one, do not pass through the clamp. However, note that the rings may be connected with each other by phonon lines, so that by no means all diagrams with several rings are necessarily disconnected diagrams (as mentioned in Section 2.4., disconnected diagrams may be ignored).

The origin of a so-called right line is placed on the contour earlier than its end, otherwise the line is a so-called wrong line. It is easy to see that any diagram contains at least one wrong line. Owing to the pairing procedure of the electron operators, the site indices of the operators connected with an electron line coincide. The corresponding site index is ascribed to the el ec tron line. All right lines are associated with a factor $1 - f_m$, and the wrong lines with a fac to f_m. Here m is the index of the corresponding line, and f_m is the equilibrium occupation probability of the mth site in the absence of a field

$$f_m = \frac{\mathrm{Tr}\left\{\exp\left[-\beta\left(\sum\limits_m \varepsilon_m a^+_m a_m - \mu\hat{N}\right)\right] a^+_m a_m\right\}}{\mathrm{Tr}\left\{\exp\left[-\beta\left(\sum\limits_m \varepsilon_m a^+_m a_m - \mu\hat{N}\right)\right]\right\}}.$$

Since the operator $\sum\limits_m \varepsilon_m a^+_m a_m - \mu\hat{N}$ is diagonal with respect to the site index, on account of the fact that $a^+_m a_m$ may take either the value 0 or 1, we readily find that

$$f_m = \{1 + \exp[\beta(\varepsilon_m - \mu)]\}^{-1}, \tag{A1.4}$$

i.e. f_m is the familiar Fermi function.

Furthermore, all diagram must be multiplied by a factor $(-1)^M$, where M denotes the number of intersections of electron lines.

Let us now turn to the phonon averages occurring in (A1.1). First, note that the following identity holds

$$\hat{\Phi}(\{b_q{}^+\}, \{b_q\}, t) \equiv \hat{\Phi}(\{b_q{}^+(t)\}, \{b_q(t)\}), \tag{A1.5}$$

where

$$b_{\boldsymbol{q}}^+(t) = \exp\left\{\frac{\mathrm{i}}{\hbar}\,\mathscr{H}_{\mathrm{ph}}t\right\} b_{\boldsymbol{q}}^+ \exp\left\{-\frac{\mathrm{i}}{\hbar}\,\mathscr{H}_{\mathrm{ph}}t\right\}, \tag{A 1.6}$$

which can be verified by expanding $\hat{\Phi}$ in a power series with respect to $b_{\boldsymbol{q}}^+$ and $b_{\boldsymbol{q}}$, subjecting each term to a unitary transformation, and summing up again this series. To get $b_{\boldsymbol{q}}^+(t)$, we differentiate $b_{\boldsymbol{q}}^+(t)$ (A 1.6) with respect to t, perform the operation of commutation between $b_{\boldsymbol{q}}^+$ and $\mathscr{H}_{\mathrm{ph}}$, and integrate the differential equation resulting for $b_{\boldsymbol{q}}^+(t)$ by making use of the initial condition $b_{\boldsymbol{q}}^+(0) = b_{\boldsymbol{q}}^+$. Then, we obtain

$$b_{\boldsymbol{q}}^+(t) = b_{\boldsymbol{q}}^+\,\mathrm{e}^{\mathrm{i}\omega_{\boldsymbol{q}}t},\ b_{\boldsymbol{q}}(t) = b_{\boldsymbol{q}}\,\mathrm{e}^{-\mathrm{i}\omega_{\boldsymbol{q}}\cdot}. \tag{A 1.7}$$

The second equation in (A 1.7) is the Hermitian conjugate of the first one. Therefore, the latter equation must not be derived separately.

Using the explicit form of the multiphonon operator $\hat{\Phi}$ (2.2.9), by means of (A 1.5) and (A 1.7), we obtain

$$\hat{\Phi}_{m'm}(t) = \exp\left\{\sum_{\boldsymbol{q}} b_{\boldsymbol{q}}^+\alpha_{\boldsymbol{q}}^*(m', m, t) - b_{\boldsymbol{q}}\alpha_{\boldsymbol{q}}(m', m, t)\right\}, \tag{A 1.8}$$

where

$$\alpha_{\boldsymbol{q}}(m', m, t) = [u_m(\boldsymbol{q}) - u_{m'}(\boldsymbol{q})]\,\mathrm{e}^{-\mathrm{i}\omega_{\boldsymbol{q}}\cdot}. \tag{A 1.8}$$

Accordingly, $\hat{\Phi}(t)$ is an exponential operator. According to (2.1.10), the commutation of the two operators occurring in the exponent of this operator yields a c-number.

$$\begin{aligned} K &= \sum_{\boldsymbol{q}\boldsymbol{q}'} \alpha_{\boldsymbol{q}}(m', m, t)\,\alpha_{\boldsymbol{q}'}^*(m', m, t)\,[b_{\boldsymbol{q}}, b_{\boldsymbol{q}'}^+] \\ &= \sum_{\boldsymbol{q}} |\alpha_{\boldsymbol{q}}(m', m, t)|^2 = \sum_{\boldsymbol{q}} |u_m(\boldsymbol{q}) - u_{m'}(\boldsymbol{q})|^2. \end{aligned} \tag{A 1.9}$$

Just the latter fact allows the phonon average over an arbitrary number of factors $\hat{\Phi}(t)$ (cf. (A 1.1)) to be performed. To evaluate such averages, we use the identity of WEYL (1928). This identity holds for any two operators A and B whose commutation yields a c-number K (or, more general, an operator K which commutes with A as well as with B). The Weyl identity reads

$$\exp(aA + bB) = \exp(aA)\exp(bB)\exp(-abK/2), \tag{A 1.10}$$

where a and b are c-numbers, and $K = [A, B]$.

We start the proof of (A 1.10), by proving the following auxiliary identity:

$$\mathrm{e}^{aA}\,\mathrm{e}^{bB} = \mathrm{e}^{bB}\,\mathrm{e}^{aA}\,\mathrm{e}^{abK}. \tag{A 1.11}$$

To this end, let us consider the operator $f(x)$:

$$f(x) = \mathrm{e}^{aAx}\,\mathrm{e}^{bB}\,\mathrm{e}^{-aAx}.$$

Differentiating $f(x)$ with respect to x, we obtain

$$\frac{\mathrm{d}f(x)}{\mathrm{d}x} = a\,\mathrm{e}^{aAx}[A, \mathrm{e}^{bB}]\,\mathrm{e}^{-aAx} = a\,\mathrm{e}^{aAx}\sum_{n=0}^{\infty}\frac{b^n}{n!}[A, B^n]\,\mathrm{e}^{-aAx}.$$

If K commutes with A as well as with B, then $[A, B^n] = nKB^{n-1}$. Inserting this expression into the above equation and performing the summation over n, we get

$$\frac{\mathrm{d}f(x)}{\mathrm{d}x} = abKf(x).$$

Integrating this equation over x, on account of the initial condition $f(0) = e^{bB}$, we find that

$$f(x) = e^{bB} e^{abKx}.$$

Setting now $x = 1$, we obtain the identity (A 1.11).

Now we may turn to the proof of the Weyl identity (A 1.10). To this end, we introduce the operator $g(x)$:

$$g(x) = \exp\{-(aA + bB)x\} \exp(aAx) \exp(bBx).$$

Differentiating $g(x)$ with respect to x, we obtain

$$\frac{dg(x)}{dx} = -b \exp[-(aA + bB)x] [B, e^{aAx}] \exp(bBx).$$

To get $[B, e^{aAx}]$, we expand (A 1.11) in a series with respect to powers of b. On comparing the coefficients of the terms linear in b, we find that $[B, e^{aA}] = -aK$. Replacing in the latter expression a by ax, from the above equation, we obtain

$$\frac{dg(x)}{dx} = abx\, Kg(x).$$

Integrating this equation, we get

$$g(x) = C \exp(abKx^2/2),$$

where C is a constant, which, owing to the condition $g(0) = 1$, is equal to unity. Setting now $x = 1$, we obtain the Weyl identity (A 1.10).

Let us now apply this identity to the product of operators $\hat{\Phi}(t)$ in (A 1.1). This product may be schematically written as

$$\exp\{A_1^+ - A_1\} \exp\{A_2^+ - A_2\} \dots \exp\{A_l^+ - A_l\}, \tag{A 1.12}$$

where the operator A_i is given by

$$A_i = \sum_q b_q \alpha_q(m_i', m_i, t). \tag{A 1.13}$$

By means of (A 1.10), we rewrite (A 1.12) as

$$\exp\left\{-\frac{1}{2}[K_1 + K_2 + \dots + K_l]\right\} \exp(A_1^+) \exp(-A_1) \exp(A_2^+) \times \exp(-A_2) \dots \exp(A_l^+) \exp(-A_l).$$

Furthermore, with the aid of (A 1.11), the latter expression may be rewritten as

$$\exp\left\{-\frac{1}{2}(K_1 + \dots + K_l) - [A_1, A_2^+ + \dots + A_l^+] - [A_2, A_3^+ + \dots + A_l^+] - \dots - [A_{l-1}, A_l^+]\right\} \exp\{A_1^+ + A_2^+ + \dots + A_l^+\} \exp\{-A_1 - A_2 - \dots - A_l\}.$$

Accordingly, the problem of performing the phonon average of operators $\hat{\Phi}(t)$ reduces to the problem of averaging operators

$$\exp\left\{\sum_q \delta_q^* b_q^+\right\} \exp\left\{-\sum_q \delta_q b_q\right\},$$

where, according to (A 1.13),

$$\delta_{\boldsymbol{q}} = \sum_{i=1}^{l} \alpha_{\boldsymbol{q}}(m_i', m_i, t).$$

It holds that

$$\left\langle \exp\left\{\sum_{\boldsymbol{q}} \delta_{\boldsymbol{q}}^* b_{\boldsymbol{q}}^+\right\} \exp\left\{-\sum_{\boldsymbol{q}} \delta_{\boldsymbol{q}} b_{\boldsymbol{q}}\right\}\right\rangle_{\text{ph}} = \exp\left\{-\sum_{\boldsymbol{q}} |\delta_{\boldsymbol{q}}|^2 N_{\boldsymbol{q}}\right\}, \tag{A 1.14}$$

where $N_{\boldsymbol{q}}$ is the phonon-distribution function

$$N_{\boldsymbol{q}} = \{\mathrm{e}^{\beta\hbar\omega_{\boldsymbol{q}}} - 1\}^{-1}. \tag{A 1.15}$$

To verify this relation, first note that we have

$$\left\langle \exp\left\{\sum_{\boldsymbol{q}} \delta_{\boldsymbol{q}}^* b_{\boldsymbol{q}}^+\right\} \exp\left\{-\sum_{\boldsymbol{q}} \delta_{\boldsymbol{q}} b_{\boldsymbol{q}}\right\}\right\rangle_{\text{ph}}$$
$$= \prod_{\boldsymbol{q}} \frac{\sum\limits_{n_{\boldsymbol{q}}=0}^{\infty} \mathrm{e}^{-\beta\hbar\omega_{\boldsymbol{q}} n_{\boldsymbol{q}}} \langle n_{\boldsymbol{q}}| \mathrm{e}^{\delta_{\boldsymbol{q}}^* b_{\boldsymbol{q}}^+} \mathrm{e}^{-\delta_{\boldsymbol{q}} b_{\boldsymbol{q}}} |n_{\boldsymbol{q}}\rangle}{\sum\limits_{n_{\boldsymbol{q}}=0}^{\infty} \mathrm{e}^{-\beta\hbar\omega_{\boldsymbol{q}} n_{\boldsymbol{q}}}}.$$

Here $|n_{\boldsymbol{q}}\rangle$ is the wave function of the phonon subsystem, when in the state $\boldsymbol{q}$ there are $n_{\boldsymbol{q}}$ phonons ($n_{\boldsymbol{q}} = 0, 1, 2, \ldots$). Expanding the exponentials in power series with respect to $\delta_{\boldsymbol{q}}^*$ and $\delta_{\boldsymbol{q}}$, and taking into account the relations

$$b_{\boldsymbol{q}}^+ |n_{\boldsymbol{q}}\rangle = \sqrt{n_{\boldsymbol{q}} + 1}\, |n_{\boldsymbol{q}} + 1\rangle, \quad b_{\boldsymbol{q}} |n_{\boldsymbol{q}}\rangle = \sqrt{n_{\boldsymbol{q}}}\, |n_{\boldsymbol{q}} - 1\rangle,$$

we obtain

$$\left\langle \exp\left\{\sum_{\boldsymbol{q}} \delta_{\boldsymbol{q}}^* b_{\boldsymbol{q}}^+\right\} \exp\left\{-\sum_{\boldsymbol{q}} \delta_{\boldsymbol{q}} b_{\boldsymbol{q}}\right\}\right\rangle_{\text{ph}}$$
$$= \prod_{\boldsymbol{q}} (1 - \mathrm{e}^{-\hbar\omega_{\boldsymbol{q}}\beta}) \sum_{n_{\boldsymbol{q}}=0}^{\infty} \sum_{i=0}^{n_{\boldsymbol{q}}} \mathrm{e}^{-\beta\hbar\omega_{\boldsymbol{q}} n_{\boldsymbol{q}}} \frac{(-1)^i |\delta_{\boldsymbol{q}}|^{2n}\, n_{\boldsymbol{q}}!}{(i!)^2 (n_{\boldsymbol{q}} - i)!}$$
$$= \prod_{\boldsymbol{q}} (1 - \mathrm{e}^{-\hbar\omega_{\boldsymbol{q}}\beta}) \sum_{n,i=0}^{\infty} \mathrm{e}^{-\beta\hbar\omega_{\boldsymbol{q}}(n+i)} \frac{(-1)^n |\delta_{\boldsymbol{q}}|^{2n} (n + i)!}{i!(n!)^2}.$$

To perform the summation over n and i, we use the integral transformation $(n + i)! = \int_0^\infty \mathrm{d}x x^{n+i} \exp(-x)$. Then, we have

$$\prod_{\boldsymbol{q}} (1 - \mathrm{e}^{-\beta\hbar\omega_{\boldsymbol{q}}}) \sum_{n,i=0}^{\infty} \mathrm{e}^{-\beta\hbar\omega_{\boldsymbol{q}}(n+i)} \frac{(-1)^n |\delta_{\boldsymbol{q}}|^{2n}}{i!(n!)^2} \int_0^\infty \mathrm{d}x x^{n+i}\, \mathrm{e}^{-x}$$
$$= \prod_{\boldsymbol{q}} (1 - \mathrm{e}^{-\beta\hbar\omega_{\boldsymbol{q}}}) \sum_{n=0}^{\infty} \mathrm{e}^{-\beta\hbar\omega_{\boldsymbol{q}} n} \frac{(-1)^n |\delta_{\boldsymbol{q}}|^{2n}}{(n!)^2} \int_0^\infty \mathrm{d}x\, x^n\, \mathrm{e}^{-x(1-\exp(-\hbar\omega_{\boldsymbol{q}}\beta))}$$
$$= \prod_{\boldsymbol{q}} \sum_{n=0}^{\infty} \mathrm{e}^{-\beta\hbar\omega_{\boldsymbol{q}} n} \frac{(-1)^n |\delta_{\boldsymbol{q}}|^{2n}}{n!(1 - \mathrm{e}^{-\hbar\omega_{\boldsymbol{q}}\beta})^n} = \exp\left\{-\sum_{\boldsymbol{q}} |\delta_{\boldsymbol{q}}|^2 N_{\boldsymbol{q}}\right\}.$$

The last equation proves the validity of (A 1.14).

Evaluating the commutators $[A_i, A_j^+]$ with $i \neq j$, and taking into account (A1.14), the phonon average over products of operators $\hat{\Phi}$ finally becomes

$$\langle \hat{\Phi}_{m_1'm_1}(t_1) \cdots \hat{\Phi}_{m_l'm_l}(t_l) \rangle_{\text{ph}} = \exp \{-S_{\text{T}}(m_1', m_1) - \cdots - S_{\text{T}}(m_l', m_l)\}$$

$$\times \exp \left\{ \sum_{\boldsymbol{q}} \frac{|\gamma(\boldsymbol{q})|^2}{N \sinh (\hbar\omega_{\boldsymbol{q}}\beta/2)} \left[a^{m_1'm_1}_{m_2'm_2} \cos \omega_{\boldsymbol{q}}(t_2 - t_1 + \mathrm{i}\hbar\beta/2) \right. \right.$$

$$+ a^{m_1'm_1}_{m_3'm_3} \cos \omega_{\boldsymbol{q}}(t_3 - t_1 + \mathrm{i}\hbar\beta/2) + \cdots + a^{m_1'm_1}_{m_l'm_l} \cos \omega_{\boldsymbol{q}}(t_l - t_1 + \mathrm{i}\hbar\beta/2)$$

$$\left. \left. + a^{m_2'm_2}_{m_3'm_3} \cos \omega_{\boldsymbol{q}}(t_3 - t_2 + \mathrm{i}\hbar\beta/2) + \cdots + a^{m_{l-1}' m_{l-1}}_{m_l'm_l} \cos \omega_{\boldsymbol{q}}(t_l - t_{l-1} + \mathrm{i}\hbar\beta/2 \right] \right\}. \tag{A1.16}$$

Here we have introduced the notation

$$a^{m_1'm_1}_{m_2'm_2} = \frac{1}{2} \{\cos \boldsymbol{q}(\boldsymbol{R}_{m_1'} - \boldsymbol{R}_{m_2}) + \cos \boldsymbol{q}(\boldsymbol{R}_{m_1} - \boldsymbol{R}_{m_2'})$$

$$- \cos \boldsymbol{q}(\boldsymbol{R}_{m_1'} - \boldsymbol{R}_{m_2'}) - \cos \boldsymbol{q}(\boldsymbol{R}_{m_1} - \boldsymbol{R}_{m_2})\}, \tag{A1.17}$$

$$S_{\text{T}}(m', m) = \frac{1}{2N} \sum_{\boldsymbol{q}} |\gamma(\boldsymbol{q})|^2 \{1 - \cos \boldsymbol{q}(\boldsymbol{R}_{m'} - \boldsymbol{R}_m)\} \coth (\hbar\omega_{\boldsymbol{q}}\beta/2). \tag{A1.18}$$

We may now formulate the phonon average in terms of the diagram language. According to (A1.16), for diagrams with l points (not counting the clamp), there are l factors of the type $\exp \{-S_{\text{T}}(m_i', m_i)\}$. It is reasonable to associate each such factor with an interaction point being the end (origin) of an incoming (outgoing) electron line $m_i(m_i')$. Furthermores according to (A1.16), there are $l(l-1)/2$ factors of the type

$$P^{m_1'm_1}_{m_2'm_2}(t) = \exp \left\{ \sum_{\boldsymbol{q}} \frac{|\gamma(\boldsymbol{q})|^2}{N \sinh (\hbar\omega_{\boldsymbol{q}}\beta/2)} a^{m_1'm_1}_{m_2'm_2} \cos \omega_{\boldsymbol{q}}(t + \mathrm{i}\hbar\beta/2) \right\}. \tag{A1.19}$$

To account for these factors, the l interaction points are pairwise connected with each other in all possible ways by wavy phonon lines. Each phonon line, connecting points i and k, is associated with a factor $P^{m_i'm_i}_{m_k'm_k}(t_k - t_i)$, where $m_k(m_k')$ is the index of the electron line running into (away from) the point k, and t_k is the time belonging to this point. The indices m_i, m_i', and t_i are the corresponding quantities for point i. On the contour C (cf. Fig. 2.1), the point k is placed ealier than the point i. The clamp is not connected with a phonon line, because it is diagonal with respect to the site indices.

According to (A1.1) and (A1.3), the diagram rules must be completed by associating each site i on the horizontal branches of C with a factor

$$\pm \frac{\mathrm{i}}{\hbar} J_{m_i'm_i}(\boldsymbol{H}) \exp \left\{ \frac{\mathrm{i}}{\hbar} \int_0^{t_i} \mathrm{d}\tau [\varepsilon_{m_i'} - \varepsilon_{m_i} - \mathrm{e}\boldsymbol{E}(\tau) (\boldsymbol{R}_{m_i'} - \boldsymbol{R}_{m_i})] \right\},$$

where the positive (negative) sign belongs to a point on the lower (upper) branch on the contour, and m_i (m_i') is the index of incoming (outgoing) electron line. Finally, it must be integrated over all times t_i, according to the arrangement of the points on the contour, and it must be summed up over all intermediate inducies m_i and m_i'. Recall that we have neglected the contribution of the polaron shift $E_{\text{p}}(m)$ (2.27), which is justified if E_{p} does not depend on the site index. In the case that this dependence must be accounted for, everywhere we have to replace ε_m by $\tilde{\varepsilon}_m = \varepsilon_m - E_{\text{p}}(m)$.

The major property of the above diagrams is the fact that (in the case of a static external electrical field) the diagrams diverge as powers of t, owing to $P(t) \to 1$ for $t \to \infty$ (in the presence of phonon dispersion). Evidently, in the absence of phonon dispersion, where for optical phonons $\omega_{\boldsymbol{q}} \equiv \omega_{\text{op}}$ (Einstein model) holds, the exponent in (A1.18) is an oscillating function of time, so that for $t \to \infty$, $P(t)$ does not go to a definite limiting value. The situation changes strongly if phonon dispersion is accounted for. Consider, for simplicity, a single optical phonon branch with the dispersion $\omega_{\boldsymbol{q}} = \omega_{\text{op}} + \Delta\omega\,(\cos q_x a + \cos q_y a + \cos q_z a)$, where a is the lattice constant, and where we have assumed a simple cubic lattice. For small dispersion, we obtain approximately

$$P(t) \approx \exp\left\{A\left(\frac{a}{2\pi}\right)^3 \int\limits_{-\pi/a}^{\pi/a} \mathrm{d}q_x\,\mathrm{d}q_y\,\mathrm{d}q_z \exp\{\mathrm{i}\omega_{\text{op}}t + \mathrm{i}\Delta\omega t\,(\cos q_x a + \cos q_y a + \cos q_z a)\right\} = \exp\{A(\cos\omega_{\text{op}}t)\,J_0^3(\Delta\omega t)\}. \quad \text{(A1.19a)}$$

Here $A \approx \overline{|\gamma|^2}\bar{a}/\sinh(\hbar\omega_{\text{op}}\beta/2)$ ($\overline{|\gamma|^2}$ and $\bar{a}$, respectively, denote $|\gamma(\boldsymbol{q})|^2$ and $a(\boldsymbol{q})$ suitably averaged over $\boldsymbol{q}$) and J_0 is the Bessel function. For large times, where $\Delta\omega t \gg 1$, according to the asymptotic behaviour of the Bessel function, the exponent in (A1.19a) oscillates and decreases as $t^{-3/2}$. Accordingly, for $t \to \infty$, $P(t)$ goes to unity. It turns out that for d dimensions, the exponent of $P(t)$ decreases as $t^{-d/2}$, independent of the concrete form of the phonon dispersion. This fact may be proved by estimating the integral over $\boldsymbol{q}$ in (A1.19) by making use of the saddle-point method for $\Delta\omega t \gg 1$.

The fact that for $t \to \infty$, $P(t)$ goes to unity, indicates the possibility of a power-like divergence with respect to t of integrals over intermediate times whose upper limit is t. In the theory of the current in the stationary regime, we are just interested in the limit $t \to \infty$. Therefore, we must sum up divergences with increasing power of t. Such summing-up also leads to the transport equation for the density matrix.

For illustration, consider the simple diagram with two interaction points shown in Fig. A1.1. According to the above diagram rules, the corresponding analytical expression reads

$$\sum_{m_1} J_{m_1 m}(\boldsymbol{H})\,J_{m m_1}(\boldsymbol{H})\,\mathrm{e}^{-2S_T(m_1,m)}\ \frac{\mathrm{i}}{\hbar}\left(-\frac{\mathrm{i}}{\hbar}\right)\int\limits_0^t \mathrm{d}t_1 \int\limits_0^t \mathrm{d}t_2\, f_{m_1}(1 - f_m)^2$$
$$\times \exp\left\{\frac{\mathrm{i}}{\hbar}\int\limits_{t_2}^{t_1} \mathrm{d}\tau[\varepsilon_{m_1} - \varepsilon_m - \mathrm{e}\boldsymbol{E}(\tau)\,(\boldsymbol{R}_{m_1} - \boldsymbol{R}_m)]\right\} P_{m m_1}^{m_1 m}(t_1 - t_2).$$

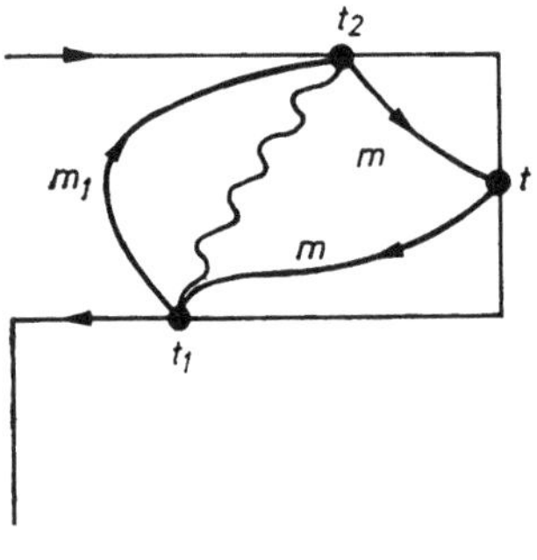

Fig. A1.1. Example of a second-order diagram for $\varrho_m(t)$

To separate clearly the divergences, we introduce the function

$$F^{m_1'm_1}_{m_2'm_2}(t) = P^{m_1'm_1}_{m_2'm_2}(t) - 1, \tag{A1.20}$$

which for $t \to \infty$ goes to zero. Writing $P(t) = \big(P(t) - 1\big) + 1$, then for $t \to \infty$, the term proportional to $F(t)$ does not increase faster than linear in t (such a linear increase occurs when, for some reason, the oscillating factor $\exp\left\{\frac{i}{\hbar}\int\limits_{t_2}^{t_1} d\tau\{\varepsilon_{m_1} - \varepsilon_m - e\boldsymbol{E}(\tau)\,(\boldsymbol{R}_{m_1} - \boldsymbol{R}_{m_2})\right\}$ is equal to unity). On the same conditions, the term containing the unity is proportional to t^2 and therefore leads to an infinitely large contribution to the current, which is proportional to $d\varrho_m(t)/dt$ (cf. (2.3.3)). With increasing complexity of the diagrams and, correspondingly, with increasing number of integrations over intermediate times, in the diverging diagrams the power of t increases. Note that the presence or absence of divergences is independent of the strength of the field. Among the diagrams are graphs for which the oscillating factors go automatically to unity, i.e. for which it is not necessary to require that the field must be weak and that certain energies ε_m must be equal to each other. In these diagrams, the indices m_1 and m, occurring in the oscillating factor, turn out to be equal to each other. An example of such a diagram is represented in Fig. A1.2, which shows a fourth order diagram. As can readily be seen, the contribution of this diagram to $\varrho_m(t)$ is proportional to t^2 if the intermediate indices m_1 and m_3, over which is summed up, are equal to each other. The general procedure of finding diverging contributions consists in introducing vertical "sections" (lines) intersecting $2l$

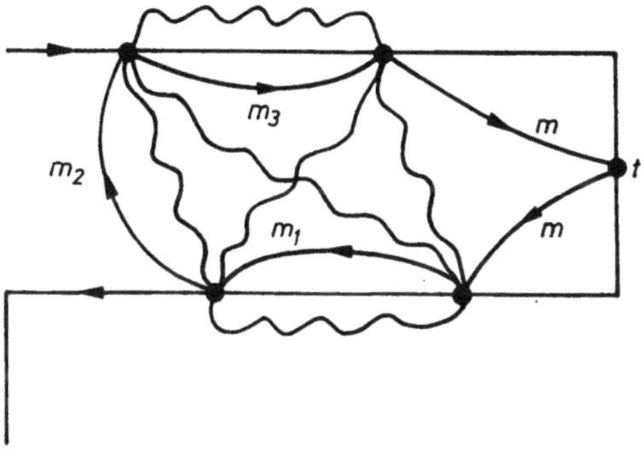

Fig. A1.2. A fourth-order diagram for $\varrho_m(t)$, which contains a divergence with respect to t if $m_1 = m_3$

electron lines where l electron lines are directed to the left and the other l lines to the right. If the site indices of the electron lines, running from the right to the left and from the left to the right, do all pairwise coincide, than the diagram contains a divergence connected with this section.

These divergences are the basic characteristic feature of transport theories basing on perturbation expansions. For interaction of arbitrary strength, the presence of such divergences requires summing-up of an infinite series of diverging diagrams, which in the final analysis leads to the transport equation. Only in the case of sufficiently high frequencies of the external electrical field, when $W/\omega \ll 1$ (W — probability for an electron transition; cf. Section 2.5), one may restrict oneself to a finite number of diagrams. For broad electron bands and weak scattering, the transport equation is the familiar Boltzmann equation, while hopping transport is governed by the rate equation for the electron density, obtained in Section 2.4 on the basis of diagram technique presented here.

The time divergences are most conveniently summed up by separating the diverging contribution by means of (A 1.20), i.e. by writing $P(t) = \big(P(t) - 1\big) + 1$. This procedure leads to the following modification of the diagram technique: now, instead of with a factor P, each phonon line must be associated with a factor F (A 1.20). In addition to these diagrams, other diagrams occur, in which phonon lines are lacking in all possible combinations. For instance, in addition to the diagram shown in Fig. A 1.1, another diagram occurs in which the phonon line is lacking. Furthermore, in addition to the fourth-order diagram shown in Fig. A 1.2, 63 modified diagrams occur: one, where all phonon lines are lacking, six having only a single phonon line, 15 with two, 12 with three, 15 with four and 6 with five such lines.

The next long step towards the summing-up of diverging diagrams is to introduce the concept of "free sections". As mentioned above, a diverging contribution may be found by means of vertical sections (dashed-dotted line in Fig. A 1.2). After introducing modified diagrams, in some diagrams, sections may be drawn which do not intersect phonon lines. A vertical section which has this property is a so-called free section. The classification of the diagrams with respect to free sections, requires further modernization of the diagram technique. So far, diagrams with different mutual arrangements of points on the upper and lower horizontal branches of the contour C have been viewed as being topologically equivalent, and for the points on different branches the integration has been performed independently. For instance, in the diagram shown in Fig. A 1.1 the point t_1 may be placed on the left of t_2 or, equivalently, on the right of t_2. The introduction of free sections requires fixation of the mutual arrangement of the points, since diagrams which differ only by the mutual arrangement of the points on the horizontal branches have different free sections. In what follows, diagrams with different mutual arrangements of the points on the two horizontal branches are considered as non-equivalent diagrams. Accordingly, for instance, we must add to the diagram shown in Fig. A 1.1 a second diagram where the point t_1 is placed on the right of the point t_2. Thereby, in the diagram in Fig. A 1.1, the integration over t_1 is from 0 to t_2, and over t_2 from 0 to t, while in the second diagram the integration over t_1 is from 0 to t, and over t_2 from 0 to t_1. Analogously, the diagram shown in Fig. A 1.2 must be replaced by six diagrams with all possible mutual arrangements of two points on the lower branch and two points on the upper branch, and the integration over the times t_1, t_2, t_3, and t_4 must be performed acording to the arrangement of the points. One of the modified diagrams originating from the diagram in Fig. A 1.2 is shown in Fig. A 1.3. The corresponding analytical expression reads

$$\sum_{m_1 m_2 m_3} \left(\frac{\mathrm{i}}{\hbar}\right)^2 \left(-\frac{\mathrm{i}}{\hbar}\right)^2 \tilde{J}_{m m_3}(\boldsymbol{H})\, \tilde{J}_{m_3 m_2}(\boldsymbol{H})\, \tilde{J}_{m_2 m_1}(\boldsymbol{H})\, \tilde{J}_{m_1 m}(\boldsymbol{H})$$

$$\times\, (1 - f_m)^2\, (1 - f_{m_1})\, f_{m_2} (1 - f_{m_3}) \int_0^t \mathrm{d}t_1 \int_0^{t_1} \mathrm{d}t_4 \int_0^{t_4} \mathrm{d}t_2 \int_0^{t_2} \mathrm{d}t_3 F^{m_3 m_2}_{m_2 m_1}(t_2 - t_3)$$

$$\times\, F^{m m_3}_{m_1 m}(t_1 - t_4) \exp \left\{ \frac{\mathrm{i}}{\hbar} \int_0^{t_1} \mathrm{d}\tau\, V_{m_1 m}(\tau) + \frac{\mathrm{i}}{\hbar} \int_0^{t_2} \mathrm{d}\tau\, V_{m_2 m_1}(\tau) \right.$$

$$\left. + \frac{\mathrm{i}}{\hbar} \int_0^{t_3} \mathrm{d}\tau\, V_{m_3 m_2}(\tau) + \frac{\mathrm{i}}{\hbar} \int_0^{t_4} \mathrm{d}\tau\, V_{m m_3}(\tau) \right\}.$$

Here we have used the notations

$$V_{m'm}(\tau) = V_{m'}(\tau) - V_m(\tau);\; V_m(\tau) = \varepsilon_m - e\boldsymbol{E}(\tau)\, \boldsymbol{R}_m, \tag{A 1.21}$$

$$\tilde{J}_{m'm}(\boldsymbol{H}) = J_{m'm}(\boldsymbol{H}) \exp\big(-S_{\mathrm{T}}(m', m)\big), \tag{A 1.22}$$

where $\tilde{J}$ has the meaning of the resonance integral renormalized by electron-phonon interaction.

Let us now point out that the above analytical expression for the diagram in Fig. A1.3 may be written as

$$\int_0^t dt_1 \int_0^{t_1} dt_4 \sum_{m_1 m_3} A_{m_1}^{m_3}(t_4)\, B_{m_1 m}^{m_3 m}(t_4, t_1), \tag{A1.23}$$

where the quantity $A_{m_1}^{m_3}(t_4)$ describes the part of the diagram placed on the left of the free section, and $B_{m_1 m}^{m_3 m}(t_4, t_1)$ the part of the diagram placed on the right of it.

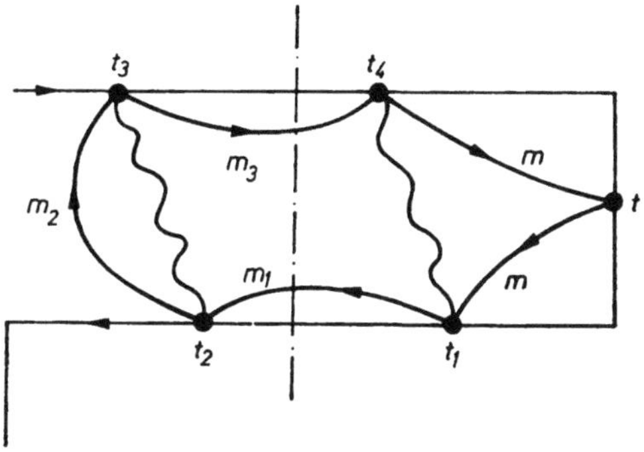

Fig. A1.3. One of the modified fourth-order diagrams for $\varrho_m(t)$, obtained from the diagram shown in Fig. A1.2 by neglection of four phonon lines. The occurring free section is indicated by a vertical dashed-dotted line

On comparing (A1.23) with the above analytical expression, we find that

$$A_{m_1}^{m_3}(t_4) = \sum_{m_2} \left(\frac{i}{\hbar}\right)\left(-\frac{i}{\hbar}\right) \tilde{J}_{m_2 m_1}(\boldsymbol{H})\, \tilde{J}_{m_3 m_2}(\boldsymbol{H})\, f_{m_2}(1 - f_{m_3})\,(1 - f_{m_1})$$

$$\times \int_0^{t_4} dt_2 \int_0^{t_2} dt_3 F_{m_2 m_1}^{m_2 m_3}(t_2 - t_3) \exp\left\{\frac{i}{\hbar}\int_0^{t_2} d\tau\, V_{m_2 m}(\tau) + \frac{i}{\hbar}\int_0^{t_3} d\tau\, V_{m_3 m_2}(\tau)\right\},$$

$$B_{m_1 m}^{m_3 m}(t_4, t_1) = \frac{i}{\hbar}\left(-\frac{i}{\hbar}\right) \tilde{J}_{m m_3}(\boldsymbol{H})\, \tilde{J}_{m_1 m}(\boldsymbol{H})\,(1 - f_m)^2$$

$$\times F_{m_1 m_3}^{m m_3}(t_1 - t_4) \exp\left\{\frac{i}{\hbar}\int_0^{t_1} d\tau\, V_{m_1 m}(\tau) + \frac{i}{\hbar}\int_0^{t_4} d\tau\, V_{m m_3}(\tau)\right\}.$$

Now it can readily be seen that A describes the part of the diagram in Fig. A1.3 which is placed on the left of the free section. Moreover, the quantity $A_{m_1}^{m_3}(t_3)$ represents the contribution to the expansion of the quantity $\varrho_{m_1}^{m_3}(t)$ in a series with respect to $\mathcal{H}'$, arising from the diagram with two points, one point on the upper branch and another point on the lower one.

Here the quantity $\varrho^m_{m'}$ is defined as[1])

$$\varrho^m_{m'}(t) = \langle U^+(t)\, a^+_{m'}\, a_m U(t)\rangle. \tag{A 1.24}$$

The quantity $B^{m_1 m}_{m_1 m}(t_4, t_1)$ is given by the part of the diagram on the right of the free section and does not include the time integration over the utmost left and right interaction points of this part of the diagram.

Hence, if a diagram is cut by a free section, then it may be represented by a formula of the type (A 1.23). Just this property allows diagrams with respect to powers of the free sections to be exactly summed up. This summing-up is the analog of the ladder summation in the Feynman technique, leading to the Dyson equation for the Green function. The main difficulty of the case in question is the possibility that a free section does not intersect only two

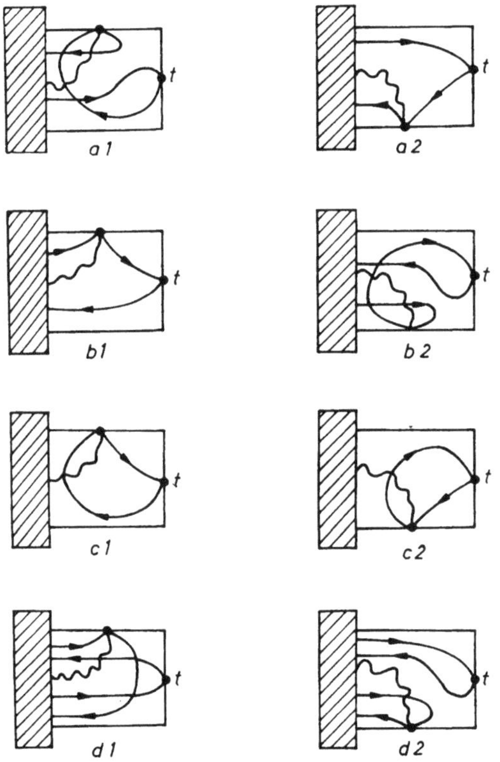

Fig. A 1.4. Diagrams for $\varrho_m(t)$, which pairwise differ by transfering the utmost right interaction point from the upper to the lower horizontal branch of the contour. The shaded blocks represent the part of the diagram without the right point. The wavy line represents here the total set of phonon lines connecting the right point with the remainder of the diagram

[1]) Here we introduce the nondiagonal elements of the density matrix after having subjected the diagonal elements to the polaron canonical transformation. Therefore, for the nondiagonal elements, the clamp is not connected with a phonon operator, i.e. the clamp is free from phonon lines. On the other hand, if one introduces the nondiagonal elements before applying the polaron canonical transformation, then after subjecting (A 1.24) to this transformation, an additional factor $\hat{\Phi}_{m'm}$ occurs, so that for the nondiagonal elements, the clamp is not free from phonon lines. Since the current contains only diagonal elements of ϱ, and the nondiagonal elements are only auxiliary quantities, we have the freedom of the choice of the definition of $\varrho^m_{m'}(t)$ with $m \neq m'$. In what follows we use the definition (A 1.24).

electron lines, but also four, six etc. To perform the summation on account of any number of electron lines, let us draw attention to an important property of the diagrams introduced. Consider the diagrams represented in Fig. A1.4, which pairwise differ only by a transfer of the utmost right interaction point from the upper branch of the contour to the lower one. Thereby, this interaction point is either connected by an electron line with the clamp (diagrams A1.4a, b, c) or it is not connected (diagrams A1.4a). The sum of the two diagrams A1.4c, and also of the two diagrams A1.4d, vanishes, i.e. diagrams of this type may be ignored. Indeed, the diagrams A1.4c 1 and A1.4c 2, and analogously the diagrams A1.4d 1 and A1.4d 2, only differ by sign, due to the transfer of the point, while the "rightness" of the electron lines and the mutual arrangement of the interaction points on the contour (from which depends the argument of the function $P(t)$ is not affected. The number of additional intersections of electrons lines is even. Therefore, such diagrams pairwise cancel. Note that the diagram A1.4c itself gives no contribution, because it contains a factor J_{mm} which is equal to zero.

Consider now the diagrams A1.4a 1 and A1.4a 2. The transfer of the utmost right interaction point leads to a factor (-1) due to one additional intersection of the electron lines, and to another factor (-1) due to the transfer of the point. Furthermore, such a transfer alters the rightness of the electron line connecting the clamp with the point quoted, i.e. f_m in the diagram A1.4a 1 changes into $1 - f_m$ in the diagram A1.4a 2. Hence, it follows that the sum of the two diagrams is equal to the contribution of the diagram A1.4a 2 without account of the factor $1 - f_m$ (which corresponds to the electron line between the clamp and the interaction point in question), because $(1 - f_m) + (-1)^2 f_m = 1$. For the diagrams A1.4b, we obtain an analogous result. From this follows the rule that one may ignore diagrams which differ from each other only by the transfer of the utmost right interaction point from one branch on the contour to other one, without changing the rightness of electron lines. A non-zero contribution may arise from diagrams whose utmost right point is connected with the clamp, whereby the corresponding line must be associated with a factor unity. In what follows, such a line will be represented by a dotted line.

It is more convenient to perform the summing-up with respect to powers of free sections from the very beginning for the whole one-particle density matrix $\varrho^m_{m'}(t)$, defined in (A1.24), whereby $\varrho^m_{m'} \equiv \varrho_m$. The introduction of $\varrho^m_{m'}$ with $m' \neq m$ is necessary because the equation for $\varrho_m(t)$ also involves nondiagonal components of the density matrix. However, before studying this problem, let us now compile the diagram rules on account of all modifications introduced above.

Construction of the diagrams: The diagrams are constructed on the contour C represented in Fig. 2.1. The contour consists of two horizontal branches in the plane of the complex time τ (from $\tau = 0$ to $\tau = t$), which are separated by a cut along the real axis, and also of a vertical branch along the imaginary axis from 0 to $-i\hbar\beta$. On the upper horizontal branch of the contour, the evolution operator $U(t)$ is expanded with respect to $\mathcal{H}'$, on the lower branch — $U^+(t)$, and on the vertical branch — the operator $\exp\{-\beta\mathcal{H}^{(0)}\}$, which is involved in the definition of the thermodynamical average (2.3.5). If we assume that at the beginning the system may be described by a Hamiltonian without interaction (i.e. in $\mathcal{H}^{(0)}$ we ignore $\mathcal{H}'$), then the vertical branch of the contour may be neglected. Such assumption is admissible if we are not interested in the transition regime occurring after suddenly turning on the electrical field at $t = 0$. The countour C has the direction indicated in Fig. 2.1 by an arrow. On studying the density matrix $\varrho^m_{m'}(t)$, the clamp $a^+_{m'}$, a_m is placed at the turning-point of the contour.

Any nth order diagram contains n interaction points, arranged on the contour. All points, including the clamp, are connected by electron lines having a direction, which is indicated by an arrow. In each point, one electron line terminates and another one starts. Any incoming electron line must be directed along the contour. Hence, all electron lines form ring-shaped closed chains. Furthermore, all interaction points are connected with each other in all possible ways by wavy phonon lines. In a nth order diagram, the number of such phonon lines is equal to $n(n-1)/2$. The clamp is not connected with phonon lines. In addition to each diagram where all interaction points are connected by phonon lines, there exist a set of diagrams in which one, two, three etc. phonon lines are lacking (up to diagrams which have no phonon lines, i.e. which only consist of "empty" interaction points). Diagrams, which differ by the mutual arrangement of the points on the two horizontal branches, are viewed as topologically inequivalent. Each electron line is labeled by a site index m_i. Accordingly, each interaction point is associated with two site indices: one arising from the incoming line, a second from the outgoing line, whereby the interaction point by all means changes the site index.

Rules for obtaining analytical expressions:

1. If the direction of an electron line agrees with the direction of the contour, then this line is a so-called right line, which is associated with a factor $1-f_m$; otherwise the line is a so-called wrong line associated with a factor f_m. The diagram is multiplied by a factor $(-1)^M$, where M is the number of intersection of electron lines with each other. The utmost right interaction point is connected with the clamp by a right line, which is dotted and which is associated with a factor unity.

2. Each interaction point is associated with a factor

$$\pm \frac{\mathrm{i}}{\hbar} \tilde{J}_{m_i'm_i} \exp\left\{\frac{\mathrm{i}}{\hbar}\int\limits_0^{t_i} \mathrm{d}\tau\, V_{m_i'm_i}(\tau)\right\},$$

where m_i' and m_i, respectively, are the indices of the incoming and outgoing electron lines connected with this point, and t_i is the time to which this point belongs. The renormalized resonance integral $\tilde{J}$ is defined in (A1.22), and V in (A1.21). The sign + refers to points on the lower horizontal branch and on the vertical branch, while the sign − to points on the upper horizontal branch.

3. Each phonon line between sites i and k, belonging, respectively, to time t_i and t_k, is associated with a factor $F^{m_i'm_i}_{m_k'm_k}(t_k-t_i)$ defined in (A1.20) and (A1.19). The indices m_k and m_k', respectively, label the incoming and outgoing electron lines connected with site k, and m_i and m_i' label the corresponding lines connected with site i. On the contour, site k is placed earlier than site i.

4. It must be summed-up over the intermediate site indices m_i' and m_i, and integrated over the intermediate times t_i. The limits of the integration over t_i are determined by the position of the corresponding point on the contour and the mutual arrangement of the points on the two horizontal branches.

5. For points on the vertical branch of the contour, we must set $E=0$, i.e. $V_{m'm}(\tau) \to \varepsilon_{m'} - \varepsilon_m$, and the integration over intermediate points must be performed in the limits from 0 to $-\mathrm{i}\hbar\beta$, in accordance with the arrangement of the points.

6. In the case of a nondiagonal clamp $a_{m'}^+, a_m$, the clamp is associated with a factor $\exp\left\{\frac{\mathrm{i}}{\hbar}\int\limits_0^{t}\mathrm{d}\tau\, V_{m'm}(\tau)\right\}$, which is a consequence of the fact that the evolution operator of a free particle $U_0(t)$ (2.4.2) does not commute with $a_{m'}^+$, a_m for $m \neq m'$.

Consider now the diagram series for the density matrix $\varrho^m_{m'}(t)$. The lowest-(zero-) order diagram does not possess an interaction point. It has only a clamp with an electron line starting and ending at it. For this diagram, we find that

$$\overset{(0)}{\varrho}{}^{m}_{m'} = \delta_{m'm} f_m \,. \tag{A1.25}$$

More complicated diagrams possess interaction points, which are arranged on the horizontal branches as well as on the vertical branch of the contour. Note that according to (A1.24), the set of diagrams, having points only on the vertical branch, yields

$$F^{m}_{m'} \exp\left\{\frac{\mathrm{i}}{\hbar}\int\limits_0^t \mathrm{d}\tau\, V_{m'm}(\tau)\right\},$$

where $F^{m}_{m'}$ denotes the thermodynamic equilibrium-density matrix, for which the following closed form can be given

$$F^{m}_{m'} = \mathrm{Tr}\{\mathrm{e}^{-\beta(\mathcal{H}^{(0)}-\mu\hat{N})} a^{+}_{m'} a_m\} / \mathrm{Tr}\{\mathrm{e}^{-\beta(\mathcal{H}^{(0)}-\mu\hat{N})}\} \,. \tag{A1.26}$$

In the zeroth approximation with respect to the interaction, it holds that $F^{m}_{m'} = f_m \delta_{m'm}$. Since on the vertical branch the limits of integration are finite (0 and $-i\hbar\beta$), such integration does not give rise to a divergence. Therefore, for sufficiently weak interaction and for studying the stationary current regime, diagrams having points on the vertical branch may be ignored. In what follows we do so.

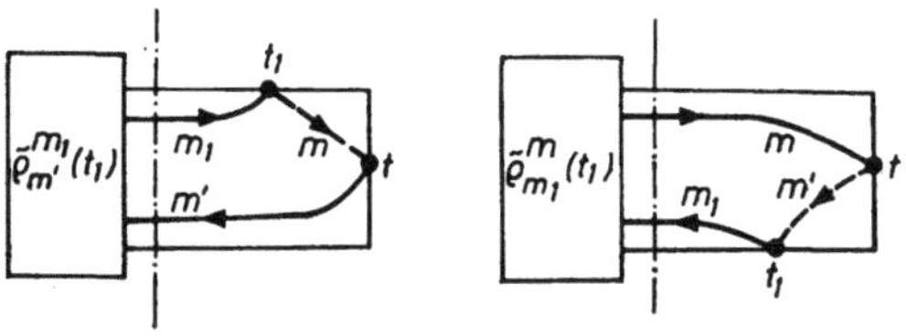

Fig. A1.5. Diagram representation of the contribution to $\varrho^{m}_{m'}(t)$ from irreducible first-order bocks. The function $\tilde{\varrho}^{m}_{m'}(t)$ agrees with $\varrho^{m}_{m'}(t)$ up to the contribution of the clamp, i.e. $\varrho^{m}_{m'}(t) = \tilde{\varrho}^{m}_{m'}(t) \exp\left\{\frac{\mathrm{i}}{\hbar}\int\limits_0^t dt\, V_{m'm}(\tau)\right\}$. The free section is indicated by a vertical dashed-dotted line

Let us now study diagrams having points only on the horizontal branches of the contour Evidently, in the presence of at least one interaction point, a diagram has at least two free sections: one between the utmost right interaction point and the clamp and another one on the left of the utmost left interaction point. The former free section may be shown to give no divergence, while the latter one, which cuts only a single wrong line, as well as all other free sections, placed between the interaction points, are associated with a divergence. Let us now take some diagram, cut it by the utmost free section (not counting the free section at the clamp) into two parts, and use the result (A1.23), where, by definition, the block B does not contain free sections, since B is enclosed between the utmost right free section and the clamp. However, unlike the concrete example on which (A1.23) is based, here the free section may intersect more than two electron lines, which requires that we generalize the above result. To do this, consider diagrams with a concrete form of the blocks B, which will be termed irreducible. In the lowest order with respect to the interaction, the block B contains a single point. The corresponding diagram is shown in Fig. A1.5. According to the diagramm technique described above, the diagram shown in Fig. A1.5 yields the following contribution to $\varrho^m_{m'}(t)$:

$$-\sum_{m_1} \frac{i}{\hbar} \int_0^t dt_1 \exp\left\{\frac{i}{\hbar}\int_0^t d\tau\, V_{m'm}(\tau)\right\} \left[\tilde{J}_{mm'}(\boldsymbol{H}) \exp\left\{\frac{i}{\hbar}\int_0^{t_1} d\tau\, V_{mm_1}(\tau)\right\} \times A^{m_1}_{m'}(t_1) - \tilde{J}_{m_1m'}(\boldsymbol{H}) \exp\left\{\frac{i}{\hbar}\int_0^{t_1} d\tau\, V_{m_1m'}(\tau)\right\} A^m_{m_1}(t_1)\right]. \tag{A1.27}$$

Here A denotes the block placed on the left of the free section (including the factor $(1 - f_{m_1}) \times (1 - f_m)$ which corresponds to the electron lines running into the block A). Note that the set of diagrams describing $A^m_{m'}(t)$ agrees with the set of diagrams for $\varrho^m_{m'}(t)$, up to a factor corresponding to the clamp, i.e. it holds that

$$A^m_{m'}(t) = \varrho^m_{m'}(t) \exp\left\{-\frac{i}{\hbar}\int_0^t d\tau\, V_{m'm}(\tau)\right\}.$$

On account of this relation, expression (A1.27) may be rewritten as

$$\sum_{m_1m_2} \int_0^t dt_1 \int_0^{t_1} dt_2\, \varrho^{m_1}_{m_2}(t_2) \overset{(1)}{\mathscr{W}}{}^{m_1m}_{m_2m'}(t_2, t_1) \exp\left\{\frac{i}{\hbar}\int_{t_1}^t d\tau\, V_{m'm}(\tau)\right\}. \tag{A1.28}$$

Here $\overset{(1)}{\mathscr{W}}$ is the first order irreducible block describing the part of the diagram placed on the right of the free section. Note that irreducible blocks do not contain integrations over the times of the utmost left and right interaction points. Furthermore, for these blocks, the factor $\exp\left\{\frac{i}{\hbar}\int_0^{t_i} d\tau\, V_{m_i'm_i}(\tau)\right\}$ associated with a point (according to the second diagram rule) must be replaced by $\exp\left\{\frac{i}{\hbar}\int_{t_2}^{t_i} d\tau\, V_{m_i'm_i}(\tau)\right\}$, where t_2 is the time of the utmost left point of this block. Owing to this transformation, the utmost left point does not contain such a factor. This rule applies to irreducible blocks of any order. For the first-order block, the analytical expression reads

$$\overset{(1)}{\mathscr{W}}{}^{m_1m}_{m_2m'}(t_2, t_1) = \frac{i}{\hbar} \{\tilde{J}_{m_1m'}(\boldsymbol{H})\, \delta_{mm_2} - \tilde{J}_{mm_1}(\boldsymbol{H})\, \delta_{m'm_2}\}\, \delta(t_1 - t_2). \tag{A1.29}$$

The presence of $\delta(t_1 - t_2)$ in (A1.29) is due to the fact that this block contains only a single point.

Let us now consider diagrams with irreducible second-order blocks (cf. Fig. A1.6). The contribution of these diagrams to $\varrho^m_{m'}(t)$ is given by

$$\sum_{\{m_i\}} \int_0^t dt_1 \int_0^{t_1} dt_2 \varrho^{m_1m_3}_{m_2m_4}(t_2) \overset{(2)}{\mathscr{W}}{}^{m_1m_3m}_{m_2m_4m'}(t_2, t_1) \exp\left\{\frac{i}{\hbar}\int_{t_1}^t d\tau\, V_{m'm}(\tau)\right\}. \tag{A1.30}$$

The second-order blocks $\overset{(2)}{\mathscr{W}}$ introduced here are described by the parts of the diagrams in Fig. A1.6 placed on the right of the free section (in this figure the irreducible blocks are indicated by a dotted line). The rule for writing down the indices of $\overset{(2)}{\mathscr{W}}$ may be taken from Fig. A1.6: the upper (lower) indices belong to electron lines running from the left to the right (from the right to the left). Since one of the electron lines always runs through the block by avoiding the points, in all diagrams the summation is not over the four indices m_1, m_2, m_3 and m_4, but only over three of them.

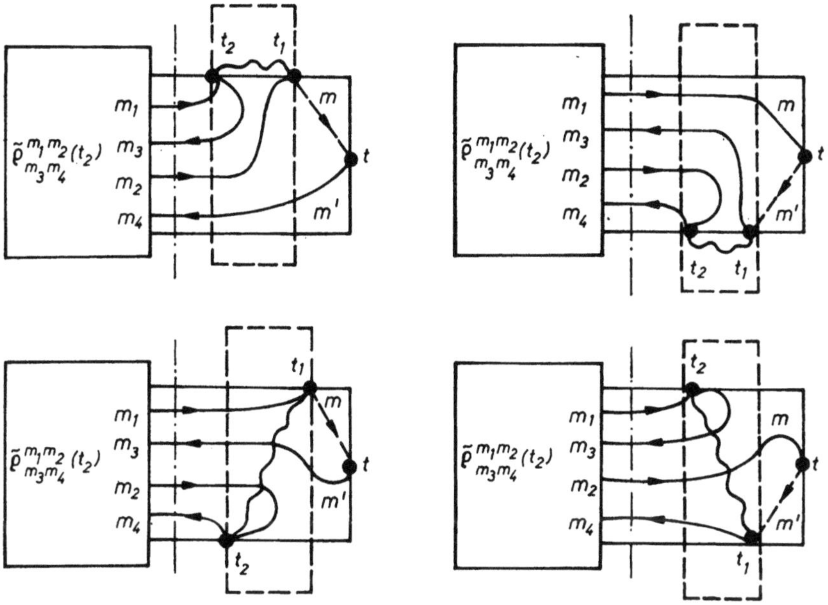

Fig. A1.6. Diagram representation of the contribution to $\varrho^{m}_{m'}(t)$ from irreducible second-order blocks. The irreducible blocks $\overset{(2)}{\mathscr{W}}{}^{m_1m_2m}_{m_3m_4m'}(t_2, t_1)$ are indicated by dashed rectangles. The function $\tilde{\varrho}^{m_1m_2}_{m_3m_4}(t)$ agrees with the two-particle density matrix up to the contribution of the clamp:

$$\varrho^{m_1m_2}_{m_3m_4}(t) = \tilde{\varrho}^{m_1m_2}_{m_3m_4}(t) \exp\left\{\frac{\mathrm{i}}{\hbar} \int_0^t \mathrm{d}\tau [V_{m_3m_1}(\tau) + V_{m_4m_2}(\tau)]\right\}$$

The block $\varrho^{m_1m_2}_{m_3m_4}(t)$ represents the two-particle density matrix defined as

$$\varrho^{m_1m_2}_{m_3m_4}(t) = \langle U^+(t)\, a^+_{m_4} a_{m_2} a^+_{m_3} a_{m_1}\, U(t)\rangle . \tag{A1.31}$$

In the language of the diagram technique, this block is an object on the contour C with the clamp $a^+_{m_4} a_{m_2} a^+_{m_3} a_{m_1}$.

Hence, on account of second-order irreducible diagrams, the one-particle density matrix turns out to be connected with the two-particle density matrix. The physical origin of this fact was discussed in Section 2.4.

On account of third-order irreducible blocks, the one-particle density matrix becomes connected with the three-particle density matrix, etc. This procedure leads to the following expression for the density matrix $\varrho_m^{m'}(t)$:

$$\varrho_{m'}^{m}(t) = \int\limits_0^t \mathrm{d}t_1 \int\limits_0^{t_1} \mathrm{d}t_2 \exp\left\{\frac{\mathrm{i}}{\hbar}\int\limits_{t_1}^{t} \mathrm{d}\tau\, V_{m'm}(\tau)\right\}$$
$$\times \sum_{\{m_i,m_i'\}} \left\{\varrho_{m_1'}^{m_1}(t_2) \overset{(1)}{\mathscr{W}}{}_{m_1'm'}^{m_1m}(t_2, t_1) + \varrho_{m_1'm_2'}^{m_1m_2}(t_2) \overset{(2)}{\mathscr{W}}{}_{m_1'm_2'm'}^{m_1m_2m}(t_2, t_1)\right.$$
$$\left. + \varrho_{m_1'm_2'm_3'}^{m_1m_2m_3}(t_2) \overset{(3)}{\mathscr{W}}{}_{m_1'm_2'm'}^{m_1m_2m}(t_2, t_1) + \cdots\right\}. \qquad \text{(A1.32)}$$

The transport equation in site-representation derived in Section 2.4. is based on the expression for the one-particle density matrix (A1.32).

After obtaining (A1.32), by means of the diagram technique, the expressions for the irreducible blocks $\overset{(n)}{\mathscr{W}}(t', t)$ must be found. Let us summarize the corresponding rules. The blocks in question are represented on the horizontal branches of the contour C in the time interval from t' (time of the utmost left interaction point) to t (time of the utmost right point). The block $\overset{(n)}{\mathscr{W}}$ is given by the set of diagrams with n points which are arbitrarily aranged on the two branches of the contour. All points are connected by phonon lines, according to the rules given above. The procedure of associating phonon lines with analytical expressions is governed by the item 3 of these rules. The most important property of the diagrams for the blocks, is the absence of free sections. The point i is associated with a factor $\pm\frac{\mathrm{i}}{\hbar} J_{m_i'm_i}(\boldsymbol{H}) \times \exp\left\{\frac{\mathrm{i}}{\hbar}\int\limits_{t'}^{t_i} \mathrm{d}\tau\, V_{m_i'm_i}(\tau)\right\}$ which, with respect to the lower limit of integration, differs from the factor indicated by the item 2 of the diagram rules. Note that unlike the diagram rules for the density matrix, for blocks $\mathscr{W}$ the initial moment of time is t', but not $t = 0$. All electron lines connected by one end with a point placed within a block run out of (or into) the block to (from) the left, except the electron line running out of the utmost right point (when it is placed on the upper branch of the contour) or incoming into it (when it is placed on the lower branch). Furthermore, one electron line passes through the block by circumventing the points. The latter two types of lines are connected with the clamp and, accordingly, labeled by indices m (upper) and m' (lower index). The electron lines leave (run into) the block in such a manner that no intersections occur between them. It is not required to give analytical expressions for the electron lines, because these lines determine only the external indices of the block $\mathscr{W}$: the indices of the lines running into the block from the left are written as upper indices whose succession (from the left to the right) agrees with the succession of the lines in the diagram (from the uppermost line downward). For such labeling, the expression for the many-particle density matrix is given by (2.4.4). Furthermore, the diagram of a block must be multiplied by $\exp\left\{\frac{\mathrm{i}}{\hbar}\int\limits_{t'}^{t} \mathrm{d}\tau\, V_{m'm}(\tau)\right\}$, analogous to the item 6 of the above diagram rules. Finally, integration must be performed over all intermediate times (i.e. the times of all points except the utmost right and left ones) according to the arrangement of the corresponding points on the segment of the contour between t' and t.

For illustration, Fig. A1.7 shows a diagram of the third-order block $\overset{(3)}{\mathscr{W}}{}_{m_1'm_2'm_3'm'}^{m_1m_2m_3m}(t', t)$, which is associated with the following analytical expression:

$$\left(-\frac{\mathrm{i}}{\hbar}\right)^2 \frac{\mathrm{i}}{\hbar} \tilde{J}_{m_1'm_1}(\boldsymbol{H})\, \tilde{J}_{mm_2}(\boldsymbol{H})\, \tilde{J}_{m_3'm_3}(\boldsymbol{H}) \int\limits_{t'}^{t} \mathrm{d}t_1 \exp\left\{\frac{\mathrm{i}}{\hbar} \int\limits_{t'}^{t_1} \mathrm{d}\tau\, V_{m_1'm_1}(\tau)\right.$$

$$\left. + \frac{\mathrm{i}}{\hbar} \int\limits_{t'}^{t} \mathrm{d}\tau \left(V_{mm_2}(\tau) + V_{m'm}(\tau)\right)\right\} F^{m_3'm_3}_{m_1'm_1}(t_1 - t')\, F^{mm_2}_{m_1'm_1}(t_1 - t)\, F^{m_3'm_3}_{mm_2}(t - t')\,. \tag{A1.33}$$

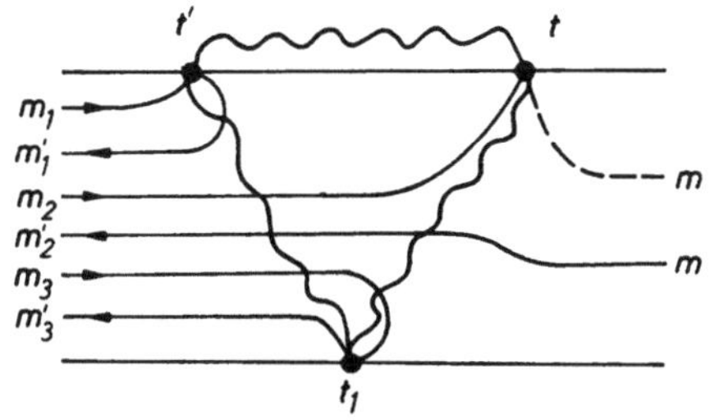

Fig. A1.7. One of the diagrams describing the irreducible third-order block $\overset{(3)}{\mathscr{W}}{}^{m_1m_2m_3m}_{m_1'm_2'm_3'm'}(t', t)$

Note that the diagram technique presented here is much more complicated than that generally used in the small-polaron theory (Firsov, 1975). Above all, this is due to the fact that the small-polaron theory is in general based on the one-electron approximation. In this approximation, all electron lines, except a single one, are directed along the contour (i.e. the are right lines), so that each free section intersects only two electron lines. This limiting case was studied in Section 2.4. It allows all many-electron correlators to be excluded from the theory, and the conception of a single irreducible block $\mathscr{W}^{m_1m}_{m_2m'}$ (cf. (2.4.7)) to be introduced.

Appendix 2. Rate equation in the presence of a magnetic field

In the presence of a magnetic field, the rate equation (2.4.23) must be generalized, because the two-site hopping probabilities $W_{m'm}$ are proportional to the square modulus of the resonance integral $|J_{m'm}(\boldsymbol{H})|^2$, which is independent of $\boldsymbol{H}$. Accordingly, at least three-site probabilities $\overset{(3)}{\mathscr{W}}$ must be accounted for in (2.4.3). Three-site processes can be most easily studied in the one-electron approximation by means of (2.4.6), on account of three-site blocks in (2.4.7).

In what follows, we derive the rate equation on account of three-site hops by means of the method used in Section 2.4 to obtain the transport equation (2.4.23). This method includes a Hartree-Fock decoupling of many-particle distribution functions, i.e. the resulting rate equation does not only apply to the case of Boltzmann statistics. To derive this equation, let us consider the third-order diagrams $\overset{(3)}{\mathscr{W}}$ contibuting to $\varrho^{m}_{m'}$ (A1.32). These diagrams are shown in Fig. A2.1. They are constructed in the same way as the diagrams in Fig. A1.6, which represent the contribution of $\overset{(2)}{\mathscr{W}}$ to $\dot{\varrho}^{m}_{m}$. Since in the following only diagonal elements of the density matrix are accounted for, in Fig. A2.1 the clamp $a^+_m a_m$ appears.

For the three-particle distribution functions $\varrho^{m_1m_2m_3}_{m_1'm_2'm_3'}$ (cf. (2.4.4) we use the Hartree-Fock decoupling, whereby we retain only one-particle density matrices being diagonal with respect

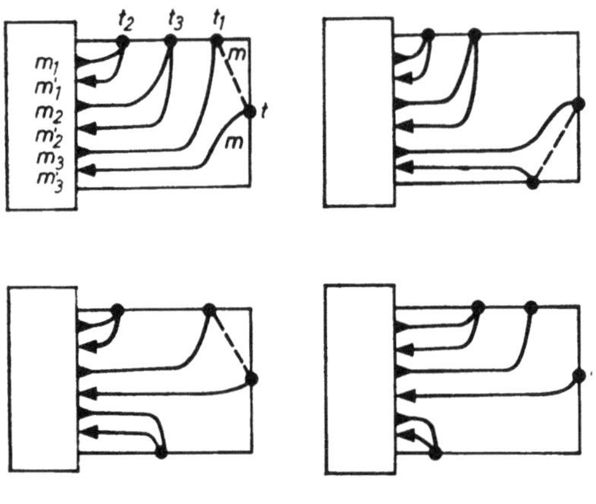

Fig. A2.1. Diagram representation of the contribution to $\varrho^{m}_{m}(t)$ (A1.32) from irreducible third-order blocks. In addition to the four diagrams represented here. there exist another four diagrams, which differ from the above ones by transfering the points from the above branch of the contour to the lower one and vice versa (complex conjugate diagrams). Phonon lines are not shown

to the site indices. Such decoupling may be performed graphically by connecting pairwise th three electron lines running into the block ϱ and the three electron lines coming out of it Thereby, each connected pair of lines m_i and m_i' is associated with a factor $\varrho_{m_i}(t_2)\,\delta_{m_i m_i'}$ (for a wrong line) or $[1 - \varrho_{m_i}(t_2)]\,\delta_{m_i m_i'}$ (for a right line). Furthermore, according to the diagrams rules, the diagram must be multiplied by an additional factor $(-1)^M$, where M is the number intersections of electron lines resulting from the pairing procedure. Of course, electron lines belonging to the same interaction point are excluded from the pairing, because $J_{mm} = 0$ holds. Consequently, in Fig. A2.1, in each diagram two pairings may be performed, e.g.: (i) m_1 and m, m_2 and m_1', m_3 and m_2'; (ii) m_1 and m_2', m_2 and m, m_3 and m_1'. The resulting eight diagrams are represented in Fig. A2.2.

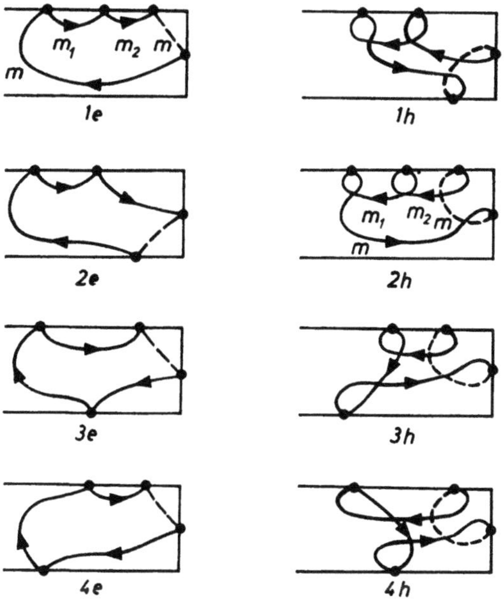

Fig. A2.2. Third-order diagrams in Hartree-Fock approximation. There exist another eight diagrams differing from the above ones by transfering the points from the above branch of the contour to the lower one and vice versa (complex conjugate diagrams). Phonon lines are not shown

The characteristic feature of diagrams of the type (e) in Fig. A2.2 is the successive connection of the points along the contour, so that there is only a single wrong electron line. These diagrams just occur in the one-electron approximation and must be accounted for solely in the limiting case of the Boltzmann statistics for electrons. Therefore, they are called the electron diagrams. On the other hand, diagrams of the type (h) in Fig. A2.2 contain only a single right line, i.e. they govern the limiting case of the Boltzmann statistics for holes $(1 - \varrho_m \ll 1)$. Therefore, they are called the hole diagrams. On comparing the diagrams (e) and (h), it turns out that the hole diagrams may be obtained from the electron ones by means of the replacements $\varrho_m \rightleftarrows (1 - \varrho_m)$, $J_{m'm}(\boldsymbol{H}) \to -J_{mm'}(\boldsymbol{H}) = -J^*_{m'm}(\boldsymbol{H})$, $\boldsymbol{E} \to -\boldsymbol{E}$, and $\varepsilon_m \to -\varepsilon_m$ and by multiplication with the factor (-1) (owing the intersection of the electron lines at the clamp in the hole diagrams). Note that the p-n-anomaly for the Hall effect (cf. Sections 3. and 8.) is a consequence of this relation.

After representing the three-particle functions by products of three one-particle functions, on the right-hand side of the equation for ϱ_m (2.4.3), the term proportional to $\overset{(3)}{\mathscr{W}}$ becomes

$$\sum_{m_1 m_2} \int_0^t \mathrm{d}t' \{\varrho_m(t') [1 - \varrho_{m_1}(t')] [1 - \varrho_{m_2}(t')] \mathscr{W}'^{(\mathrm{e})}_{m_1 m_2 m}(t', t)$$
$$+ \varrho_{m_1}(t') [1 - \varrho_{m_2}(t')] [1 - \varrho_m(t')] \mathscr{W}^{(\mathrm{e})}_{m_1 m_2 m}(t', t)$$
$$-[1 - \varrho_m(t')] \varrho_{m_1}(t') \varrho_{m_2}(t') \mathscr{W}'^{(\mathrm{h})}_{m_1 m_2 m}(t', t)$$
$$-[1 - \varrho_{m_1}(t)] \varrho_{m_2}(t') \varrho_m(t') \mathscr{W}^{(\mathrm{h})}_{m_1 m_2 m}(t', t)\} \,. \tag{A 2.1}$$

The probability $\mathscr{W}'^{(e)}_{m_1m_2m}$ is described by the diagram 1e in Fig. A2.2 (plus the corresponding complex conjugate diagram), and $\mathscr{W}^{(e)}_{m_1m_2m}$ by the sum of the diagrams 2e, 3e and 4e (plus the corresponding complex conjugate diagrams). The diagrams $\mathscr{W}'^{(h)}_{m_1m_2m}$ and $\mathscr{W}^{(h)}_{m_1m_2m}$, respectively, may be obtained from $\mathscr{W}'^{(e)}_{m_1m_2m}$ and $\mathscr{W}^{(e)}_{m_1m_2m}$ by means of the replacement $J_{m'm}(\boldsymbol{H}) \to -J_{mm'}(\boldsymbol{H})$, $\varepsilon_m \to -\varepsilon_m$ and $\boldsymbol{E} \to -\boldsymbol{E}$ (the factor (-1) is already accounted for in (A2.1)).

The transition to the Markovian limit is performed analogously as in the case of the two-site contribution (cf. Section 2.4). In this limit, the rate equation in the presence of a magnetic field reads

$$\begin{aligned}\frac{\mathrm{d}\varrho_m}{\mathrm{d}t} &= \sum_{m_1} \{\varrho_{m_1}(1-\varrho_m)\, W_{m_1m} - \varrho_m(1-\varrho_{m_1})\, W_{mm_1}\} \\ &+ \sum_{m_1m_2} \{\varrho_m(1-\varrho_{m_1})(1-\varrho_{m_2})\, W'^{(e)}_{m_1m_2m} + \varrho_{m_1}(1-\varrho_{m_2})(1-\varrho_m)\, W^{(e)}_{m_1m_2m} \\ &- (1-\varrho_m)\,\varrho_{m_1}\varrho_{m_2} W'^{(h)}_{m_1m_2m} - (1-\varrho_{m_1})\,\varrho_{m_2}\varrho_m W^{(h)}_{m_1m_2m}\}. \end{aligned} \tag{A2.2}$$

The three-site probabilities $W^{(e,h)}$ are related to $\mathscr{W}^{(e,h)}$ as (cf. (2.4.19))

$$W^{(e,h)}_{m_1m_2m}(t) = \int_0^\infty \mathrm{d}t'\, \mathscr{W}_{m_1m_2m}(t-t', t). \tag{A2.3}$$

Let us write down the explicit expression for the three-site blocks $W^{(e,h)}$, by making use of the diagram rules presented in Appendix 1. The contribution of the diagram 1e (plus its complex conjugate) reads

$$\begin{aligned} W'^{(e)}_{m_1m_2m} &= 2\mathrm{Re}\left(-\frac{\mathrm{i}}{\hbar}\right)^3 \tilde{J}_{mm_2}(\boldsymbol{H})\, \tilde{J}_{m_2m_1}(\boldsymbol{H})\, \tilde{J}_{m_1m}(\boldsymbol{H}) \int_0^\infty \mathrm{d}t_1 \int_0^{t_1} \mathrm{d}t_2 \\ &\times \exp\left\{\frac{\mathrm{i}}{\hbar}[\varepsilon_{m_1} - \varepsilon_m - \mathrm{e}\boldsymbol{E}(\boldsymbol{R}_{m_1} - \boldsymbol{R}_m)] + \frac{\mathrm{i}}{\hbar}[\varepsilon_m - \varepsilon_{m_2} \right. \\ &\left. - \mathrm{e}\boldsymbol{E}(\boldsymbol{R}_m - \boldsymbol{R}_{m_2})]\right\} \{F^{m_1m}_{mm_2}(t_1)\, F^{m_1m}_{m_2m_1}(t_2)\, F^{m_2m_1}_{mm_2}(t_1-t_2) + F^{m_1m}_{mm_2}(t_1)\, F^{m_1m}_{m_2m_1}(t_2) \\ &+ F^{m_1m}_{mm_2}(t_1)\, F^{m_2m_1}_{mm_2}(t_1-t_2) + F^{m_1m}_{m_2m_1}(t_2)\, F^{m_2m_1}_{mm_2}(t_1-t_2) + F^{m_1m}_{mm_2}(t_1)\}. \end{aligned}$$

The function F is defined in (A1.20) and (A1.19), and $\tilde{J}$ in (A1.22). The contribution of the product with three F's arises from diagrams with three phonon lines (the corresponding transitions are completely nondiagonal with respect to phonons), that with two F's from diagrams with two phonon lines, and that with one F from diagrams with one phonon line. The latter two contributions contain diagonal as well as nondiagonal transitions with respect to phonons (in diagonal transitions, the number of phonon is conserved). Note that diagrams with empty points must be summed up (cf. Firsov, 1975). This procedure leads to an oscillating factor of the type $\exp\left(\frac{\mathrm{i}}{\hbar}\,\Delta\varepsilon_p t_1\right)$, which describes the contribution of a finite width of the polaron band. In the case of hopping transport, these diagrams may be neglected.

For strong electron-phonon coupling, where $S_\mathrm{T} \gg 1$, the main contribution to (A2.4) comes from the product of three F's, since, after performing the time integration, only in this term the exponentially small factor $\exp(-3S_\mathrm{T})$ is completely compensated, while the contribution of terms with two factors F is proportional to $\exp(-\gamma S_\mathrm{T})$, where γ is of order unity. For weak interaction with phonons in disordered systems, the situation is more complicated. In this case, the contribution from products of two F's is proportional to $(|\gamma|^2)^2$, while that from products of three F's is proportional to $(|\gamma^2|)^3$, so that the former contribution is formally larger than the latter one ($\gamma \ll 1$). However, a summing-up of diagrams with a different number of phononless transitions leads possibly to the result that this contribut ion is small with the factor J/Γ (Γ — width of the spread of the electron levels) compared with the contribution from the totally nondiagonal phonon transitions (i.e. with the product of three F's). So far, however, this important problem has not been solved.

The three-site probabilities are in general assumed to be small compared with the two-site ones. Nevertheless, in the presence of a magnetic field, they must be accounted for, because they yield a contributions linear in $\boldsymbol{H}$ and they determine the Hall transport. From (A2.4), it can readily be seen that

$$W'^{(e,h)}_{m_1m_2m}(\boldsymbol{H}) = W'^{(e,h)}_{m_2m_1m}(-\boldsymbol{H}) \tag{A2.5}$$

holds. Hence, it follows that the rate equation (A2.2) involves only the combination $W'^{(e,h)}_{m_1m_2m} + W'^{(e,h)}_{m_2m_1m} = W^{(e,h)}_{m_1m_2m}(\boldsymbol{H}) + W^{(e,h)}_{m_1m_2m}(-\boldsymbol{H})$, which is symmetrical with respect to $\boldsymbol{H}$. Accordingly, if one is interested in effects being linear in $\boldsymbol{H}$, then in the rate equation the terms $W'^{(e,h)}$ may be ignored, and from $W^{(e,h)}$, only contributions being linear in $\boldsymbol{H}$ must be accounted for. On account of this fact, the rate equation in the presence of a magnetic field (A2.2) takes the form (3.5.4).

The summing-up of the diagrams 2e, 3e and 4e (plus the complex conjugate diagrams) leads to an expression for $W^{(e)}_{m_1m_2m}$, which agrees with the expression for $W'^{(e)}_{m_1m_2m}$ (A2.4), up to a change of the sign and the following change of the limits of integration over t_2: instead from 0 to t_1, the integration must be performed from $-\infty$ to $+\infty$. Neglecting all diagonal phonon contributions proportional to F^2, in the linear approximation with respect to $\boldsymbol{H}$, we obtain

$$\begin{aligned}
W^{(e)}_{m_1m_2m} = {} & \frac{e\boldsymbol{H}}{2\hbar^4 c} [(\boldsymbol{R}_m - \boldsymbol{R}_{m_1}) \times (\boldsymbol{R}_{m_1} - \boldsymbol{R}_{m_2})] \, J_{mm_2} J_{m_2m_1} J_{m_1m} \\
& \times \exp\{-S_T(m, m_2) - S_T(m_1, m_2) - S_T(m_1, m)\} \int_{-\infty}^{\infty} dt_1 \int_{-\infty}^{\infty} dt_2 \\
& \times \exp\left\{\frac{it_1}{\hbar} [\varepsilon_{m_1} - \varepsilon_{m_2} - e\boldsymbol{E}(\boldsymbol{R}_{m_1} - \boldsymbol{R}_{m_2})]\right. \\
& \left. + \frac{it_2}{\hbar} [\varepsilon_m - \varepsilon_{m_1} - e\boldsymbol{E}(\boldsymbol{R}_m - \boldsymbol{R}_{m_1})]\right\} \\
& \times \left\{\exp\left[\sum_{\boldsymbol{q}} \frac{|\gamma(\boldsymbol{q})|^2}{2N \sinh(\hbar\omega_{\boldsymbol{q}}\beta/2)} \left\{\left(1 - \cos \boldsymbol{q}(\boldsymbol{R}_{m_1} - \boldsymbol{R}_m)\right.\right.\right.\right. \\
& \left. + \cos \boldsymbol{q}(\boldsymbol{R}_{m_1} - \boldsymbol{R}_{m_2}) - \cos \boldsymbol{q}(\boldsymbol{R}_{m_2} - \boldsymbol{R}_m)\right) \cos \omega_{\boldsymbol{q}} \left(t_1 + \frac{i\hbar\beta}{2}\right) \\
& + \left(1 - \cos \boldsymbol{q}(\boldsymbol{R}_{m_1} - \boldsymbol{R}_m) - \cos \boldsymbol{q}(\boldsymbol{R}_{m_1} - \boldsymbol{R}_{m_2}) + \cos \boldsymbol{q}(\boldsymbol{R}_{m_2} - \boldsymbol{R}_m)\right) \\
& \times \cos \omega_{\boldsymbol{q}} \left(t_2 + \frac{i\hbar\beta}{2}\right) \\
& + \left(1 + \cos \boldsymbol{q}(\boldsymbol{R}_{m_1} - \boldsymbol{R}_m) - \cos \boldsymbol{q}(\boldsymbol{R}_{m_1} - \boldsymbol{R}_{m_2}) - \cos \boldsymbol{q}(\boldsymbol{R}_{m_2} - \boldsymbol{R}_m)\right) \\
& \left.\left.\left.\left. \times \cos \omega_{\boldsymbol{q}} \left(t_1 - t_2 + \frac{i\hbar\beta}{2}\right)\right\}\right] - 1\right\}.
\end{aligned} \tag{A2.6}$$

In (A 2.6), the time integration may be performed analogously as in Section 3.5., where such an integral was studied for an ordered system (all $\varepsilon_m = 0$) with strong electron-phonon coupling. We find that

$$W^{(e)H}_{m_1m_2m} = \frac{eH}{\hbar^2 c} \frac{\pi}{4\sqrt{3}} [(\boldsymbol{R}_m - \boldsymbol{R}_{m_1}) \times (\boldsymbol{R}_{m_2} \times \boldsymbol{R}_{m_1})]$$

$$\times \frac{J_{mm_1}J_{m_1m_2}J_{m_2m}}{E_a kT} \exp\left\{-\frac{1}{kT}\left[\frac{4}{3}E_a + \frac{1}{24E_a}(\varepsilon^2_{mm_1} + \varepsilon^2_{mm_2} + \varepsilon^2_{m_1m_2}) - \frac{1}{3}(\varepsilon_{m_1m} + \varepsilon_{m_1m_2})\right]\right\}, \quad \text{(A 2.7)}$$

where $\varepsilon_{mm'} = \varepsilon_m - \varepsilon_{m'}$. For $\varepsilon_m = 0$, (A 2.7) agrees with the corresponding expression given in Section 3.5. Note that here we have not accounted for the effect of disorder on E_a. This may be justified, in particular, in the case of short-range electron-phonon interaction, where in the expression for E_a (3.2.8) the term $\cos \boldsymbol{qg}$ is neglected, i.e. where $E_a = \sum_{\boldsymbol{q}} \hbar\omega_{\boldsymbol{q}} |\gamma_{\boldsymbol{q}}|^2/4N$ holds.

For weak electron-phonon coupling, the integrand of (A 2.6) is expanded in a series of powers of the coupling constant γ. Restricting ourself to the lowest nonvanishing approximation, then, on account of the contribtuion from F^2 in (A 2.4), we obtain

$$W^{(e)H}_{m_1m_2m} = \frac{e\boldsymbol{H}}{8c} [(\boldsymbol{R}_m - \boldsymbol{R}_{m_1}) \times (\boldsymbol{R}_{m_2} - \boldsymbol{R}_{m_1})]$$

$$\times \left\{\frac{\exp(\varepsilon_{m_1m_2}/2kT)}{\cosh\left(\frac{\varepsilon_{m_1m}}{2kT}\right)\cosh\left(\frac{\varepsilon_{m_2m}}{2kT}\right)} \frac{J_{m_1m_2}}{J_{m_1m}J_{mm_2}} \tilde{W}_{m_1m}\tilde{W}_{mm_2} + m \rightleftarrows m_2 + m \rightleftarrows m_1\right\}, \quad \text{(A 2.8)}$$

where

$$\tilde{W}_{m_1m} = \pi \frac{|J_{m_1m}|^2}{\hbar^2} \coth\frac{|\varepsilon_{m_1m}|}{2kT} \sum_{\boldsymbol{q}} \frac{1}{N} |\gamma_{\boldsymbol{q}}|^2 \,\delta\left(\omega_{\boldsymbol{q}} - \frac{|\varepsilon_{m_1m}|}{\hbar}\right) \quad \text{(A 2.9)}$$

is equal to the symmetrized two-site hopping probability for $\boldsymbol{E} = 0$ (2.4.19) for weak electron-phonon coupling, $\tilde{W}_{m_1m} = \sqrt{W^{(0)}_{m_1m} W^{(0)}_{mm_1}}$ (the upperscript (0) indicates that $\boldsymbol{E} = 0$).

Appendix 3. Percolation problems and the Potts model

The Potts model (POTTS, 1952, for a review, see WU, 1982, 1984) (often called Ashkin-Teller-Potts model) is a generalization of the Ising model to more than two components. In this model each site of a regular lattice is occupied by a spin which can take on Q possible values. Nearest neighbour like and unlike spins have an interaction energy 0 and J, respectively.

The Hamiltonian of the Q-state Potts model may be written as

$$\mathcal{H} = -J \sum_{\langle m,n \rangle} (\delta_{\sigma_m \sigma_n} - 1) - 2\mu H \sum_{m} (\delta_{\sigma_m 1} - 1), \tag{A3.1}$$

where J is the exchange integral, μ the magnetic moment, H the magnetic field; the variable σ_m assumes the values $1, 2, \ldots, Q$, and the sum $\langle m, n \rangle$ is over nearest-neighbour pairs. For $Q = 2$, (A3.1) is the Hamiltonian of the Ising model.

Let us consider a simple cubic lattice with N sites and $L = 3N$ nearest neighbour bonds. For the Hamiltonian (A3.1), the following expression for the partition function can readily be found

$$Z = q^L \sum_{\{\sigma_m\}} \left\{ \prod_{\langle m,n \rangle} \left(1 + \frac{p}{q} \delta_{\sigma_m \sigma_n}\right) \prod_{m} \exp\left[h(\delta_{\sigma_m 1} - 1)\right] \right\}, \tag{A3.2}$$

where $h = 2\mu H/kT$, $q = e^{-2J/kT}$, $p = 1 - q$; $\prod_{\langle m,n \rangle}$ denotes the product over all L bonds and $\prod_{m}$ the product over all N sites of the lattice.

Analogous to the Ising problem, the partition function (A3.2) may be evaluated by expanding it in powers of p/q and summing up the expressions thus obtained over all possible configurations of the lattice. In this way we find that

$$Z = \sum_{\text{conf}} q^L (p/q)^R \prod_{s=1}^{\infty} \left[1 + (Q - 1) e^{-sh}\right]^{\mathcal{N}_s}$$

$$\equiv \left\langle \prod_{s=1}^{\infty} \left[1 + (Q - 1) e^{-sh}\right]^{\mathcal{N}_s} \right\rangle. \tag{A3.3}$$

Here $\sum_{\text{conf}}$ denotes the sum over all possible configurations, whereby in the given configuration there are R bonds and $(L - R)$ deleted bonds; $\mathcal{N}_s$ is the number of clusters of connected bonds of size s, and $\langle \ldots \rangle$ indicates the configuration average. The conception of a configuration average may be introduced here, since $p + q = 1$ and $q < 1$, so that in (A3.3) the factor $p^R q^{L-R}$ may be considered as the probability of finding a configuration having R bonds and $(L - R)$ deleted bonds.

To relate the Potts model with the (bond) percolation problem, let us now consider the limit $Q \to 1$, $N \to \infty$, whereby we assume that $(Q - 1) N \ll 1$. By means of (A3.3) (and the rule of de l'Hospital), we get

$$F(p, h) \equiv \lim_{N\to\infty} \lim_{Q\to 1} \frac{\ln Z}{N(s - 1)} = \sum_s n_s \, e^{-sh}, \tag{A3.4}$$

where

$$n_s = \lim_{N\to\infty} N^{-1} \langle \mathcal{N}_s \rangle \tag{A3.5}$$

denotes the mean number of clusters per site of size s. $F(p, h)$ may be used as generating function for the (bond) percolation problem (see Section 3.2.2).

On the basis of the Potts model the bond percolation problem was first formulated by Kasteleyn and Fortuin (1969). A reformulation of the Kasteleyn-Fortuin result in a more compact form was given by Stephen (1977). As pointed out by Giri et al. (1977), Kunz and Wu (1978) and Murata (1979), the site percolation problem can also be formulated as the limit of a Potts model. Furthermore, according to Klein (1982), a continuum version of the Potts model can be used to describe continuum percolation. The Potts model has been used in renormalization-group studies of the percolation problem (Harris et al., 1975, Dasgupta, 1976). The Potts-model formulations of percolation are essential in our understanding of the relation between thermal phase transitions and percolation (see, e.g., Heerman and Stauffer, 1981, Coniglio and Peruggi, 1982).

References

Abkowitz, M., P. G. Le Comber, and W. E. Spear (1976) Commun. Phys. **1**, 175

Abou-Chacra, R., P. W. Anderson, and D. J. Thouless (1973), J. Phys. **C 6**, 1734

Abrahams, E., P. W. Anderson, P. A. Lee, and T. V. Ramakrishnan (1981), Phys. Rev. **B 24**, 6783

Abrahams, E., P. W. Anderson, D. C. Licciardello, and T. W. Ramakrishnan (1979), Phys. Rev. Letters **42**, 693

Abram, R. A., and S. F. Edwards (1972), J. Phys. **C 5**, 1183, 1196

Adkins, C. J. (1978), J. Phys. **C 11**, 851

Adler, J., and M. Silver (1982), Phil. Mag. **B 45**, 307

Ahmed, G., and J. A. Blackman (1979), J. Phys. **C 12**, 837

Ahmed, G., and J. A. Blackman (1980a), Phil. Mag. **B 42**, 517

Ahmed, G., and J. A. Blackman (1980b), Phil. Mag. **B 42**, 529

Alefeld, G., and J. Völkl (editors) (1978), Hydrogen in Metals I, Basic Properties; Hydrogen in Metals II, Application-Oriented Properties, Springer-Verlag; Berlin, Heidelberg, New York

Alexander, S. (1981), Phys. Rev. **B 23**, 2951

Alexander, S. (1982), Phys. Rev. **B 26**, 2956

Alexander, S., J. Bernasconi, W. R. Schneider, and R. Orbach (1981), Revs. Mod. Phys. **53**, 175

Alexander, S., and P. Pincus (1978), Phys. Rev. **B 18**, 2011

Alexandrov, A., and J. Ranninger (1981), Phys. Rev. **B 23**, 1796, **B 24**, 1164

Allen, R. F., and D. J. Adkins (1972), Phil. Mag. **26**, 1027

Altshuler, B. L., A. G. Aronov, and P. A. Lee (1980), Phys. Rev. Letters **44**, 1288

Ambegaokar, V., S. Cochran, and J. Kurkijärvi (1973), Phys. Rev. **B 8**, 3682

Ambegaokar, V., B. I. Halperin, and J. S. Langer (1971), Phys. Rev. **B 4**, 2612

Amitay, M., and M. Pollak (1966), J. Phys. Soc. Japan Suppl. **21**, 549

Anderson, P. W. (1958), Phys. Rev. **109**, 1492

Anderson, P. W. (1975), Phys. Rev. Letters **34**, 953

Anderson, P. W. (1978), Revs. Mod. Phys. **50**, 191

Anderson, P. W. (1981), Phys. Rev. **B 23**, 4828

Anderson, P. W., B. I. Halperin, and C. M. Varma (1972) Phil. Mag. **25**, 1

Anderson, P. W., D. J. Thouless, E. Abrahams, and D. S. Fisher (1980), Phys. Rev. **B 22**, 3519

Aoki, H. (1979), J. Phys. **C 12**, 633

Appel, J. (1966), Phys. Rev. **141**, 506

Appel, J. (1968), Solid State Phys. **21**, 193

APSLEY, N., and H. P. HUGHES (1974), Phil. Mag. **30**, 963
APSLEY, N., and HUGHES, H. (1975), Phil. Mag. **31**, 1327
ARNOLD, G. B., and T. HOLSTEIN (1974), Phys. Rev. Letters **33**, 1547
ARNOLD, G. B., and T. HOLSTEIN (1981), Ann. Phys. (N. Y.) **132**, 163
ASAI, S., T. TOYABE, and M. HIRAO (1970), Proc. X. Int. Conf. Phys. of Semicond., Boston, 1970, p. 578
AUSTIN, I. G. (1972), J. Phys. **C 5**, 1687
AUSTIN, I. G., and N. F. MOTT (1969), Adv. Phys. **18**, 41
AUSTIN, I. G., and SAYER, M. (1974), J. Phys. **C 7**, 905
BANYAI, L., and A. ALDEA (1979), Fortschr. Phys. **27**, 435
BANYAI, L., A. ALDEA, and P. GARTNER (1981), Rev. Roum. Phys. **26**, 923
BARANOVSKII, S. D., A. L. EFROS, B. L. GELMONT, and B. I. SHKLOVSKII (1979), J. Phys. **C 12**, 1023
BARBANEL, V. I., and V. N. BOGOMOLOV (1975), Fiz. tverd. Tela **17**, 1782
BARI, R. A. (1974), Phys. Rev. **B 9**, 4328
DE BELL, K., and J. W. ESSAM (1981), J. Phys. **A 14**, 1993
DE BELL, K., and J. ESSAM (1983), J. Phys. **A 16**, 385
BEREZIN, A. A. (1980), J. Phys. **C 13**, L 947
BEREZIN, A. A. (1981), Phys. Letters **86 A**, 480
BEREZIN, A. A. (1983), Phys. Rev. Letters **50**, 1520
BEREZINSKII, V. B. (1973), Zh. eksper. teor. Fiz. **65**, 1251
BERGMAN, D. J., and KANTOR, Y. (1981), J. Phys. **C 14**, 3365
BERNASCONI, J. (1978), Phys. Rev. **B 18**, 2185
BERNASCONI, J., W. R. SCHNEIDER, and W. WYSS (1980), Z. Phys. **B 37**, 175
BERNASCONI, J., and H. J. WIESMANN (1976), Phys. Rev. **B 13**, 1131
BISHOP, D. J., D. C. TSUI, and R. C. DYNES (1980) Phys. Rev. Letters **44**, 1153
BLACKMAN, J. A. (1976), J. Phys. **C 9**, 2049
BLATT, F. J. (1957), Theory of Mobility of Electrons in Solids, Academic Press Inc., New York
BLOCH, A. N., V. K. SHANTE, and C. M. VARMA (1973), Phys. Rev. **B 8**, 4885
BLUMEN, A., J. KLAFTER, and G. ZUMOFEN (1983), Phys. Rev. **B 27**, 3429
BOGOMOLOV, V. N., E. K. KUDINOV, and YU. A. FIRSOV (1967), Fiz. tverd. Tela **9**, 3175
BONCH-BRUEVICH, V. L. (1977), phys. stat. sol. (b) **84**, 99
BONCH-BRUEVICH, V. L., A. G. MIRONOV, and I. P. ZVYAGIN (1973), La Resista del Nuovo Cimento **3**, 321
BONCH-BRUEVICH, V. L., I. P. ZVYAGIN, R. KEIPER, A. G. MIRONOV, R. ENDERLEIN, and B. ESSER (1981), Electron Theory of Disordered Solids (in Russisch), Nauka, Moscow
BÖTTGER, H. (1983), Principles of the Theory of Lattice Dynamics, Akademie-Verlag, Berlin; Physik-Verlag, Weinheim
BÖTTGER, H., and V. V. BRYKSIN (1974), phys. stat. sol. (b) **64**, 449
BÖTTGER, H., and V. V. BRYKSIN (1975a), phys. stat. sol. (b) **68**, 285
BÖTTGER, H., and V. V. BRYKSIN (1975b), Fiz. tverd. Tela **17**, 3208
BÖTTGER, H., and V. V. BRYKSIN (1975c), phys. stat. sol. (b) **71**, 93
BÖTTGER, H., and V. V. BRYKSIN (1976a), phys. stat. sol. (b) **78**, 9
BÖTTGER, H., and V. V. BRYKSIN (1976b), phys. stat. sol. (b) **78**, 415
BÖTTGER, H., and V. V. BRYKSIN (1976c), Fiz. tverd. Tela **18**, 88
BÖTTGER, H., and V. V. BRYKSIN (1976d), Fiz. tverd. Tela **18**, 1888
BÖTTGER, H., and V. V. BRYKSIN (1976e), Proc. VI. Internat. Conf. Conf. Amorph. Liquid Semicond., Leningrad 1975, Vol. Electronic Processes in Non-Cryst. Semicond., Nauka Leningrad, p. 98

BÖTTGER, H., and V. V. BRYKSIN (1976f), phys. stat. sol. (b) **74**, K 151

BÖTTGER, H., and V. V. BRYKSIN (1977a), phys. stat. sol. (b) **80**, 569

BÖTTGER, H., and V. V. BRYKSIN (1977b), phys. stat. sol. (b) **81**, 97

BÖTTGER, H., and V. V. BRYKSIN (1977c), phys. stat. sol. (b) **81**, 433

BÖTTGER, H., and V. V. BRYKSIN (1977d), Solid State Commun. **23**, 227

BÖTTGER, H., and V. V. BRYKSIN (1979), phys. stat. sol. (b) **96**, 219

BÖTTGER, H., and V. V. BRYKSIN (1980), Phil. Mag. **B 42**, 297

BÖTTGER, H., and V. V. BRYKSIN (1982), phys. stat. sol. (b) **113**, 9

BÖTTGER, H., V. V. BRYKSIN, and P. KLEINERT (1978) phys. stat. sol. (b) **89**, 675

BÖTTGER, H., V. V. BRYKSIN, and P. KLEINERT (1979a) Fiz. tverd. Tela **21**, 808

BÖTTGER, H., V. V. BRYKSIN, and P. KLEINERT (1979b) Fiz. tverd. Tela **21**, 594

BÖTTGER, H., V. V. BRYKSIN, and G. YU. YASHIN (1979a), J. Phys. **C 12**, 2797

BÖTTGER, H., V. V. BRYKSIN, and G. YU. YASHIN (1979b), J. Phys. **C 12**, 3951

BÖTTGER, H., V. V. BRYKSIN, and G. YU. YASHIN (1979c), Fiz. tverd. Tela **21**, 1256

BÖTTGER, H., H. RÖLL, D. WEGENER, and L. ZIEGLER (1982), Wiss. Z. Techn. Hochsch. Magdeburg **26**, 67

BÖTTGER, H., and D. WEGENER (1984a), Wiss. Z. Techn. Hochsch. Magdeburg **28**, 101

BÖTTGER, H., and D. WEGENER (1984b), Phil. Mag. **B 50**, 409

BÖTTGER, H., and D. WEGENER (1984c), phys. stat. sol. (b) **121**, 413

BOYCE, J. B., and B. A. HUBERMAN (1979), Phys. Rep. **51**, 189

BRENIG, W., G. DÖHLER, and P. WÖLFE (1971), Z. Physik **246**, 1

BRENIG, W., G. DÖHLER, and P. WÖLFE (1973a), Z. Physik **258**, 381

BRENIG, W., G. H. DÖHLER, and H. HEYZENAU (1973b) Phil. Mag. **27**, 1093

BROADBENT, S. R., and J. M. HAMMERSLEY (1957), Proc. Cambridge Phil. Soc. **53**, 628

BRUGGEMAN, D. A. G. (1935), Ann. Phys. (Leipzig) **24**, 636

BRYKSIN, V. V. (1980), Fiz. tverd. Tela **22**, 2048, 2441

BRYKSIN, V. V. (1982), Fiz. tverd. Tela **24**, 1110

BRYKSIN, V. V., M. N. DYAKONOV, and S. D. HANIN (1980), Fiz. tverd. Tela **22**, 1403

BRYKSIN, V. V., M. N. DYAKONOV, V. M. MUZHDABA, and S. D. HANIN (1981), Fiz. tverd. Tela **23**, 1516

BRYKSIN, V. V., M. N. DYAKONOV, V. M. MUZHDABA, and S. D. HANIN (1982), Fiz. tverd. Tela **24**, 2682

BRYKSIN, V. V., and YU. A. FIRSOV (1968), Fiz. tverd. Tela **10**, 1314

BRYKSIN, V. V., and YU. A. FIRSOV (1969), Fiz. tverd. Tela **11**, 539

BRYKSIN, V. V., and YU. A. FIRSOV (1970a) Proc. X. Intern. Conf. Phys. Semicond. Cambridge (USA), US Atomic Energy Commission, p. 767

BRYKSIN, V. V., and YU. A. FIRSOV (1970b), Fiz. tverd. Tela **12**, 627

BRYKSIN, V. V., and YU. A. FIRSOV (1971a), Zh. eksper. teor. Fiz. **61**, 2373

BRYKSIN, V. V., and YU. A. FIRSOV (1971b), Fiz. tverd. Tela. **13**, 3246

BRYKSIN, V. V., and YU. A. FIRSOV (1972a), Fiz. tverd. Tela **14**, 463

BRYKSIN, V. V., and YU. A. FIRSOV (1972b), Fiz. tverd. Tela **14** 3599

BRYKSIN, V. V., and YU. A. FIRSOV (1972c) Solid State Commun. **10**, 471

BRYKSIN, V. V., and YU. A. FIRSOV (1973), Fiz. tverd. Tela **15**, 3235, 3344

BRYKSIN, V. V., and YU. A. FIRSOV (1974a), Fiz. tverd. Tela **16**, 811

BRYKSIN, V. V., and YU. A. FIRSOV (1974b), Fiz. tverd. Tela **16**, 1941

BRYKSIN, V. V., YU. A. FIRSOV, and S. A. KTITOROV (1981), Sol. State Commun. **39**, 385

BRYKSIN, V. V., and G. YU. YASHIN (1981), Fiz. tverd. Tela **23**, 3063

BURKHARDT, T. W., and J. M. J. VAN LEEUWEN (eds.) (1982), Real-Space Renormalization, Springer-Verlag, Heidelberg, Berlin, New York

BUTCHER, P. N. (1972), J. Phys. **C 5**, 1817
BUTCHER, P. N. (1974), J. Phys. **C 7**, 879
BUTCHER, P. N. (1978), Phil. Mag. **B 37**, 653
BUTCHER, P. N. (1980), Phil. Mag. **B 42**, 799
BUTCHER, P. N., and J. D. CLARK (1980), Phil. Mag. **B 42**, 191
BUTCHER, P. N., and J. D. CLARK (1981), Phil. Mag. **43**, 1029
BUTCHER, P. N., K. J. HAYDEN, and J. A. MC INNES (1977) Phil. Mag. **36**, 19
BUTCHER, P. N., K. J. HAYDEN, and J. A. MC INNES (1979), Phil. Mag. **B 39**, 357
BUTCHER, P. N., and J. A. MC INNES (1978), Phil. Mag. **B 37**, 249
BUTCHER, P. N., and J. A. MC INNES (1981), Phil. Mag. **B 44**, 595
BUTCHER, P. N., and A. A. KUMAR (1980), Phil. Mag. **B 42**, 201
BUTCHER, P. N., and S. SUMMERFIELD (1981), J. Phys. **C 14**, L 1099
BUTCHER, P. N., and L. SWIERKOWSKI (1980), Solid State Commun. **33**, 1151
CARDY, J. L., and R. L. SUGAR (1980), J. Phys. **A 13**, L 423
CARMELI, B., and A. NITZAN (1983), J. Chem. Phys. **79**, 393
CHAKRAVERTY, B. K., and C. SCHLENKER (1976), J. Physique (Paris), Collog. **37**, C 4—353
CHANDRASEKHAR (1947), Stochastic Processes in Physics and Astronomy (in Russian), Nauka, Moscow
CHRISTOV, S. G. (1982), Phys. Rev. **B 26**, 6918
CHROBOCZEK, J. A. (1980), Phil. Mag. **B 42**, 933
CHROBOCZEK, J., H. FRITZSCHE, C.-L. JIANG, M. POLLAK, and R. L. WILD, Phil. Mag. **B 44**, 685 (1981)
CHROBOCZEK, J. A., A. KLOKOCKI, and K. KOPALKO (1974), J. Phys. **C 7**, 3042
CHROBOCZEK, J. A., and R. J. SLADEK (1966), Phys. Rev. **151**, 595
CLARK, A. H., M. H. COHEN, M. CAMPI, and H. P. D. LANYON (1974) J. noncrystall. Solids **16**, 117
CLARKE, P. S., J. W. ORTON, and A. J. GUEST (1978), Phys. Rev. **B 18**, 1813
COHEN, M. H. (1970), J. noncrystall. Solids **2**, 432
COLL, C. F., and G. BENI (1974), Sol. State Commun. **15**, 997
CONIGLIO, A. (1981), Phys. Rev. Letters **46**, 256
CONIGLIO, A. (1982), J. Phys. **A 15**, 3829
CONIGLIO, A., C. NAPPI, F. PERRUGI, and L. RUSSO (1977), J. Phys. **A 10**, 205
CONIGLIO, A., and F. PERUGGI (1982), J. Phys. **A 15**, 1873
CONIGLIO, A., H. E. STANLEY, W. KLEIN (1979), Phys. Rev. Letters 42, 518
COX, M. A. A., and J. W. ESSAM (1976), J. Phys. **C 9**, 3985
CUTLER, M., and N. F. MOTT (1969), Phys. Rev. **181**, 1336
DASGUPTA, C. (1976), Phys. Rev. **B 14**, 1221
DASGUPTA, C., A. B. HARRIS, and T. C. LUBENSKY (1978), Phys. Rev. **B 17**, 1375
DAVIES, J. H., P. A. LEE, and T. M. RICE (1982), Phys. Rev. Letters **49**, 758
DAVIES, R. A., M. PEPPER, and M. KAVEH (1983), J. Phys. **C 16**, L 285
DERSCH, U., B. POHLMANN, and P. THOMAS (1983), J. Phys. **C 16**, 3725
DIETERICH, W. (1981), Festkörperprobleme (Advances in Solid State Physics), Vol. XXI, ed. J. TREUTSCH, Vieweg, Braunschweig, p. 325
DIETERICH, W., P. FULDE, and I. PESCHEL (1980), Adv. Phys. **29**, 527
DIETERICH, W., I. PESCHEL, and W. R. SCHNEIDER (1977) Z. Physik **B 27**, 177
DOLAN, D. J., and D. D. OSHEROFF (1979), Phys. Rev. Letters **43**, 721
DOMANY, E., and S. SARKER (1979), Phys. Rev. **B 20**, 4726
DUNN, A. G., J. W. ESSAM, and J. M. LOVELUCK (1975). J. Phys. **C 8**, 743
DYSON, F. J. (1953), Phys. Rev. **92**, 1331

EAGLES, D. M. (1963), Phys. Rev. **130**, 1381
EAGLES, D. M. (1966), Phys. Rev. **145**, 645
ECONOMOU, E. N., and M. H. COHEN (1972), Phys. Rev. **B 5**, 2931
EDDOLS, D. V. (1966), phys. stat. sol. **17**, 67
EDWARDS, S. F. (1970), J. Phys. **C 3**, L 30
EDWARDS, J. T., and D. J. THOULESS (1972), J. Phys. **C 5**, 807
EFROS, A. L. (1967), Fiz. tverd. Tela **9**, 1152
EFROS, A. L. (1976), J. Phys. **C 9**. 2021
EFROS, A. L. (1981), Phil. Mag. **B 43**, 829
EFROS, A. L., NGUEN VAN LIEN and B. I. SHKLOVSKII (1979) J. Phys. **C 12**, 1869
EFROS, A. L., B. I. SHKLOVSKII, and I. Y. YANCHEV (1972), phys. stat. sol. (b) **50**, 45
EFROS, A., and B. SHKLOVSKII (1975), J. Phys. **C 8**, L 49
ELLIOTT, S. R. (1977), Phil. Mag. **36**, 1291
ELLIOTT, S. R. (1978), Phil. Mag. **B 38**, 325, 553
ELLIOTT, R. J., J. A. KRUMHANSL and P. L. LEATH (1974), Rev. Mod. Phys. **46**, 465
ELLIOTT, R. J., and R. J. LOUDON (1960), J. Phys. Chem. Solids **15**, 196
EMEL'JANENKO, O. V., T. S. LAGUROVA, D. N. NASLEDOV, D. D. NEDEOGLO, and I. N. TIMCHENKO (1973), Fiz. Tekh. Poluprov. **7**, 1919
EMEL'JANENKO, O. V., K. G. MASAGUTOV, D. N. NASLEDOV, and I. N. TIMCHENKO (1975), Fiz. Tekh. Poluprov. **9**, 503
EMIN, D. (1970), Phys. Rev. Letters **25**, 1751
EMIN, D. (1971a), Phys. Rev. **B 3**, 1321
EMIN, D. (1971b), Phys. Rev. **B 4**, 3639
EMIN, D. (1971c), Ann. Phys. (N. Y.) **64**, 336
EMIN, D. (1973) in: Electrons and Structural Properties of Amorphous Semiconductors. Ed. P. G. LE COMBER and J. MORT, Academic Press, New York, p. 261
EMIN, D. (1974), Phys. Rev. Letters **32**, 303
EMIN, D. (1975a), Adv. Phys. **24**, 305
EMIN, D. (1975b), Phys. Rev. Letters **35**, 882
EMIN, D. (1977), Phil. Mag. **B 35**, 1189
EMIN, D. (1983), Comments Solid State Phys. **11**, 35, 59
EMIN, D. and T. HOLSTEIN (1969), Ann. Phys. (N. Y.) **53**, 439
ENDERLEIN, R., and K. PEUKER (1977), in: 75 Jahre Quantentheorie. Ed. W. BRAUER, H. W. STREITWOLF, and K. WERNER, Akademie-Verlag, Berlin, p. 193
ESSAM, J. W. (1972), Phase Transitions and Critical Phenomena, Vol. 2. Ed. C. DOMB and M. S. GREEN, Academic. Press, New York, chap. 6, p. 197
ESSAM, J. W. (1980), Rep. Prog. Phys. **43**, 833
ESSAM, J. W., and K. M. GWILYM (1971), J. Phys. **C 4**, L 228
DA FANO, A., and G. JACUCCI (1977), Phys. Rev. Letters **39**, 950
FEDDERS, P. A. (1978), Phys. Rev. **B 17**, 2098
FEDDERS, P. A., and O. F. SANKEY (1977), Phys. Rev. **B 15**, 3580
FEDDERS, P. A., and O. F. SANKEY (1978), Phys. Rev. **B 18**, 5938
FERTIS, A. C., A. N. ANDRIOTIS, and E. N. ECONOMOU (1981), Phys. Rev. **B 24**, 5806
FIRSOV, YU. A. (1963), Fiz. tverd. Tela **5**, 2149
FIRSOV, YU. A. (1968), Fiz. tverd. Tela **10**, 1950
FIRSOV, YU. A. (1975), Polarons (in Russian), Mir, Moskow
FISCH, R., and A. B. HARRIS (1978), Phys. Rev. **B 18**, 416
FISHCHUK, I. I. (1982a), phys. stat. sol. (b) **113**, 549
FISHCHUK, I. I. (1982b), phys. stat. sol. (b) **111**, K 17

FISHCHUK, I. I., and V. N. RUDKO (1979), Solid State Commun. **29**, 99
FISHCHUK, I. I., and V. N. RUDKO (1978), phys. stat. sol. (b) **89**, 61
FISHCHUK, I. I., and V. N. RUDKO (1980a), J. Phys. **C 13**, 2703
FISHCHUK, I. I., and V. N. RUDKO (1980b), phys. stat. sol. (b) **99**, 477
FISHCHUK, I. I., and V. N. RUDKO (1980c) J. Phys. **C 13**, L 493
FISHER, M. E. (1964), Proceedings of the IBM Sience and Computer Symposium on Combinatoric Problems. IBM Data Processing Division, White Plains, New York, p. 179
FITZPATRICK, J. P., R. B. MALT, and F. SPAEPEN (1974), Phys. Letters **47 A**, 207
FLEISHMAN, L., and P. W. ANDERSON (1980), Phys. Rev. **B 21**, 2366
FLEISHMAN, L., D. C. LICCIARDELLO, and P. W. ANDERSON (1978), Phys. Rev. Letters **40**, 1341
FLYNN, C. P., and G. JACUCCI (1982), Phys. Rev. **B 25**, 6225
FLYNN, C. P., and A. M. STONEHAM (1970), Phys. Rev. **B 1**, 3966
FREED, K. F. (1972), Phys. Rev. **B 5**, 4802
FREMLIN, D. H. (1976), J. Physique **37**, 813
FRIEDMAN, L. (1963), Phys. Rev. **131**, 2445
FRIEDMAN, L. (1971), J. non-crystall. Solids **6**, 329
FRIEDMAN, L. (1977), Phil. Mag. **36**, 553
FRIEDMAN, L. (1980), Phil. Mag. **B 41**, 374
FRIEDMAN, L., and T. HOLSTEIN (1963), Ann. Phys. (N. Y.) **21**, 494
FRIEDMAN, L., and M. POLLAK (1978), Phil. Mag. **B 38**, 173
FRIEDMAN, L., and M. POLLAK (1981), Phil. Mag. **B 44**, 487
FRITZSCHE, H. (1955), Phys. Rev. **99**, 406
FRITZSCHE, H. (1962), Phys. Rev. **125**, 1552
FRITZSCHE, H. (1971), Solid State Commun. **9**, 1813 (1971)
FRITZSCHE, H., and M. CUEVAS (1960), Phys. Rev. **119**, 1238
FUCITO, F., and G. PARISI (1981), J. Phys. **A 14**, L 507
FUJII, S. (1979), J. Phys. Soc. Japan **46**, 1833, 1843
FUJITA, T., and J. HORI (1972), J. Phys. **C 5**, 1059
FUKUYAMA, H., and K. YOSHIDA (1979), J. Phys. Soc. Japan **46**, 102
FULDE, P., L. PIETRONERO, W. R. SCHNEIDER, and S. STRÄSSLER (1975), Phys. Rev. Letters **26**, 1776
GAUNT, D. S., A. J. GUTTMANN, and S. G. WHITTINGTON (1979), J. Phys. **A 12**, 75
GAVRILYUK, A. I., V. G. PROHVATILOV, and F. A. CHUDNOVSKII (1982), Fiz. tverd. Tela **24**, 982
GAWLINSKI, E. T., and S. REDNER (1983), J. Phys. **A 16**, 1063
GAWLINSKI, E. T., and H. E. STANLEY (1981), J. Phys. **A 14**, L 291
GEFEN, Y., A. AHARONY, and S. ALEXANDER (1983), Phys. Rev. Letters **50**, 77
GEFEN, Y., A. AHARONY, B. B. MANDELBROT, and S. KIRKPATRICK (1981), Phys. Rev. Letters **47**, 1771
GEFEN, Y., A. AHARONY, Y. SHAPIR, and N. BERKER (1982), J. Phys. **C 15**, L 801
DE GENNES, P. G. (1976), J. Physique **37**, L 1
GERSHENSON, E. M., V. A. IL'JIN, and L. B. LITVAK-GORSKAJA (1974), Fiz. Tekh. Poluprov. **8**, 295
GILBERT, M. H., and C. J. ADKINS (1976), Phil. Mag. **34**, 143
GILROY, K. S., and W. A. PHILLIPS (1981), Phil. Mag. **B 43**, 735
GINZBURG, S. L. (1976), Zh. eksper. teor. Fiz. **71**, 1145
GIRI, M. R., M. J. STEPHEN, and G. S. GREST (1977), Phys. Rev. **B 16**, 4971
GIRVIN, S. M. JONSON, and P. A. LEE (1982), Phys. Rev. **B 26**, 1651
GOCHANOUR, C. R., H. C. ANDERSEN, and M. D. FAYER (1979), J. Chem. Phys. **70**, 4254

GORDY, L. (1972), Bull. Am. Phys. Soc. **17**, 305
GORHAM-BERGERON, E., and D. EMIN (1977), Phys. Rev. **B 15**, 3667
GORKOV, L., D. KHMELNITZKIJ, and A. I. LARKIN (1979), Pisma Zh. eksper. teor. Fiz. **30**, 248
GOSAR, P. (1980), J. Phys. **C 13**, L 111
GOSAR, P. (1983), phys. stat. sol. (b) **118**, 853; Physica **A 120**, 339
GÖTZE, W. (1981), Phil. Mag. **B 43**, 219
GRAENER, H., M. ROSENBERG, T. E. WHALL, and M. R. B. JONES (1979), Phil. Mag. **B 40**, 389
GRAF, H., G. BALZER, E. RECKNAGEL, A. WEIDINGER, and R. I. GRYNSZPAN, (1980), Phys. Rev. Letters **44**, 1333
GRASSBERGER, P., and I. PROCACCIA (1982), Phys. Rev. **A 26**, 3686
GRUENEWALD, M., H. MUELLER, P. THOMAS, and D. WUERTZ (1981), Sol. State Commun. **38**, 1011
GRÜNEWALD, M., B. POHLMANN, L. SCHWEITZER, and D. WÜRTZ (1982), J. Phys. **C 15**, L 1153
GRÜNEWALD, M., B. POHLMANN, D. WÜRTZ, and B. MOVAGHAR (1983), J. Phys. **C 16**, 3739
GRÜNEWALD, M., and P. THOMAS (1979), phys. stat. sol. (b) **93**, K 17
GUTMANN, F., and L. E. LYONS (1967), Organic Semiconductors. John Wiley and Sons, Inc., New York, London
HAAN, S. W., and R. ZWANZIG (1977), J. Phys. **A 10**, 1547
HACKER, K., and G. OBERMAIR (1970), Z. Physik **234**, 1
HALEY, S. B. (1982), Phys. Rev. **B 25**, 5824
HALPERN, V. (1981), J. Phys. **C 14**, 3195
HAMILTON, E. M. (1972), Phil. Mag. **26**, 1043
HARRIS, A. B., and R. FISCH (1977), Phys. Rev. **B 15**, 5674
HARRIS, A. B., and T. C. LUBENSKY (1981), Phys. Rev. **B 24**, 2656
HARRIS, A. B., T. C. LUBENSKY, W. K. HOLCOMB, and C. DASGUPTA (1975), Phys. Rev. Letters **35**, 327
HASEGAWA, H., and R. E. HOWARD (1961), J. Phys. Chem. Solids **21**, 179
HARTMANN, O., E. KARLSSON, L. O. NORLIN, T. P. NIINIKOSKI, K. W. KEHR, D. RICHTER, J. M. WELTER, A. YAOUANC, and J. LE HERICY (1980), Phys. Rev. Letters **44**, 337
HAUBENREISSER, (1961), phys. stat. sol. **1**, 619
HAUS, J. W., K. W. KEHR, and K. KITAHARA (1982), Phys. Rev. **B 25**, 4918
HAUS, J. W., K. W. KEHR, and K. KITAHARA (1983) Z. Physik **B 50**, 161
HAUSER, J. J. (1975), Solid State Commun. **17**, 1577
HAUSER, J. J. (1977), J. non-crystall. Solids **23**, 21
HEDGCOCK, F. T., and T. W. RAUDORF (1970), Solid State Commun. 8, 1819
HEERMAN, D. W., and D. STAUFFER (1981), Z. Physik **B 44**, 339
HEIKES, R., and W. JONSTON (1957), J. Chem. Phys. **26**, 582
HEINRICHS, J. (1977), Phys. Rev. **B 16**, 4365
HEINRICHS, J. (1980), Phys. Rev. **B 22**, 3093
HEINRICHS, J. (1982), Phys. Rev. **B 25**, 1388
HERLACH, D. M., E. F. WASSERMANN, and R. WILLNECKER (1983), Phys. Rev. Letters **50**, 529
HILL, R. M. (1971), Phil. Mag. **24**, 1307
HILL, R. M. (1976a), phys. stat. sol. (a) **34**, 601
HILL, R. M. (1976b), phys. stat. sol. (a) **35**, K 29
HILL, R. M. (1977), phys. stat. sol. (a) **39**, 615
HILL, R. M., and A. K. JONSCHER (1979), J. non-crystall. Solids **32**, 53
HILLEBRAND, K., and R. M. NIEMINEN (1981), J. Phys. **C 14**, L 133
HINKELMANN, H., and H. G. REIK (1975) Sol. State Commun. **16**, 567

HOHLER, B., and H. SCHREYER (1982), J. Phys. **F 12**, 857
HOLSTEIN, T. (1959), Ann. Phys. (N. Y.) **8**, 325, 343
HOLSTEIN, T. (1961), Phys. Rev. **124**, 1329
HOLSTEIN, T. (1973), Phil. Mag. **27**, 255
HOLSTEIN, T. (1978a), Phil. Mag. **B 37**, 49
HOLSTEIN, T. (1978b), Phil. Mag. **B 37**, 499
HOLSTEIN, T. (1981), Ann. Phys. (N. Y.) **132**, 212
HOLSTEIN, T., S. K. LYO, and R. ORBACH (1976), Phys. Rev. Letters **36**, 891
HORI, M., and F. YONEZAWA (1974), J. Math. Phys. **15**, 2177
HORI, M., and F. YONEZAWA (1975), J. Math. Phys. **16**, 352, 365
HORI, M., and F. YONEZAWA (1977), J. Phys. **C 10**, 229
HOSHINO, K., and M. WATABLE (1978), J. Phys. **C 11**, 1381
HSU, C., and R. C. POWELL (1975), Phys. Rev. Letters **35**, 734
HUBERMAN, B. A., and J. B. BOYCE (1978), Solid State Commun. **25**, 759, 943
ICHE, G., and P. NOZIÈRES (1976), J. Physique **37**, 1313
MC INNES, J. A., P. N. BUTCHER, and J. C. CLARK (1980), Phil. Mag. **B 41**, 1
IOFFE, A. F. (1956), Canad. Jour. Phys. **34**, 1393
IONOV, A. N., and I. S. SHLIMAK (1972), Pis'ma Zh. eksper. teor. Fiz. **16**, 374
ISHIDA, S., and E. OTSUKA (1977), J. Phys. Soc. Japan **43**, 124
JÄCKLE, J., and K. W. KEHR (1983), J. Phys. **F 13**, 753
JACOBSON, S. H., M. A. RATNER, and A. NITZAN (1983), J. Chem. Phys. **78**, 4154
KADANOFF, L. P. (1971), Proc. Enrico Fermi Summer School of Physics, Varenna, 1970. Ed. M. S. GREEN, Academic Press, New York
KAGAN, YU. M. (1981), in: Defects in Insulating Crystals, Proc. Int. Conf. Riga, 1981. Eds. V. M. TUCHKEVICH, K. K. SHVARTS, „Zinatne" Publishing House, Riga; Springer-Verlag, Berlin, Heidelberg, New York, p. 17
KAGAN, YU., and M. I. KLINGER (1974), J. Phys. **C 7**, 2791
KAGAN, YU., and L. A. MAKSIMOV (1983), Phys. Letters **95 A**, 242
KAHLERT, H., G. LANDWEHR, A. SCHLACHETZKI, and H. SALOW (1976), Z. Physik **B 24**, 361
KAMIMURA, H. (1980), Phil. Mag. **B 42**, 763
KAMIMURA, H., A. KUROBE, and T. TAKEMORI (1983) Physica **117 B** & **118 B**, 652
KAMIMURA, H., T. TAKEMORI, and A. KUROBE (1982), in: Anderson Localization. Springer Series in Solid State Sciences, Vol. 39. Eds. Y. NAGAOKA and H. FUKUYAMA, Springer Verlag, Berlin, Heidelberg, New York, p. 156
KANERT, O. (1982), Phys. Rep. **91**, 183
KANEYOSHI, T. (1976), Phil. Mag. **33**, 11
KARIORIS, F. G., and K. S. MENDELSON (1981), J. Phys. **C 14**, 321
KARLSSON, E. (1982), Phys. Rep. **82**, 271
KASKI, K., R. A. TAHIR-KHELI, and R. J. ELLIOTT (1982), J. Phys. **C 15**, 209
KASTELEYN, P. W., and C. M. FORTUIN (1969), J. Phys. Soc. Japan **26** (Suppl.) 11
KAVEH, M. (1982), J. Phys. **C 30**, L 181
KAVEH, M., and N. F. MOTT (1981a), J. Phys. **C 14**, L 177
KAVEH, M., and N. F. MOTT (1981b), J. Phys. **C 14**, L 183
KAWABATA, A. (1980), J. Phys. Soc. Japan, Suppl. **A 49**, 375
MC KEE, R. A. (1980), Phys. Rev. **B 21**, 4269
KEHR, K. W., D. RICHTER, J. M. WELTER, O. HARTMANN, E. KARLSSON, L. O. NORLIN, T. O. NIINIKOSKI, and A. YAOUANC (1982), Phys. Rev. **B 26**, 567
KERTESZ, J., B. K. CHAKRABARTI and J. A. M. S. DUARTE (1982), J. Phys. **A 15**, L 13
KHOSLA, R. F., and J. R. FISCHER (1970), Phys. Rev. **B 2**, 4084

KILLIAS, H. R. (1966), Phys. Letters **20**, 5
KIMBALL, J. C., and L. W. ADAMS, Jr. (1978), Phys. Rev. **B 18**, 5851
KINZEL, W., and J. M. YEOMANS (1981), J. Phys. **A 14**, H 163
KIRKPATRICK, S. (1971), Phys. Rev. Letters **27**, 1722
KIRKPATRICK, S. (1973), Revs. Mod. Phys. **45**, 574
KIRKPATRICK, S. (1976), Phys. Rev. Letters **36**, 69
KIRKPATRICK, S. (1977), Phys. Rev. **B 15**, 1533
KIRKPATRICK, S. (1978), AIP Conference Proceedings No. 40. American Institute of Physics, New York, p. 99
KITTEL, C. (1963), Quantum Theory of Solids, John Wiley and Sons, Inc., New York, London
KIVELSON, S. (1980), Phys. Rev. **B 21**, 5755
KIVELSON, S. (1981), Phys. Rev. Letters **46**, 1344
KIVELSON, S. (1982), Phys. Rev. **B 25**, 3798
KLAFTER, J., and R. SILBEY (1980a), Surface Sci. **92**, 393
KLAFTER, J., and R. SILBEY (1980b), Phys. Rev. Letters **44**, 55
KLAFTER, J., and R. SILBEY (1983), Phil. Mag. **B 47**, 337
KLEIN, W. (1982), Phys. Rev. **B 26**, 2677
KLEIN, W., and W. KINZEL, J. Phys. **A 14**, L 405
KLINGER, M. I. (1963a), Phys. Letters **7**, 67
KLINGER, M. I. (1963b), Phys. Letters **7**, 102
KNOTEK, M. L. (1975), Solid State Commun. **17**, 1431
KNOTEK, M. L., and M. POLLAK (1972), J. non-crystall. Solids **8—10**, 505
KNOTEK, M. L., and M. POLLAK (1974), Phys. Rev. **B 9**, 664
KNOTEK, M. L., M. POLLAK, T. M. DONOVAN, and H. KURTZMAN (1973), Phys. Rev. Letters **30**, 853
KOGUT, P. M., and J. P. STRALEY (1979), J. Phys. **C 12**, 2151
KOHN, W., and J. M. LUTTINGER (1975), Phys. Rev. **108**, 590
KONSTANTINOV, O. V., and V. I. PEREL (1960) Zh. eksper. teor. Fiz. **39**, 197
KOPLIK, J. (1981), J. Phys. **C 14**, 4821
KOSAREV, V. V. (1974), Fiz. Tekh. Poluprov. **7**, 1378
KOSAREV, V. V., R. V. PARFENEV, V. V. POPOV, and S. S. SHALYT (1976), Fiz. tverd. Tela **18**, 489
KRAMERS, H. A. (1940), Physica (Utrecht) **7**, 284
KRENKE, V. M., and P. E. PARRIS (1983), Phys. Rev. **B 27**, 3221
KUBO, R. (1957), J. Phys. Soc. Japan **12**, 570
KUBO, R., M. YOKOTA, and S. NAKAJIMA (1957), J. Phys. Soc. Japan **12**, 1203
KUDINOV, E. K., and YU. A. FIRSOV (1964) Zh. eksper. teor. Fiz. **47**, 601
KUDINOV, E. K., and YU. A. FIRSOV (1965) Zh. eksper. teor. Fiz. **49**, 867
KUDINOV, E. K., and YU. A. FIRSOV (1966), Fiz. tverd. Tela **8**, 666
KUDINOV, E. K., D. N. MIRLIN, and YU. A. FIRSOV (1969), Fiz. tverd. Tela **11**, 2789
KUIVALAINEN, P., J. HELESKIVI, M. LEPPIHALME, U. GYLLENBERG-GÄSTRIN, and H. ISOTALO (1982), Phys. Rev. **B 26**, 2041
KUMAR, A. A., and J. HEINRICHS (1980), J. Phys. **C 13**, 2131
KUNZ, H., and F. Y. WU (1978), J. Phys. **C 11**, L 1
KURKIJÄRVI, J. (1973), Phys. Rev. **B 8**, 922
KURKIJÄRVI, J. (1974), Phys. Rev. **B 9**, 770
KUROBE, A., and H. KAMIMURA (1982), J. Phys. Soc. Japan **51**, 1904
LAGENDIJK, A. (1982), Phys. Rev. **B 25**, 4691
LAM, L., and A. BUNDE (1978), Z. Physik **B 30**, 65

Landau, L. D., and E. M. Lifshitz (1963), Quantum Mechanics (in Russian), Moscow
Lang, I. G., and Yu. A. Firsov (1962), Zh. eksper. teor. Fiz. **43**, 1843
Lang, I. G., and Yu. A. Firsov (1963), Fiz. tverd. Tela **5**, 2799
Lang, I. G., and Yu. A. Firsov (1967), Fiz. tverd. Tela **9**, 3422
Lang, I. G., and Yu. A. Firsov (1968), Zh. eksper. teor. Fiz. **54**, 826
Larkin, A. I., and P. A. Lee (1978), Phys. Rev. **B 17**, 1596
Last, B. J. and D. J. Thouless (1971), Phys. Rev. Letters **27**, 1719
Lee, P. A. (1982), in: Anderson Localization, Springer Series in Solid State Sciences, Vol. 39. Eds. Y. Nagaoka and H. Fukuyama, Springer Verlag, Berlin, Heidelberg, New York, p. 78
Lee, Y. C., C. S. Chu, and E. Castano (1983), Phys. Rev. **B 27**, 6136
Lee, P. A., and D. S. Fisher (1981), Phys. Rev. Letters **47**, 882
Lee, P. A., and T. V. Ramakrishnan (1982), Phys. Rev. **B 26**, 4009
Levinshtein, M. E. (1977), J. Phys. **C 10**, 1895
Levinshtein, M. E., B. I. Shklovskij, M. S. Shur, and A. L. Efros (1975), Zh. eksper. teor. Fiz. **69**, 386
Levinson, I. B. (1969), Zh. eksper. teor. Fiz. **57**, 660
Lifshitz, I. M., S. A. Gredeskul, and L. A. Pastur (1982), Introduction to the Theory of Disordered Systems (in Russian), Nauka, Moscow
Licciardello, D. C., and E. N. Economou (1975), Phys. Rev. **B 11**, 3697
Licciardello, D. C., D. L. Stein, and F. D. M. Haldane (1981), Phil. Mag. **B 43**, 189
Lobb, C. J., and D. J. Frank (1979), J. Phys. **C 12**, L 827
Long, A. R. (1982), Adv. Phys. **31**, 553
Long, A. R., K. Long, and N. Balkan (1980), J. non-crystall. Solids **35—36**, 415
Loring, R. F., H. C. Andersen, and M. D. Fayer (1983), Phys. Rev. Letters **50**, 1324
Lyo, S. Y., and T. Holstein (1973), Phys. Rev. **B 8**, 682
Lyo, S. K., T. Holstein, and R. Orbach (1978), Phys. Rev. **B 18**, 1637
Ma, M. (1982), Phys. Rev. **B 26**, 5097
Machta, J. (1981), Phys. Rev. **B 24**, 5260
Machta, J. (1983), J. Stat. Phys. **30**, 305
Maekawa, S. (1970), Phys. Rev. Letters **24**, 1175
Makumel, S. (1983), J. Stat. Phys. **30**, 179
Mansfield, R., and L. Kusztelan (1978), J. Phys. C **11**, 4157
Marshall, J. M. (1978), Phil. Mag. **B 38**, 335
Marshall, J. M. (1981), Phil. Mag. **B 43**, 401
Marshall, J. M. (1983), Phil. Mag. **B 47**, 323
Marshall, J. M., H. Michiel, and G. J. Adriaenssens (1983), Phil. Mag. **B 47**, 211
Maschke, K., H. Overhof, and P. Thomas (1974), phys. stat. sol. (b) **62**, 113
van der Meer, M., R. Schuchardt, and R. Keiper (1982), phys. stat. sol. (b) **110**, 571
Mehra, R. M., S. C. Agarwal, S. Rani, R. Shyam, and P. C. Mathur (1981), Thin Solid Films **76**, 379
Mell, H. (1974), in: Proceedings of the Fifth International Conference on Amorphous and Liquid Semiconductors, Garmisch-Partenkirchen, 1973. Ed. J. Stuke and W. Brenig, Taylor and Francis, London, p. 203
Mell, H., L. Schweitzer, and U. Voget-Grote (1980), J. noncrystall. Solids **35—36**, 639
Mell, H., and J. Stuke (1970), J. noncrystall. Solids **4**, 304
Mellander, B. E. (1982), Phys. Rev. **B 26**, 5886
Mendelson, K. S., and G. F. Karioris (1980), J. Phys. **C 13**, 6197
Metropolis, N., A. W. Rosenbluth, M. N. Rosenbluth, A. H. Teller, and E. Teller (1953), J. Chem. Phys. **21**, 1087

MICHEEV, V. A., V. A. MAYDANOV, and M. P. MICHIN (1982), Fiz. nizk. Temp. 8, 1000
MIKOSHIBA, N. (1962), Phys. Rev. **127**, 1962
MIKOSHIBA, N., and S. GONDA (1962), Phys. Rev. **127**, 1954
MILLER, A., and B. ABRAHAMS (1960), Phys. Rev. **120**, 745
MITESCU, C. D., M. ALLAIN, E. GUYON, and J. P. CLERC (1982), J. Phys. **A 15**, 2523
MONTROLL, E. W., and G. H. WEISS (1965), J. Math. Phys. **6**, 167
MOORE, E. (1974), J. Phys. **C 7**, 1840
MORIGAKI, K. (1980), Phil. Mag. **B 42**, 979
MORIN, F. J. (1958), Bell Syst. tech. J. **37**, 1047
MOTT, N. F. (1956), Canad. J. Phys. **31**, 1356
MOTT, N. F. (1967), Adv. Phys. **16**, 49
MOTT, N. F. (1968), J. non-crystall. Solids **1**, 1
MOTT, N. F. (1969a), Phil. Mag. **19**, 835
MOTT, N. F. (1969b), Festkörperprobleme **9**, 22
MOTT, N. F. (1970), Phil. Mag. **22**, 7
MOTT, N. F. (1971), Phil. Mag. **24**, 911
MOTT, N. F. (1972), J. non-crystall. Solids **8/10**, 1
MOTT, N. F. (1975), J. Phys. **C 8**, L 239
MOTT, N. F. (1976a), Phil. Mag. **34**, 643
MOTT, N. F. (1976b), Commun. Phys. **1**, 203
MOTT, N. F. (1979), Festkörperprobleme **19**, 331
MOTT, N. F. (1980) J. Phys. **C 13**, 5433
MOTT, N. F. (1981a), Phil. Mag. **B 43**, 941
MOTT, N. F. (1981b), Phil. Mag. **44**, 265
MOTT, N. F., and E. A. DAVIS (1979), Electronic Processes in Non-Crystalline Materials, 2nd edition. Clarendon Press, Oxford
MOTT, N. F., and J. H. DAVIES (1980), Phil. Mag. **B 42**, 845
MOTT, N. F., E. A. DAVIS, and R. A. STREET (1975), Phil. Mag. **32**, 961
MOTT, N. F., and W. D. TWOSE (1961), Adv. Phys. **10**, 107
MOVAGHAR, B., and N. A. CADE (1982), J. Phys. **C 16**, L 807
MOVAGHAR, B., M. GRÜNEWALD, B. POHLMANN, D. WÜRTZ, and W. SCHIRMACHER (1983a) J. Stat. Phys. **30**, 315
MOVAGHAR, B., B. POHLMANN, and G. W. SAUER (1980), phys. stat. sol. (b) **97**, 533
MOVAGHAR, B., B. POHLMANN, and W. SCHIRMACHER (1980), Phil. Mag. **B 41**, 49
MOVAGHAR, B., P. POHLMANN, and D. WÜRTZ (1981), J. Phys. **C 14**, 5127
MOVAGHAR, B., B. POHLMANN, and D. WÜRTZ (1983b) J. Phys. **C 16**, 3755
MOVAGHAR, B., and G. W. SAUER (1980), Solid State Commun. **35**, 841
MOVAGHAR, B., G. W. SAUER, and D. WÜRTZ (1982), J. Stat. Phys. **27**, 473
MOVAGHAR, B., and W. SCHIRMACHER (1981), J. Phys. **C 14**, 859
MOVAGHAR, B., and L. SCHWEITZER (1977), phys. stat. sol. (b) **80**, 491
MOVAGHAR, B., and L. SCHWEITZER (1978), J. Phys. **C 11**, 125
MOVAGHAR, B., L. SCHWEITZER, and H. OVERHOF (1978), Phil. Mag. **B 37**, 683
MURATA, K. K. (1979), J. Phys. **A 12**, 81
MURCH, G. E. (1982), Phil. Mag. **A 45**, 685
NAGAOKA, Y., and H. FUKUYAMA (Eds.) (1982), Anderson Localization, Springer Series in Solid State Sciences, Vol. 39, Springer-Verlag, Berlin, Heidelberg, New York
NAGATANI, T. (1981), J. Phys. **C 14**, 3383
NAGATANI, T. (1982), J. Phys. **C 15**, 5987
NAIR, K., and S. S. MITRA (1977), J. Noncrystall. Solids **24**, 1

NAKAMURA, M. (1982), J. Phys. **C 15**, L 749
NAKANISHI, H., and H. E. STANLEY (1978), J. Phys. **A 11**, L 189
NARAYANAMURTI, V., and R. O. POHL, Revs. Mod. Phys. **42**, 201
NGUEN VAN LIEN, and B. I. SHKLOVSKIJ (1981), Sol. State Commun. **38**, 99
DEN NIJS, M. P. M. (1979), J. Phys. **A 12**, 1857
NOOLANDI, J. (1977), Phys. Rev. **B 16**, 4474
OBUKHOV, S. P. (1980), Physica **101 A**, 145
ODAGAKI, T., and M. LAX (1981a), Phys. Rev. **B 24**, 5284
ODAGAKI, T., and M. LAX (1981b), Phys. Rev. Letters **47**, 1496
ODAGAKI, T., and M. LAX (1982), Phys. Rev. **B 26**, 6480
ODAGAKI, T., N. OGITA, and H. MATSUDA (1980), J. Phys. **C 13**, 189
ODAGAKI, T., and F. YONEZAWA (1978), Solid State Commun. **27**, 1203
OHTSUKI, T., and K. OKANO (1982), J. Chem. Phys. **77**, 1443
OLEŚ, B., A. M. OLEŚ, and K. A. CHAO (1982), Phys. Rev. **B 27**, 2504 (1983)
OLIVER, D. (1962), Phys. Rev. **127**, 1045
ONIZUKA, K. (1975), J. Phys. Soc. Japan **39**, 527
ORENSTEIN, J., M. A. KASTNER, and V. VANINOV (1982), Phil. Mag. **B 46**, 23
ORTUNO, M. (1981), J. Phys. **C 14**, L 421
ORTUNO, M., and M. POLLAK (1983), J. Phys. **C 16**, 1459
OSAKA, Y. (1979), J. Phys. Soc. Japan **47**, 729
OTTAVI, H., J. CLERC, G. GIRAUD, J. ROUSSENQ, E. GUYON, and C. D. MITESCU (1978), J. Phys. **C 11**, 1311
OVERHOF, H. (1975), phys. stat. sol. (b) **67**, 709
PAALANEN, M. A., and G. A. THOMAS (1983), Helv. Phys. Acta **56**, 27
PANT, M. M., and B. Y. TONG (1980), J. Phys. **C 13**, 1237
PAYANDEH, B. (1980), Revista Nuovo Cimento **3**, N. 3, p. 1
PEACOCK-LÓPEZ, E., and H. SUHL (1982), Phys. Rev. **B 26**, 3774
PEPPER, M., and M. J. UREN (1982), J. Phys. **C 15**, L 617
PETZINGER, K. G. (1982), Phys. Rev. **B 26**, 6530
PFEILSTICKER, R., S. KALBITZER, and G. MÜLLER (1978), Z. Physik **31**, 233
PFISTER, G. (1976), Phys. Rev. Letters **36**, 271
PFISTER, G., and H. SCHER (1978), Adv. Phys. **27**, 747
PHILLIPS, W. A. (1972), J. Low Temp. Phys. **7**, 351
PIKE, G. E., and C. H. SEAGER (1974), Phys. Rev. **B 10**, 1421
PIKE, G. E., and C. H. SEAGER (1976), phys. stat. sol. (b) **75**, 289
POLLAK, M. (1970), Disc. Faraday Soc. **50**, 13
POLLAK, M. (1971), Proc. Roy. Soc. **A 325**, 383
POLLAK, M. (1972), J. non-crystall. Solids **11**, 1
POLLAK, M. (1974), Proc. V. Intern. Conf. Amorph. Liquid Semicond., Garmisch-Partenkirchen 1973, Vol. 1. Taylor and Francis, London, p. 127
POLLAK, M. (1977), Phil. Mag. **36**, 1157
POLLAK, M. (1980), Phil. Mag. **B 42**, 781
POLLAK, M. (1981), J. Phys. **C 14**, 2977
POLLAK, M., and T. M. GEBALLE (1961), Phys. Rev. **122**, 1742
POLLAK, M., and M. L. KNOTEK (1979), J. non-crystall. Solids **32**, 141
POLLAK, M., and I. RIESS (1976), J. Phys. **C 9**, 2339
POTTS, R. B. (1952), Proc. Camb. Phil. Soc. **48**, 106
POWELL, M. J. (1979), Phys. Rev. **B 20**, 4194
POWELL, M. J. (1980), Phys. Rev. **B 21**, 3725

PRIGODIN, V. N. (1980), Zh. eksper. teor. Fiz. **79**, 2338
PRIGODIN, V. N. (1984), Zh. eksper. teor. Fiz. **86**, 242
PURI, A., and T. ODAGAKI (1981), Phys. Rev. **B 24**, 5541
RAY, R. K., and H. Y. FAN (1961), Phys. Rev. **121**, 768
REDFIELD, D. (1975), Adv. Phys. **24**, 463
REDNER, S. (1982), Phys. Rev. **B 25**, 5646
REIK, H. G. (1963), Sol. State Commun. **1**, 67
REYNOLDS, P. J., W. KLEIN, and H. E. STANLEY (1977), J. Phys. **C 10**, L 167
REYNOLDS, P. J., H. E. STANLEY, and W. KLEIN (1978), J. Phys. **A 11**, L 199
REYNOLDS, P. J., H. E. STANLEY, and W. KLEIN (1980), Phys. Rev. **B 21**, 1223
REZAYI, E. H., and H. SUHL (1982), Phys. Rev. Letters **45**, 1115
RICHARDS, P. M. (1977), Phys. Rev. **B 16**, 1393
RICHARDS, P. M. (1982), Phys. Letters **88 A**, 202
RITALA, R. K., and J. KURKIJÄRVI (1982), J. Phys. **C 15**, 3101
ROBASZKIEWICZ, S., R. MICNAS, and K. A. CHAO (1981), Phys. Rev. **B 24**, 4018
ROBASZKIEWICZ, S., R. MICNAS, and K. A. CHAO (1982), Phys. Rev. **B 26**, 3915
ROLLIN, B. V., and J. P. RUSSEL (1963), Proc. Phys. Soc. **81**, 571
ROSMAN, R., and B. SHAPIRO (1977), Phys. Rev. **B 16**, 5117
ROUSSENQ, J., J. CLERK, G. GIRAUD, E. GUYON, and H. OTTAVI (1976), J. Physique Letters **37**, 99
RUDENKO, A. I., and V. I. ARKHIPOV (1982), Phil. Mag. **B 45**, 177, 189, 209
SANKEY, O. F., and P. A. FEDDERS (1977), Phys. Rev. **B 15**, 3586
SANKEY, O. F., and P. A. FEDDERS (1980), Phys. Rev. **B 22**, 5135
SARKER, S., and E. DOMANY (1980), J. Phys. **C 13**, L 273
SARKER, S., and E. DOMANY (1981), Phys. Rev. **B 23**, 6018
SARYCHEV, A. K., and A. P. VINOGRADOFF (1981), J. Phys. **C 14**, L 487
SAWAKI, N., and T. NISHINAGA (1980), J. Phys. **C 13**, 5543
SCHER, H., and T. HOLSTEIN (1981), Phil. Mag. **B 44**, 343
SCHER, H., and M. LAX (1973), Phys. Rev. **7**, 4491, 4502
SCHER, H., and E. W. MONTROLL (1975), Phys. Rev. **B 12**, 2455
SCHER, H., and R. ZALLEN (1970), J. Chem. Phys. **53**, 3759
SCHIRMACHER, W. (1981), Solid State Commun. **39**, 893
SCHNAKENBERG, J. (1968), phys. stat. sol. **28**, 623
SCHMIDLIN, F. W. (1977), Phys. Rev. **B 16**, 2362
SCHMIDLIN, F. (1980), Phil. Mag. **B 41**, 535
SCHOBER, H. R., and A. M. STONEHAM (1982), Phys. Rev. **B 26**, 1819
SCHOMMERS, W. (1980), Phys. Rev. **B 21**, 847
SCHÖNHERR, G., H. BÄSSLER, and M. SILVER (1981), Phil. Mag. **B 44**, 47
SCHOTTE, K. (1966), Z. Physik **196**, 393
SCHWINGER, J. (1962), Brownian Motion of a Quantum Oscillator (in Russian), Inostr. Literat., Moscow
SEAGER, C. H., D. EMIN, and R. K. QUINN (1973), Phys. Rev. **B 10**, 4746
SEAGER, C. H., and G. E. PIKE (1974), Phys. Rev. **B 10**, 1435
SEEGER, A., and P. MONACHESI (1982), Phil Mag. **B 46**, 283
SHALITIN, D. (1981), J. Phys. **A 14**, 1983
SHANTE, V. K. S. (1977), Phys. Rev. **B 16**, 2597
SHANTE, V. K. S., and S. KIRKPATRICK (1971), Adv. Phys. **20**, 325
SHIMAKAWA, K. (1982), Phil. Mag. **B 46**, 123
SHIMAKAWA, K., and A. KONDO (1983), Phys. Rev. **B 27**, 1136

SHKLOVSKII, B. I. (1971), Zh. eksper. teor. Fiz. **61**, 2033
SHKLOVSKII, B. I. (1972a), Fiz. Tekh. Poluprov. **6**, 1197
SHKLOVSKII, B. I. (1972b), Fiz. Tekh. Poluprov. **6**, 2335
SHKLOVSKII, B. I. (1973), Fiz. Tekh. Poluprov. **8**, 416
SHKLOVSKII, B. I. (1975), Phys. Letters **51 A**, 289
SHKLOVSKII, B. I. (1976), Fiz. Tekh. Poluprov. **10**, 1440
SHKLOVSKII, B. I. (1977a), Fiz. Tekh. Poluprov. **11**, 2142
SHKLOVSKII, B. I. (1977b), phys. stat. sol. (b), **83**, K 11
SHKLOVSKII, B. I. (1978), phys. stat. sol. (b) **85**, K 111
SHKLOVSKII, B. I., and A. L. EFROS (1971), Zh. eksper. teor. Fiz. **60**, 867
SHKLOVSKII, B. I., and A. L. EFROS (1975a), Pis'ma Zurn. eksp. teor. Fiz. **1**, 174
SHKLOVSKII, B. I., and A. L. EFROS (1975b), Uspekh. Fiz. Nauk **117**, 401
SHKLOVSKII, B. I., and A. L. EFROS (1979), Electronic Properties of Doped Semiconductors (in Russian) Nauka, Moscow; English translation: Springer-Verlag, Berlin (1984)
SILVER, M., K. RISKO, and H. BÄSSLER (1979), Phil. Mag. **B 40**, 247
SILVER, M., G. SCHOENHERR, and H. BAESSLER (1982), Phys. Rev. Letters **48**, 352
SKAL, A. S., and B. I. SHKLOVSKII (1973), Fiz. Tekh. Poluprov. **7**, 1589
SKAL, A. S., and B. I. SHKLOVSKII (1974), Fiz. Tekh. Poluprov. **8**, 1586
SKAL, A. S., and B. I. SHKLOVSKII (1976), Fiz. tverd. Tela **16**, 1820
SKAL, A. S., B. I. SHKLOVSKII, and A. L. EFROS (1975), Fiz. tverd. Tela **17**, 506
SLADEK, R. J. (1958), J. Phys. Chem. Solids **5**, 157
SMYTHE, R. T., and J. C. WIERMAN (1978), Lecture Notes in Math. **671**, 1
SOUKOULIS, C. M., and E. N. ECONOMOU (1980), Phys. Rev. Letters **45**, 1590
SOUKOULIS, C. M., J. V. JOSÉ, E. N. ECONOMOU, and P. SHENG (1983), Phys. Rev. Letters **50**, 764
SPEAR, W. E. (1974), Adv. Phys. **23**, 523
SRINIVASAN, C. (1971), Phys. Rev. **B 4**, 2581
STAUFFER, D. (1976), Z. Physik **B 25**, 391
STAUFFER, D. (1979), Phys. Rev. **54**, 1
STEPHEN, M. J. (1977), Phys. Letters **56 A**, 149
STEPHEN, M. J. (1978), Phys. Rep. **B 17**, 4444
STEPHEN, M. J., and R. KARIOTIS (1982), Phys. Rev. **B 26**, 2917
STINCHCOMBE, R. B. (1973), J. Phys. **C 6**, L 1
STINCHCOMBE, R. B. (1974), J. Phys. **C 7**, 179
STINCHCOMBE, R. B., and B. P. WATSON (1976), J. Phys. **C 9**, 3221
STRALEY, J. P. (1976), J. Phys. **C 9**, 783
STRALEY, J. P. (1977a), Phys. Rev. **B 15**, 5733
STRALEY, J. P. (1977b), J. Phys. **C 10**, 3009
STRALEY, J. P. (1979), J. Phys. **C 12**, 3711
STRALEY, J. P. (1980), J. Phys. **C 13**, 819
STRALEY, J. P. (1982a), J. Phys. **C 15**, 2333
STRALEY, J. P. (1982b), J. Phys. **C 15**, 2343
STREET, R. A. (1977), Adv. Phys. **25**, 397
STREET, R. A., and N. F. MOTT (1975), Phys. Rev. Letters **35**, 1293
SUPRAPTO, B. B., and P. N. BUTCHER (1975), J. Phys. **C 8**, L 517
SUR, A., J. L. LEBOWITZ, J. MARRO, M. H. KALOS, and S. KIRKPATRICK (1976), J. Stat. Phys. **15**, 345
SYKES, M. F., and J. W. ESSAM (1964), Phys. Rev. **133**, A 310

SZPILKA, A. M., and P. VISCOR (1982), Phil. Mag. **B 45**, 485
TAHIR-KHELI, R. A., and R. J. ELLIOTT (1983), Phys. Rev. **B 27**, 844
TANAKA, S., and H. Y. FAN (1963), Phys. Rev. **132**, 1516
THOMAS, G. A., A. KAWABATA, Y. OOTUKA, S. KATSUMOTO, S. KOBAYASHI, and W. SASAKI (1982), Phys. Rev. **B 26**, 2113
THOULESS, D. J. (1972), The Quantum Mechanics of Many-Body Systems. Academic Press, New York
THOULESS, D. J. (1974), Phys. Rep. **13**, 93
THOLUESS, D. J. (1980), Phys. Rep. **67**, 5
THOULESS, D. J. (1982), Physica **109 & 110 B**, 1523
TITEICA, S. (1935), Ann. Phys. (Leipzig) **22**, 129
TOKUMOTO, H., R. MANSFIELD, and M. J. LEA (1982), Phil. Mag. **B 46**, 93
TONKS, D. L., and R. N. SILVER (1982), Phys. Rev. **B 26**, 6455
TOULOUSE, G. (1974), Nuovo Cim. **23**, 234
TOYOZAWA, Y. (1961), Prog. Theor. Phys. **26**, 29
TOYAZAWA, Y. (1962), J. Phys. Soc. Japan **17**, 986
TRIBERIS, G. P., and L. R. FRIEDMAN (1981), J. Phys. **C 14**, 4631
TRIFONOV, E. D., and K. PEUKER (1965), Fiz. tverd. Tela **8**, 2345
TUNALEY, J. K. E. (1974), Phys. Rev. Letters **33**, 1037
TURBAN, L. (1979), J. Phys. **C 12**, 1479
TURBAN, L., and P. GUILMIN (1979), J. Phys. **C 12**, 961
TUTHILL, G. F., and W. KLEIN (1982), J. Phys. **A 15**, L 377
TYABLIKOV, S. V. (1952), Zh. eksper. teor. Fiz. **23**, 381
UREN, M. J., R. A. DAVIES, and M. PEPPER (1980), J. Phys. **C 13**, L 985
VANDERBILT, D., and J. D. JOANNOPOULOS (1982), Phys. Rev. Letters **49**, 823
VICSEK, T. (1981), Z. Physik **45**, 153
VICSEK, T., and A. KERTESZ (1981), J. Phys. **A 14**, L 31
VILLAIN, J. (1977), J. Phys. **C 10**, 4793
VINYARD, G. C. (1957), J. Phys. Chem. Solids **3**, 121
VOLLHARDT, D., and P. WÖLFLE (1982), Phys. Rev. Letters **48**, 699
VOLLMER, H. D., and H. RISKEN (1980), Z. Physik **B 37**, 343
WANNIER, G. H. (1960), Phys. Rev. **117**, 423
WATSON, B. P., and P. L. LEATH (1974), Phys. Rev. **B 9**, 4893
WEBMAN, I., J. JORTNER, and M. H. COHEN (1977), Phys. Rev. **B 16**, 2593
WEBMAN, I., and KLAFTER, J. (1982), Phys. Rev. **B 26**, 5950
WEGNER, C. (1979), Z. Physik **B 35**, 207
WEYL, H. (1928), Z. Physik **46**, 1
WHALL, T. E. (1981), J. Phys. **C 14**, L 887
WHALL, T. E., M. O. RIGO, M. R. B. JONES, and A. J. POINTON (1977), J. Physique **C 1**, Suppl. 4, 229
WILSON, K. G., and J. KOGUT (1974), Phys. Rep. **12 C**, 75
WINTLE, H. J. (1975), Solid State Commun. **17**, 755
WINTLE, H. J., and P. A. PUHACH (1978), J. Stat. Phys. **18**, 557
WOLF, D. (1980), J. Phys. Chem. Solids **41**, 1053
WU, F. Y. (1982), Revs. Mod. Phys. **54**, 235
WUERTZ, D., and P. THOMAS (1978), phys. stat. sol. (b) **88**, K 73
YAFET, Y., R. W. KEYES, E. N. ADAMS (1956), J. Phys. Chem. Solids **1**, 137
YAMAGUCHI, E., H. AOKI, and H. KAMIMURA (1979), J. Phys. **C 12**, 4801

YAMASHITA, J., and T. KUROSAWA (1958), J. Phys. Chem. Sol. **5**, 34
YEOMANS, J. M., and R. B. STINCHCOMBE (1978), J. Phys. **C 11**, 4095
YONEZAWA, F. (Ed.) (1981), Fundamental Physics of Amorphous Semiconductors. Springer Series in Solid State Sciences, Vol. 25. Springer-Verlag, Berlin, Heidelberg, New York
YONEZAWA, F., I. WEBMAN, and M. H. COHEN (1982), in: Anderson Localization. Springer Series in Solid State Sciences, Vol. 39. Eds. Y. NAGAOKA and H. FUKUYAMA, Springer Verlag, Berlin, Heidelberg, New York, p. 166
YOSHIDA, A., S. NAKASHIMA, and M. IEDA (1980), J. noncrystall. Solids **35–36**, 421
YOUNG, A. P., and R. B. STINCHCOMBE (1975), J. Phys. **C 8**, L 535
ZABRODSKII, A. G. (1977), Fiz. Tekh. Poluprov. **11**, 595
ZABRODSKII, A. G. (1980), Fiz. Tekh. Poluprov. **14**, 1130
ZABRODSKII, A. G., and I. G. SHLIMAK (1977), Fiz. Tekh. Poluprov. **11**, 736
ZIMAN, J. M. (1961), Phil. Mag. **6**, 1013
ZIMAN, J. M. (1968), J. Phys. **C 1**, 1532
ZIMAN, J. M. (1979), Models of Disorder, Cambridge University Press, Cambridge, London, New York, Melbourne
ZIMAN, T. A. L. (1979), J. Phys. **C 12**, 2645
ZUBAREV, D. N. (1971), Nonequilibrium Statistical Thermodynamics (in Russian), Nauka, Moscow; German translation: D. N. Subarew, Statistische Thermodynamik des Nichtgleichgewichts, Akademie-Verlag, Berlin 1976
ZVYAGIN, I. P. (1973), phys. stat. sol. (b) **58**, 443
ZVYAGIN, I. P. (1977), phys. stat. sol. (b) **83**, 63
ZVYAGIN, I. P. (1978), Vestnik Moskovsk. Univ. Fiz./Astr. **19**, 82
ZVYAGIN, I. P. (1979), phys. stat. sol. (b) **95**, 227
ZVYAGIN, I. P. (1980a), phys. stat. sol. (b) **101**, 9
ZVYAGIN, I. P. (1980b), phys. stat. sol. (b) **97**, 143
ZVYAGIN, I. P., and R. KEIPER (1972), Wiss. Zeitschr. Humboldt-Univ. Berlin, Math.-Nat., R XXI, p. 459
ZWANZIG, R. W. (1961), Phys. Rev. **124**, 983

References added in proof

ADLER, J., A. AHARONY, and D. STAUFFER (1985), J. Phys. **A 18**, L 129
AHARONY, A., S. ALEXANDER, O. ENTIN-WOHLMAN, and R. ORBACH (1985), Phys. Rev. **B 31**, 2565
AHARONY, A., Y. GEFEN and A. KAPITULNIK (1984), J. Phys. **A 17**, L 197
AHARONY, A., and D. STAUFFER (1984), Phys. Rev. Letters **52**, 2368
ALEXANDER, S., J. BERNASCONI, and R. ORBACH (1978), Phys. Rev. **B 17**, 4311
ALEXANDER, S., G. S. GREST, H. NAKANISHI, and T. A. WITTEN, Jr. (1984), J. Phys. **A 17**, L 185
ALEXANDER, S., C. LAERMANS, R. ORBACH, and H. M. ROSENBERG (1983), Phys. Rev. **B 28**, 4615
ALEXANDER, S., and R. ORBACH (1982), J. Phys. (Paris), Lettres 43, L 625
ANDERSON, P. W. (1983), Physica **117 B** & **118 B**, 30
ANDERSON, P. W., K. A. MUTTALIB, T. V. RAMAKRISHNAN (1983), Phys. Rev. **B 28**, 117

ANDO, T. (1984), J. Phys. Soc. Japan **53**, 3101, 3126
ANGLES, D'AURIAC, J. C., A. BENOIT, and R. RAMMAL (1983), J. Phys. **A 16**, 4039
ANGLES D'AURIAC, J. C., and R. RAMMAL (1984), J. Phys. **A 17**, L 15
ANLAUF, J. K. (1984), Phys. Rev. Letters **52**, 1845
AOKI, H. (1983), J. Phys. **C 16**, L 205
AOKI, H., and T. ANDO (1981), Solid State Commun. **38**, 1079
AOKI, H., and T. ANDO (1985), Phys. Rev. Letters **54**, 831
ARKHIPOV, V. I., J. A. POPOVA, and A. I. RUDENKO (1983), Phil. Mag. **B 48**, 401
ARONOV, A. G. (1984), Physica **126 B**, 314
AZBEL, M. YA. (1984), Phil. Mag. **B 50**, 229
AZBEL, M. YA., and D. P. DI VENZENZO (1984), Phys. Rev. **B 30**, 6877
BALAKRISHNAN, V., and M. KHANTHA (1985), J. Phys. **C 18**, L 351
BALBERG, I. (1985), Phys. Rev. **B 31**, 4053
BALBERG, I., C. H. ANDERSON, S. ALEXANDER, and N. WAGNER (1984b) Phys. Rev. **B 30**, 3933
BALBERG, I., N. BIRNENBAUM, and N. WAGNER (1984a), Phys. Rev. Letters **52**, 1465
BANAVAR, J. R., and M. CIEPLAK (1983), Phys. Rev. **B 28**, 3813
BARMA, M. (1985), J. Phys. **A 18**, L 277
VAN BEIJEREN, H. (1982), Revs. Mod. Phys. **54**, 195
VAN BEIJEREN, H., K. W. KEHR, and R. KUTTNER (1983), Phys. Rev. **B 28**, 5711
BELITZ, D., and W. GÖTZE (1982), J. Phys. **C 15**, 981
BELITZ, D., and W. SCHIRMACHER (1983), J. Phys. **C 16**, 913
DE BELL, K., and J. W. ESSAM (1983), J. Phys. **A 16**, 385, 3145, 3553
BEN-AVRAHAM, D., and S. HAVLIN (1984), Phys. Rev. **A 29**, 2309
BENZAQUEN, M., and D. WALSH (1984), Phys. Rev. **B 30**, 7287
BENZONI, J. (1984), J. Phys. **A 17**, 2651
BENZONI, J., and J. L. CARDY (1984), J. Phys. **A 17**, 179
BEREZIN, A. (1984), J. Phys. **C 17**, L 393
BERGMANN, G. (1984), Phys. Rep. **107**, 1
BERNASCONI, J., and W. R. SCHNEIDER (1982), J. Phys. **A 15**, L 729
BIDAUX, R., and G. FORGÀCS (1984), J. Phys. **A 17**, 1853
BILLER, R. (1984), Z. Physik **B 55**, 7
BILLER, R. (1985), J. Phys. **A 18**, 989
BISHOP, D. J., R. C. DYNES, B. J. LIN and D. C. TSUI (1984), Phys. Rev. **B 30**, 3539
BISKUPSKI, G., H. DUBOIS, J. L. WOJKIEWICZ, A. BRIGGS, and G. REMENYI (1984), J. Phys. **C 17**, L 411
BLUMEN, A., J. KLAFTER, B. S. WHITE, and G. ZUMOFEN (1984), Phys. Rev. Letters **53**, 1301
BODUNOV, E. N., and V. A. MALYSHEV (1984), Fiz. tverd. Tela **26**, 2990
BÖTTGER, H., P. SZYLER, and D. WEGENER (1985a), phys. stat. sol (b) **128**, K 179
BÖTTGER, H., P. SZYLER, and D. WEGENER (1985b), phys. stat. sol. (b), in press
BOUGHALEB, Y., and J. F. GOUYET (1983), Sol. State Ionics **9** & **10**, 1401
BRAZOVSKII, S. A., and N. N. KIROVA (1981), Pis'ma Zh. eksper. teor. Fiz. **33**, 6
BRÉDAS, J. L., R. R. CHANCE, and R. SILBEY (1982b), Phys. Rev. **B 26**, 5843
BRÉDAS, J. L., B. THÉMANS, and J. M. ANDRÉ (1982a), Phys. Rev. **B 26**, 6000
BREZINI, A., and G. OLIVIER (1983), Phil. Mag. **B 47**, 461
BRYKSIN, V. V. (1983a), Fiz. tverd. Tela **25**, 1780
BRYKSIN, V. V. (1983b), Fiz. tverd. Tela **25**, 2431
BRYKSIN, V. V. (1984), Fiz. tverd. Tela **26**, 1362
BRYKSIN, V. V. (1985), Fiz. tverd. Tela **27**, in press
BRYKSIN, V. V., and V. S. VOLOSHIN (1984), Fiz. tverd. Tela **26**, 2357

BRYKSIN, V. V., V. S. VOLOSHIN, and A. V. RAJTSEV (1983), Fiz. tverd. Tela **25**, 1427
BRYKSIN, V. V., and G. YU. YASHIN (1983), Fiz. tverd. Tela **25**, 3025
BUG, A. L. R., S. A. SAFRAN, and I. WEBMAN (1985), Phys. Rev. Letters **54**, 1412
BUNDE, A., D. K. CHATURVEDI, and W. DIETERICH (1983), Sol. State Ionics **9 & 10**, 1415
BUNDE, A., A. CONIGLIO, D. C. HONG, and H. E. STANLEY (1985), J. Phys. **A 18**, L 137
CUTCHER, P. N., J. A. MC INNES, and S. SUMMERFIELD (1983), Phil. Mag. **B 48**, 551
CADE, N. A., and B. MOVAGHAR (1983), J. Phys. **C 16**, 539
CAI, W., and C. S. TING (1985), J. Phys. **C 18**, 1827, 1843
CALDEIRA, A. O., and A. J. LEGGETT (1981), Phys. Rev. Letters **46**, 211
CALDEIRA, A. O., and A. J. LEGGETT (1983), Ann. Phys. (N.Y.) **149**, 374
CAMPBELL, D. K., A. R. BISHOP, and K. FESSER (1982), Phys. Rev. **B 26**, 6862
CARDY, J. L. (1983), J. Phys. **A 16**, L 709
CARDY, J. L., and P. GRASSBERGER (1985), J. Phys. **A 18**, L 267
CARINI, J. P., S. R. NAGEL, L. K. VARGA, and T. SCHMIDT (1983), Phys. Rev. **B 27**, 7589
CARMELI, B., and A. NITZAN (1984), Phys. Rev. **A 29**, 1481
CARTON, J. P. (1983), J. Phys. **A 16**, L 219
CASTELLANI, C., C. DI CASTRO, P. A. LEE, M. MA, S. SORELLA, and E. TABET (1984), Phys. Rev. **B 30**, 1596
CHAKRABARTI, B. K., and S. S. MANNA (1983), J. Phys. **A 16**, L 113
CHAKRAVERTY, B. K. (1982), Solar En. Mater. 8, 71
CHAKRAVERTY, B. K., J. M. SIENKO, and J. BONNEROT (1978), Phys. Rev. **B 17**, 3781
CHANCE, R. R., J. L. BRÉDAS, and R. SILBEY (1984), Phys. Rev. **B 29**, 4491
CHATURVEDI, D. K. (1983), Phys. Rev. **B 28**, 6868
CHATURVEDI, D. K. (1984), J. Phys. **C 17**, L 449
CHATURVEDI, D. K., and M. P. TOSI (1984), Z. Physik **58**, 49
COHEN, M. H. (1982), in: Melting, Localization and Chaos, eds. R. K. KALIA and P. VASHISHTA, Elsevier, New York, p. 125
COHEN, M. H., E. N. ECONOMOU, and C. M. SOUKOULIS (1983), Phys. Rev. Letters **51**, 1202
COHEN, M. H., E. N. ECONOMOU, and C. M. SOUKOULIS (1984), Phys. Rev. **B 29**, 4496, 4500
CONIGLIO, A. (1983), in: Magnetic Phase Transitions, Sol. State Ser. Vol. **48**, Springer-Verlag, Berlin, p. 195
DAVIES, J. H. (1984), J. Phys. **C 17**, 3031
DELYON, F., B. SIMON and B. SOUILLARD (1984), Phys. Rev. Letters **52**, 2187
DENTENEER, P. J. H., and M. H. ERNST (1983), J. Phys. **C 16**, L 961; Phys. Rev. **B 29**, 1755
DERRIDA, B. (1983), J. Stat. Phys. **31**, 433
DERRIDA, B., and J. M. LUCK (1983), Phys. Rev. **B 28**, 7183
DERRIDA, B., and R. ORBACH (1983), Phys. Rev. **B 27**, 4694
DERRIDA, B., R. ORBACH, and KIN-WAH YU (1984), Phys. Rev. **B 29**, 6645
DEUTSCHER, G., A. M. GOLDMAN, and H. MICKLITZ (1985), Phys. Rev. **B 31**, 1679
DEUTSCHER, G., R. ZALLEN, and J. ADLER (eds.) (1983), Percolation Structures and Processes, Adam Hilger, Bristol
DOMANY, E., S. ALEXANDER, D. BENSIMON, and L. P. KADANOFF (1983), Phys. Rev. **B 28**, 3110
DRUGER, S. D., M. A. RATNER, and A. NITZAN (1985), Phys. Rev. **B 31**, 3939
DUBOIS, H., G. BISKUPSKI, J. P. SPRIET, and A. BRIGGS (1985), J. Phys. **C 18**, L 195
DUXBURY, P. M., and S. L. A. DE QUEIROZ (1985), J. Phys. **A 18**, 661
DYNES, R. C., T. H. GEBALLE, G. W. HULL, JR., and J. P. GARNO (1983), Phys. Rev. **B 27**, 5188
DYRE, J. C. (1985), Phys. Letters **108 A**, 457

EAVES, L., P. S. S. GUIMARAES, P. C. MAIN, I. P. ROCHE, J. A. CHROBOCZEK, H. MITTER, J. PORTAL, and G. HILL (1984), J. Phys. **C 17**, L 345
ECONOMOU, E. N. (1974), Physica **127 B**, 246
EFETOV, K. B. (1983), Adv. Phys. **32**, 53
ELAM, W. T., A. R. KERSTEIN, and J. J. REHR (1984), Phys. Rev. Letters **52**, 1516
EMIN, D., and T. HOLSTEIN (1976), Phys. Rev. Letters **36**, 323
ENDERLEIN, R., K. PEUKER, and F. BECHSTEDT (1979), phys. stat. sol. (b) **92**, 149
ENTIN-WOHLMAN, O., S. ALEXANDER, and KIN-WAH YU (1984), Phys. Rev. **B 29**, 4588
ERNST, M. H., and J. R. DORFMAN (1972), Physica **61**, 157
EVANS J. W., and D. K. HOFFMAN (1984), Phys. Rev. **B 30**, 2704
FAMILY, F. (1984), J. Stat. Phys. **36**, 881
FAMILY, F., and A. CONIGLIO (1985), J. Physique **46**, L 9
FEINBERG, D., and J. RANNINGER (1984), Physica **14 D**, 29
FESSER, K., A. R. BISHOP, and D. K. CAMPBELL (1983), Phys. Rev. **B 27**, 4804
FINKEL'SHTEIN, A. M. (1983), Zh. eksper. teor. Fiz. **84**, 168
FINKEL'SHTEIN, A. M. (1984), Z. Physik **B 56**, 189
FISHER, D. S. (1984), Phys. Rev. **A 30**, 960
FRÖHLICH, J., and T. SPENCER (1984), Phys. Rep. **103**, 9; Physica **124 A**, 303
GEFEN, Y., and Y. IMRY (1983), Phys. Rev. **B 28**, 3569
GEFEN, Y., D. J. THOULESS, and Y. IMRY (1983), Phys. Rev. **B 28**, 6677
GEISEL, T., and S. THOMAE (1984), Phys. Rev. Letters **52**, 193
DE GENNES, P. G. (1976), J. Phys. (Paris) Lettres **37**, L 1
GHIBAUDO, G. (1984), J. Phys. **C 17**, 3067
GIRVIN, S. M., A. H. MAC DONALD, and P. M. PLATZMAN (1985), Phys. Rev. Letters **54**, 581
GIRVIN, S. M., and T. JACH (1984), Phys. Rev. **B 29**, 5617
GIRVIN, S. M., and M. JONSON (1980), Phys. Rev. **B 22**, 3583
GÖTZE, W. (1981), Phil. Mag. **B 43**, 219
GRABERT, H., and U. WEISS (1984), Phys. Rev. Letters **53**, 1787
GRASSBERGER, P. (1983), Math. Biosci. **63**, 157
GRASSBERGER, P. (1985), Phys. Letters **107 A**, 101
GREST, D. S., and P. A. LEE (1983), Phys. Rev. Letters **50**, 693
GRÜNEWALD, M., B. POHLMANN, B. MOVAGHAR, and D. WÜRTZ (1984), Phil. Mag. **B 49**, 341
HALPERIN, B. I. (1982), Phys. Rev. **B 25**, 2185
HÄNGGI, P. (1984), Phys. Letters **103 A**, 97
HÄNGGI, P., and F. MOJTABAI (1983), J. Stat. Phys. **30**, 401
HARRIS, A. B. (1983), Phys. Rev. **B 28**, 2614
HARRIS, A. B., S. KIM, and T. C. LUBENSKY (1984), Phys. Rev. Letters **53**, 743
HARRIS, A. B., and T. C. LUBENSKY, (1983), J. Phys. **A 16**, L 365
HAUS, J. W., and K. W. KEHR (1983), Phys. Rev. **B 28**, 3573
HAVLIN, S., M. DISHON, J. E. KIEFER, and G. H. WEISS (1984a), Phys. Rev. Letters **53**, 407
HAVLIN, S., R. NOSSAL, B. TRUS, and G. H. WEISS (1984b), J. Phys. **A 17**, L 957
HAVLIN, S., G. H. WEISS, D. BEN-AVRAHAM, and D. MOVSHOVITZ (1984c), J. Phys. **A 17**, L 849
HAVLIN, S., B. TRUS, G. H. WEISS, and D. BEN-AVRAHAM (1985), J. Phys. **A 18**, L 247
HERRMANN, H. J., F. F. FAMILY, and H. E. STANLEY (1983), J. Phys. **A 16**, L 375
HERRMANN, H. J., and H. E. STANLEY (1984), Phys. Rev. Letters **53**, 1121
HIKAMI, S. (1981), Phys. Rev. **B 24**, 2671
HIKAMI, S. (1984), Prog. Theor. Phys. **72**, 722
HIKAMI, S., A. I. LARKIN, Y. NAGAOKA (1980), Prog. Theor. Phys. **63**, 5142
HIRAMOTO, H., and TOYOZAWA (1985), J. Phys. Soc. Japan **54**, 245

DEN HOLLANDER, W. TH. F. (1984), J. Stat. Phys. **37**, 331
HOLSTEIN, T. (1981), Mol. Cryst. Liq. Cryst. **77**, 235
HONG, D. H., S. HAVLIN, H. J. HERRMANN and H. E. STANLEY (1984), Phys. Rev. **B 30**, 4083
HOWARD, I. A., and E. M. CONWELL (1984), Phys. Rev. **B 29**, 6879
HOWSON, M. A. (1984), J. Phys. **F 14**, L 25
HOWSON, M. A., and D. GREIG (1984), Phys. Rev. **B 30**, 4805
HUBER, D. L. (1984), J. Chem. Phys. **81**, 689
IHLE, D. (1985), Z. Physik **B 58**, 91
IMRY, Y. (1980), Phys. Rev. Letters **44**, 469
IMRY, Y. (1981), J. Appl. Phys. **52**, 1817
JACUCCI, G., M. TOLLER, G. DE LORENZI, and C. P. FLYNN (1984), Phys. Rev. Letters **52**, 295
JAN, N., D. L. HUNTER, and T. LOOKMAN (1984), Phys. Rev. **B 29**, 6356
JONSON, M., and S. M. GIRVIN (1979), Phys. Rev. Letters **43**, 1447
KAGAN, YU., and L. A. MAKSIMIV (1983), Phys. Letters **95 A**, 242
KANTOR, Y. (1984), J. Phys. **A 17**, L 843
KAPITULNIK, A., A. AHARONY, G. DEUTSCHER, and D. STAUFFER (1983), J. Phys. **A 16**, L 269
KARIOTIS, R. (1984), J. Phys. **A 17**, 2889
KARIOTIS, R., and M. J. STEPHEN (1983), Phys. Letters **98 A**, 349
KAUFMAN, J. H., N. COLANERI, J. C. SCOTT, and G. B. STREET (1984), Phys. Rev. Letters **53**, 1005
KAVEH, M. (1984), Physica **127 B**, 292
KAVEH, M., and N. F. MOTT (1982), J. Phys. **C 15**, L 707
KAVEH, M., and N. F. MOTT (1983), Phil. Mag. **B 47**, L 9
KAVEH, M., and N. WISER (1984), Adv. Phys. **33**, 257
MAC KAY, G., and N. JAN (1984), J. Phys. **A 17**, L 757
KEHR, K. W. (1983), J. Stat. Phys. **30**, 509
KEHR, K. W., R. KUTNER, and K. BINDER (1983), Phys. Rev. **B 23**, 4931
KERSTEIN, A. R. (1983), J. Phys. **A 16**, 3071
KERTÉSZ, J. (1983), J. Phys. **A 16**, L 471
KERTÉSZ, J., and J. METZGER (1984), J. Phys. **A 17**, L 501
KHANTHA, M., and V. BALAKRISHNAN (1984) Phys. Rev. **B 29**, 4679
KHMEL'NITSKII, D. E. (1984), Physica **126 B**, 235
KHOLODENKO, A. L., and K. F. FREED (1984), J. Phys. **A 17**, L 55
MAC KINNON, A. (1983), J. Phys. **C 16**, L 1155
KIRKPATRICK, S. (1978), AIP Conf. Proc. **40**, 99
KIVELSON, S., and A. J. EPSTEIN (1984), Phys. Rev. **B 29**, 3336
KLAFTER, J., A. BLUMEN, and G. ZUMOFEN (1984), J. Stat. Phys. **36**, 561
KLAFTER, J., and R. SILBEY (1983), Phil. Mag. **B 47**, 337
KLEIN, R. S. (1985), Phys. Rev. **B 31**, 2014
VON KLITZING, K. (1984), Physica **126 B**, 242
VON KLITZING, K., G. DORDA, and M. PEPPER (1980), Phys. Rev. Letters **45**, 494
VON KLITZING, K., and G. EBERT (1983), Physica **117 B & 118 B**, 682
KORZENIEWSKI, G., R. FRIESNER, and R. SILBEY, J. Stat. Phys. **31**, 451
KRENKRE, V. M., V. ERN, and A. FORT (1983), Phys. Rev. **B 28**, 598
KUTNER, R., H. VAN BEIJEREN, and K. W. KEHR (1984), Phys. Rev. **B 30**, 4382
KUTNER, R., K. BINDER, and K. W. KEHR (1983), Phys. Rev. **B 28**, 1846
KUROBE, A., T. TAKEMORI, and H. KAMIMURA (1984), Phys. Rev. Letters **52**, 1457
LANGER, J., and T. NEAL (1966), Phys. Rev. Letters **16**, 984
LAUGHLIN, R. B. (1983), Phys. Rev. Letters **50**, 1395

LEE, P. A. (1984), Phys. Rev. Letters **53**, 2042
LEHR, W., J. MACHTA, and M. NELKIN (1984), J. Stat. Phys. **36**, 15
LEVIN, E. I. (1984), Fiz. Tekh. Poluprov. **18**, 255
LEVIN, E. I., and B. I. SHKLOVSKII (1984), Fiz. Tekh. Poluprov. **18**, 856
LEVINE, H., S. B. LIBBY, and A. M. M. PRUISKEN (1983), Phys. Rev. Letters **51**, 1915
LIN, B. J. F., M. A. PAALANEN, A. C. GOSSARD, and D. C. TSUI (1984), Phys. Rev. **B 29**, 927
LIU, Y., and Z. Q. ZHANG (1985), J. Phys. **A 18**, 1027
LOBB, C. J., and D. J. FRANK (1984), Phys. Rev. **B 30**, 4090
LOGAN, D. E., and P. G. WOLYNES (1985), Phys. Rev. **B 31**, 2437
DE LORENZI, G., G. P. FLYNN, and G. JACUCCI (1984), Phys. Rev. **B 30**, 5430
LORING, R. F., H. C. ANDERSEN, and M. D. FAYER, J. Chem. Phys. **80**, 5731
LUCK, J. M. (1983), Nucl. Phys. **B 225**, 169
LUCK, J. M. (1984), J. Phys. **A 17**, 2069
LUQUE, J. J., and A. CORDOBA (1982), J. Chem. Phys. **76**, 6393
MA, M., and E. FRADKIN (1983), Phys. Rev. **B 28**, 2990
MACHTA, J., M. H. ERNST, H. VAN BEIJEREN, and J. R. DORFMAN (1984), J. Stat. Phys. **35** 413
MANDELBROT, B. B., Y. GEFEN, A. AHARONY, and J. PEYRIÈRE (1985), J. Phys. **A 18**, 335
MANSFIELD, R., and H. TOKUMOTO (1983), Phil. Mag. **B 48**, L 1
MARCHESONI, F., and J. K. VIJ (1985), Z. Physik **58**, 187
MARGOLINA, A., and H. J. HERRMANN (1984), Phys. Letters **104 A**, 295
MARINARI, E., G. PALADIN, G. PARISI, and A. VULPIANI (1985), J. Physique **45**, 657
MARINARI, E., G. PARISI, D. RUELLE, and P. WINDEY (1983), Phys. Rev. Letters **50**, 1223
MARKIEWICZ, R. S., and C. J. ROLLINS (1984), Phys. Rev. **B 29**, 735
MARRO, J., and J. MASOLIVER (1985), Phys. Rev. Letters **54**, 731
MARSHALL, J. M. (1983), Rep. Prog. Phys. **46**, 1235
MARSHALL, J. M., and C. MAIN (1983), Phil. Mag. **B 47**, 471
MARTIN, J. F. (1985), J. Phys. **A 18**, L 207
MARTINELLI, F., and H. HOLDEN (1984), Commun. Math. Phys. **93**, 197
MEL'NIKOV, V. I. (1985), Physica **130 A**, 606
MICHIEL, H., J. M. MARSHALL, and G. J. ADRIAENSSENS (1983), Phil. Mag. **B 48**, 187
MC MILLAN, W. L. (1981), Phys. Rev. **B 24**, 2739
MC MILLAN, W. L. (1968), Phys. Rev. **167**, 331
MONROE, D. (1985), Phys. Rev. Letters **54**, 146
MONROE, D., and M. A. KASTNER (1983), Phil. Mag. **B 47**, 605
MOOIJ, J. H. (1973), phys. stat. sol. (a) **17**, 521
MORAWECK, M., and U. BRANDT (1984), Z. Physik **B 56**, 327
MOTT, N. F. (1984), Rep. Prog. Phys. **47**, 909, Phil. Mag. **B 49**, L 75
MOTT, N. F., and M. KAVEH (1983), Phil. Mag. **B 47**, 577, L 17
MOVAGHAR, B., D. W. MURRAY, B. POHLMANN, and D. WÜRTZ (1984), J. Phys. **C 17**, 1677
MÜLLER, H., and P. THOMAS (1983), Phys. Rev. Letters **51**, 702
MÜLLER, H., and P. THOMAS (1984), J. Phys. **17**, 5337
MUNAKATA, T. (1985), Prog. Theor. Phys. **73**, 570, 826
MUTTALIB, K. A. (1983), Phys. Rev. **B 27**, 4628
NADAL, J. P., and J. VANNIMENUS (1985), J. Physique **46**, 17
NAKAMURA, M. (1984), Phys. Rev. **B 29**, 3691
NAKANISHI, H. (1984), J. Phys. **A 17**, L 329
NARA, S. (1983), Physica **121 B**, 169
NAUGLE, D. G. (1984), J. Phys. Chem. Solids **45**, 367
NIEUWENHUIZEN, TH. M., and M. H. ERNST (1985), Phys. Rev. **B 31**, 3518

NIGHTINGALE, P. (1982), J. Appl. Phys. **53**, 7927
NIU, Q., D. J. THOULESS, and Y. S. WU (1985), Phys. Rev. **B 31**, 3372
ODAGAKI, T., and K. C. CHANG (1984), Phys. Rev. **B 30**, 1612
ODAGAKI, T., M. LAX, and D. S. DAY (1984), Phys. Rev. **B 30**, 6911
ODAGAKI, T., M. LAX, and A. PURI (1983), Phys. Rev. **B 28**, 2755
OHTSUKI, T., and T. KEYES (1984a), Phys. Rev. Letters **52**, 1177
OHTSUKI, T., and T. KEYES (1984b), J. Phys. **A 17**, L 267
OHTSUKI, T., and T. KEYES (1985a), Phys. Letters **108 A**, 203
OHTSUKI, T., and T. KEYES (1985b), J. Phys. **A 18**, L 171
ONODERA, Y., and S. OKUNO (1983), J. Phys. Soc. Jap. **52**, 2478
PALEVSKI, A., and G. DEUTSCHER (1984), J. Phys. **A 17**, L 895
PANDEY, R. B. (1984), Phys. Rev. **B 30**, 489
PANDEY, R. B., D. STAUFFER, A. MARGOLINA, and J. G. ZABOLITZKY (1984), J. Stat. Phys. **34**, 427
PERUGGI, F., F. DI LIBERTO, and G. MOUROY (1984), Physica **123 A**, 175
PHANI, M. K., and D. DHAR (1984), J. Phys. **A 17**, L 645
POLLAK, M. (1984), Phil. Mag. **B 50**, 265
POON, S., K. M. WONG, and A. J. DREHMAN (1985), Phys. Rev. **B 31**, 1668
PREGENZER, A. L., and H. SUHL (1985), Phys. Rev. **A 31**, 1718
PRAMANIK, M. H. A., P. N. BUTCHER, and I. D. COX (1983), Phil. Mag. **B 47**, 437
PRIGODIN, V. N. (1984a), Sol. St. Commun. **51**, 371
PRIGODIN, V. N. (1984b), Sol. St. Commun. **51**, 607
PRIGODIN, V. N., and A. N. SAMUKHIN (1983a), Fiz. tverd. Tela **25**, 1720
PRIGODIN, V. N., and A. N. SAMUKHIN (1983b), Sol. St. Commun. **46**, 379; Zh. eksper. teor. Fiz. 84, 301
PRIGODIN, V. N., CH. SEIDEL, and A. P. NAKHMEDOV (1985), phys. stat. sol. (b) **128**, 109
PUECH, L., and R. RAMMAL (1983), J. Phys. **C 16**, L 1197
PURI, A., and T. ODAGAKI (1984), Phys. Rev. **B 29**, 1707
DE RAEDT, H., and A. LAGENDIJK (1984), Phys. Rev. **B 30**, 1671
RAMMAL, R. (1983), Phys. Rev. **B 28**, 4871
RAMMAL, R. (1984a), J. Physique **45**, 191
RAMMAL, R. (1984b), Phys. Letters **102 A**, 117
RAMMAL, R., J. C. ANGLES D'AURIAC, and A. BENOIT (1984), Phys. Rev. **B 30**, 4087
RAMMAL, R., C. TANNOUS, and A. M. S. TREMBLAY (1985), Phys. Rev. **A 31**, 2662
RAMMAL, R., and G. TOULOUSE (1983), J. Physique Lettres **44**, L 13
RIES, B., G. SCHÖNHERR, H. BÄSSLER, and M. SILVER (1984), Phil. Mag. **B 49**, 27, 259, 597
RISEBOROUGH, P. S. (1983), Z. Physik **B 51**, 173
RISEBOROUGH, P. S. (1984), Ann. Phys. (N.Y.) **153**, 1
ROSENBAUM, T. F., R. F. MILLIGAN, M. A. PAALANEN, G. A. THOMAS, R. N. BHATT, and W. LIN (1983) Phys. Rev. **B 27**, 7509
SAHIMI, M. (1983), in: The Mathematics and Physics of Disordered Media, Lecture Notes in Mathematics **1035**, Springer-Verlag, Berlin, p. 314
SAHIMI, M. (1984a), J. Phys. **A 17**, 3073
SAHIMI, M. (1984b), J. Phys. **A 17**, L 601; J. Phys. **C 17**, L 355
SAHIMI, M. (1985), J. Phys. **A 18**, 83
SAHIMI, M., B. D. HUGHES, L. E. SERIVEN, and H. T. DAVIS (1983), Phys. Rev. **B 28**, 307
SAHIMI, M., L. E. SERIVEN, and H. T. DAVIS (1984), J. Phys. **C 17**, 1941
SANTHANAM, P., S. WIND, and D. E. PROBER (1984), Phys. Rev. Letters **53**, 1179
SARKER, S. K. (1983), J. Phys. **C 16**, L 991

Saso, I. (1984), J. Phys. **C 17**, 2905
Schäfer, L., and F. J. Wegner (1980), Z. Physik **B 38**, 113
Schüttler, H. B., and T. Holstein (1983), Phys. Rev. Letters **51**, 2337
Seifert, E., and M. Suessenbach (1984), J. Phys. **A 17**, L 703
Seiferheld, U., H. Bässler, and B. Movaghar (1983), Phys. Rev. Letters **51**, 813
Shapir, Y. (1984), J. Physique Lettres **45**, L 895
Shapir, Y., A. Aharony, and A. B. Harris (1982), Phys. Rev. Letters **49**, 486
O'Shaughnessy, B., and I. Procaccia (1985), Phys. Rev. Letters **54**, 455
Shie, J. S. (1984), Phys. Rev. **B 30**, 5566
Shlifer, G., W. Klein, P. J. Reinolds, and H. E. Stanley (1979), J. Phys. **A 12**, L 169
Skocpol, W. J., L. D. Jackel, R. E. Howard, E. L. Hu, and L. A. Fetter (1983), Physica **117 B & 118 B**, 667
Sommer, E., and H. J. Kreuzer (1982), Phys. Rev. **B 26**, 4094
Soonpaa, H. H., and W. A. Schwalm (1984) Phys. Letters **100 A**, 156
Soukoulis, C. M., and E. N. Economou (1984), Phys. Rev. Letters **52**, 565
Stanley, H. E., and A. Coniglio (1984), Phys. Rev. **B 29**, 522
Stephen, M. J. (1984), Phys. Rev. **B 29**, 374
Stone, A. D., and P. A. Lee (1985), Phys. Rev. Letters **54**, 1196
Stormer, H. L., A. Chang, D. C. Tsui, J. C. M. Hwang, A. C. Gossard, and W. Wiegmann (1983) Phys. Rev. Letters **50**, 1953
Straley, J. P., and S. W. Kenkel (1984), Phys. Rev. **B 29**, 6299
Su, W. P., J. R. Schrieffer, and A. J. Heeger (1980), Phys. Rev. **B 22**, 2099
Summerfield, S., and P. N. Butcher (1983) J. Phys. **C 16**, 295
Summerfield, S., and P. N. Butcher (1984), Phil. Mag. **B 49**, L 65
Summerside, P., E. Sommer, R. Teshima, and H. J. Kreuzer (1982), Phys. Rev. **B 25**, 6235
Suhrke, M., and B. Esser (1984), phys. stat. sol.(b) **126**, 393
Tahir-Kheli, R. A. (1983), Phys. Rev. **B 27**, 7229
Tahir-Kheli, R. A. (1984), Phys. Rev. **B 29**, 1349
Takemori, T., and H. Kamimura (1983), Adv. Phys. **32**, 715
Tanaka, F. (1980), J. Phys. **C 13**, L 1
Theumann, A. (1983), Phys. Rev. **B 28**, 6453
Thouless, D. J. (1984), Physica **126 B**, 18
Triberis, G. P., and L. R. Friedman (1985), J. Phys. **C 18**, 2281
Trugman, S. A., and A. Weinrib (1985), Phys. Rev. **B 31**, 2974
Tsui, D. C., H. L. Stormer, and A. C. Gossard (1982) Phys. Rev. Letters **48**, 1559
Tuthill, G. F., and W. Klein (1983), J. Phys. **A 16**, 3561
Vannimenus, J., and J. P. Nadal (1984), Phys. Rep. **103**, 47
Vannimenus, J., J. P. Nadal, and H. Martin (1984), J. Phys. **A 17**, L 351
Verbaarschot, J. J. M., and M. R. Zirnbauer (1985), J. Phys. **A 18**, 1093
Visscher, P. B. (1984), Phys. Rev. **B 29**, 5472
Vomvas, V., and M. Roilos (1984), Phil. Mag **B 49**, 143
Voss, F. R. (1984), J. Phys. **A 17**, L 373
Webman, I. (1984), J. Stat. Phys. **36**, 603
Wegner, F., and S. Grossman (1985), Z. Physik **59**, 197
White, S. R., and M. Barma (1984), J. Phys. **A 17**, 2995
Willemsen, J. F. (1984), Phys. Rev. Letters **52**, 2197
Williams, J. K., and N. D. Mackenzie (1984), J. Phys. **A 17**, 3343
Wilson, E. G. (1983), J. Phys. **C 16**, 1039, 6739
Wu, F. Y., (1984), J. Appl. Phys. **55**, 2421

Wysokinski, K. I., and W. Brenig (1983), Z. Physik **B 54**, 11
Wysokinski, K. I., and W. Brenig (1985), Z. Physik **59**, 127
Yamada, K. (1984), Prog. Theor. Phys. **72**, 195
Zabolitzky, J. G. (1984), Phys. Rev. **B 30**, 4077
Zhang, Z. Q., and Y. S. Yang (1984), J. Phys. **A 17**, 1267
Zumofen, G., A. Blumen, and J. Klafter (1984), J. Phys. **A 17**, L 479
Zvyagin, I. P. (1984), Kinetic Phenomena in Disordered Semiconductors (in Russian), Moscow University Press, Moscow

Index

In this series original contributions of physical research in all fields are contained, such as
— research monographs
— collections of papers to a single topic
— reports on congresses of exceptional interest for physical research.

This series is aimed at promoting quick information and communication between physicists of the various special branches.

In diese Reihe werden Originalbeiträge zu allen Gebieten der physikalischen Forschung aufgenommen wie
— Forschungsmonographien
— Sammlungen von Arbeiten zu einem speziellen Thema
— Berichte von Tagungen, die für die physikalische Forschung besonders aktuell sind.

Die Reihe soll die schnelle Information und gute Kommunikation zwischen den Physikern der verschiedenen Fachgebiete fördern.

Manuscripts
in English comprising at least 100 pages and not more than 500 pages can be admitted to this series. With respect to a quick publication in future the manuscripts are reproduced photomechanically. Authors who are interested in this series please turn directly to the 'Akademie-Verlag'. Here you will get more detailed information about the form of the manuscripts and the modalities of publication.

Manuskripte
in englischer Sprache, die mindestens 100 Seiten und nicht mehr als 500 Seiten umfassen, können in diese Reihe aufgenommen werden. Im Interesse einer schnellen Publikation werden die Manuskripte künftig auf fotomechanischem Weg reproduziert. Autoren, die an der Veröffentlichung entsprechender Arbeiten in dieser Reihe interessiert sind, wenden sich bitte direkt an den Akademie-Verlag. Sie erhalten dort genauere Informationen über die Gestaltung der Manuskripte und die Modalitäten der Veröffentlichung.

Bei Fragen zur Produktsicherheit wenden Sie sich bitte an:
If you have any questions regarding product safety,
please contact:

Walter de Gruyter GmbH
Genthiner Straße 13
10785 Berlin
productsafety@degruyterbrill.com